Fundamentals of Acoustics: Volume 1

Fundamentals of Acoustics: Volume I

Triston Rodriguez

Larsen & Keller
www.larsen-keller.com

Fundamentals of Acoustics: Volume 1
Triston Rodriguez
ISBN: 979-8-88836-012-5 (Hardback)

Larsen & Keller
Published by Larsen and Keller Education,
5 Penn Plaza,
19th Floor,
New York, NY 10001, USA

Cataloging-in-Publication Data

Fundamentals of acoustics : Volume 1 / Triston Rodriguez.
p. cm.
Includes bibliographical references and index.
ISBN 979-8-88836-012-5
1. Sound. 2. Sound--Transmission. 3. Acoustical engineering. I. Rodriguez, Triston.
QC225.15 .I58 2023
534--dc23

For more information regarding Larsen and Keller Education and its products, please visit the publisher's website www.larsen-keller.com

Table of Contents

Preface

Sound is a vibration that propagates as an acoustic wave through a transmission medium such as a gas, liquid or solid. It travels in the form of longitudinal waves that involve a succession of compressions or rarefactions in the elastic medium. Acoustics is the branch of physics that deals with the science of sound. This branch is mainly concerned with the production, control, transmission, reception and effects of sound. The fundamental properties of sound are amplitude, frequency, wavelength, velocity and loudness. There are various types of sound waves based on different frequencies which include audible sound waves, ultrasonic waves and infrasonic sound waves. Audible sound waves can be heard by humans and animals whereas ultrasound and infrasound waves cannot be heard by normal human ears. Acoustics has important applications in almost every area of life. It plays a fundamental role in the development of music and musical instruments. It has also been crucial in the study of geologic, atmospheric and underwater phenomena. The topics included in this book on acoustics are of utmost significance and bound to provide incredible insights to the readers. It will help new researchers by foregrounding their knowledge in this branch of physics.

The information contained in this book is the result of intensive hard work done by researchers in this field. All due efforts have been made to make this book serve as a complete guiding source for students and researchers. The topics in this book have been comprehensively explained to help readers understand the growing trends in the field.

I would like to thank the entire group of writers who made sincere efforts in this book and my family who supported me in my efforts of working on this book. I take this opportunity to thank all those who have been a guiding force throughout my life.

Triston Rodriguez

1

Understanding Vibrational and Acoustical Systems

"The discussion of any problem in science or engineering has two aspects: the physical side, the statement of the facts of the case in everyday language and of the results in a manner that can be checked by experiment; and the mathematical side, the working out of the intermediate steps by means of the symbolized logic of calculus. These two aspects are equally important and are used side by side in every problem, one checking the other." [1]

The difference between engineering and science, and all other human activity, is the fact that engineers and scientists make quantitative predictions about measurable outcomes and can specify their uncertainty in such predictions. Because those predictions are quantitative, they must employ mathematics. This chapter is intended as an introduction to some of the more useful mathematical concepts, strategies, and techniques that are employed in the description of vibrational and acoustical systems and the calculations of their behavior.

It is not necessary to master the content of this chapter before working through this textbook. Other chapters will refer back to specific sections of this chapter as needed. If you are unsure of your competence or confidence with mathematics, it may be valuable to read through this chapter before going on.

1.1 The Five Most Useful Math Techniques

Below is a list of the five most useful mathematical techniques for the study of acoustics and vibration based on my experience. Techniques number one and number five are self-explanatory. The other three will be introduced in more detail in this section.

- Substitution
- Taylor series
- The product rule or integration by parts
- Logarithmic differentiation
- Garrett's First Law of Geometry: "Angles that look alike are alike."

1.1.1 Taylor Series

Acoustics and vibration are the "sciences of the subtle." Most of our attention will be focused on small deviations from a state of stable equilibrium. For example, a sound pressure level[1] of 115 dB_{SPL} is

[1] Do not worry if "sound pressure level" is not yet a familiar term. It will be defined in the fluid part of this textbook when intensity is explained in Sect. 10.5.1. For this example, it is only meant to specify a very loud sound.

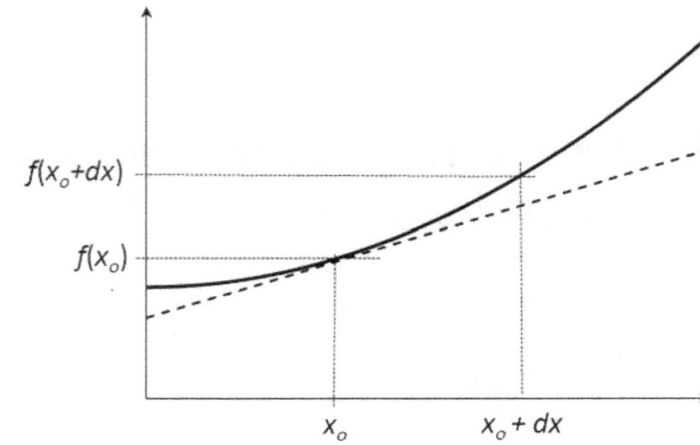

Fig. 1.1 Graph of an arbitrary function of position, $f(x)$. The straight dashed line is tangent to $f(x)$ at x_o. It represents the first derivative (slope) of $f(x)$ evaluated at x_o

capable of creating permanent damage to your hearing with less than 15 min of exposure per day [2]. That acoustic pressure level corresponds to a peak excess pressure of $p_1 = 16$ Pa (1 Pa = 1 N/m^2). Since "standard" atmospheric pressure is $p_m = 101{,}325$ Pa [3], that level corresponds to a relative deviation from equilibrium that is less than 160 parts per million (ppm) or $p_1/p_m = 0.016\%$.

If we assume that any parameter of interest (e.g., temperature, density, pressure) varies smoothly in time and space, we can approximate the parameter's value at a point (in space or time) if we know the parameter's value at some nearby point (typically, the state of stable equilibrium) and the value of its derivatives evaluated at that point.[2] The previous statement obscures the true value of the Taylor series because it is frequently used to permit substitution of the value of the derivative, as we will see throughout this textbook.

Let us start by examining the graph of some arbitrary real function of position, $f(x)$, shown in Fig. 1.1. At position x_o, the function has a value, $f(x_o)$. At some nearby position, $x_o + dx$, the function will have some other value, $f(x_o + dx)$, where we will claim that dx is a small distance without yet specifying what we mean by "small."

The value of $f(x_o + dx)$ can be approximated if we know the first derivative of $f(x)$ evaluated at x_o.

$$f(x_o + dx) \cong f(x_o) + \left.\frac{df}{dx}\right|_{x_o} dx \tag{1.1}$$

As can be seen in Fig. 1.1, the approximation of Eq. (1.1) produces a value that is slightly less than the actual value $f(x_o + dx)$ in this example. That is because the actual function has some curvature that happens to be upward in this case. The differential, dx, is used to represent both finite and infinitesimal quantities, depending upon context. For approximations, dx is assumed to be small but finite. For derivation of differential equations, it is assumed to be infinitesimal.

We can improve the approximation by adding another term to the *Taylor series expansion* of $f(x)$ that includes a correction proportional to the second derivative of $f(x)$, also evaluated at x_o. For the example in Fig. 1.1, the curvature is upward so the second derivative of $f(x)$, evaluated at x_o, is a positive number, so $(d^2f/dx^2)_{x_o} > 0$.

[2] What I am calling a "smooth" function is specified by mathematicians as being "infinitely differentiable," meaning that all of the function's derivatives are finite, remembering that zero is also finite.

If the curve had bent downward, the second derivative would have been negative. A more accurate approximation than Eq. (1.1) is provided by the following expression containing a correction proportional to $(dx)^2$:

$$f(x_o + dx) \cong f(x_o) + \left.\frac{df}{dx}\right|_{x_o} dx + \left.\frac{d^2f}{dx^2}\right|_{x_o} \frac{(dx)^2}{2} \tag{1.2}$$

Since $(dx)^2$ is intrinsically positive and the upward curvature makes d^2f/dx^2 positive, we can see that this *second-order correction* improves our estimate of $f(x_o + dx)$. Had the curvature been downward, making $(d^2f/dx^2)_{x_o} < 0$, then the second-order correction would have placed the estimated value of $f(x_o + dx)$ the first-order (linear) estimate, as required, since $(dx)^2$ would still be positive.

In principle, we can continue to improve the Taylor series approximation by adding higher and higher-order derivatives, although it is rare to extend such a series extended beyond the first three terms in Eq. (1.2). The full generic form of the Taylor series is provided in Eq. (1.3).

$$f(x_o + dx) = f(x_o) + \sum_{n=1}^{\infty} \left.\frac{f^n(x)}{(dx)^n}\right|_{x_o} \frac{(dx)^n}{n!} = f(x_o) + \sum_{n=1}^{\infty} f^{(n)}(x_o)\frac{(x - x_o)^n}{n!} \tag{1.3}$$

The Taylor series approach of Eq. (1.3) can be used to express continuous functions in a *power series*. Below are some functions that have been expanded about $x_o = 0$. As we will see, they will be particularly useful when $|x| \ll 1$.

$$e^x = 1 + x + \frac{x^2}{2!} + \frac{x^3}{3!} + \cdots = 1 + x + \frac{x^2}{2} + \frac{x^3}{6} + \cdots \tag{1.4}$$

$$\sin(x) = x - \frac{x^3}{3!} - \frac{x^5}{5!} - \cdots = x - \frac{x^3}{6} + \frac{x^5}{120} - \cdots \tag{1.5}$$

$$\cos(x) = 1 - \frac{x^2}{2!} + \frac{x^4}{4!} - \cdots = 1 - \frac{x^2}{2} + \frac{x^4}{24} - \cdots \tag{1.6}$$

$$\tan(x) = x + \frac{x^3}{3} + \frac{2x^5}{15} + \frac{17x^7}{315} + \frac{62x^9}{2835} \cdots \tag{1.7}$$

$$\ln(1 + x) = x - \frac{x^2}{2} + \frac{x^3}{3} - \frac{x^4}{4} + \cdots \tag{1.8}$$

$$(1 + x)^\alpha = 1 + \alpha x + \frac{\alpha(\alpha - 1)}{2!}x^2 + \frac{\alpha(\alpha - 1)(\alpha - 2)}{3!}x^3 + \cdots \tag{1.9}$$

1.1.2 The Product Rule or Integration by Parts

If we have two functions of x, say $u(x)$ and $v(x)$, the *product rule* tells us how we can take the derivative of the product of those functions.

$$\frac{d(uv)}{dx} = \frac{du}{dx}v + u\frac{dv}{dx} \tag{1.10}$$

This rule can be extended, for example, to the product of three functions of a single variable, where we have now added $w(x)$.

$$\frac{d(uvw)}{dx} = \frac{du}{dx}vw + u\frac{dv}{dx}w + uv\frac{dw}{dx} \tag{1.11}$$

Although it is not the intent of this chapter to derive all the quoted results, it is instructive, in this context, to introduce the concept of a *differential*.[3] If we say that a small change in x, by the amount dx, produces a small change in $u(x)$, by the amount du, and that $|du/u| \ll 1$, then we can show that Eq. (1.10) is correct by taking the product of the small changes.

$$d(uv) = (u + du)(v + dv) - uv = u(dv) + v(du) + (du)(dv) \tag{1.12}$$

Since both du and dv are small, the product of du times dv must be much less than $u\,(dv)$ or $v\,(du)$. If we make the changes small enough, then $(du)\,(dv)$ can be neglected in Eq. (1.12). Throughout this textbook, we will make similar assumptions regarding our ability to ignore the products of two very small quantities [4].

The *Fundamental Theorem of Calculus* states that integration and differentiation are inverse processes. By rearranging Eq. (1.10) and integrating each term with respect to x, we can write an expression that allows for integration of the product of one function and the derivative of another function.

$$\int u(x)\frac{dv}{dx}dx = u(x)v(x) - \int v\frac{du}{dx}dx \tag{1.13}$$

This result is known as the method of *integration by parts*.

1.1.3 Logarithmic Differentiation

The natural logarithm can be defined in terms of the integral of x^{-1}.

$$\int \frac{dx}{x} = \ln(x) + C \tag{1.14}$$

C is a constant. If we again apply the Fundamental Theorem of Calculus, we can write an expression for the differential, dx, in terms of the derivative of ln (x), remembering that the derivative of any constant must vanish.

$$d(\ln x) = \frac{dx}{x} \tag{1.15}$$

We will frequently be confronted by mathematical expressions that contain products and exponents. We can simplify our manipulations of such terms by taking the natural logarithms of those expressions and then differentiating. One typical example is the determination of the temperature change, dT, that accompanies a pressure change, dp, in a sound wave propagating under adiabatic conditions based on the Ideal Gas Law in Eq. (1.16) and the Adiabatic Gas Law in Eq. (1.17).

[3] Because the concept of a "differential" is so useful, it has multiple definitions that correspond to its usage in different contexts. For this discussion, the definition of the differential as a "small change" is adequate. In thermodynamic contexts, there are "inexact" differentials that produce changes in functions that are dependent upon the "path" of the change taken by the variable. A detailed discussion of the difference is presented by F. Reif, *Fundamentals of Statistical and Thermal Physics* (McGraw-Hill, 1965), Ch. 2, §11.

$$pV = n\Re T \tag{1.16}$$

$$pV^\gamma = \text{constant} \tag{1.17}$$

If we substitute the Ideal Gas Law into the Adiabatic Gas Law, we can write the absolute (kelvin) temperature, T, and pressure, p, in terms of a constant, C, related to the initial pressure and volume; the universal gas constant, $\Re$; and the polytropic coefficient, γ. (Do not worry if none of this is familiar at this point, it is only meant to illustrate the mathematics and will be treated explicitly in Sect. 7.1.4.)

$$pV^\gamma = (n\Re \mathrm{T})^\gamma p^{1-\gamma} \quad \Rightarrow \quad T^\gamma p^{1-\gamma} = \frac{p_o V_o^\gamma}{(n\Re)^\gamma} \equiv C \tag{1.18}$$

Again, C is just a constant that depends upon the initial conditions. The natural logarithm of both sides of the right-hand version of Eq. (1.18) produces Eq. (1.19).

$$\gamma \ln T + (1-\gamma)\ln p = \ln C \tag{1.19}$$

Using *logarithmic differentiation*, as expressed in Eq. (1.15),

$$\gamma \frac{dT}{T} + (1-\gamma)\frac{dp}{p} = 0 \quad \Rightarrow \quad \frac{dT}{T} = \frac{(\gamma-1)}{\gamma}\frac{dp}{p} \tag{1.20}$$

Rearranging terms and recalling that we substituted into the Adiabatic Gas Law, this result provides an expression for the change in the temperature of an ideal gas in terms of the change in the pressure due to a sound wave that is propagating under adiabatic conditions.

$$\left(\frac{\partial T}{\partial p}\right)_s = \frac{(\gamma-1)}{\gamma}\frac{T}{p} \tag{1.21}$$

The subscript "s" has been placed along the partial derivative to remind us that the change in temperature with pressure is evaluated for an adiabatic process in which the specific entropy (per unit mass), s, is held constant. As will be illustrated in our investigations of the coupling between sound waves and temperature changes, this is a particularly useful result. It has only been previewed here to illustrate how simple it is to do some calculations by taking a logarithm of an expression before differentiation. This will also be a rather convenient approach for calculation of the uncertainty of a result based on the uncertainty in the parameters that were used to calculate the result (i.e., "error propagation"; see Sect.1.8.4).

1.2 Equilibrium, Stability, and Hooke's Law

Pause for a moment as you are reading and look around. Close your eyes, count to ten, and then open your eyes. Probably not much has changed. This is because most of the material we observe visually is in a state of *stable equilibrium*.[4] For an object to be in equilibrium, the vector sum of all the forces acting on that body (i.e., the net force) must be zero, and the first derivative of the object's *potential energy* with respect to its position must be zero. For that equilibrium state to be stable, the second derivative of the object's potential energy must be positive.

[4] Here, we are referring to what is known as Lyapunov stability, which applies to dynamical systems in the vicinity of their equilibrium configuration.

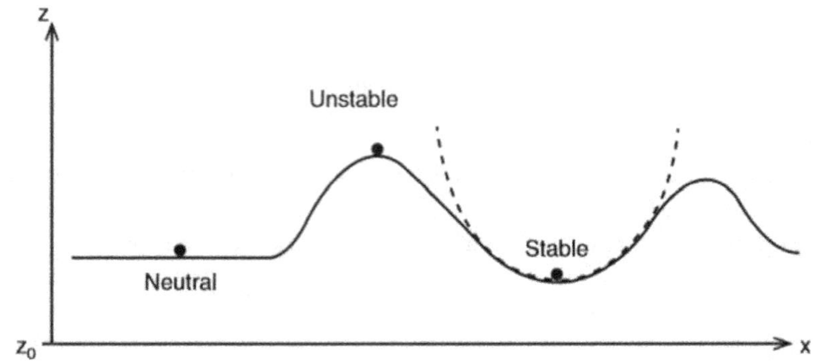

Fig. 1.2 Potential energy (proportional to height) is shown as a function of position for three objects that are at rest in an equilibrium state within a gravitational field. The ball at the left is in a state of *neutral equilibrium*, since it can move to the left or the right by a small distance and not change its potential energy. The ball at the center is in a state of *unstable equilibrium*. Although the net force (gravitational downward and the force of the surface upward) is zero, if it is displaced by an infinitesimal distance to either the right or the left, it will leave its equilibrium position. The ball at the right is in a position of *stable equilibrium*. If it is displaced by an infinitesimal distance to either the *right* or to the *left*, the net force on the ball will tend to return it toward its equilibrium position. The dashed line is a parabola that is fit to match the curvature of the line near the position of stable equilibrium corresponding to the quadratic contribution in the Taylor series expansion of Eq. (1.23)

Figure 1.2 illustrates three possible equilibrium conditions based on the rate of change of a body's potential energy with position. For this illustration, let us assume that the solid curve represents the height, z, above some reference height, z_o, in a uniform gravitational field. The (gravitational) potential energy, $PE\ (z)$, of each of the three balls shown in Fig. 1.2 is therefore proportional to their height, z, where the mass of each ball is m and g is the acceleration due to gravity that is assumed to be independent of z: $PE\ (z) = mg\ (z{-}z_o)$.

The three balls in Fig. 1.2 each respond to a small displacement from their equilibrium positions differently. We can think of the solid curve as representing a flat surface on the left, two peaks at the center and the right, and a valley between two peaks. All three balls are in a state of mechanical equilibrium because the vector sum of the force of gravity (down) and of the surface (up) is zero. In all three cases, the first derivative of the potential energy vanishes at all three locations: $d(PE)/dx = 0$. The ball on the left is in a state of *neutral equilibrium* because it can be moved to the left or to the right by a small distance and it will still be at equilibrium, even at its displaced position. The curve at that location is flat and horizontal.

The other two balls are located at local extrema. The slopes of the tangents to the curves at those extrema are just both as horizontal and as the flat region on the left.[5] The ball near the center of Fig. 1.2

[5] The inability of some to connect mathematics with reality is illustrated by this short story told by Richard Feynman in his most entertaining autobiography entitled *Surely You're Joking, Mr. Feynman!* (W. W. Norton, 1985); ISBN 0-393-01921-7.

"I often liked to play tricks on people when I was at MIT. Once, in mechanical drawing class, some joker picked up a French curve (a piece of plastic for drawing smooth curves – a curly, funny-looking thing) and said 'I wonder if the curves on this thing have some special formula?'"

"I thought for a moment and said, 'Sure they do. The curves are very special curves. Lemme show ya,' and I picked up my French curve and began to turn it slowly. 'The French curve is made so that at the lowest point on each curve, no matter how you turn it, the tangent is horizontal.'"

"All the guys in the class were holding their French curve up at different angles, holding their pencil up to it at the lowest point and laying it along, and discovering that, sure enough, the tangent is horizontal. They were all excited by this 'discovery' – even though they had already gone through a certain amount of calculus and had already 'learned' that the derivative (tangent) of the minimum (lowest point) of any curve is zero (horizontal). They did not put two and two together. They didn't even know what they 'knew.'"

is at the top of the "hill," so $d(PE)/dx = 0$ and $d^2(PE)/dx^2 < 0$. The ball near the right in that figure is at the lowest point in the valley, so $d(PE)/dx = 0$, but down in the valley, $d^2(PE)/dx^2 > 0$.

The ball at the top of the peak is in a state of *unstable equilibrium* because if it is displaced by an infinitesimal amount in either direction, it will run away from its equilibrium position. In general, objects in states of unstable equilibrium do not remain in those states for very long.

The ball at the lowest point in the valley is in a state of stable equilibrium. If it is displaced in either direction, the potential energy increases, and the ball can reduce its energy by returning to its position of stable equilibrium. A dashed parabola has been drawn which matches the second derivative of the actual potential energy curve. The greater the distance from the point of stable equilibrium, the greater the difference between the actual curve representing the potential energy and the dashed parabola, but for small displacements from equilibrium, the real potential energy curve and the dashed parabola will be indistinguishable. The dashed parabola represents the second-order term in a Taylor series expansion of the potential energy about the equilibrium position provided in Eq. (1.23).

With this understanding of equilibrium and stability, we are in a position to formalize the application of potentials and forces to develop the formalism that is used to characterize the oscillations of systems that obey Hooke's law.

1.2.1 Potentials and Forces

The relationship between forces and potential energy can be extended beyond our simple stability example. In general, the net (vector) force, $\vec{F}_{net}$, is the negative of the gradient of the (scalar) potential energy: $\vec{\nabla}(PE) = -\vec{F}_{net}$.[6] This is consistent with the definitions of work and energy.

$$W_{1,2} \equiv \int_1^2 \vec{F} \bullet d\vec{x} = (PE)_1 - (PE)_2 = -\Delta(PE) \tag{1.22}$$

The right-hand side assumes that $\vec{F}$ is a conservative force[7] and the work, $W_{1,2}$, done in moving an object from position 1 to position 2, over some distance along the direction of the force (indicated by the "dot product" under the integral), leads to a change in potential energy, $-\Delta(PE)$. Again, application of the Fundamental Theorem of Calculus leads to the desired relationship between the gradients of the potential energy and the net force: $\vec{\nabla}(PE) = -\vec{F}_{net}$.

If we limit ourselves to the current example of a ball in the valley, we can expand the potential energy about the stable equilibrium position that is identified as x_o.

---

[6] The sign convention becomes obvious if we consider work against a gravitational field. In that case, if we raise an object, $\vec{F}$ and $d\vec{x}$ are antiparallel, and the integral is negative, although we have increased the mass's potential energy by lifting it.

[7] The work done against a "conservative" force depends only upon the endpoints so that no work is done if the path brings the body back to its original position. The application of Stokes' Theorem converts the closed path integral into a requirement that $\text{curl}(\vec{F}) = \vec{\nabla} \times \vec{F} = 0$, so conservative forces can be derived from the gradient of a scalar potential, U: $\vec{F} = -\text{grad}(U) = -\vec{\nabla}U$. For a more complete discussion, see H. Goldstein, *Classical Mechanics* (Addison-Wesley, 1950).

$$PE(x_o + dx) = PE(x_o) + \left.\frac{d^2(PE)}{dx^2}\right|_{x_o} \frac{(dx)^2}{2} + \left.\frac{d^3(PE)}{dx^3}\right|_{x_o} \frac{(dx)^3}{6} + \ldots \tag{1.23}$$

Note that the first derivative of *PE* is missing from Eq. (1.23) because it is zero if x_o is the equilibrium position.[8] The term proportional to $(dx)^3$ corresponds to the contribution to the difference between the actual curve and the dashed parabola in Fig. 1.2. If the deviation between the two curves is symmetric, then the leading correction term would be proportional to $(dx)^4$. If the deviation is not symmetric, the leading correction term will be proportional to $(dx)^3$.

In our one-dimensional example, the gradient of the potential energy will simply be the derivative of Eq. (1.23) with respect to *x*.

$$F_{net}(dx) = -\frac{d(PE)}{dx} = -\left.\frac{d^2(PE)}{dx^2}\right|_{x_o} (dx) - \left.\frac{d^3(PE)}{dx^3}\right|_{x_o} \frac{(dx)^2}{2} - \cdots \tag{1.24}$$

For sufficiently small displacements from equilibrium, the parabolic approximation to the potential energy curve (i.e., the dashed parabola in Fig. 1.2) provides an adequate representation, and therefore the series of Eq. (1.24) that describes the force can be truncated after the first term.

$$F_{net}(dx) \cong -\left.\frac{d^2(PE)}{dx^2}\right|_{x_o} (dx) \equiv -\mathrm{K}(dx) \tag{1.25}$$

The result in Eq. (1.25) is known as *Hooke's law.*[9] It states that the net force is proportional to the displacement from equilibrium, *dx*. Because of the minus sign, the net force is directed opposite to that displacement if $\mathrm{K} > 0$. The new constant, K, is known as the "stiffness" when the potential energy is created by the extension or compression of a spring. It is equal to the negative of the second derivative of the potential energy, evaluated at the equilibrium position, x_o.

In this form, it is obvious that a positive curvature in the potential energy (i.e., $d^2(PE)/dx^2 > 0$) leads to a stable equilibrium but that a negative curvature in the potential energy causes the system to become unstable. In an unstable state of equilibrium, a small displacement from equilibrium creates a force that causes a small displacement to grow, regardless of the sign of the original displacement; at a position of unstable equilibrium, the stiffness constant, K, is a negative number.

The approximation represented by Hooke's law leads to a *linear relationship* between net force on a body and the displacement of that body from its position of stable equilibrium. It is normally introduced in elementary mechanics courses as the law which describes the force exerted by a spring when the spring is displaced from its equilibrium length, x_o. Two such springs are showed schematically in Fig. 1.3.

Both of the springs in Fig. 1.3 obey Hooke's law if the displacements, *dx*, from the equilibrium position at x_o are small: $(dx)/x_o \ll 1$. In both cases, large displacements from equilibrium will lead to nonlinear behavior. For the coil spring, excess compression will cause the coils to close up, and the force necessary for further compression will be much larger than predicted by Hooke's law. Similarly, for excess extension, the coils will straighten out, again requiring excess force, until the material fractures and the broken end can be moved an infinite distance without requiring any additional force.

[8] Any constant added to the potential energy has no effect on any observable quantity. This independence of any constant offset in the potential is known as "gauge invariance."

[9] This law is named after Robert Hooke, FRS, who first stated it as a Latin anagram in 1660 and then published the solution in 1678 as *Ut tensio, sic vis*; literally translated, it means "As the extension, so the force."

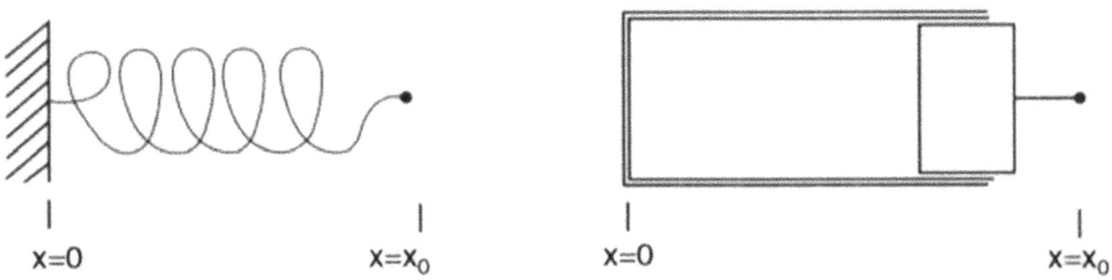

Fig. 1.3 Schematic representation of two springs that obey Hooke's law for small displacements from their equilibrium positions, x_o. The spring at the *left* represents a coil of wire with one end fixed (represented by the hatched wall) at $x = 0$ and an equilibrium length of x_o. The spring at the right represents an air-filled cylinder that is closed at the end located at $x = 0$ and is sealed at the opposite end by a close-fitting, frictionless piston. (It is possible to purchase a piston and cylinder combination that behaves very much like this leak-tight, frictionless idealization. It is called an Airpot® and consists of a glass cylinder and a matched graphite piston: www.airpot.com. One such Airpot® is shown in Prob. 4 at the end of Chap. 7, in Fig. 7.6.) The equilibrium position of the gas spring at x_o occurs when the mean gas pressure inside the cylinder is equal to the mean gas pressure outside the cylinder

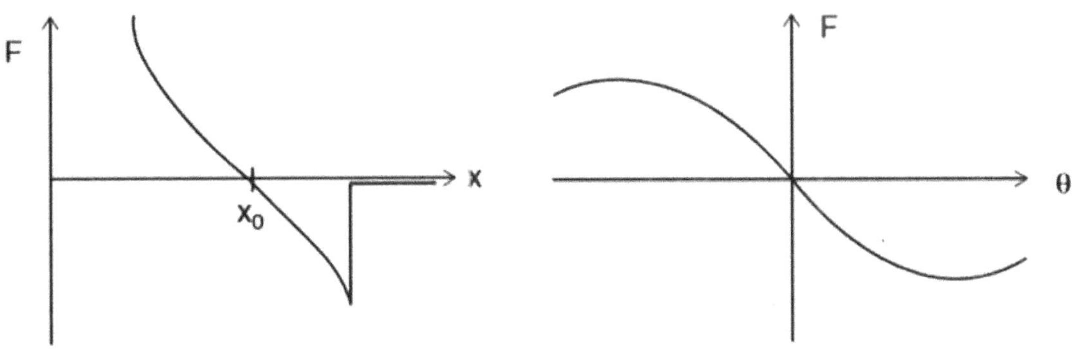

Fig. 1.4 Two graphs of force vs. displacement which illustrate deviations from linear Hooke's law behavior for two systems that each possess a position of stable equilibrium. At the left is a graph representing springs like those shown in Fig. 1.3. For small displacements from the equilibrium position, x_o, the force is a linear function of displacement. As x becomes significantly smaller than x_o, the spring "stiffens," producing significantly larger forces for a given displacement, presumably because adjacent coils have made contact. As x becomes significantly greater than x_o, the spring again stiffens. In this limit, the coils have begun to uncoil, and the force is increasing due to the tensile strength of the coil spring's material. Eventually, the material fractures or the piston is removed from the cylinder, and the force goes to zero. The graph at the right might represent the restoring force of a pendulum that is displaced from its equilibrium vertical position hanging down at $\theta_o = 0$ (see Fig. 1.5). Again, there is a linear portion of the force vs. angular displacement for small values of $\theta \ll 1$, but as the angle increases, less force is required for each increment of angle. This nonlinear spring behavior is called "softening"

Similar deviations from the linear behavior predicted by Hooke's law occur for the gas spring. Two examples of the force vs. displacement that go beyond the linear regime are illustrated in Fig. 1.4.

1.2.2 A Simple Pendulum

To emphasize the fact that "linear" behavior is not restricted to linear motion, consider the spherical mass suspended from a rigid ceiling by a (massless) string of length, L, in a uniform gravitational field, as shown schematically in Fig. 1.5. Its position of stable equilibrium corresponds to the mass hanging

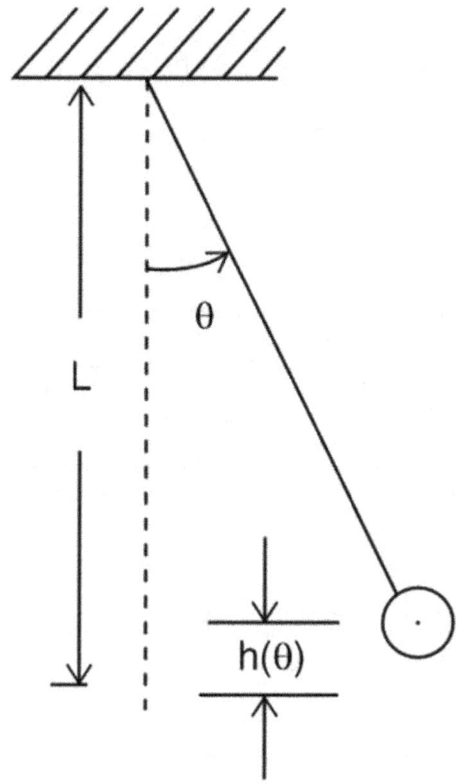

Fig. 1.5 The pendulum consists of a sphere of mass, *m*, which is supported by a massless string. The distance from the point of support to the center of mass of the sphere is *L*. When the sphere is displaced from its equilibrium position, $\theta_o = 0$, this angular displacement, θ, causes the sphere's center of mass to rise above its equilibrium height by a vertical distance, *h*. The associated increase in the gravitational potential energy of the sphere is given by Eq. (1.26). The restoring torque on the pendulum is given by the gradient in the potential energy and is shown schematically in the graph of Fig. 1.4 (*right*)

vertically downward beneath the point of suspension; hence $\theta_o = 0$. In this case, the gravitational potential energy is a function of the angular displacement corresponding to the change in vertical height of the sphere's center of mass, $h(\theta)$, within a gravitational field that is presumed to produce a constant gravitational acceleration, g. The relationship between the displacement angle, θ, and the height change, $h(\theta)$, can be calculated from trigonometry.

$$PE(\theta) = mgL[1 - \cos\theta] \cong mgL\left[\frac{\theta^2}{2} - \frac{\theta^4}{24} + \cdots\right] \tag{1.26}$$

The above approximation uses a power (Taylor) series expansion of the cosine function from Eq. (1.6). The magnitude of the corresponding restoring torque, $N(\theta)$, is given by the negative of the gradient of the potential energy with respect to angle.

$$N(\theta) = -\frac{\partial[PE(\theta)]}{\partial\theta} \cong -mgL\left[\theta - \frac{\theta^3}{6} + \cdots\right] \tag{1.27}$$

For $\theta \ll 1$, the restoring torque is proportional to the angular displacement from stable equilibrium. In this limit, this results in a torsional stiffness that is a constant. For larger displacements from equilibrium, additional effects generated by the θ^3 term in Eq. (1.27) reduce the torque producing the "softening" behavior shown in Fig. 1.4 (Right).

1.3 The Concept of Linearity

At least 95% (by weight?) of the systems analyzed in this textbook will assume that the dynamical processes occur within the linear regime. This assumption brings with it amazing mathematical and physical simplifications. As will be discussed here, linearity guarantees that:

1. The frequency of oscillation of a "free" system is amplitude-independent (*isochronism*).
2. The steady-state response of a driven linear system occurs only at the driving frequency.
3. We can superimpose an infinite number of responses to simple stimuli to determine the linear system's response to a very complicated stimulus and vice versa.

The founder of modern science, Galileo Galilei (1564–1642), was the first to record his observation that the period of a pendulum was independent of the amplitude of its displacement. Legend has it that he made that observation in the Cathedral of Pisa when he was a student living in that city. Being bored in church, he noticed that when the wind moved the chandeliers, the period of oscillation, measured against his own pulse rate, was independent of the amplitude of chandeliers' swing. Although Galileo recognized the value of this amplitude independence for time-keeping devices, it was not until 17 years after Galileo's death that the great Dutch physicist, Christiaan Huygens (1629–1695), described his "pendulum clock" and other time pieces with errors of less than 1 min/day (later, less than 10 s/day) whose accuracy relied on the amplitude independence of the period of the pendulum.

The behavior of nonlinear springs is dramatically different than the Hooke's law behavior [5]. When driven at a frequency, ω, the system can respond at multiples, or sub-multiples of the driving frequency [6], and can produce non-zero time-averaged changes in the equilibrium position. Nonlinear elastic behavior in solids is responsible for the fact that the thermal expansion coefficient of solids is non-zero. As temperature increases, the thermal vibrational amplitude of the solid's atoms causes their equilibrium separation to change [7].

Although we have thus far only considered the behavior of systems displaced a small distance from their position of stable equilibrium (thus producing a linear restoring force), we will combine that result with the behavior of masses and dampers in the next chapter on simple harmonic oscillators. The dynamics of masses and dampers also exhibit a substantial linear regime. Newton's Second Law of Motion, $\vec{F} = m\ddot{\vec{x}}$, is linear because it relates the acceleration of a mass to the net force acting on that mass. Acceleration is the second time derivative of displacement and does not involve higher powers of displacement. Similarly, there is a regime where the damping force can be linear in velocity, $v(t) = \dot{x}(t) \equiv dx(t)/dt$, which is the first time derivative of displacement. Although we will investigate the behavior of the damped simple harmonic oscillator in the next chapter, we can write down the equation for a linear operator, $\underline{L}\,(x)$, that describes the response of such an oscillator to an externally applied time-dependent force, $F\,(t)$.

$$\underline{L}(x) = m\frac{d^2x}{dt^2} + R_m\frac{dx}{dt} + \mathrm{K}x = F(t) \tag{1.28}$$

To simplify our discussion of linear response, we can lump all of these linear operations performed on x into a general linear operator, $\underline{L}(x)$, where $\underline{L}$ is underlined to remind us that it is an "operator," not a function.[10] That operator defines a combination of mathematical procedures that are applied to an independent variable, x in this case. The operator in Eq. (1.28) is a second-order linear differential

[10] This "linear operator" approach is taken from *The Feynman's Lectures on Physics*, R. P. Feynman, R. B. Leighton, and M. Sands (Addison-Wesley, 1963), Vol. I, Ch. 25.

equation with constant coefficients, those coefficients being the mass, m; the mechanical resistance, R_m; and the stiffness, K.

By substitution into Eq. (1.28), it is easy to demonstrate that all linear operations exhibit both an additive and a multiplicative property where a is an arbitrary scalar.

$$\underline{L}(x+y) = \underline{L}(x) + \underline{L}(y) \tag{1.29}$$

$$\underline{L}(ax) = a\underline{L}(x) \tag{1.30}$$

In fact, these two properties of linear operators can be extended indefinitely.

$$\underline{L}(ax + by + cz + \cdots) = a\underline{L}(x) + b\underline{L}(y) + c\underline{L}(z) + \cdots \tag{1.31}$$

These properties demonstrate that if we have a solution to Eq. (1.28), that any scalar multiple of that solution is also a solution; if we double the force, we double the response. Similarly, if both $x_a(t)$ and $x_b(t)$ are two responses to two forces, $F_a(t)$ and $F_b(t)$, possibly acting at two different frequencies, then their sum is the response to the sum of those forces.

$$\underline{L}(x_a + x_b) = \underline{L}(x_a) + \underline{L}(x_b) = F_a(t) + F_b(t) \tag{1.32}$$

1.4 Superposition and Fourier Synthesis

"Superposition is our compensation for enduring the limitations of linearity". [8]

The ability exhibited by Eq. (1.31) to combine solutions to linear equations is so significant that it has its own name: *The Principle of Superposition*. It guarantees that if we can decompose a complicated excitation into the sum of simpler excitations, we can calculate the responses to the simple excitations, and then we can combine those simple solutions to determine the response to the more complicated excitation.

We all have experience with this concept when we locate a point on a two-dimensional surface using Cartesian coordinates. The x and y axes form an orthogonal basis that specifies the horizontal and vertical directions. When we want to specify a particular point on that surface, we draw a vector, $\vec{r}$, from the origin of the coordinate system to the point and project that vector on to our orthogonal axes by taking the dot product of the vector, $\vec{r}$, with the unit vectors, $\hat{e}_x$ and $\hat{e}_y$, along each axis.

$$(x, y) = \left(\vec{r} \bullet \hat{e}_x, \vec{r} \bullet \hat{e}_y\right) \tag{1.33}$$

In that way, the ordered sequence of two numbers, (x, y), produced by those two operations, uniquely specifies a point on the plane.

This simple concept can be extended to a representation of a function as a superposition of *orthogonal functions*. A general n^{th}-order, linear differential equation with constant coefficients, a_i, will have the form expressed below:

$$a_n \frac{d^n x}{dt^n} + a_{n-1} \frac{d^{n-1} x}{dt^{n-1}} + \cdots + a_1 \frac{dx}{dt} + a_o = F(t) \tag{1.34}$$

There will be n linearly independent solutions to such an equation, where we mean that the solutions are "independent" if it is not possible to express one of the linearly independent solutions in terms of linear combinations of the others.[11] With a few minor exceptions (such as flexural waves on bars and plates), our focus in this textbook will be on systems described by second-order differential equations with constant coefficients. For that case $n = 2$, so there will be two linearly independent solutions. One convenient choice of independent solutions for vibration and acoustics is the sine and cosine functions.

The representation of periodic functions as the sum of simple oscillating functions dates back to the third century BC when ancient astronomers proposed models of planetary motion based on epicycles. The first use of trigonometric functions as a basis for solving a linear differential equation was made by Jean-Baptiste Joseph Fourier (1768–1830). In his *Treatise on the Propagation of Heat in Solid Bodies*, published in 1807, he used the technique that now bears his name to analyze the diffusion of heat through a metal plate.

As we will see, the solutions to equations describing vibration and sound are particularly simple if the disturbances are sinusoidal in space and/or time. For more complicated disturbances, which are periodic with period, T_1, it is possible to decompose the disturbance into the sum of sinusoidal disturbances with the same period as the complicated waveform. If the complicated waveform is a continuous function, the periodic waveform can be represented by an infinite number of sines and cosines, each with the correct amplitude, that have frequencies that are integer multiples of the fundamental frequency of the complicated waveform, $f_1 = T_1^{-1} = \omega/2\pi$.

$$f(t) = \frac{A_o}{2} + \sum_{n=1}^{\infty} A_n \cos(n\omega t) + B_n \sin(n\omega t)]. \tag{1.35}$$

The coefficients, A_n and B_n, are determined in the same way we selected the x and y coordinates for a point on a Cartesian plane; we take the equivalent of the dot product of the complicated function with the linearly independent basis functions over an entire period, T. Such a "dot product," also known as the "inner product," is represented by the following integrals:

$$A_0 = \frac{1}{T}\int_0^T f(t)\ dt \text{ and } A_n = \frac{2}{T}\int_0^T f(t)\cos(n\omega t)dt \text{ for } n \geq 1 \tag{1.36}$$

$$B_n = \frac{2}{T}\int_0^T f(t)\sin(n\omega t)dt \tag{1.37}$$

Since $\sin(0) = 0$, there is no need for B_o. Since $\cos(0) = 1$, A_o is just the average of $f(t)$ over one period,[12] divided by the period, T. The sine and cosine functions are linearly independent, and the orthogonality of the harmonics of the individual sine and cosine functions can be expressed in terms of the *Kronecker delta* function, $\delta_{m,n}$, which is zero if $m \neq n$ and one if $m = n$.

$$\begin{aligned} \frac{2}{T}\int_0^T \cos(m\omega t)\cos(n\omega t)dt &= \delta_{m,n} \\ \frac{2}{T}\int_0^T \sin(m\omega t)\sin(n\omega t)dt &= \delta_{m,n} \end{aligned} \tag{1.38}$$

Depending upon the nature of $f(t)$, it is possible that many of the A_n and B_n coefficients may be zero. If $f(t)$ is symmetric about the time axis, then $A_o = 0$. If $f(t) = f(-t)$, then $f(t)$ is an even function of t and

[11] For our example of a Cartesian coordinate system, the x and y axes are independent because it is impossible to specify some non-zero location on the y axis using any x value.

[12] In the parlance of electrical engineers, A_o is the "DC offset."

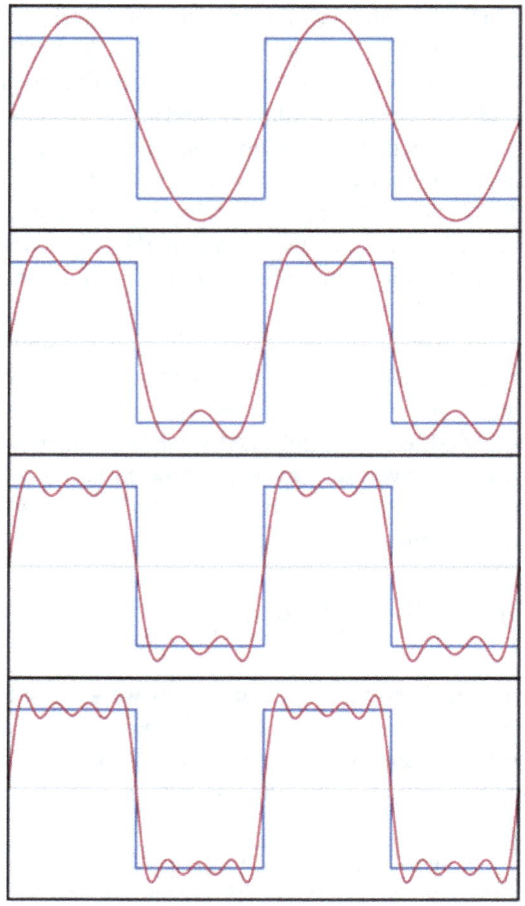

Fig. 1.6 The square wave defined in Eq. (1.39) is shown along with the sum of the first through fourth non-zero Fourier components. The first term is simply a sinusoidal waveform that is synchronized with the square wave. The third harmonic contribution starts to deform the sum into a shape that more closely resembles the square wave. Addition of the fifth harmonic, and then the seventh harmonic, improves the fidelity of the superposition and makes the slope of the zero-crossing transitions more nearly vertical. Clearly superposition of an infinite number of harmonics would be required to simulate the infinite slope of those transitions and reduce the "ripple" on the plateaus, although the small (9%) overshoot will remain. It does not disappear as more terms are added; it simply becomes narrower [10]. Since the square wave is not a "continuous" function but has a discontinuous slope, its Fourier series representation will have an "overshoot" (known as "ringing") about those transitions. This is known as the "Gibbs phenomenon"

$B_n = 0$ for all n, since sine is an odd function. Similarly, if $f(t) = -f(-t)$, then $f(t)$ is an odd function of t, and $A_n = 0$ for all n, since cosine is an even function.

The process of assembling a "complex" waveform from simpler sinusoidal components is known as *Fourier synthesis*. To illustrate Fourier synthesis, let us represent a square wave, $f_{\text{square}}(t)$, of unit amplitude and period, T, that is defined for one period below and is shown in Fig. 1.6:

$$f_{\text{square}}(t) = \begin{cases} +1 & 0 \leq t < T/2 \\ -1 & T/2 \leq t < T \end{cases} \tag{1.39}$$

Since $f_{square}(t)$ is symmetric, $A_o = 0$ and because it is an odd function of t, only the B_n coefficients will be non-zero. Substitution of Eq. (1.39) into Eq. (1.37) breaks the integral into two parts.

$$B_n = \frac{2}{T}\left[\int_0^{T/2} \sin(n\omega t)dt - \int_{T/2}^{T} \sin(n\omega t)dt\right] \tag{1.40}$$

The integral of the sin (ax) over x is available from any table of integrals. My favorite is the Russian compilation by Gradshteyn and Ryzhik [9].

$$\int \sin(ax)\,dx = \frac{-1}{a}\cos(ax) + C \tag{1.41}$$

For this problem $\omega = 2\pi/T$, so $a = 2\pi n/T$.

$$B_n = \frac{1}{n\pi}\left[-\cos\left(\frac{2\pi nt}{T}\right)\Big|_0^{T/2} + \cos\left(\frac{2\pi nt}{T}\right)\Big|_{T/2}^{T}\right] \tag{1.42}$$

For odd values of n, $B_n = (4/n\pi)$, and for even values of n, $B_n = 0$, so the Fourier series representation of our unit symmetric square wave can be written as an infinite sum.

$$f_{square}(t) = \frac{4}{\pi}\left[\sin\left(\frac{2\pi t}{T}\right) + \frac{1}{3}\sin\left(\frac{6\pi t}{T}\right) + \frac{1}{5}\sin\left(\frac{10\pi t}{T}\right) + \cdots\right] \tag{1.43}$$

Again, the fact that $\omega = 2\pi/T$ allows us to rewrite this result in terms of angular frequencies.

$$f_{square}(t) = \frac{4}{\pi}\left[\sin(\omega t) + \frac{\sin(3\omega t)}{3} + \frac{\sin(5\omega t)}{5} + \cdots\right] \tag{1.44}$$

The sum of the first four terms is shown in Fig. 1.6, illustrating that the addition of successive terms makes the Fourier series superposition an ever more faithful representation of $f_{square}(t)$.[13]

We could repeat the process to determine the Fourier representation of a triangle wave $f_{triangle}(t)$, but if we simply recognize that a triangular wave is the integral of a square wave, then we can integrate Eq. (1.44) term by term.

$$f_{triangle}(t) = \frac{4}{\pi^2}\left[\cos(\omega t) + \frac{\cos(3\omega t)}{9} + \frac{\cos(5\omega t)}{25} + \ldots\right] \tag{1.45}$$

Since the shape of the triangle wave is much closer to that of a sine or cosine, the high-frequency components decay more quickly (as n^{-2}) than those of the square wave (as n^{-1}).

1.5 Convenience (Complex) Numbers

"The most beautiful formula in mathematics: $e^{j\pi} + 1 = 0$." [11]

[13] Since the square wave is not a "continuous" function but has a discontinuous slope, its Fourier series representation will have an "overshoot" of about 9% with oscillations (known as "ringing") about those transitions, even with the inclusion of an infinite number of terms. This is known as the "Gibbs phenomenon."

Calculus is a set of rules that allow us to convert expressions that involve integrals and derivatives into algebraic equations. The simplest of these rules are those which allow us to write down the integrals and derivatives of polynomials and exponentials:

$$\frac{d(ax^n)}{dx} = anx^{n-1} \ ; \qquad \int ax^n dx = \frac{ax^{n+1}}{n+1} + C \tag{1.46}$$

$$\frac{d(ae^{bx})}{dx} = abe^{bx} \ ; \qquad \int ae^{bx} dx = \frac{ae^{bx}}{b} + C \tag{1.47}$$

We have already exploited the simplicity of Eq. (1.46) because our Taylor series expansions resulted in polynomial representations of various functions (see Sect. 1.1.1). In this section, we will exploit the simplicity of Eq. (1.47) in a way that allows us to encode the two linearly independent solutions to a second-order differential equation into a single function by simply multiplying one of those two solutions by the square root of (-1). That special linear coefficient will be designated as $j \equiv \sqrt{-1}$.

The fact that j allows us to combine (i.e., superimpose) the solutions to differential equations of interest in a way that permits the use of Eq. (1.47) to do our calculus is the reason I call j the "convenience number." Unfortunately, I am the only person who calls j the convenience number. Everyone else calls j the unit "imaginary number."[14] Numbers that contain an imaginary component are called "complex numbers." This poor historical choice of nomenclature suggests that there is something "complex" about the use of j when the contrary is true; j makes computations simpler.

To appreciate the convenience of j, we can start by examining the solution to the simplest homogeneous second-order differential equation with a constant coefficient that has been designated ω_o^2 for reasons that will become obvious in the next chapter.

$$\frac{d^2x}{dt^2} + \omega_o^2 x = 0 \qquad \text{or} \qquad \frac{d^2x}{dt^2} = -\omega_o^2 x \tag{1.48}$$

When written as the right-hand version of Eq. (1.48), it is clear that we seek a solution which produces the negative of itself when differentiated twice. Sine and cosine functions have exactly that property, being linearly independent.[15]

$$\frac{d\sin(at)}{dt} = a\cos(at) \ ; \qquad \int \sin(at)dt = -\frac{\cos(at)}{a} + C \tag{1.49}$$

$$\frac{d\cos(at)}{dt} = -a\sin(at) \ ; \qquad \int \cos(at)dt = \frac{\sin(at)}{a} + C \tag{1.50}$$

Double differentiation of either sine or cosine regenerates the negative of the original function, as required by Eq. (1.48).[16] Superposition allows us to write the complete solution to Eq. (1.48) by introducing two arbitrary scalar constants, A and B:

---

[14] This textbook uses j to designate the unit imaginary number. That choice is more common among engineers. Physicists and mathematicians prefer to let $i = (-1)^{1/2}$. Of course, the choice is arbitrary.

[15] The orthogonality of the sine and cosine functions can be expressed by integrating their product over one period, T: $\int_0^T \cos(mt)\sin(nt)dt = 0$, where m and n are any integers, including zero.

[16] We could have obtained the same result in a more general way by assuming an infinite-order polynomial solution and then identifying the resulting series using the Taylor series expansions of Eqs. (1.5) and (1.6). For this approach, see P. M. Morse, *Vibration and Sound* (Acoustical Society of America, 1976); ISBN 0-88318-287-4.

$$x(t) = A\cos(\omega_o t) + B\sin(\omega_o t) \tag{1.51}$$

Using our definition of $j \equiv \sqrt{-1}$ and the rules for differentiating exponentials in Eq. (1.47), we see that an exponential with an imaginary exponent will also obviously satisfy Eq. (1.48), where again A and B are scalar constants that facilitate superposition of the two solutions.

$$x(t) = Ae^{j\omega_o t} + Be^{-j\omega_o t} \tag{1.52}$$

Thanks to our earlier investment in the Taylor series expansions of sine, cosine, and exponential functions, we see that these solutions are identical if we use Eq. (1.4) to expand $e^{j\theta}$ in a power series and group together terms that are multiplied by j.

$$e^{j\theta} = 1 - \frac{\theta^2}{2!} + \frac{\theta^4}{4!} - \cdots + j\left[\theta - \frac{\theta^3}{3!} + \frac{\theta^5}{5!} - \cdots\right] = \cos\theta + j\sin\theta \tag{1.53}$$

This result is known as *Euler's formula* named after Swiss mathematician and physicist Leonhard Euler (1707–1783). By similar means, we see that $e^{-j\theta} = \cos\theta - j\sin\theta$. It is also easy to demonstrate the relationship between exponential functions of imaginary arguments and trigonometric functions by taking sums and differences of these expressions of Euler's identity.

$$\cos\theta = \frac{e^{j\theta} + e^{-j\theta}}{2} \quad ; \qquad \sin\theta = \frac{e^{j\theta} - e^{-j\theta}}{2j} \tag{1.54}$$

These algebraic manipulations certainly build confidence in the equivalence of the two solutions of Eq. (1.48), but for our purposes, a geometric interpretation will have significant utility in our study of both vibrations and waves.

1.5.1 Geometrical Interpretation on the Argand Plane

To develop and exploit this geometric interpretation of exponential functions, which contain complex numbers within their arguments (hereafter referred to as *complex exponentials*), we can represent a complex number on a two-dimensional plane known as the "complex plane" or the *Argand plane*. In that representation, we define the x axis as the "real axis" and the y axis as the "imaginary axis." This is shown in Fig. 1.7. In this geometric interpretation, multiplication by j would correspond to making a "left turn" [12], that is, making a 90° rotation in the counterclockwise direction. Since $j*j = j^2 = -1$ would correspond to two left turns, a vector pointing along the real axis would be headed backward, which is the equivalent of multiplication by -1.

In this textbook, complex numbers will be expressed using **bold** font. A complex number, $\mathbf{z} = x + jy$, where x and y are real numbers, would be represented by a vector of length, $|\vec{r}| = \sqrt{x^2 + y^2}$, from the origin to the point, z, on the Argand plane, making an angle with the positive real axis of $\theta = \tan^{-1}(y/x)$. The complex number could also be represented in polar coordinates on the Argand plane as $\mathbf{z} = Ae^{j\theta}$, where $A = |\vec{r}|$. The geometric and algebraic representations can be summarized by the following equation:

$$\mathbf{z} = x + jy = |\mathbf{z}|(\cos\theta + j\sin\theta) = |\mathbf{z}|e^{j\theta} \tag{1.55}$$

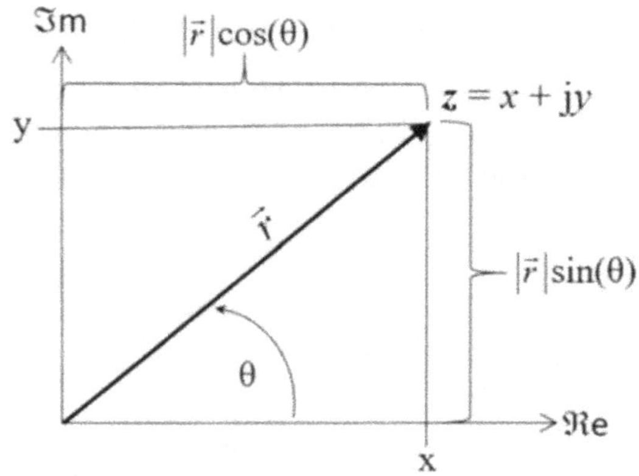

Fig. 1.7 Representation of a complex number, z, on the Argand plane. The projection of z on the Cartesian x axis represents the real part of z, and the y axis projection represents the imaginary part of z. The projections of the vector, $\vec{r}$, onto the real and imaginary axes provide $\Re e(z) = x = |\vec{r}| \cos\theta$ and $\Im m(z) = y = |\vec{r}| \sin\theta$. The ordinary rules of plane geometry and trigonometry apply. For example, $|\vec{r}| = (x^2 + y^2)^{1/2}$ and $\theta = \tan^{-1}(y/x)$

1.5.2 Phasor Notation

In this textbook, much of our analysis will be focused on response of a system to a single-frequency stimulus. We will use complex exponentials to represent time-harmonic behavior by letting the angle θ increase linearly with time, $\theta = \omega_o t + \phi$, where ω_o is the frequency (angular velocity) which relates the angle to time and ϕ is a constant that will accommodate the incorporation of initial conditions (see Sect. 2.1.1) or the phase between the driving stimulus and the system's response (see Sect. 2.5). As the angle, θ, increases with time, the projection of the uniformly rotating vector, $\vec{x} = |\vec{x}|e^{j\omega_o t + j\phi t} \equiv \widehat{\mathbf{x}}e^{j\omega_o t}$, traces out a sinusoidal time dependence on either axis. This choice is also known as *phasor* notation. In this case, the phasor is designated $\widehat{\mathbf{x}}$, where the "hat" reminds us that it is a phasor and its representation in **bold** font reminds us that the phasor is a complex number.

$$\widehat{\mathbf{x}} = |\widehat{\mathbf{x}}|e^{j\theta} \tag{1.56}$$

Although the projection on either the real or imaginary axis generates the time-harmonic behavior, the traditional choice is to let the real component (i.e., the projection on the real axis) represents the physical behavior of the system. For example, $x(t) \equiv \Re e[\widehat{\mathbf{x}}e^{j\omega_o t}]$.

1.5.3 Algebraic Operations with Complex Numbers

The rectangular representation and the polar representation have complementary utility, as the following review of algebraic operations will demonstrate. Addition and subtraction of two complex numbers allow a vector operation to be converted to an algebraic operation, as it would if our concern were analytic geometry instead of complex arithmetic. Starting with two complex vectors, $\mathbf{A_1} = a_1 + jb_1 = |\mathbf{A_1}|e^{j\theta_1}$ and $\mathbf{A_2} = a_2 + jb_2 = |\mathbf{A_2}|e^{j\theta_2}$, where a and b are real scalars, their sum and difference are just the sums and differences of their real and imaginary components:

$$\mathbf{A_1} + \mathbf{A_2} = (a_1 + a_2) + j(b_1 + b_2) \text{ and } \mathbf{A_1} - \mathbf{A_2} = (a_1 - a_2) + j(b_1 - b_2) \tag{1.57}$$

The utility of the polar representation becomes manifest for operations involving multiplication and division. In terms of their real and imaginary components, a and b, multiplication of two complex numbers proceeds just as in the case of the multiplication of two binomials.

$$\mathbf{A_1 A_2} = (a_1 + jb_1)(a_2 + jb_2) = (a_1 a_2 - b_1 b_2) + j(a_1 b_2 + b_1 a_2) \tag{1.58}$$

Multiplication by anything other than a scalar is not a linear process, so the real and imaginary components become mixed by multiplication. The equivalent operation expressed in polar coordinates is both simpler to execute and easier to visualize.

$$\mathbf{A_1 A_2} = |\mathbf{A_1}|e^{j\theta_1}|\mathbf{A_2}|e^{j\theta_2} = |\mathbf{A_1}||\mathbf{A_2}|e^{j(\theta_1+\theta_2)} \tag{1.59}$$

The division of two complex numbers in component form introduces the procedure known as *rationalization*. This is accomplished by multiplying both the numerator and denominator by the *complex conjugate* of the denominator. The complex conjugate of a complex number is just that number with the sign of the imaginary part changed. The complex conjugate is usually designated by an asterisk: $z* = x - jy$; hence, $\Im m[(\mathbf{z})(\mathbf{z}*)] = \Im m[(\mathbf{z}*)(\mathbf{z})] = 0$ (see Sect. 1.5.5). When faced with division by a complex number, the rationalization process makes the denominator a real number and allows the result to be separated into its real and imaginary components.

$$\frac{\mathbf{A_1}}{\mathbf{A_2}} = \frac{(a_1 + jb_1)}{(a_2 + jb_2)} = \frac{(a_1 + jb_1)}{(a_2 + jb_2)}\frac{(a_2 - jb_2)}{(a_2 - jb_2)} = \frac{(a_1 a_2 + b_1 b_2)}{a_2^2 + b_2^2} + j\frac{b_1 a_2 - a_1 b_2}{a_2^2 + b_2^2} \tag{1.60}$$

Again, the process of division is nonlinear, so the real and imaginary components are comingled. The polar representation provides both greater insight and simplified execution.

$$\frac{\mathbf{A_1}}{\mathbf{A_2}} = \frac{|\mathbf{A_1}|e^{j\theta_1}}{|\mathbf{A_2}|e^{j\theta_2}} = \frac{|\mathbf{A_1}|}{|\mathbf{A_2}|}e^{j(\theta_1-\theta_2)} \quad \text{if} \quad |\mathbf{A_2}| \neq 0 \tag{1.61}$$

Raising a number to a real power or extracting a real root is also simplified using a polar representation.

$$\mathbf{A}^n = \left(|\mathbf{A}|e^{j\theta}\right)^n = |\mathbf{A}|^n e^{jn\theta} \tag{1.62}$$

Since the extraction of the nth root generates n solutions, it is helpful to include the periodicity of the complex exponential explicitly, since addition of integer multiples of 2π to the angle θ does not change its value.

$$\sqrt[n]{\mathbf{A}} = \mathbf{A}^{(1/n)} = \left(\mathbf{A}e^{j(\theta+2\pi\nu)}\right)^{(1/n)} = |\mathbf{A}|^{(1/n)}e^{j[(\theta/n)+2\pi(\nu/n)]} \tag{1.63}$$

This will generate n roots with $\nu = 0, 1, 2, \ldots, (n-1)$. Inclusion of larger integer values of $\nu \geq n$ just regenerates the same roots. For example, if we calculate the square root of 4,

$$\sqrt{4} = 2e^{j(0/2)} = 2 \quad \text{and} \quad \sqrt{4} = 2e^{j(2\pi/2)} = 2e^{j\pi} = -2. \tag{1.64}$$

The cube root of 8 has three solutions:

$$\begin{aligned}
\sqrt[3]{8} &= 2e^{j(0/3)} = 2e^{j0} = 2 \\
\sqrt[3]{8} &= 2e^{j(2\pi/3)} = 2[\cos(2\pi/3) + j\sin(2\pi/3)] = -1 + j\sqrt{3} \\
\sqrt[3]{8} &= 2e^{j(4\pi/3)} = 2[\cos(4\pi/3) + j\sin(4\pi/3)] = -1 - j\sqrt{3}
\end{aligned} \tag{1.65}$$

In polar representation, the three roots are three two-unit-long vectors separated by 120°, with the first of those three solutions aligned along the real axis.

The calculation of the natural logarithm of a complex number is also simplified by the use of the polar form.[17]

$$\ln[\mathbf{A}] = \ln\left[|\mathbf{A}|e^{j\theta}\right] = \ln[|\mathbf{A}|] + j\theta \tag{1.66}$$

As for the usual case with real numbers, $\log_{10}[\mathbf{A}] = (\log_{10}e)\ \ln\ [|\mathbf{A}|e^{j\theta}] \cong 0.4343\ \ln\ [\mathbf{A}]$.

1.5.4 Integration and Differentiation of Complex Exponentials

The primary motivation for the introduction of complex exponentials is the ease with which it is possible to write their integrals or derivatives as was expressed in Eq. (1.47).

1.5.5 Time Averages of Complex Products (Power)

Two complex numbers are said to be complex conjugates if the imaginary parts are negatives of each other.

$$\text{If } \mathbf{A} = a + jb = |\mathbf{A}|e^{j\theta}, \text{ then } \mathbf{A}* = a - jb = |\mathbf{A}|e^{-j\theta} \tag{1.67}$$

The polar version shows that the complex conjugate, $\mathbf{A}*$, of a complex number, $\mathbf{A}$, is just the reflection of $\mathbf{A}$ about the real axis. The complex conjugate provides a direct method for calculation of the modulus $|\mathbf{A}|$ of a complex number as well as for the extraction of the real and imaginary parts of a complex number..

$$|\mathbf{A}| \equiv \sqrt{\mathbf{A}\mathbf{A}^*} = [(a+jb)(a-jb)]^{1/2} = \left(a^2+b^2\right)^{1/2} \tag{1.68}$$

$$\Re e[\mathbf{A}] = \left(\frac{1}{2}\right)(\mathbf{A}+\mathbf{A}^*) = \left(\frac{1}{2}\right)[(a+jb)+(a-jb)] = a \tag{1.69}$$

$$\Im m[\mathbf{A}] = (1/2j)(\mathbf{A}-\mathbf{A}^*) = (1/2j)[(a+jb)-(a-jb)] = b \tag{1.70}$$

Complex conjugates are particularly useful for the calculation of time-averaged power, $\langle\Pi\rangle_t$, since that calculation requires evaluation of the "dot product" between two variables that characterize a potential and a flow. For a simple harmonic oscillator, those complex variables are force, $\mathbf{F}$, and velocity, $\mathbf{v}$; for a fluid, they are pressure, $\mathbf{p}$, and volume flow rate, $\mathbf{U}$.[18] The dot product guarantees that only the components of the two variables which are in-phase contribute to the result.

It is important to realize that the product of the real parts of two variables is not equal to the real part of their product.

$$\Re e[\mathbf{F}] \times \Re e[\mathbf{v}] \neq \Re e[\mathbf{F}\mathbf{v}] \tag{1.71}$$

Using the definition of the complex conjugate in Eq. (1.67),

[17] The natural logarithm of a complex number is not unique, as was the case for roots, but for the applications addressed in this textbook, it can be expressed as shown in Eq. (1.66) and can be treated as being unique.

[18] In electrical circuit theory, the potential and flow are the voltage and the electrical current, respectively.

$$\Re e[\mathbf{F}] = \frac{1}{2}\Re e[\mathbf{F} + \mathbf{F}^*] \quad \text{and} \quad \Re e[\mathbf{v}] = \frac{1}{2}\Re e[\mathbf{v} + \mathbf{v}^*] \tag{1.72}$$

The instantaneous power, $\Pi(t)$, can be written using the results of Eq. (1.72).

$$\begin{aligned} \Pi(t) &= F \bullet v = \frac{1}{4}(\mathbf{F} + \mathbf{F}^*)(\mathbf{v} + \mathbf{v}^*) = \frac{1}{4}[(\mathbf{F}\mathbf{v} + \mathbf{F}^*\mathbf{v}^*) + (\mathbf{F}\mathbf{v}^* + \mathbf{v}\mathbf{F}^*)] \\ \Pi(t) &= \frac{1}{2}(\Re e[\mathbf{F}\mathbf{v}] + \Re e[\mathbf{F}\mathbf{v}^*]) = \frac{1}{2}\Re e\left[|\mathbf{F}||\mathbf{v}|e^{j(2\omega t + \phi_F + \phi_v)} + |\mathbf{F}||\mathbf{v}|e^{j(\phi_F - \phi_v)}\right] \end{aligned} \tag{1.73}$$

The second version of the instantaneous power assumes simple harmonic time variation of both **F** and **v** at frequency, ω, with a phase difference of $\phi_F - \phi_v$. If we take the real part of the term in square brackets, the instantaneous power can be expressed in terms of two cosines.

$$\Pi(t) = \frac{1}{2}|\mathbf{F}||\mathbf{v}|[\cos(2\omega t + \phi_F + \phi_v) + \cos(\phi_F - \phi_v)] \tag{1.74}$$

The time-averaged power is determined by integration of Eq. (1.74) over a complete cycle with period, $T = 2\pi/\omega$, or if the integral is over a time much longer than one period. The first term is oscillatory, so the time average is zero. The second term is just a constant, so the time integral over the second term is non-zero.

$$\langle \Pi(t) \rangle_t \equiv \frac{1}{T}\int_0^T \Pi(t)dt = \frac{1}{2}|\mathbf{F}||\mathbf{v}|\cos(\phi_F - \phi_v) = \frac{1}{2}\Re e[\mathbf{F}^*\mathbf{v}] \tag{1.75}$$

Taking the complex conjugate of the argument on the right-hand side of Eq. (1.75) does not change the result, since we are extracting only the real part.

$$\langle \Pi(t) \rangle_t = \frac{1}{2}\Re e[(\mathbf{F}^*\mathbf{v})^*] = \frac{1}{2}\Re e[\mathbf{v}^*\mathbf{F}] \tag{1.76}$$

1.6 Standard (SI) Units and Dimensional Homogeneity

We take for granted that every term in any equation must have the same units as all the other terms in that same equation. This concept is known as *dimensional homogeneity*. From the time of Isaac Newton (1642–1727) through the early nineteenth century, mathematical theories representing physical phenomena were expressed as ratios. For example, Ohm's law was expressed by Georg Ohm as "The force of the current in a galvanic circuit is directly as the sum of all the tensions (i.e., voltages), and inversely as the entire reduced length of the circuit." Similarly, Hooke's law was expressed as "Stress is proportional to strain." The idea that laws could be written as equations with consistent units for all terms was introduced by Jean-Baptiste Joseph Fourier. He had to fight for 20 years before this radical concept gained widespread acceptance [13].

By advocating this approach, Fourier also necessarily introduced dimensional constants. At that time, his concentration was on the transfer of heat, so he introduced both the thermal conductivity, κ [W/m-K], and the convective heat transfer coefficient, h [W/m^2-K].[19] Not only did this make it possible to express physical laws as equations, such "constants" characterized the properties of specific materials. The American Society of Metals publishes the *ASM Handbook* that has 23 volumes that are

[19] In this textbook, when it is appropriate to designate units, they will be shown within square brackets [ ].

filled with such constants for a variety of metals, alloys, and composites. The printed version of the *ASM Handbook* takes over a meter of shelf space, although it is now also available in electronic form.

Throughout this textbook, we will use the *SI System of Units* exclusively.[20] Although this system is also commonly known as "MKS" units, because it uses meter [m], kilogram [kg], and second [s], it is actually based on seven units, of which those are just the first three. The remaining "base units" that complete the system are ampere [A], kelvin [K], mole [mol], and candela [cd].[21] These seven form the basis of other "derived" units, many of which are commonly used in acoustics and vibration: hertz, Hz [s^{-1}]; newton, N [kg-m/s^2]; pascal, Pa [N/m^2]; joule, J [N-m]; watt, W [J/s]; coulomb, C [A-s]; volt, V [W/A]; farad, F [C/V]; ohm, Ω[V/A]; siemens, S [Ω^{-1}]; weber, Wb [V-s]; tesla, T [Wb/m^2]; and henry, H [Wb/A]. Although I was not pleased by the change of "cycles per second" to Hz and the conversion of the conductance unit from mho [℧] (the inverse of ohm) to siemens [S], this system is very convenient because it is easy to integrate mechanical and electrical effects, an important feature for acoustical transduction systems (e.g., speakers, microphones, accelerometers, geophones, seismometers, etc.).

As of the date this book is being written, only the United States, Liberia, and Burma (Myanmar) have yet to accept SI as their national standard.[22] We are frequently required either to accept data that is expressed in other systems of units or to report results in other units. This textbook includes problems that require results reported in "customary units," but it is recommended that all computations be done using SI units and then converted as required. It follows that any result that is not a pure number must be reported along with its units.

The other recommendation regarding units is that any equation should be checked to ensure dimensional homogeneity before any numerical values are placed into that equation. An equation that is dimensionally homogeneous may be wrong, but one that is not dimensionally homogeneous is meaningless. An equation that is not dimensionally homogeneous is not even wrong!

1.7 Similitude and the Buckingham Π-Theorem (Natural Units)

> *"I have often been impressed by the scanty attention paid even by original workers in physics to the great principle of similitude. It happens not infrequently those results in the form of 'laws' are put forward as novelties on the basis of elaborate experiments, which might have been predicted a priori after a few minutes of consideration."* (J. W. Strutt (Lord Rayleigh) [14])

I will be the first to admit that "a few minutes of consideration" by Lord Rayleigh might be the equivalent of many hours of my personal effort, but the idea that physical units can dictate the form of mathematical expressions describing physical laws is important at a very fundamental level. Those of you who have studied Mechanical Engineering and/or Fluid Mechanics can attest to the significance of dimensionless groups like the Reynolds number for description of viscous fluid flow and the Nusselt number for description of convective heat transfer.

Although most attention to the discussion of units focuses on the topics in the previous section and we rely on the "standard" SI System of Units to report results and execute calculations, there is an equally important role played by the use of "natural units" that is often overlooked. Because its

[20] The International System of Units is designated "SI" from the French *Le Système international d'unités*.

[21] When the names of units are spelled out, they are not capitalized, even though many are based on proper names. When used as abbreviations, the first letter of those abbreviations are capitalized.

[22] The United States is proceeding to the metric system inch by inch.

exploitation is rather subtle, it will be introduced first by example; then it will be generalized within a more formal (and powerful) structure.

1.7.1 Three Simple Examples

The simplest vibratory system is the mass-spring oscillator. Although that topic will be covered in detail in the next chapter, there are significant results that can be obtained without even writing down the equations governing the oscillations of a mass on a spring (i.e., Newton's Second Law of Motion and Hooke's law). Following Fourier's introduction of dimensional constants, we assume that the "mass" that is attached to the spring is characterized only by its mass, m. We also know that the spring is characterized by its "stiffness," K, as defined in Eq. (1.25). Mass happens to be one of the SI base units. According to Eq. (1.25), the dimensions of K must be force divided by displacement.

$$[\mathrm{K}] = \frac{[\text{Force}]}{[\text{displacement}]} = \left[\frac{N}{m}\right] = \left[\frac{kg - m}{s^2 - m}\right] = \left[\frac{kg}{s^2}\right] \equiv \left[\frac{M}{T^2}\right] \tag{1.77}$$

For the discussions that employ dimensional analysis to extract functional relations (a part of a process called *similitude*), it is common to abbreviate the "base units" as M for mass, L for length or distance, T for time, and Θ for temperature.

Instead of thinking of this problem in terms of SI units, we can use m and K as our "natural units." First, notice that the definitions of m and K involve two base units, M and T, and that both M and T appear in K. Let's say we are interested in the calculation of the frequency of oscillation, f, of this simple mass-spring combination. Since frequency has the units of reciprocal time ($1/T$), we can ask what combination of K and m, the natural units for the description of this specific system, will have the units of frequency? Although it is not too difficult to simply guess the correct result, let's employ a more formal procedure that will become useful when we analyze more complicated applications.

To express the natural frequency of vibration, f_o, in terms of K and m, we can let a and b be exponents and write an equation to guarantee dimensional homogeneity:

$$m^a \mathrm{K}^b = \frac{M^a M^b}{T^{2b}} = \frac{1}{T} = T^{-1} M^0 \tag{1.78}$$

The term at the far right is just included to emphasize that we are seeking a result that does not include M. Now the values of a and b can be determined by inspection: $a + b = 0$ so $a = -b$. Equating the time exponents, $b = ½$ so $a = -½$. Based solely on the units of K and m, we can write the required functional form of our desired result.

$$f_o \propto \sqrt{\frac{\mathrm{K}}{m}} \tag{1.79}$$

This approach cannot determine the numerical constant that would change this proportionality into an equation. As we will see in the next chapter, if we choose to use the radian frequency, $\omega_o = 2\pi f_o$, then no numerical constant is required: $\omega_o = \sqrt{\mathrm{K}/m}$. If we were deriving this relationship for a more complicated system, for which an analytic solution may not exist, then an experiment or a numerical calculation could be used to determine the constant. Even more importantly, if the experiments or the numerical calculations did not obey Eq. (1.79), then there must be other parameters required to specify the dynamics, or the experiments or calculations are flawed.

What if we added damping to the mass-spring system? The damping is characterized by a mechanical resistance, R_m, that has units of [kg/s]. Now we have three parameters (m, K, and R_m), but those parameters still only involve two base units. When the number of parameters exceeds the number of base units, it is then possible to combine those parameters to form *dimensionless groups*.

Using the same approach as employed in Eq. (1.78), we can identify this dimensionless group.

$$m^a \mathrm{K}^b R_m^c = \frac{M^a M^b M^c}{T^{2b} T^c} = M^0 T^0 \tag{1.80}$$

The resulting equations for the exponents show that $a + b + c = 0$ and $2b + c = 0$, so $c = -2b$; thus $a = b$. These dimensionless groups are called *Π-groups*, based on the classic paper by Buckingham [15]. In this case, we have created one Π-group that is designated Π_1.

$$\Pi_1 = \frac{\mathrm{K}m}{R_m^2} = \frac{\omega_o^2 m^2}{R_m^2} \equiv Q^2 \tag{1.81}$$

Substitution for K, using $\omega_o{}^2 = (\mathrm{K}/m)$, produced an expression for Π_1 that will appear frequently in the next chapter as the dimensionless quality factor, Q. When damping is present, our result for frequency can only depend upon R_m through some function of Q^2, hence the damped frequency, $\omega_d = f$ (K, m, Q^2). Again, that function f (K, m, Q^2) can be determined either by experiment or by solution of the fundamental Eq. (1.28) with $F(t) = 0$.

One additional simple example might be helpful before introducing the general result. Equation (1.27) expresses the torque on a simple pendulum. The parameters in that expression are the mass of the bob, m; the length of the string, L; and the acceleration due to gravity, $g \cong 9.8$ m/s^2. We have three parameters and three "base units," so there are no dimensionless groups. As before, recognizing that frequency, either f_o or ω_o, has units of T^{-1}, the dimensional equation below can be solved for the appropriate exponents.

$$m^a L^b g^c = \frac{M^a L^b L^c}{T^{2c}} = \frac{1}{T} = T^{-1} M^o L^o \tag{1.82}$$

The first result is that $a = 0$, so the frequency is independent of the mass of the bob. Since time only enters through the acceleration due to gravity, $2c = 1$ and $b + c = 0$, so $c = -b = ½$.

$$f_o \propto \sqrt{\frac{g}{L}} \tag{1.83}$$

Even though we included the mass of the pendulum bob as a parameter, that turned out to have no effect on the pendulum's frequency; our dimensional analysis was able to exclude that parameter from the final result even if our intuition did not.

1.7.2 Dimensionless Π-Groups

The general idea can be summarized by expressing some dependent variable, q_o, in terms of the independent variables, q_1 through q_m. Those independent variables characterize the physical process that defines the dependent variable through some function, F, of the independent variables.

$$q_o = F(q_1, q_2, \ldots q_m) \tag{1.84}$$

q_1 through q_m define m independent parameters. If those independent parameters can be expressed with n "base units," then there will be $(m-n)$ dimensionless groups. If we rearrange Eq. (1.84) into a different function, G, which includes all the variables and set G equal to zero, then we obtain one more dimensionless group that results in a homogeneous expression.

For our simple oscillator examples, we solved for q_o (f_o or ω_o in our examples) based on the independent parameters. If we would have included f_o, then we would have generated one dimensionless group that would have been equal to zero. With either approach, we would have arrived at the same result.

Those three examples that dictated the functional dependence of the frequencies for three oscillators were particularly simple. We will be able to obtain these results (including the required numerical pre-factor) by solving the differential equations directly. In acoustics and vibration, there are problems for which we cannot obtain closed-form analytical solutions, even when we know the fundamental governing equations [16]. There are occasions when a numerical answer for a particular case can be obtained using digital computation, but an "answer" for a single case is not the same as a "solution." The use of dimensional analysis can provide an independent check on an "answer" and can provide the functional form of a solution (where the "answer" might be used to produce the numerical pre-factor). Also, our dimensional solution can be used to test answers for different cases (e.g., flow velocity, object sizes, etc.), since the parameters in the numerical solution have presumably been identified.

1.7.3 Windscreen Noise*

We will complete this section by examining an acoustical example that does not easily yield to conventional mathematical analysis. Windscreens are placed around microphones to reduce the noise that such microphones detect when they measure sound in the presence of flowing air (wind). Figure 1.8 shows measurements of the sound pressure levels in one-third octave bands, $p_{1/3}$, as a function of the band-centered frequencies for air flow velocities from 2 m/s to 14 m/s.

To apply our dimensional analysis, we first need to determine the relevant parameters. Some thought is required to select a physically significant set. Obviously, frequency, f; one-third octave-band pressure level, $p_{1/3}$; flow velocity, v; and windscreen diameter, D, seem to be potentially relevant. We also might include the properties of the fluid. Three possible candidates are the density, ρ; the sound speed, c; and the shear viscosity, μ, of the flowing air. As we saw with the pendulum frequency, we can choose the wrong (or suboptimal) parameter set and still end up with a valid solution. Our practical interest is in understanding the noise pressure levels.

If we examine the first four parameters that are identified in Table 1.1, we notice that there is no other parameter that includes mass other than $p_{1/3}$. To nondimensionalize $p_{1/3}$, we have to include a fluid parameter that has units of mass. Since the flow speeds are much less than the sound speed (which does not contain units of mass) and we expect that the windscreen noise is generated by turbulence around the screen, I will select density rather than viscosity.[23]

We have identified five parameters that can be expressed using three basic units, so we will be able to form two dimensionless Π-groups: $(m-n) = 5\text{–}3 = 2$. Again, since our interest is in the pressure levels at different frequencies, the choice suggested by the raw data plotted in Fig. 1.8 is to create a

[23] Here I have used my knowledge that the effects of density dominate those of viscosity for turbulent solutions to the Navier-Stokes equation (which presumably generate the turbulent flow noise the windscreens are intended to suppress). The success of dimensional analysis is frequently enhanced by what Rayleigh called "a few minutes of consideration" [14], which I interpret as the application of additional physical insight related to the specific problem.

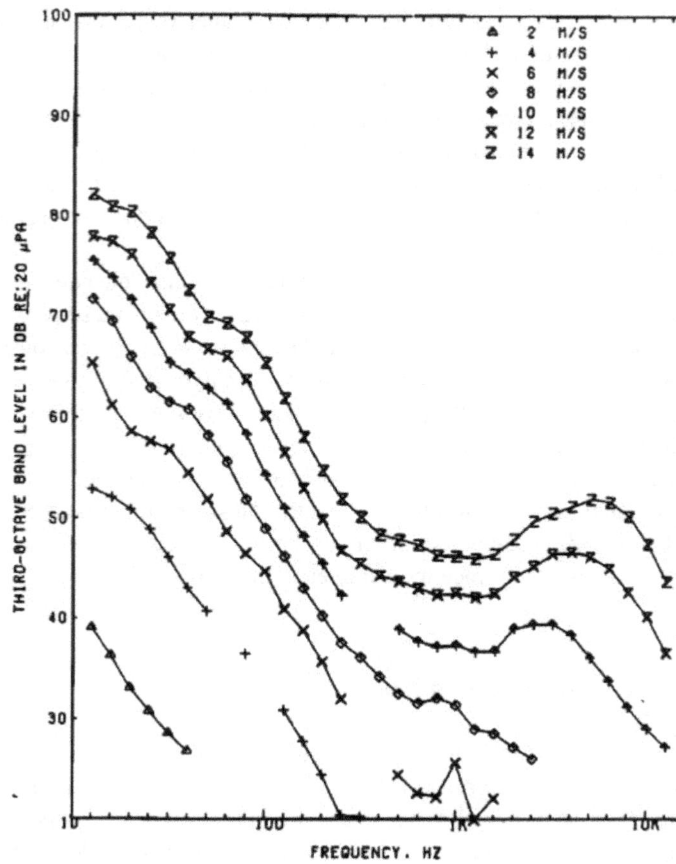

Fig. 1.8 Measured one-third octave-band levels of noise generated by a microphone with a 9.5-cm-diameter spherical windscreen in air flows that ranges from 2 m/s to 14 m/s [18]. The band center frequencies of the one-third octave-band levels are plotted on a logarithmic horizontal axis. The magnitudes of the one-third octave bands are plotted as dB *re:* 20 μPa$_{rms}$

Table 1.1 Five parameters that are being tested to determine the dimensionless groups that will characterize the one-third octave-band pressure, $p_{1/3}$, detected by a microphone protected by a windscreen of characteristic diameter, D, in air of density, ρ, that is flowing past the microphone with a velocity, v, as a function of the band center frequency, f, of each one-third octave band

Parameter	f	$p_{1/3}$	v	D	ρ
Basic units	s^{-1}	$N/m^2 = kg/m \cdot s^2$	m/s	m	kg/m^3
Dimensions	T^{-1}	$ML^{-1}T^{-2}$	LT^{-1}	L	ML^{-3}

With five parameters selected and three base units, there will be two dimensionless Π-groups

dimensionless frequency, Π_f, and a dimensionless pressure, Π_p. For the calculation of Π_f, we know that the units for f do not include mass, so we can eliminate ρ from the Π_f group.

$$\Pi_f = f^a v^b D^c = \frac{1}{T^a}\frac{L^b}{T^b}L^c = T^0 L^0 \tag{1.85}$$

The length terms suggest $b + c = 0$, or $b = -c$, and the time terms make $a + b = 0$, so $a = -b = c = 1$; hence $\Pi_f = fD/v$. This dimensionless group is known as the Strouhal number.

The Π-group for pressure will not include the frequency, since Π_p will be used to replace the results plotted with reference to the vertical axis of the graph in Fig. 1.8 with the dimensionless axis in Fig. 1.9.

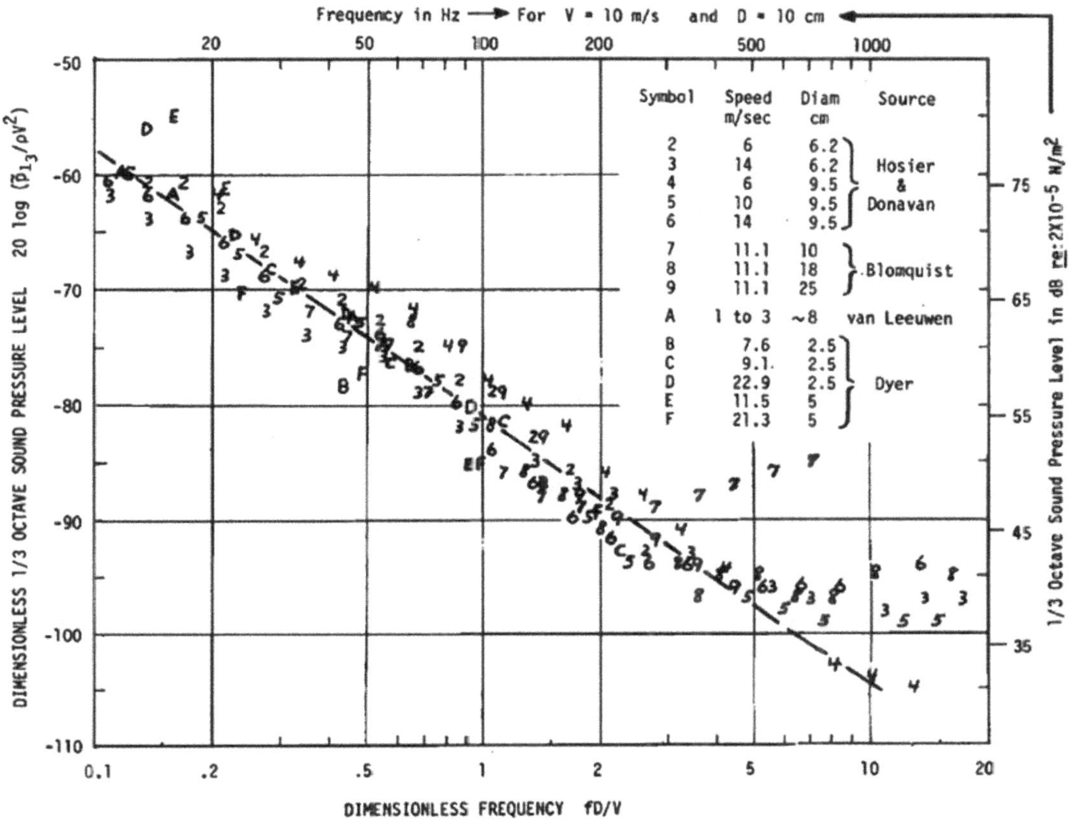

Fig. 1.9 Plot of the dimensionless one-third octave-band pressures, $\Pi_p = p_{1/3}/\rho v^2$, versus the dimensionless frequencies, $\Pi_f = f D/v$ (Strouhal number), for four different groups of measurements, performed by four different researchers, using windscreens of diameters ranging from 2.5 cm to 25 cm, in air flows with velocities ranging from 1 m/s to 23 m/s [17]. The vertical axis at the left is Π_p scaled in decibels. The vertical axis at the right is the one-third octave-band sound pressure levels in dB *re:* 20 μPa$_{rms}$, if $v = 10$ m/s. The lower horizontal axis is logarithmic in dimensionless frequency, Π_f, and the upper axis is scaled in frequency for $v = 10$ m/s and $D = 10$ cm. The systematic deviation of some data points from the dashed best-fit straight line above $\Pi_f > 2$ indicates that another parameter may have become important, possibly the windscreen pore size

$$\Pi_p = \left(p_{1/3}\right)^a v^b D^c \rho^d = \frac{M^a}{L^a T^{2a}} \frac{L^b}{T^b} L^c \frac{M^d}{L^{3d}} = M^0 L^0 T^0 \tag{1.86}$$

The mass terms require $a + d = 0$, so $a = -d$; the time terms require $2a + b = 0$ or $b = -2a$; and the length terms require $-a + b + c - 3d = 0$. Substituting time and mass results into length, we obtain $-a - 2a + c + 3a = 0$, so D did not need to be part of $\Pi_p = p_{1/3}/\rho v^2$; one-third octave-band pressure levels are therefore normalized by the kinetic energy density of the fluid.

Strasberg has taken the windscreen noise data from Fig. 1.8 and plotted it, along with measurements made by three other investigators in three other laboratories, as $p_{1/3}/\rho v^2$ versus fD/v in Fig. 1.9 [17]. This choice of nondimensional variables causes all of the data in Fig. 1.8, which were represented by seven different curves, to fall on *a single straight line* along with three other sets of measurements made by three other researchers using windscreens of eight different diameters in Fig. 1.9.

The fact that Fig. 1.9 shows systematic deviation of some data sets from the best-fit straight line for $\Pi_f > 2$ could indicate that some other parameter that was not included in the dimensional analysis has become important in that regime. For these measurements, the windscreen pore size might start to play a role at higher frequencies.

1.7.4 Similitude Summary

Fourier's requirement for dimensional homogeneity introduced dimensional constants that can provide sets of "natural units." Such "derived units" lead to important restrictions on the functional form (to within a multiplicative constant) of mathematical expressions for the behavior of physical systems. We have used a formalism introduced by Buckingham [15] to extract those functions and to determine how many independent dimensionless groups can be formed that control the behavior of the physical system under the assumption that we have identified all the controlling parameters correctly. Although we may not be able to derive or even guess the form of the functions, we know that those functions can only depend on those dimensionless Π-groups.

Figure 1.9 demonstrated how the use of Π-groups as the plotting variables can produce a universal relationship that unifies seemingly diverse sets of data, like those in Fig. 1.8, to produce a single correlation across a large variation of experimental parameters.

The Π-groups also provide guidance in the design and planning of experiments. Using the Π_f-group, we could select flow velocities and windscreen diameters to optimize our variation of Strouhal number. The Π_ρ-group introduces the option of using gases of different densities to further increase the range of an experimental investigation, either by using less dense gases like H_2 or He or a denser gas like SF_6 (sulfur hexafluoride), in addition to air, or by changing the mean pressure to modify the density of a single gas.

Experiments are expensive! It is very valuable to be able to span the largest range of parameters with the fewest measurements.

Since the use of similitude is not commonly taught as a powerful technique for acoustics and vibration (although it is indispensable to the education of mechanical engineers in fields like fluid flow and heat transfer), the steps in determining Π-groups will be summarized again below:

1. List all of the parameters involved and their SI "base units." An arrangement like Table 1.1 can be useful.
2. The difference between the number of parameters, m, and the required number of SI units, n, needed to express the m parameters will determine the number of dimensionless Π-groups. The resultant number of Π-groups equals $(m - n)$.
3. Select n of the m parameters you will use as the "natural units," keeping in mind that you cannot use parameters that differ by only an exponent, for example, a base unit like length, L, and the volume, L^3.
4. Express the other $(m - n)$ parameters, one at a time, in terms of the "natural units" using the requirement of dimensional homogeneity (e.g., Eqs. (1.78), (1.80), (1.81), (1.82), (1.85), and (1.86)) to produce $(m - n)$ dimensionless Π-groups.
5. Check to be sure that the Π-groups you have formed are dimensionless.

1.8 Precision, Accuracy, and Error Propagation

"The best instrument for detecting systematic error is between your ears." [19]

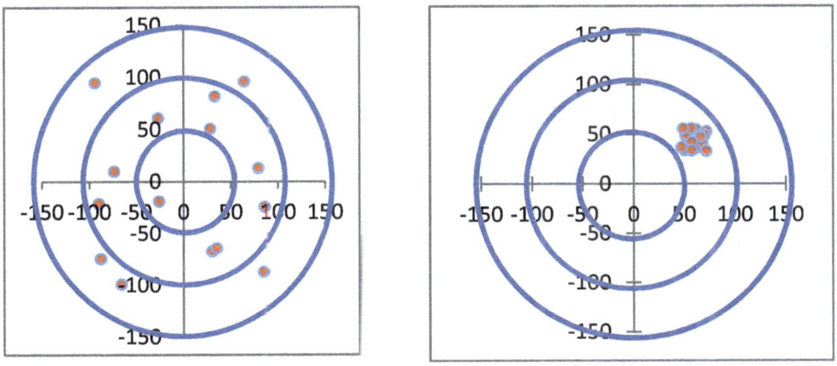

Fig. 1.10 The target patterns created by 16 bullets fired from two hypothetical guns illustrating the difference between random error and systematic (bias) error. If we assume that the numbering on the targets corresponds to centimeters, the holes in the target at the left have an average position at the exact center of the target, but the standard deviation of the distance of those holes from the center is 94 cm. The average distance of the 16 holes in the target at the right is 60 cm to the right of the center and 45 cm above the center, corresponding to an average "miss" distance from the center of 75 cm, but the standard deviation about that average is 12 cm. The target on the left exhibits large random error (possibly due to large fluctuations in the wind), and the target on the right is dominated by systematic error, possibly due to a misaligned gun sight

This textbook will dedicate a lot of space to the development and derivation of mathematical results that summarize the behavior of vibratory and acoustical systems. All of these equations will contain parameters that either specify the properties of a material or component (e.g., stiffness, mass density, elastic modulus, sound speed, heat capacity, thermal conductivity, viscosity, etc.) or the value of some geometric feature of the system under consideration (e.g., length, perimeter, diameter, surface area, volume, etc.). Each of these parameters will have some uncertainty in its specification and those uncertainties will be reflected in the error estimate of the result. For that reason, this section as well as the subsequent section of this chapter will address some basic concepts that are used to quantify those uncertainties and to combine them to provide an estimate of the uncertainty in the final result.

It is important to first recognize that there are two distinct types of errors (other than blunders): (1) *random errors* due to noise-induced statistical fluctuations and (2) *systematic errors*, also known as *bias errors*, which arise because of calibration errors or an oversimplification of the equation that connects the measurements to the desired result. The reason that it is important to distinguish between those errors is that they need to be addressed in two very different ways. Figure 1.10 illustrates this difference.

1.8.1 Random Errors (Noise) and Relative Uncertainty

We are usually able to quantify random errors by making N measurements of a scalar quantity, x_i, under (presumably!) identical conditions and then using the average (mean) of those measurements, $\bar{x}$, as the best estimate of the value of the parameter being measured.

$$\bar{x} = \frac{\sum_{i=1}^{N} x_i}{N} \tag{1.87}$$

The variability of such a collection of N measurements can be quantified by defining the deviation of each individual measurement from the mean value: $\delta x_i = x_i - \bar{x}$. Based on the definition of the mean, the sum of all the deviations will be zero, being equally positive and negative. If all of the deviations, δx_i, are squared, then the square root of the sum is an indication of the *statistical fluctuations* about the mean value.

The square root of the sum of the squares of the deviations is generally not used as a statistical measure of the fluctuation since it is widely recognized that if the measurement were repeated and another set of N measurements were collected, both the average, $\bar{x}$, and the deviations from the average, δx_i, would be different, although both samples would represent the same phenomenon.[24] Careful statistical analysis distinguishes between the "population" and the "sample" and suggests that the *standard deviation*, σ_x, be slightly larger than the root-mean-squared deviation [20]. Since our values of N are usually fairly large, the difference is usually not significant.

$$\sigma_x = \sqrt{\frac{\sum_{i=1}^{N} (x_i - \bar{x})^2}{N-1}} = \sqrt{\frac{\sum_{i=1}^{N} \delta x_i^2}{N-1}} \qquad (1.88)$$

This variability can be attributed to a number of different sources depending upon the experimental conditions, sensors, signal-conditioning electronics, cabling, connectors, and instrumentations. Since many measurements today are made with instruments that employ an analog-to-digital conversion, some variability (noise) might be due to digitization errors, round-off errors, or simple truncation.

Electronic measurements are also susceptible to electromagnetic interference (EMI) that may be correlated to the power line frequency (e.g., 60 Hz in the North America, 50 Hz in Europe) and its harmonics[25] or higher-frequency radiation from switch-mode power supplies, motor controllers, or other power electronics. Sometimes the fluctuations in the readings might be due to fluctuating conditions such as background noises and vibrations, wind, etc. For our purposes, the sources of these fluctuations are not the immediate concern, although recognizing the magnitude and importance of such fluctuations is the reason for introducing a standard means of quantifying such random influences.

In this textbook (and in my own research), we will use the ratio of the standard deviation to the mean to define a *relative uncertainty* of the mean value. This normalization of the standard deviation is convenient both because the relative uncertainty is dimensionless and because it makes it very easy to combine errors from different sources to produce an uncertainty in a result that depends upon simultaneous errors in the measurement of several different parameters (see Sect. 1.8.4). I designate this relative uncertainty as $\delta x/|\bar{x}|$. In this simple case, involving a single variable, x, $\delta x/|\bar{x}| \equiv \sigma_x/|\bar{x}|$. We typically use the relative uncertainty to specify the *precision* of a measurement.

This concept is extended to two dimensions in Fig. 1.10, which simulates randomness using a pattern of $N = 16$ bullets fired at targets by two different weapons. The holes in the target on the left were produced by generating two columns of uncorrelated random numbers between -100 and $+100$ representing the x_i and y_i for $i = 1$ through 16. The average values of both components were zero: $\bar{x} = \bar{y} = 0$. If the numbers on the axes in Fig. 1.10 represent centimeters, the standard deviations of both variables are also nearly equal: $\sigma_x \cong \sigma_y \cong 66$ cm. The average of those 16 shots is a "bull's-eye," although only one of the bullets landed within 50 cm of the origin. Only 11 of the 16 bullets (69%) were within $|\vec{r}| = \left(\sigma_x^2 + \sigma_y^2\right)^{1/2} = 94\,\text{cm}$ of the center.

[24] The *ergodic hypothesis* is a statement of the assumption that the average of a parameter over time in a single system and the average of the same parameter at a single time in a number of similarly prepared systems yield the same average value.

[25] The best modern electronic instruments are designed to integrate the measured signal over an integer number of power line cycles so that any variations that are synchronized to the line frequency will average to zero.

The target on the right-hand side of Fig. 1.10 has a much smaller variation in the positions of the bullets, $\sigma_x \cong \sigma_y \cong 8$ cm and $\left(\sigma_x^2 + \sigma_y^2\right)^{1/2} = 12\,\text{cm}$, but the average position is displaced; $\overline{y} = 60$ cm and $\overline{y} = 45$ cm, so the center of the distribution is located = 75 cm from the origin of the coordinates at an average angle of 37° above the horizontal to the right of center.

1.8.2 Normal Error Function or the Gaussian Distribution

"Everyone believes that the Gaussian distribution describes the distribution of random errors; mathematicians because they think physicists have verified it experimentally, and physicists because they think mathematicians have proved it theoretically." [21]

When we think about random processes, a common example is the coin toss. If we flip a fair coin, there is a 50% probability that the coin will land "heads" up and an equal probability that it will land "tails" up. If we toss *N* coins simultaneously, the probability of any particular outcome, say *h* heads and $t = N - h$ tails, is given by a binomial distribution. The average of that distribution will still be $\overline{h} = \overline{t} = N/2$, but the likelihood of getting exactly *N*/2 heads in any given toss is fairly small and grows smaller with increasing *N*.

The probability, P_B (*h*, *p*, *N*), of obtaining *h* heads and $t = N - h$ tails is given by a binomial distribution where the probability of obtaining a head is $p = 0.5$. Of course, the probability of a tail is $q = 0.5$.

$$P_B(h, p, \text{N}) = \frac{N!}{h!(N-h)!} p^h q^{(N-h)} \tag{1.89}$$

For the binomial distribution, the average outcome with the largest probability is the mean, $\overline{h} = Np$, and the standard deviation about that mean is $\sigma = [Np(1-p)]^{1/2}$. It is worthwhile to recognize that $\overline{h}$ is proportional to the number of trials *N*, whereas σ is proportional to $\sqrt{N}$. Therefore, the relative width of the distribution function, $\sigma/\overline{h}$, decreases in proportion to $\sqrt{N}$. This "sharpening" of the distribution with increasing *N* is evident in Fig. 1.11.

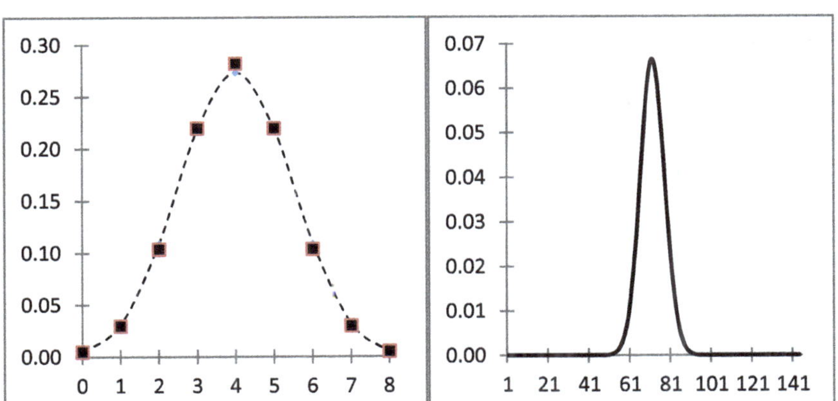

Fig. 1.11 (*Left*) The squares are the probabilities of obtaining a number of heads when tossing $N = 8$ coins. The highest probability is for $h = 4$ heads, P_B (4, 0.5, 8) = 0.273. The standard deviation is $\sigma = \sqrt{2}$. Superimposed over the discrete results obtained from the binomial distribution is the dashed line representing a Gaussian distribution with the same mean and standard deviation, P_G (4, 4, √2) = 0.282. (*Right*) With $N = 144$, the probabilities determined by both the Gaussian and the binomial distribution functions are indistinguishable. Here the highest probability P_G (72, 72, 6) = 0.0665 is for $h = 72$ heads, and the standard deviation is $\sigma = 6$

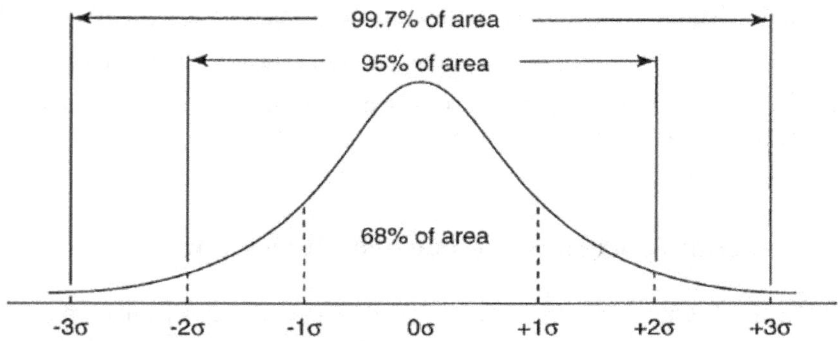

Fig. 1.12 Integration of the Gaussian distribution makes it possible to determine the area corresponding to the probability of a measurement being within a chosen distance from the mean in terms of the standard deviation of the distribution

If we are interested in a continuous variable instead of a variable with a discrete number of possible outcomes, we can use a Gaussian distribution that provides the probability $P_G(x, \bar{x}, \sigma)$ of obtaining a value, x.

$$P_G(x, \bar{x}, \sigma) = \frac{1}{\sigma\sqrt{2\pi}} e^{\left[-\frac{1}{2}\left(\frac{x-\bar{x}}{\sigma}\right)^2\right]} \tag{1.90}$$

Both of these distribution functions are normalized. The sum of the probabilities for the binomial distribution must equal one, and the integral under the Gaussian distribution must also equal one. These two probability distribution functions are superimposed in Fig. 1.11 for $N = 8$ coins and $N = 144$ coins.

The purpose of introducing these distribution functions is to provide some convenient interpretation of the standard deviation. If the measurement errors are truly random, then it is probable that 68% of the measurements will fall within one standard deviation from the mean. For a normal (Gaussian) distribution, 95% of the results will be within two standard deviations from the mean. This is illustrated in Fig. 1.12. In the "target" example on the left of Fig. 1.10, we have already shown that 69% of the holes were within one standard deviation of the distance from the center of the target, so that example quite accurately exhibits Gaussian "randomness."[26]

In the two cases treated in Fig. 1.11, the relative uncertainty for the $N = 8$ case is $\delta x/|\bar{x}| = (2)^{1/2}/4 = 0.354$ and for the $N = 144$ case is $\delta x/|\bar{x}| = 6/72 = 0.083$. Their ratio is 4.243, which is $\sqrt{144/8}$. This is not a coincidence. For random errors, the relative uncertainty decreases as the square root of the number of measurements. This is a result that is worth keeping in mind when you have to decide whether to reduce the random error by increasing the number of measurements or trying to improve the measurement in a way that reduces the noise (e.g., provide better shielding against EMI, use differential inputs to provide common-mode rejection, etc.).

[26] This is only a rough check of the "Gaussianity" of the distribution. More sophisticated statistical tests, such as the Kolmogorov-Smirnov test, can be applied to determine a quantitative estimate of the distribution's similarity to a Gaussian distribution.

1.8.3 Systematic Errors (Bias)

Systematic error is not reduced by increasing the number of measurements. In the "target" example on the right-hand side of Fig. 1.10, taking more shots will not bring the average any closer to the bull's-eye. On the other hand, an adjustment of the sighting mechanism could produce results that are far better than those shown in the left-hand side of Fig. 1.10 by bringing the much tighter cluster of holes on the right-hand side toward the center of the target.

The right-hand side of the target example in Fig. 1.10 represents a type of systematic error that I call a "calibration error." These can enter a measurement in a number of ways. If a ruler is calibrated at room temperature but used at a much higher temperature, the thermal expansion will bias the readings of length. In acoustic and vibration experiments, often each component of the measurement system may be calibrated, but the calibration could be a function of ambient pressure and temperature. The "loading" of the output of one component of the system by a subsequent component can reduce the output or provide a gain that is load-dependent.

For example, a capacitive microphone capsule with a capacitance of 50 pF has an electrical output impedance at 100 Hz of $Z_{el} = (\omega C)^{-1} = 32\ \text{M}\Omega$. If it is connected to a preamplifier with an input impedance of 100 MΩ, then the signal amplified by that stage at 100 Hz is reduced by (100/132) $\cong$ 0.76. Even though the capsule may be calibrated with a sensitivity of 1.00 mV/Pa, it will present a sensitivity to the preamplifier of 0.76 mV/Pa. At 1.0 kHz, the capsule's output impedance drops to 3.2 MΩ, so the effective sensitivity at that higher frequency will be (100/103) $\times$ 1.00 mV/Pa $=$ 0.97 mV/Pa. A typical acoustic measurement system may concatenate many stages from the sensor to its preamplifier, through the cabling, to the input stage of the data acquisition system, through some digital signal processing, and finally out to the display or some recording (storage) device.

As important as it is to know the calibration or gain (i.e., transfer function) of each stage in the measurement system, it is also imperative that the entire system's overall behavior be tested by an *end-to-end calibration* that can confirm the calculation of the overall system's sensitivity. This is usually accomplished by providing a calibrated test signal to the sensor and reading the output at the end of the signal acquisition chain. Commercial calibrators are available for microphones, hydrophones, and accelerometers. If a commercial calibrator is not available for a particular sensor, some end-to-end calibration system should be designed as part of the test plan for every experiment.

Another common source of systematic error is an oversimplification of the equation that is employed to relate the measured quantities to the determination of the parameter of interest. For example, let us say that we wanted to determine the local value of the acceleration due to gravity, g, by measuring the period, T, of a pendulum that suspends a bob of mass, m, from a string of length, L. The simplest expression relating those parameters to the gravitational acceleration involves only the length of the pendulum and its period of oscillation, T.

$$g = \frac{4\pi^2 L}{T^2} \tag{1.91}$$

It is very easy to make a precise determination of the period, T, by timing 100 cycles of the pendulum and dividing by 100. The length, L', of the string, between the support and the mass is also easy to determine accurately, but the length that appears in Eq. (1.91) should be the distance from the support point to the center of mass of the pendulum bob. Also, Eq. (1.91) does not include the effects of air that provides some buoyancy and also must accelerate and decelerate to get out of the path of the bob's motion. Since the amplitude of the oscillations is not zero, there are nonlinear effects that are introduced because Eq. (1.91) assumes the small angle approximation by truncating all higher-order terms in the potential energy of Eq. (1.26). Because of the viscous drag of the air, the amplitude of

Table 1.2 There are several physical effects that influence the period of a pendulum which are not captured by Eq. (1.91) relating the gravitational acceleration, *g*, to the period, *T*, of a pendulum. Several of the sources of these systematic errors are listed that shift the period by an amount ΔT and are comparable to the statistical uncertainty of ± 170 μs in the determination of the period [22]

Effect	ΔT(μs)
Non-zero amplitude (i.e., second order)	596
Buoyancy	292
Added mass due to air acceleration	346
Mass of the wire	–463
Decay of oscillations	–77
Moment of inertia of the bob	72

those swings decrease over time and that damping increases the period (see Sect. 2.4). Table 1.2 summarizes some of these systematic errors for a pendulum experiment that had $L \cong 3.00$ m and a spherical bob with diameter $D \cong 6.01$ cm and $m \cong 0.857$ kg. The pendulum had an average period of $T \cong 3.47880 \pm 0.00017$ s [22]. Although the statistical uncertainty in the determination of the period is only ± 170 μs, there are several systematic errors, listed in Table 1.2, that have a comparable effect on the relationship between the period and the gravitational acceleration that are not captured by Eq. (1.91).

Finally, there are errors that are just simply blunders. If you are writing results in a laboratory notebook or typing them into a computer, it is possible to enter the wrong number, skip an entry, lose count, forget to write down the model and serial number of a sensor, or transcribe a number incorrectly from the notebook into your calculator or into a spreadsheet or other data analysis software package. The best defense against blunders is to be well-prepared before actually taking the data and be sure you have arranged a place that will be comfortable and well-lit and affords easy access to the knobs that need twisting, the displays that must be read, and the keyboard if you are entering data or observations directly into a computer.

You are really not ready to start an experiment if you have not labeled the columns into which the data will be entered and included the units with that label. Before taking data, you should have a good idea of how the data will be plotted, at least to the level that you have identified the axes and their units. Plotting some data early in an experiment and calculating results based on that early data are always a good way to be sure you are not wasting hours or days acquiring data that will later be recognized as useless.

1.8.4 Error Propagation and Covariance

Our interest in the quantification of errors is motivated by the need to specify the uncertainty in our measurement results. Since most calculations require more than a single input parameter, we also need a means to combine those errors in a way that provides the overall uncertainty. The critical question for making such a combination is whether the errors in the determination of the parameters are *correlated* or *uncorrelated*. In the previous example, which used Eq. (1.91) to determine *g* from measurements of *L* and *T*, it is obvious that the errors in determination of *T* were uncorrelated to the errors in the determination of *L*. If a stopwatch was used to determine *T* and a ruler was used to determine *L*, then errors introduced by the stopwatch do not have any influence on errors made in reading the length of the string using a ruler.

The correlation, or lack of correlation, between sources of error can be formalized by the introduction of the *covariance*. If a particular parameter, *x*, is a function of several other variables, $x = f(u, v,$

...), we assume that the most probable value of the dependent variable, $\bar{x}$, is the same function of the most probable values of the independent variables: $\bar{x} = f(\bar{u}, \bar{v}, \ldots)$. Using Eq. (1.88) in the limit of very large N, it is possible to approximate the variance, σ_x^2.

$$\sigma_x^2 \cong \lim_{N\to\infty} \frac{\sum (x_i - \bar{x})^2}{N} \tag{1.92}$$

Limiting ourselves to only two independent variables, u and v, and using a Taylor series, the deviation (also known as the *residual*), $\delta x_i = x_i - \bar{x}$, can be expanded in terms of the deviations of those two independent variables.

$$\delta x_i = x_i - \bar{x} \cong \left(\frac{\partial x}{\partial u}\right)(u_i - \bar{u}) + \left(\frac{\partial x}{\partial v}\right)(v_i - \bar{v}) \tag{1.93}$$

Substitution of Eq. (1.93) into Eq. (1.92) generates three contributions to σ_x^2.

$$\begin{gathered}\sigma_x^2 \cong \lim_{N\to\infty} \frac{1}{N} \sum \left[\left(\frac{\partial x}{\partial u}\right)(u_i - \bar{u}) + \left(\frac{\partial x}{\partial v}\right)(v_i - \bar{v})\right]^2 = \\ \lim_{N\to\infty} \frac{1}{N} \sum \left[\left(\frac{\partial x}{\partial u}\right)^2 (u_i - \bar{u})^2 + \left(\frac{\partial x}{\partial v}\right)^2 (v_i - \bar{v})^2 + 2\left(\frac{\partial x}{\partial u}\right)\left(\frac{\partial x}{\partial v}\right)(u_i - \bar{u})(v_i - \bar{v})\right]\end{gathered} \tag{1.94}$$

The last term introduces the covariance σ_{uv}^2.

$$\sigma_{uv}^2 \equiv \lim_{N\to\infty} \frac{\sum (u_i - \bar{u})(v_i - \bar{v})}{N} \tag{1.95}$$

This allows Eq. (1.94) to be re-written in a more compact and intuitive form.

$$\sigma_x^2 \cong \left(\frac{\partial x}{\partial u}\right)^2 \sigma_u^2 + \left(\frac{\partial x}{\partial v}\right)^2 \sigma_v^2 + 2\left(\frac{\partial x}{\partial u}\right)\left(\frac{\partial x}{\partial v}\right)\sigma_{uv}^2 \tag{1.96}$$

If the fluctuations $\delta u_i = u_i - \bar{u}$ and $\delta v_i = v_i - \bar{v}$ are uncorrelated, then $\sigma_{uv}^2 = 0$, and the standard deviation of the dependent variable, x, is related to the variances in the individual independent variables.

$$\sigma_x \cong \left[\left(\frac{\partial x}{\partial u}\right)^2 \sigma_u^2 + \left(\frac{\partial x}{\partial v}\right)^2 \sigma_v^2\right]^{1/2} \tag{1.97}$$

We can apply this formalism directly to a weighted sum. If $x = au \pm bv$, then $(\partial x/\partial u) = a$ and $(\partial x/\partial v) = \pm b$, so Eq. (1.96) becomes

$$\sigma_x^2 \cong a^2\sigma_u^2 + b^2\sigma_v^2 \pm 2ab\sigma_{uv}^2. \tag{1.98}$$

For products, ratios, and powers of independent variables that are uncorrelated, it is more convenient to deal with the relative uncertainties and to combine the uncertainties using logarithmic differentiation (see Sect. 1.1.3). We can take the natural logarithm of Eq. (1.91).

$$\ln g = \ln 4\pi^2 + \ln L - 2\ln T \tag{1.99}$$

Using Eq. (1.15), differentiation of that expression leads to a relationship between the relative uncertainties.

$$\frac{\delta g}{g} = \frac{\delta L}{L} - 2\frac{\delta T}{T} \tag{1.100}$$

Since we have already established the statistical independence of the errors in L and T (i.e., $\sigma_{LT}^2 = 0$), Eq. (1.97) dictates that the relative uncertainty in g becomes the Pythagorean sum of the individual contributions of the relative uncertainties of the independent variables.

$$\frac{\delta g}{g} = \left[\left(\frac{\delta L}{L}\right)^2 + \left(-2\frac{\delta T}{T}\right)^2\right]^{1/2} \tag{1.101}$$

1.8.5 Significant Figures

It is important that the overall uncertainty in a result is part of the number that reports that result. The first consideration should be the number of significant digits used to report the result. For example, if my calculation of sound speed produces a result of 1483.4 m/s and my uncertainty in that result is ± 30 m/s, it makes no sense to report the result as $c = 1483.4 \pm 30$ m/s. In that example, only three significant digits are justified, so the result should have been reported as $c = 1480 \pm 30$ m/s.[27]

The specification of the expected uncertainty in a result differs among users. In my laboratory notebook, my publications, and my technical reports, I usually report a result as the mean plus or minus one standard deviation, thus indicating that I have a 68% confidence in the accuracy of the result. Standards organizations, such as the National Institutes of Standards and Technology (formerly the National Bureau of Standards) in the United States, the National Research Council in Canada, or the National Physical Laboratory in the United Kingdom, customarily specify their uncertainties with errors that are $\pm 2\sigma$ to provide a 95% confidence level in the result. In every case, it is the responsibility of the experimentalist to specify the meaning of his or her stated uncertainty.

1.9 Least-Squares Fitting and Parameter Estimation

To this point, our discussion of measurement error has focused on repeated measurement of individual parameters. A more common occurrence in the determination of parameters in an acoustical or vibrational experiment is the measurement of some response, y (the dependent variable), that is a function of some stimulus, x (the independent variable). In that case, the goal is to find the function, f, such that $y = f(x)$. Of course, this can be extended to functions of several variables. Instead of measuring y several times for a given value of x, we would like to treat the measurement of y_i values corresponding to x_i inputs, where $i = 1, 2, \ldots, N$, to produce N data pairs (x_i, y_i). These pairs are typically displayed graphically [23].

[27] There are different standards for the number of significant digits that should be included in the representation of a measurement. For example, the ASTM International (known prior to 2001 as the American Society for Testing and Materials) publishes "Standard Practice for Using Significant Digits in Test Data to Determine Conformance with Specifications," ASTM Designation E 29 – 08. For certain applications, there are legal ramifications associated with misrepresenting a result by supplying more significant figures than are justified.

The simplest possible relationship between the pairs of data is the linear relationship: $y = mx + b$. In a graphical representation, m corresponds to the slope of the line, and b is the y-axis intercept. This section will address the question of how one chooses m and b so that the line specified by those two parameters minimizes the deviations (residuals), δy_i, between the experimental measurements, y_i, and the values generated by the equation for the straight line when x_i is substituted into that equation: $\delta y_i = y_i - y(x_i) = y_i - (mx_i + b)$.

As before, the line that is considered a "best-fit" to the data will minimize the sum of squares of the residuals. If we assume that the values of x_i are known exactly, then the Gaussian probability of measuring a value y_i is given by $P_i(y_i)$.

$$P_i(y_i) = \frac{1}{\sigma_i\sqrt{2\pi}} \exp\left[\frac{1}{2}\left(\frac{y_i - y(x_i)}{\sigma_i}\right)^2\right] \tag{1.102}$$

For N measurement pairs, the overall probability will be the product of the probability of the individual measurements.

$$P(m,b) = \prod_{i=1}^{N} P_i(y_i) = \prod_{i=1}^{N}\left(\frac{1}{\sigma_i\sqrt{2\pi}}\right)\exp\left[\sum_{i=1}^{N}\frac{1}{2}\left(\frac{\delta y_i}{\sigma_i}\right)^2\right] \tag{1.103}$$

We can abbreviate the argument in the sum as χ^2.

$$\chi^2 = \sum_{i=1}^{N}\left(\frac{\delta y_i}{\sigma_i}\right)^2 = \sum_{i=1}^{N}\left[\frac{(y_i - mx_i - b)^2}{\sigma_i^2}\right] \tag{1.104}$$

Determination of the best values of m and b is therefore equivalent to finding the minima of χ^2 by setting the derivatives of χ^2 with respect to m and b equal to zero.

$$\frac{\partial(\chi^2)}{\partial m} = 0 \quad \text{and} \quad \frac{\partial(\chi^2)}{\partial b} = 0 \tag{1.105}$$

Using these criteria, it is possible to express the best values of m and b in terms of various sums of the measurements.

$$b = \frac{\left(\sum_{i=1}^{N} x_i^2\right)\left(\sum_{i=1}^{N} y_i\right) - \left(\sum_{i=1}^{N} x_i\right)\left(\sum_{i=1}^{N} x_i y_i\right)}{N\left(\sum_{i=1}^{N} x_i^2\right) - \left(\sum_{i=1}^{N} x_i\right)^2} \tag{1.106}$$

$$m = \frac{N\left(\sum_{i=1}^{N} x_i y_i\right) - \left(\sum_{i=1}^{N} x_i\right)\left(\sum_{i=1}^{N} y_i\right)}{N\left(\sum_{i=1}^{N} x_i^2\right) - \left(\sum_{i=1}^{N} x_i\right)^2} \tag{1.107}$$

Although the expressions look intimidating, they are exactly the types of expressions that a digital computer can evaluate nearly instantaneously. Practically any software package will automatically determine m and b and several other features of such a *least-squares fit*.

1.9.1 Linear Correlation Coefficient

"If your experiment needs statistics, you should have done a better experiment." (E. Rutherford, F.R.S [24])

The calculation of m and b assumed that the values, x_i, were known exactly and all of the statistical fluctuations were due to the y_i values. In some experimental circumstances, that assumption might be valid. For mathematically perfect data (i.e., values of y_i that are generated by substitution of x_i into an equation like $y_i = mx_i + b$), we could just as well describe the line which fits $x = m'y + b'$, where m' and b' would be different from m and b in both their numerical values and their units.

Solving the inverse relation, $x = m'y + b'$ for y, we obtain the relationships between the primed quantities and the unprimed fit parameters, m and b.

$$y = \frac{x}{m'} - \frac{b'}{m'} = mx + b \quad \Rightarrow \quad b = \frac{-b'}{m'} \quad \text{and} \quad m = \frac{1}{m'} \tag{1.108}$$

For mathematically perfect data, $(m)(m') \equiv R^2 = 1$; the product of the slope for a line that plots y vs. x is the reciprocal of the slope for the plot of x vs. y. By reversing the plotting axes, we are now also assuming that the y_i values are exact and all of the uncertainty is caused by the x_i values. If there are errors, the value of R^2 will be less than unity. R^2 is called the square of the *correlation coefficient* that can also be calculated in terms of the same sums used in Eqs. (1.106) and (1.107).

$$R = \sqrt{mm'} = \frac{N\sum x_i y_i - (\sum x_i)(\sum y_i)}{\left[N\sum x_i^2 - (\sum x_i)^2\right]^{1/2}\left[N\sum y_i^2 - (\sum y_i)^2\right]^{1/2}} \tag{1.109}$$

If the (x_i, y_i) data pairs are uncorrelated, then $R = 0$; if the data are perfectly correlated and noise-free, then $R^2 = 1$. If we use this formalism to fit the data that produced the left-hand "target" in Fig. 1.10, the best-fit gives the following equation for the y_i values in terms of the x_i values: $y = 0.0393x + 0.397$. The square of the correlation coefficient for that fit is $R^2 = 0.0015$. We expect that slope and R^2 for a truly random set of points should both be very close to zero. It is difficult to say how close is "close enough" to declare that the data are truly random and that the slope is zero, since the answer to such a question is inherently probabilistic.

A strict interpretation of the meaning of the correlation coefficient as a measure of randomness and the relationship between the correlation coefficient and the relative uncertainty in the slope, $\delta m/|m|$, requires the evaluation of a two-dimensional Gaussian distribution with careful attention to the distinction between the population and the sample, especially if N is small (see Table 1.4) [25]. Again, it is the type of result that can be generated by the proper software or by consulting Table 1.3. That table provides the smallest value of R that would rule out randomness with some specified level of confidence. For example, there is a 10% chance that a set of $N = 8$ data pairs would produce a correlation coefficient of $R \geq 0.621$, even if the data were uncorrelated. At the 1% probability level, $R \geq 0.834$ would be required to imply causality and $R \geq 0.925$ to ensure a 0.1% confidence level.

For the least-squares fit to the data plotted on the left-hand side of Fig. 1.10, the correlation coefficient $R = (0.0015)^{1/2} = 0.039$. Linear interpolation within Table 1.3 for $N = 16$ points requires that $R \geq 0.428$ to rule out, with a 10% probability, that the data were not random and $R \geq 0.744$ for a 0.1% probability that the data were not random. Those data appear to be truly random, as claimed.

Before leaving this subject, it is important to recognize that in physical and engineering acoustics, controlled experiments that are supposed to be represented by a linear relationship routinely produce $R^2 > 0.99$, while in areas like community response to noise or other environmental, architectural, psychological, or physiological acoustical investigations, the square of the correlation coefficients can

Table 1.3 Values of the correlation coefficient, R, that would have to be exceeded for N data pairs to have the probability, given in the top row of the table, that the correlation coefficient, R, was produced by N causally related pairs

	Probability				
N	0.10	0.05	0.02	0.01	0.001
3	0.988	0.997	0.999	1.000	1.000
4	0.900	0,950	0.980	0.990	0.999
5	0.805	0.878	0.934	0.959	0.992
6	0.729	0.811	0.882	0.917	0.974
7	0.669	0.754	0.833	0.874	0.951
8	0.621	0.707	0.789	0.834	0.925
10	0.549	0.632	0.716	0.765	0.872
12	0.497	0.576	0.658	0.708	0.823
15	0.441	0.514	0.592	0.641	0.760
20	0.378	0.444	0.516	0361	0.679
30	0.307	0.362	0.423	0.464	0.572
40	0.264	0.312	0.367	0.403	0.502
60	0.219	0.259	0.306	0.337	0.422
80	0.188	0.223	0.263	0.291	0.366
100	0.168	0.199	0.235	0.259	0.327

Adapted from Fisher and Yates [26]
For the example on the left-hand side of Fig. 1.10, linear interpolation for $N = 16$ points requires that $R \geq 0.428$ to guarantee with a 10% probability that the data were not random and $R \geq 0.744$ for a 0.1% probability that the data were not random

Table 1.4 Comparison of the error estimate in slope and intercept using Eq. (1.112) or their Higbie equivalents in Eqs. (1.113) and (1.114)

Topic	N	R^2	Higbie slope (%)	Small N slope (%)	Higbie intercept (%)	Small N intercept (%)
Language test	24	0.254	36.9	37.2	18.3	18.6
Frequency	8	0.999937	0.32	0.35		
Decay time	7	0.99973	0.77	0.80		
Phase speed	10	0.99989	0.37	0.40	0.30	0.20
Bevington [24]	9	0.9437	9.23	9.88	101.6	108.6
Beers [28]	6	0.99988	0.55	0.63	239	281

The examples include Fig. 1.14 (language), Fig. 1.15 (frequency and decay time), and Fig. 1.17 (phase speed), as well as the example in Table 6.1 of Bevington [25] and Table XI in Beers [29]. The very large (i.e., >100%) relative errors in the intercepts from the Bevington and Beers examples arise because the value of the intercept, $|b|$, is very close to zero. For all entries, the reported errors are based on $\pm 1\sigma$

be one-half or less (see Fig. 1.14). This is not the result of inferior scientific methods but a consequence of the inability to control the large number of *confounding variables* that influence specific responses of animals and particularly humans.

Recognition of this challenge is not a valid justification for the reliance solely on statistical tests, particularly when a stated "result" can trigger regulatory remediation that may require enormous financial burdens.

The four graphs shown in Fig. 1.13 are all fit with a best-fit straight line. All four lines have the same slope and intercept, and all four have the same squared correlation coefficient. The fit designated "y_1" (upper left) is what is usually assumed when we think of a linear relationship that describes fairly noisy data. The data in the graph fit by "y_2" (upper right) is much smoother but clearly would be better represented by a quadratic equation rather than a linear one. In fact, those points were generated by the following equation: $y_2 = -0.127\, x^2 + 2.78\, x + 6$.

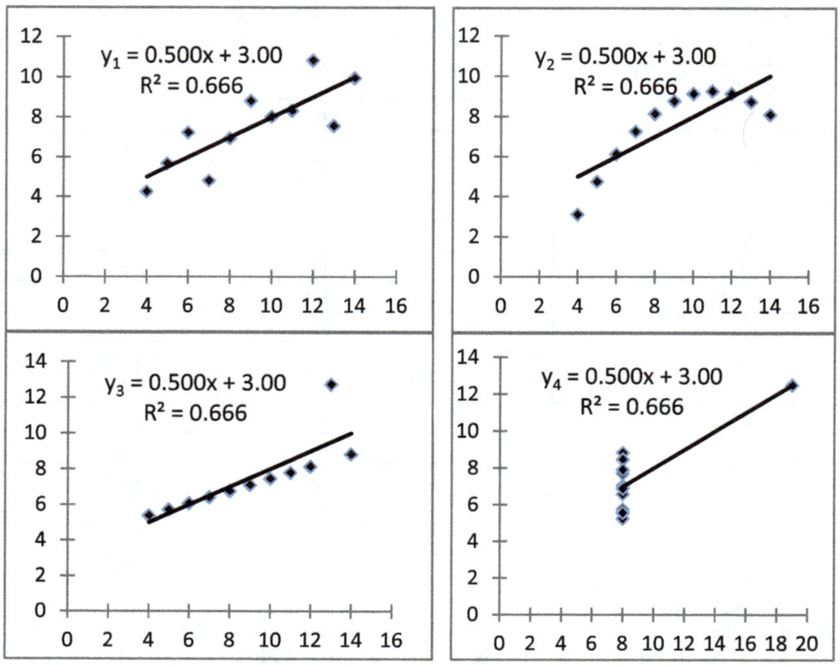

Fig. 1.13 These four data sets each contain $N = 11$ data pairs that are plotted and include solid lines that were determined by a best-fit to each data set. All lines have the same slope, $m = 0.500$; intercept, $b = 3.00$; and (diabolical) squared correlation coefficient, $R^2 = 0.666$. (Data taken from Anscombe [23].)

The third and fourth examples in Fig. 1.13 are intentionally pathological. The fit designated "y_3" (lower left) is dominated by one "outlier," but all the other points were generated by a linear equation with a different slope and intercept: $y_3 = 0.346x + 4$. Similarly, the slope of the line that is the best-fit to produce "y_4" is entirely dominated by one "outlier" which, if removed, would leave the remaining points with an infinite slope since the variety of all y_i values corresponds to a single value for all the remaining $x_i = 8$.

Although I've called "y_3" and "y_4" least-squares fits in the lower half of Fig. 1.13 "pathological," they are not uncommon. Figure 1.14 is taken from Fig. 9 in an article on the relationships between unoccupied classroom acoustical conditions and a metric for elementary school student achievement [27]. The authors fit their data to a straight line and claim "a significant negative correlation between the Terra Nova language test score and the temperature-weighted average BNL [background noise level] . . . ."

Using the analysis of Table 1.3, it is true that there is only about a 5% probability that such a data set incorporating those 24 points would have been generated by a random set of ordered pairs, but if only the 21 points, corresponding to $L_{eqA} \leq 45$ dB_A,[28] are fit to a straight line, then *percentile rank* [from Fig. 1.14] $= -0.135\ L_{eqA} + 70.95$ with $R^2 = 0.0014$. Using techniques introduced in the next section, that "best-fit" slope of $m = -0.135$ could range anywhere from $m = -0.97$ to $m = +0.70$ if the fit were good to $\pm 1\sigma$. This correlation coefficient is just as "random" as the intentionally random data set on the left of Fig. 1.10 (i.e., $R^2 = 0.0015$). The reported "effect" is due only to the three additional points corresponding to $45 \leq L_{eqA} \leq 52$ dB_A.

---

[28] The unit dB_A is a version of a sound pressure level (see footnote 1) that compensates for the frequency dependence of the sensitivity of human hearing.

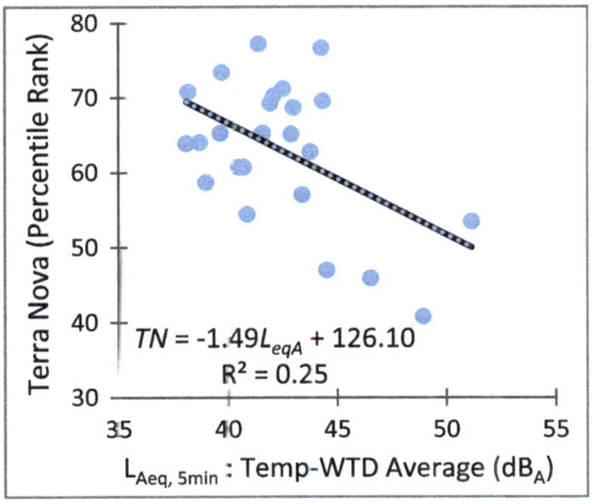

Fig. 1.14 Graph taken from an article that claims a "statistically significant" relationship between background noise levels and language test achievement scores [27]. The line is the authors' least-squares fit to a straight line with a squared correlation coefficient $R^2 = 0.25$. A fit to all the data with $L_{Aeq} \leq 45$ dB$_A$ has a slope of -0.135 and $R^2 = 0.0014$. The addition of those three "outliers" produces a result similar to that shown as the "y_3" fit in the lower-left quadrant of Fig. 1.13

The message that I hope these examples provide is that both graphical and statistical interpretations of data, and the models used to fit data, must be applied simultaneously to avoid erroneous conclusions. Once confidence in the analysis that might involve a least-squares fit is established, there are further techniques that are invaluable in the determination of the uncertainty of the slope and intercept that are generated by such a fit.

1.9.2 Relative Error in the Slope

For most data sets generated by experiments in physical or engineering acoustics, causality is not in question. In such cases, it is far more important to determine the relative error in the slope $\delta m/|m|$ and the intercept $\delta b/|b|$. These error estimates are related to the variance determined by the squared sum of the residuals, δy_i, produced by the fitting function.

$$\sigma^2 \cong \frac{\sum (y_i - m x_i - b)^2}{N - 2} \tag{1.110}$$

Unlike the definition of the standard deviation in Eq. (1.88), the denominator above is $N - 2$ instead of $N - 1$. This is due to the fact that the best-fit straight line had two adjustable parameters, m and b, whereas the standard deviation had only one: $\bar{x}$.

The estimate of the error in m and b is related to Eq. (1.110) through combinations of sums for the x_i and y_i measurements like those in Eqs. (1.106), (1.107), and (1.109).

$$\Delta = N\left(\sum_{i=1}^{N} x_i^2\right) - \left(\sum_{i=1}^{N} x_i\right)^2 \tag{1.111}$$

The relative errors in the slope and the intercept can be expressed in terms of Δ.

$$\begin{aligned}\frac{\delta m}{|m|} &= \frac{\sigma_m}{|m|} = \frac{N}{|m|}\frac{\sigma}{\Delta} \\ \frac{\delta b}{|b|} = \frac{\sigma_b}{|b|} &= \frac{\sqrt{\sum x_i^2}}{|b|}\frac{\sigma}{\Delta} = \frac{N^{1/2} x_{rms}}{|b|}\frac{\sigma}{\Delta} = x_{rms}\frac{\sigma_m}{|b|}\end{aligned} \tag{1.112}$$

Again, these combinations are commonly provided automatically by many software packages when data is fit to a straight line.

I have found an expression suggested by Higbie [28] to be a convenient way to calculate the relative uncertainties in slope, $\delta m/|m|$, and intercept, $\delta b/|b|$, directly from the R^2 obtained by a straight-line least-squares fit, which is usually available even when data are analyzed on a handheld calculator in the laboratory.

$$\frac{\delta m}{|m|} = \frac{\tan\left(\cos^{-1} R\right)}{\sqrt{N-2}} = \sqrt{\frac{\frac{1}{R^2} - 1}{N-2}} \tag{1.113}$$

The relative uncertainty in the intercept, $\delta b/|b|$, can be calculated from substitution of Eq. (1.113) into Eq. (1.112) where x_{rms} is the root-mean-squared value of the x_i values.

$$\frac{\delta b}{|b|} = \delta m \frac{x_{rms}}{|b|} = \left(\frac{\delta m}{|m|}\right)\frac{|m|}{|b|} x_{rms} \tag{1.114}$$

1.9.3 Linearized Least-Squares Fitting

The formalism developed in the previous section can easily be extended to many other models that have two adjustable parameters but are not plotted as straight lines. Most books on the data analysis demonstrate that exponential growth or decay, power law relationships, and logarithmic relaxation can be transformed into straight lines. These will be demonstrated, along with more specific applications, in acoustic problems that will be analyzed later in this textbook that also fix the values of two adjustable parameters.

The first example is the envelope of an exponentially decaying damped simple harmonic oscillator (Sect. 2.4). For a damping force that is proportional to velocity, we expect the amplitude to decay with time according to $V(t) = V_o e^{-t/\tau}$, where τ is the characteristic exponential decay time (i.e., the time required for the amplitude to decrease by a factor of $e^{-1} \cong 36.8\%$). By taking the natural logarithm of that expression, we can see that the expected behavior can be plotted as a straight line if the natural logarithm of the amplitude envelope is plotted against time as shown in Fig. 1.15.

$$\ln\left[V(t)\right] = \ln\left[V_o\right] - \frac{t}{\tau} \tag{1.115}$$

Sample data taken from the recording of a digital oscilloscope is shown in Fig. 1.15 along with the least-squares line fit to those data. The use of Eq. (1.113) provides the relative uncertainty in the slope: $\delta\tau/|\tau| = \pm 0.73\%$. The slope of the line it provides is the negative reciprocal of the characteristic exponential decay time: $\tau = 22.18 \pm 0.17$ s. For this type of measurement, we are not interested in the intercept, V_o, since it is just a measure of the amplitude extrapolated back to an arbitrary time designated $t = 0$.

The reverse of the transformation in Eq. (1.115) can be used to fit a straight line to a logarithmic decay. This is illustrated in Fig. 1.16 showing the creep of an elastomeric loudspeaker surround. The

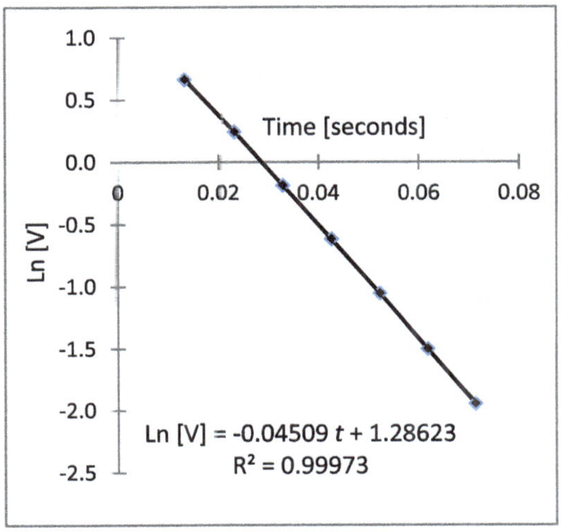

Fig. 1.15 Free-decay data that plots the natural logarithm of the peak-to-peak amplitude of the waveform vs. the average time between successive peaks and troughs. The slope of the line provides the negative reciprocal of the characteristic exponential decay time, $\tau = (22.18 \pm 0.17) \times 10^{-3}$ s, with the uncertainty determined from R^2 by way of Eq. (1.113). A plot of the time for zero-crossings vs. the cycle number provides $f_d = 51.82 \pm 0.17$ Hz

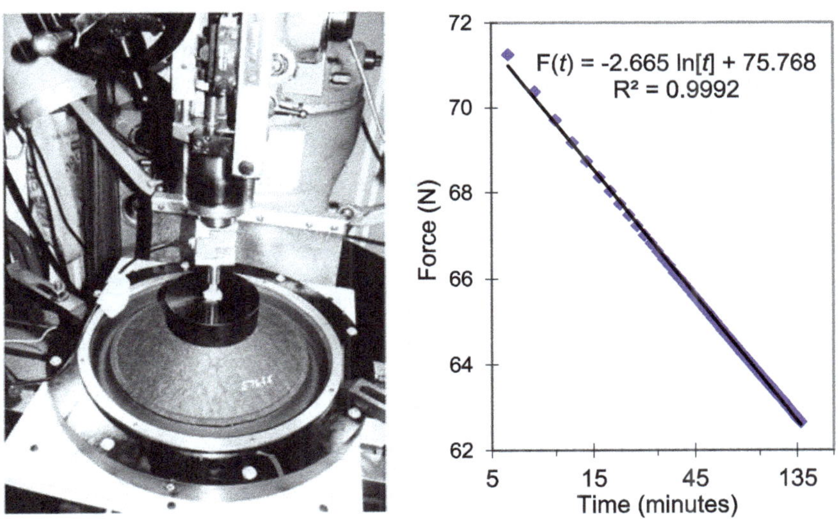

Fig. 1.16 *(Left)* The force necessary to displace a loudspeaker cone from its equilibrium position is applied by the quill of a milling machine. Between the quill and the fixture applying the force to the loudspeaker cone is a load cell that measures the force. *(Right)* Plot of the force necessary to displace an elastomeric loudspeaker surround as a function of the logarithm of the time it was held in a fixed position. The least-squares fit to the force vs. the logarithm of the holding time yields a straight line. The force decreases by 10% from its value at the first second for each multiple of 17.2 s. The force is reduced by 20% at $(17.2)^2 = 295$ s and by 30% at $(17.2)^3 = 5060$ s = 1.4 h and would reach half its value after 1.49×10^6 s = 17.3 days

surround was displaced from its equilibrium position, and the force needed to hold it in that position is plotted vs. the logarithm of the holding time.

The same approach can be used to transform power-law expressions, $V(t) = bt^m$, where b and m can be determined by again taking logarithms, typically using base 10 to make it easy to read a graph.

$$\log_{10}[V(t)] = \log_{10}[b] + m \log_{10}[t] \tag{1.116}$$

This linearization technique should not be limited only to the two-parameter fits that are treated in the data analysis textbooks. There are several instances in this textbook where proper plotting of the results of acoustical or vibrational measurements simplifies the extraction of important parameters and their uncertainties. I will use one example that occurs when we investigate the propagation of sound in waveguides and introduce the concepts of group speed, c_{gr}, and phase speed, c_{ph}, and their relationship to the square of the thermodynamic sound speed, $c_o^2 = c_{ph}c_{gr}$. In the following equation, f is the frequency of the sound propagating within the waveguide in a mode with a cut-off frequency, f_{co}.

$$c_{ph} = \frac{c_o}{\sqrt{1 - \left(\frac{f_{co}}{f}\right)^2}} \tag{1.117}$$

It is possible to make very accurate measurements of phase speed, and with a synthesized function generator, the frequency can be selected with relative accuracies on the order of a few parts per million. Equation (1.117) has two adjustable parameters: c_o and f_{co}. By squaring Eq. (1.117) and inverting the result, it can be cast into the form of a straight line.

$$\frac{1}{c_{ph}^2} = \frac{1}{c_o^2} - \frac{f_{co}^2}{c_o^2}\frac{1}{f^2} = \frac{1}{c_o^2} - \frac{f_{co}^2}{c_o^2}T^2 \tag{1.118}$$

The results of measurements of phase speed vs. frequency in a water-filled waveguide are shown on the left-hand side of Fig. 1.17 (also see Prob. 5 in Ch. 13). One could attempt to locate a vertical asymptote to determine the cut-off frequency, f_{co} (where the phase speed would have become infinite), and a horizontal asymptote that would intersect the y axis at a value of $c_{ph} = c_o$. Plotting the same data using Eq. (1.118) makes such guesswork totally unnecessary. The right-hand side of Fig. 1.17 shows a plot of the reciprocal of the square of the phase speed $(c_{ph})^{-2}$ vs. the square of the period of the sound,

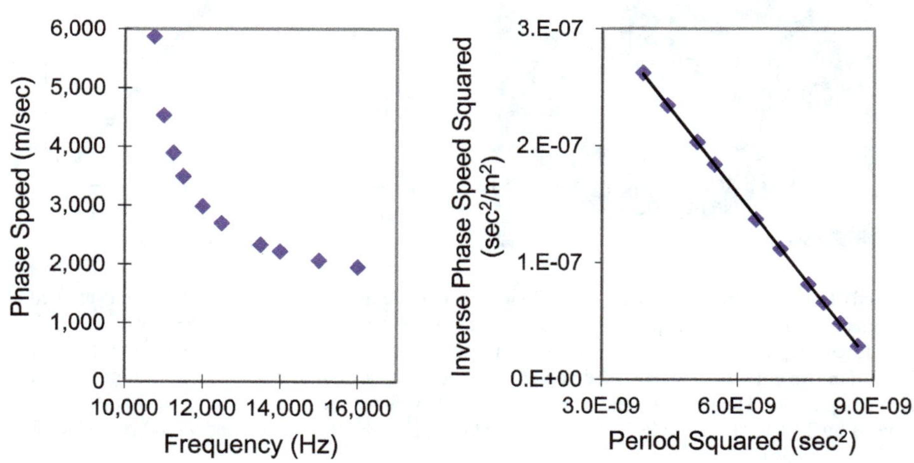

Fig. 1.17 The same data plotted in two different ways. (*Left*) The raw data consisting of phase speed, c_{ph}, in a waveguide vs. the frequency of the sound propagating through the waveguide. (*Right*) The same data plotted using Eq. (1.118) that is fit with the straight line: $c_{ph}^{-2} = -49.04\,T^2 + 4.531 \times 10^{-7}$ s^2/m^2. The correlation coefficient of the fit is $R = -0.999944$, corresponding to an uncertainty in the slope of $\delta m/|m| = 0.37\%$. The uncertainty in the intercept is $\delta b/|b| = 0.30\%$. The combined uncertainty gives $c_o = 1485.6 \pm 2.2$ m/s and $f_{co} = 10{,}404 \pm 25$ Hz

T^2. The thermodynamic sound speed, c_o, is the square root of the y axis intercept of the best-fit straight line. The square root of the ratio of the slope to the intercept is the cut-off frequency, f_{co}.

1.9.4 Caveat for Data Sets with Small *N**

The uncertainties in the slopes and intercepts reported for the graphs in the previous section were based on Eq. (1.112) or the Higbie equivalents of Eq. (1.113) and Eq. (1.114). For smaller data sets, these estimated uncertainties are not exactly correct, even though those estimates incorporate the number of measurements through the $(N-2)^{1/2}$ in the denominator or Eq. (1.113). In fact, the error estimates are not necessarily symmetric (i.e., instead of $\pm$, the + and – errors may differ slightly). Most software packages that can calculate least-squares fits can also provide estimates for the errors. Table 1.4 provides a comparison between those simpler error estimates and the correct estimates that incorporate the actual number of points. In general, those error estimates are slightly larger than obtained by use of the Higbie expression. In both cases, the quoted errors correspond to variation by $\pm 1\sigma$.

1.9.5 Best-Fit to Models with More Than Two Adjustable Parameters

The availability of digital computers makes it rather easy to optimize the fit of nearly any function to a data set, although it should always be remembered that the number of fitting (free) parameters must be less than the number of data points. The extension of the least-squares fitting to a polynomial of arbitrary order is deterministic but algebraically messy [29]. Again, most software packages are capable of executing a fit to an nth-order polynomial automatically if the number of data pairs $N \geq n+1$.

Figure 1.18 shows 2000 data points acquired by a digital oscilloscope and imported to a spreadsheet program. The goal was to fit the entire data set to a free-decay expression that is related to Eq. (1.115) used to produce the fit shown in Fig. 1.15. Since all the measurements were used, instead of just the amplitudes of the peaks and the troughs, the equation had also to incorporate the time dependence explicitly.

$$V(t_i) = V_o e^{-t_i/\tau} \sin(2\pi f_d t_i + \phi) \tag{1.119}$$

This expression has four adjustable parameters: V_o, τ, f_d, and ϕ. Those parameters have to be adjusted simultaneously to minimize the sum of the square of the residuals.

The fit shown in Fig. 1.18 was made by taking the time and amplitude data columns and creating a column that took each time and plugged it into Eq. (1.119) using guesses for the values of each parameter that seemed reasonable based on inspection of the plot of $V(t)$ vs. t. For example, the initial amplitude was approximately $|V(t)| \cong 0.8$ volts, so one initial guess was $V_o = 0.8$ volts. We see approximately three complete cycles between $t \cong 0.018$ s and $t \cong 0.082$ s, so a reasonable guess for frequency is $f_d \cong 3$ cycles/0.064 s $= 47$ Hz. Just considering the first four peaks, it appears that amplitude decreases by about $e^{-1} \cong 37\%$, from about 0.6 volts to about 0.2 volts, and also in about 0.064 s, so $\tau = 0.064$ s is a reasonable guess. Finally, we assumed sinusoidal time dependence, but a sine wave starts from zero amplitude at $t = 0$. If the data in Fig. 1.18 were actually a sine wave, we would have to translate the origin of our time axis backward by about 0.007 s. Using the period based on the frequency guess, $T = f_d^{-1} \cong 0.021$ s, this means we need to retard the phase by about one-third

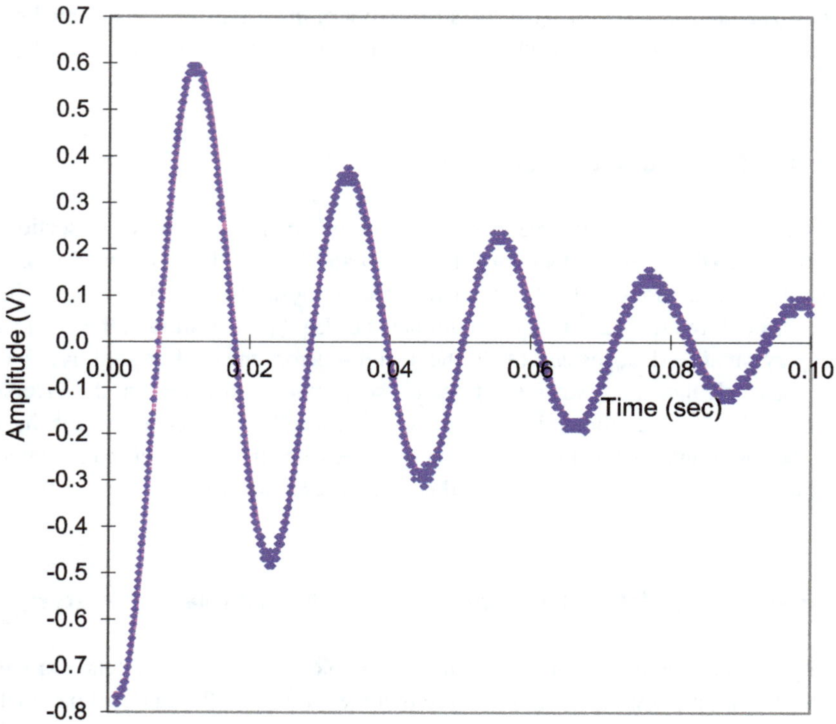

Fig. 1.18 The 2000 plotted points were acquired by a digital oscilloscope and were imported to a spreadsheet program. By allowing the program to vary the four parameters in Eq. (1.119), the program attempted to minimize the sum of the squared residuals, $(\delta V_i)^2 = (V_i - V(t_i))^2$, based on the initial guesses provided. The result is the characteristic exponential decay time, $\tau = 0.0445$ s, with a root-mean-squared relative error of $\pm 0.15\%$. The quality of that fit is so good that the line representing Eq. (1.119) is almost entirely obscured by the individual data points. The other parameters that were optimized by the program are $f_d = 46.141$ Hz, $V_o = 0.788$ volts, and $\phi = -2.032$ radians

of a cycle or advance it by two-thirds of a cycle corresponding to $\phi \cong -2\pi/3$ or $+ 4\pi/3$ radians. I choose to use $\phi \cong -2$ radians.

With those initial guesses, it is possible to calculate the residuals, $\delta V_i = V_i - V(t_i)$, square those residuals, and add them together for all 2000 points. (Ain't computers wonderful!) To create the fit shown in Fig. 1.18, I used MS Excel's "solver," which I let vary those four adjustable parameters from their guessed values in an attempt to minimize the sum of the squared residuals. The final values are provided in the caption of Fig. 1.18.

1.10 The Absence of Rigorous Mathematics

"In the mathematical investigations I have usually employed such methods as present themselves naturally to a physicist. The pure mathematician will complain, and (it must be confessed) sometimes with justice, of deficient rigor. But to this question there are two sides. For, however important it may be to maintain a uniformly high standard in pure mathematics, the physicist may occasionally do well to rest content with arguments which are fairly satisfactory and conclusive from his point of view. To his mind, exercised in a different order of ideas, the more severe procedures of the pure mathematician may appear not more but less demonstrative. And further, in many cases of difficulty, to insist upon the highest standard would mean the exclusion of the subject altogether in view of the space that would be required." [30]

Talk Like an Acoustician

In addition to thinking like an acoustician and calculating like an acoustician, you need to be able to speak like an acoustician. You should understand the meaning of the technical terms below that have been introduced in this chapter. (Note: The first time a significant new term is introduced in the notes, it is *italicized*.) If a term is descriptive, you should have your own definition (e.g., isotropic medium → a medium whose properties are the same in every direction). If it is a physical quantity, you should know its definition and its units (mechanical work, $dW = F \cdot dx$ or $dW = P \cdot dV$[Joules]).

Taylor series expansion	Phasor notation
Second-order correction	Rationalization
Power series	Complex conjugate
Product rule	Dimensional homogeneity
Differential	SI System of Units
The Fundamental Theorem of Calculus	Similitude
Integration by parts	Π-groups
Logarithmic differentiation	Dimensionless groups
Stable equilibrium	Random errors
Potential energy	Systematic (bias) errors
Neutral equilibrium	Statistical fluctuations
Unstable equilibrium	Relative uncertainty
Linear relationship	Precision
Hooke's law	Standard deviation
Isochronism	Ergodic hypothesis
Harmonic generation	Residual
The Principle of Superposition	End-to-end calibration
Orthogonal functions	Correlated or uncorrelated errors
Fourier synthesis	Covariance
Kronecker delta	Least-squares fit
Euler's identity	Correlation coefficient
Complex exponentials	Confounding variables
Argand plane	

Exercises

1. **Taylor series**. It is easy to evaluate trigonometric functions using a handheld calculator. In this exercise, you will estimate the value of sin 1.10 = 0.8921 using a Taylor series based on the value at $x_o = 1.0$ radians, sin 1.0 = 0.8415, and $dx = 0.10$ radians.
   (a) Using Eq. (1.1) and the fact that $d(\sin x)/dx = \cos x$, evaluate the first approximation to sin (1.10), and compare that result to the exact result. Express that difference as a percentage relative to the exact result.
   (b) Using Eq. (1.2) and the fact that $d^2(\sin x)/dx^2 = -\sin x$, evaluate the second approximation to sin (1.10), and compare that result to the exact result. Express that difference as a percentage relative to the exact result.

(c) Using the Taylor series expansion of $\sin x$ about $x_o = 0$ in Eq. (1.5), approximate sin (1.10) using the first three terms. Express the difference between the exact result and the result using the x, x^3, and x^5 terms as a percentage relative to the exact result.

2. **Product rule**. For a simple mass-spring system, the sum of the kinetic energy of the mass, m, and the potential energy of the spring having a spring constant, K, is conserved.

$$\frac{1}{2}m\left(\frac{dx}{dt}\right)^2 + \frac{1}{2}\mathrm{K}x^2 = \text{constant} \tag{1.120}$$

Use the product rule to show that time differentiation of Eq. (1.120) leads to Newton's Second Law of Motion.

$$\left[m\frac{d^2x}{dt^2} + \mathrm{K}x\right]\frac{dx}{dt} = 0 \quad \Rightarrow \quad m\frac{d^2x}{dt^2} + \mathrm{K}x = 0$$

3. **Logarithmic differentiation**. The volume of a cylinder is $V = \pi a^2 h$, where a is the cylinder's radius and h is its height. Calculate the relative change in the volume of the cylinder, $\delta V/V$, in terms of the relative change in radius, $\delta a/a$, and the relative change in height, $\delta h/h$.

4. **Complex roots**. Find the value of z that solves the following equation:

$$e^{2z} = 2j$$

5. **Convenience numbers**. Evaluate the following expressions where $j = (-1)^{1/2}$:

(a) $\sqrt[3]{-j}$

(b) $(32)^{1/5}$

(c) $\ln[3 + 4j]$

(d) $\left|\frac{a-jb}{b-ja}\right|$

(e) $\Re e\left[\frac{e^{-jx}}{1+e^{a+jb}}\right]$

(f) $\Re e\left[\frac{1}{1-j}\right]$

(g) $\left(j + \sqrt{3}\right)^2$

(h) $\left(1 - j\sqrt{3}\right)^{1/2}$

(i) $\left|\frac{\sqrt{5}+3j}{1-j}\right|$

6. **Lennard-Jones potential**. This potential function, sometimes known as the "6–12 potential," describes the interaction of two inert gas atoms as a function of their separation and is useful in calculations of scattering or determination of the temperature of condensation from a gas to a liquid. It was first proposed by John Lennard-Jones in 1924 [31] (Fig. 1.19).

$$V(r) = 4\varepsilon\left[\left(\frac{\sigma}{r}\right)^{12} - \left(\frac{\sigma}{r}\right)^{6}\right]$$

Fig. 1.19 Lennard-Jones interatomic potential

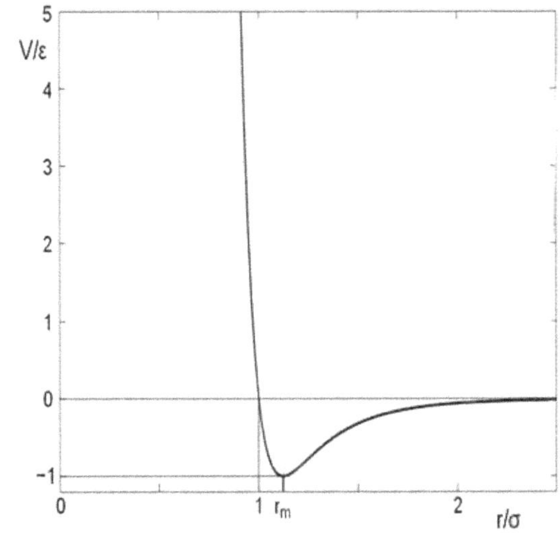

The term proportional to r^{-12} represents an empirical fit to the hard-core repulsion produced by the Pauli exclusion at short ranges when the atoms are close enough that the electron orbital overlaps. The r^{-6} term represents the van der Waals attraction due to the mutual interactions of the fluctuating dipole moments. Determine the equilibrium separation of two atoms, r_m, in terms of σ and ε.

7. **Work**. Determine the work done by the specified force over the specified paths where e_x is the unit vector in the x-direction and e_y is the unit vector in the y-direction. The units of F are newtons and x and y are in meters.

$$\vec{F} = x\hat{e}_x + 3j\hat{e}_y \quad \text{and} \quad W = \int_S \vec{F} \bullet d\vec{s}$$

   (a) Evaluate the work done by the force along Path S_1, (0, 0) to (2, 0) to (2, 1), with all motion along straight lines.
   (b) Evaluate the work done by the force along Path S_2, which is one straight line directly from (0, 0) to (2,1).
   (c) Evaluate the work along the straight-line Path S_3: (0, 0) to (2, 0) to (2, 1) to (0, 0).

8. **Similitude**. Use dimensional analysis to determine the functional dependence of the variables that describe the following physical systems:
   (a) *Boleadoras*. A mass, m, at the end of a string of length, ℓ, is swung overhead. Its tangential velocity is v. Determine the tension in the string. (You may neglect gravity.)
   (b) *Droplet oscillations*. A liquid drop of radius, a, and mass density, ρ [kg/m^3], may oscillate with its shape being restored to a sphere if the liquid has a surface tension, σ[N/m]. Determine how the period of oscillation could depend upon these three parameters.
   (c) *Deep water ocean waves*. Waves on a very deep ocean travel with a speed, v, that depends upon their wavelength, λ, but not upon their amplitude. How should this speed depend upon the wavelength, λ; the acceleration due to gravity, g [m/s^2]; and the density of water, ρ [kg/m^3]?
   (d) *Shallow-water gravity waves*. If the wavelength of a disturbance on the surface of water is much greater than the depth of the water, $h_o \ll \lambda$, the speed of the surface wave depends only upon depth and the acceleration due to gravity. How does the shallow-water gravity wave speed depend upon h_o and g?

(e) *Capillary waves*. At very short wavelengths, λ, the speed of surface waves on water depends upon wavelength; the fluid mass density, ρ; and the surface tension, σ [N/m].

(f) *Piano string*. How should the frequency of a string depend upon its tension, T [N]; its length, ℓ; and its linear mass density, ρ_L [kg/m]?

9. **Oscillator output impedance**. One way to determine the output impedance of an electronic oscillator or an amplifier is to measure the output voltage as a function of load resistance. The oscillator's internal resistance, R_{int}, and the load resistance, R_{load}, form a voltage divider circuit that relates V_{out} to V_{osc}, as shown in Fig. 1.20.

$$V_{\text{out}} = V_{\text{osc}}\left(\frac{R_{\text{load}}}{R_{\text{int}} + R_{\text{load}}}\right) \quad \Rightarrow \quad \frac{1}{V_{\text{out}}} = \frac{R_{\text{int}}}{V_{\text{osc}}}\frac{1}{R_{\text{load}}} + \frac{1}{V_{\text{osc}}}$$

Table 1.5 provides values of the load resistance, R_{load}, and the measured output voltage, V_{out}, corresponding to that load. Linearize the equation, and make a least-squares plot to determine the oscillator's internal resistance, R_{int}, and the oscillator's internal voltage, V_{osc}.

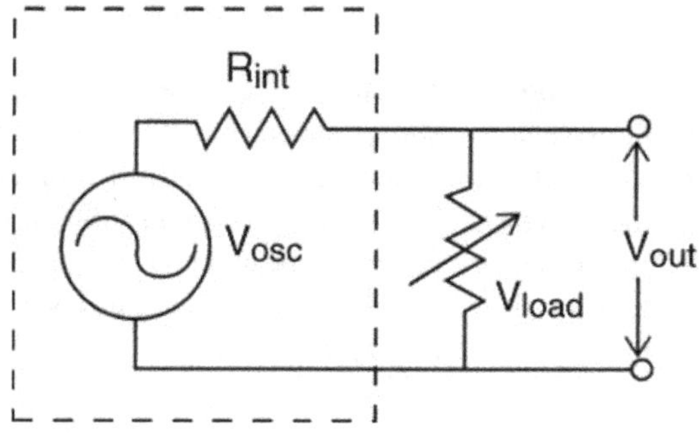

Fig. 1.20 An oscillator which produces a constant output voltage, V_{osc}, will produce voltage, V_{out}, across a load resistor, R_{load}, that depends upon the load resistance and the oscillator's own internal output resistance, R_{int}

Table 1.5 Measured values of output voltage, V_{out}, vs. load resistance, R_{load}

R_{load} [kΩ]	V_{out} [V_{ac}]
∞	1.97650
11.0	1.81276
9.0	1.77892
8.0	1.75779
7.0	1.73038
6.0	1.69524
5.0	1.64860
4.0	1.58302
3.0	1.48438
2.0	1.32063
1.0	0.99272
0.9	0.94760
0.8	0.88300

10. **Pendulum data analysis**. An experiment proceeds as follows: The pendulum is clamped at the 50th mark, approximately 1 m from the attachment point between wire and the pendulous mass. The oscillation period of the pendulum is recorded. The attachment point is raised by 2.00 ± 0.01 cm to the 49th mark (so the wire is shortened by 2.00 cm), and the period of oscillation is measured again. This same procedure is repeated nine more times, producing an 11-element data set containing the approximate wire length, ℓ_i, and oscillation period, T_i, for $i = 1, 2, 3, \ldots, 11$.
    (a) Write an algebraic expression for the acceleration due to gravity, g, in terms of the period, T_i; the wire length, l_i; and an undetermined constant, a, that accounts for the unknown distance between the last mark on the wire and the center of gravity of the mass.
    (b) Rearrange the equation you wrote in part (a) so that the data set can be plotted as a straight line and the slope and intercept of the line can be used to determine the local acceleration due to gravity, g, and the undetermined offset distance, a. Write expressions for g and for a in terms of the slope and the intercept of the straight line.
11. **Energy released by the first atomic bomb**. The first atomic bomb was detonated in Alamogordo, New Mexico, on 16 July 1945. The explosion was captured in a series of photographs like the one shown in Fig. 1.21. Using these photos, G. I. Taylor [32] was able to use similitude to estimate the energy released by the blast based on the rate at which the fire ball expanded, assuming that the fire ball was hemispherical and that the physical size of the source of the energy was very small.
    (a) Use dimensional analysis to determine the radius of the hemisphere if that radius depends upon time, t; the energy released by the explosion, E; and the density, ρ, of the air. You may assume that the proportionality constant is unity.
    (b) The blast radius (in meters) as a function of time (in milliseconds) is given in Table 1.6. Plot the radius vs. time using log-log axes, and determine the power law of the best-fit to those data. How close was the exponent that relates time to radius to the value predicted in part (a)?
    (c) Express your result in part (a) in terms of the energy release, E.
    (d) Plot R^5 vs. t^2 and set the intercept to zero.

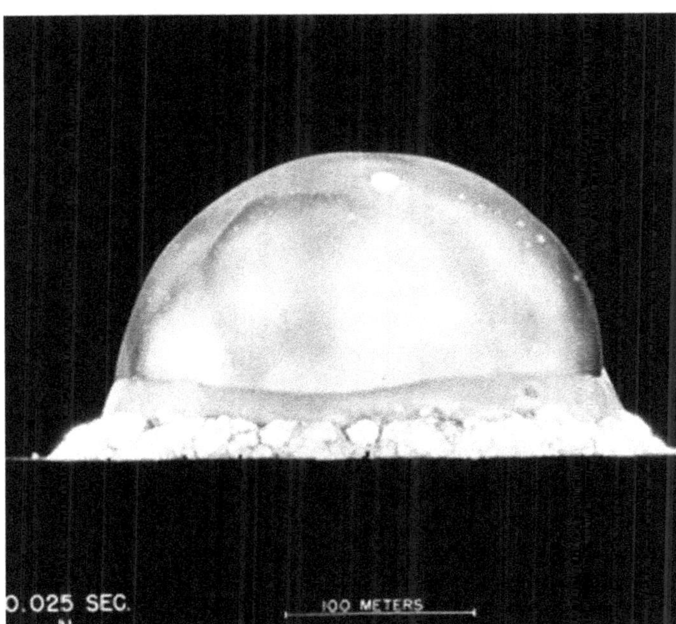

Fig. 1.21 Photo of the first atomic bomb explosion 25 milliseconds after detonation. The diameter of the blast hemisphere is approximately 260 m

Table 1.6 Blast radius

Time[ms]	Radius [m]
0.38	25.4
0.52	28.8
0.66	31.9
0.80	34.2
0.94	36.3
1.08	38.9
1.22	41.0
1.36	42.8
1.50	44.4
1.65	46.0
1.79	46.9
1.93	48.7
3.26	59.0
3.53	61.1
3.80	62.9
4.07	64.3
4.34	65.6
4.61	67.3
15.00	106.5
25.00	130.0
34.00	145.0
53.00	175.0
62.00	185.0

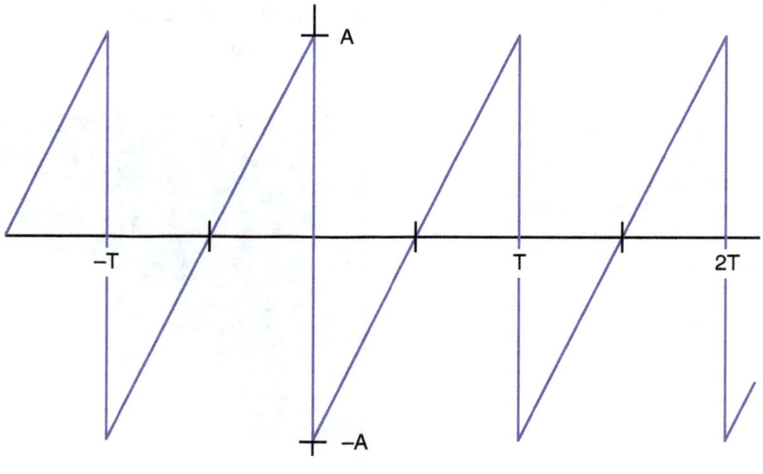

Fig. 1.22 Sawtooth waveform

(e) Using the slope determined in part (d), estimate the energy released by the explosion assuming the density of air is 1.2 kg/m^3. Report your result in joules and in kilotons of TNT. (1 kiloton of TNT = 4.184×10^9 J.)

12. **Fourier synthesis of a sawtooth waveform**. The sawtooth waveform occurs when nonlinear effects distort a sinusoidal sound wave into a fully developed shock wave (Fig. 1.22).

One cycle of a sawtooth wave, with period T and amplitude A, be expressed algebraically.

$$y(t) = \frac{2At}{T} - A \tag{1.121}$$

Write an expression for the sawtooth as a superposition of Fourier components with frequencies $f_n = n/T$, where $n = 1,2,3,\ldots$

References

1. P.M. Morse, K.U. Ingard, *Theoretical Acoustics* (McGraw-Hill, New York, 1968); ISBN 07-043330-5
2. U.S. Department of Labor, Occupational Safety & Health Administration (OSHA), Standard No. 1910.95(b)(2): https://www.osha.gov/pls/oshaweb/owadisp.show_document?p_table=standards&p_id=9735
3. U.S. Standard Atmosphere (National Oceanic and Atmospheric Administration, National Aeronautics and Space Administration, and United States Air Force, Washington, DC); NOAA-S/T 76-1562. U.S. Gov't. Printing Office, Stock No. 003-17-00323-0 (1976)
4. R. Courant, *Differential & Integral Calculus*, 2nd edn. (Blackie & Son, Glasgow, 1937), Ch. 2, §3.9
5. J. J. Stoker, *Nonlinear Vibrations in Mechanical and Electrical Systems* (Interscience, New York, 1950); ISBN 978-04-71570-332
6. A. B. Pippard, *The Physics of Vibration* (Cambridge University Press, Cambridge, 1989). See Ch. 9; ISBN 0 521 37200 3
7. L.D. Landau, E.M. Lifshitz, *Theory of Elasticity*, 2nd edn. (Pergamon, Oxford, 1970). See §7; ISBN 0 08 006465 5
8. B. Kinsman, *Wind Waves* (Prentice-Hall, Englewood Cliffs, 1965)
9. I.S. Gradshteyn, I.M. Ryzhik, *Tables of Integrals, Series, and Products*, 4th edn. (Academic, New York, 1980); ISBN 0-12-294760-6
10. M.L. Boas, *Mathematical Methods in the Physical Sciences*, 2nd edn. (Wiley, New York, 1983). See Ch. 7; ISBN 0-471-04409-1
11. D. Wells, Are these the most beautiful? Math. Intell. **12**(3), 37–41 (1990)
12. H. Lamb, *Dynamical Theory of Sound*, 2nd edn. (Edward Arnold & Co., London, 1931) or (Dover, 1960), pg. 54
13. E.F. Adiutori, Fourier. Mech. Eng. **127**(8), 30–31 (2005)
14. J.W. Strutt (Lord Rayleigh), The principle of similitude. Nature **95**, 66–68 (1915)
15. E. Buckingham, On physically similar systems; illustrations of the use of dimensional equations. Phys. Rev. **4**(4), 345–376 (1914)
16. J.R. Olson, G.W. Swift, Similitude in thermoacoustics. J. Acoust. Soc. Am. **95**(3), 1405–1412 (1994)
17. M. Strasberg, Dimensional analysis of windscreen noise. J. Acoust. Soc. Am. **83**(2), 544–548 (1988)
18. N.R. Hosier, P.R. Donovan, "Microphone Windscreen Performance," National Bureau of Standards Report NBSIR 79-1559 (Jan. 1979)
19. Private communication from Thomas J. Hofler
20. S. Rabinovich, *Measurement Errors: Theory and Practice* (American Institute of Physics, New York, 1995); ISBN 1-56396-323-X
21. H.D. Young, *Statistical Treatment of Experimental Data* (McGraw-Hill, New York, 1962); ISBN 07-072646-9.
22. R.A. Nelson, M.G. Olsson, The pendulum – Rich physics from a simple system. Am. J. Phys. **54**(2), 112–121 (1985). A better, though less readily accessible treatment, is that of G. G. Stokes, *Mathematical and Physical Papers* (Macmillan, 1905)
23. F.J. Anscombe, Graphs in statistical analysis. Am. Stat. **27**(1), 17–21 (1973)
24. Ernest Rutherford, F.R.S., 1st Baron Rutherford of Nelson (1871–1937)
25. P.R. Bevington, *Data Reduction and Error Analysis for the Physical Sciences* (McGraw-Hill, New York, 1969)
26. R.A. Fisher, F. Yates, *Statistical tables for biological, agricultural and medical research*, 6th edn. (Oliver and Boyd, London, 1963)
27. L.M. Ronsse, L.M. Wang, Relationships between unoccupied classroom acoustical conditions and elementary student achievement in eastern Nebraska. J. Acoust. Soc. Am. **133**(3), 1480–1495 (2013)
28. J. Higbie, Uncertainty in the linear regression slope. Am. J. Physiol. **59**(2), 184–185 (1991)
29. Y. Beers, *Introduction to the Theory of Error*, 2nd edn. (Addison-Wesley, Reading, 1957)
30. J. W. Strutt (Lord Rayleigh), *The Theory of Sound*, vol I, 2nd edn. (Macmillan, London, 1894) reprinted (Dover, 1945)
31. J.E. Lennard-Jones, On the determination of molecular fields. Proc. R. Soc. Lond. A **106**, 463–477 (1924)
32. G.I. Taylor, The formation of a blast wave by a very intense explosion. II. The atomic explosion of 1945. Proc. Roy. Soc. Lond **201**(1065), 175–186 (1950)

Part 1
Bars and Plates: Vibrations

2

An Integrated Study of Harmonic Oscillator

This chapter will introduce a system that is fundamental to our understanding of more physical phenomena than any other. Although the "simple" harmonic oscillator seems to be only the combination of the most mundane components, the formalism developed to explain the behavior of a mass, spring, and damper is used to describe systems that range in size from atoms to oceans.

The treatment of the harmonic oscillator in this chapter goes beyond the "traditional" treatments found in the elementary physics textbooks. For example, the introduction of damping will open a two-way street: a damping element (i.e., a mechanical resistance, R_m) will dissipate the oscillator's energy, thus reducing the amplitudes of successive oscillations, but it will also connect the oscillator to the surrounding environment (where the dissipated energy must go to "leave" the system). That surrounding environment will also return energy into the oscillator. The oscillation amplitude does not decay to zero but to a value that represents the oscillator's thermal equilibrium with its surroundings.

The excitation of such a harmonic oscillator by an externally applied force, displacement, or combination of the two will result in a response that is critically dependent upon the relationship between the frequency of excitation and the natural frequency of the oscillator, which is the oscillation of the undamped free oscillations. We will pay special attention to both the magnitude and phase of the driven response and will introduce the critical concepts of mechanical impedance, resonance, and quality factor.

Finally, the harmonic oscillator model will be extended to coupled oscillators that are represented by combinations of several masses and several springs. As the number of coupled masses and springs increases, several new features of such combinations will preview the behavior of continuous systems (like strings, bars, membranes, and plates). Ultimately, solid matter consists of a nearly infinite number of masses (atoms and molecules) that are coupled together by three times that number of stiffnesses arising from their mutual attraction or repulsion.

2.1 The Undamped Harmonic Oscillator

We begin this exploration by simply joining one *point mass* (meaning we neglect the spatial distribution of the mass which might require the specification of a moment of inertia), m, to one end of a spring with stiffness constant, K, where the concept of a stiffness constant was addressed in Sect. 1.2.1 and the stiffness was related to potential gradients that produced forces or vice versa. The other end of the spring is assumed to be rigidly immobilized (i.e., fixed, as shown in Fig. 1.3). We further restrict the junction of the spring and the mass to move only along a single direction: the x axis. This system is called a one-dimensional, single degree-of-freedom, undamped simple harmonic oscillator. (Damping will be added in Sect. 2.4.) Why does such a simple combination of two idealized components have such a complicated designation? It is an indication that there will be numerous variations on this theme.

At this stage, we will assume that the spring is massless, $m_s = 0$, but we will see soon that a non-zero spring mass can be easily accommodated under some simplifying assumptions. Our analysis of the motion of this simple mass-spring system will begin by combining a *dynamical equation* with an *equation of state*. The dynamical equation is Newton's Second Law of Motion. It relates the acceleration of the mass to the net external force acting on the mass. The net force will be that which the spring exerts on the mass, determined by the displacement of the spring, x, from its equilibrium extension, x_o, according to Hooke's law in Eq. (1.25). You may be more familiar with thermodynamic equations of state that relate the pressure of a gas to its density or temperature, but Hooke's law is a simple example of an equation of state that relates some deformation of the system (a stimulus) to its response (the net force). The dynamical equation is generic (i.e., system independent), but the equation of state is particular to a specific system, the spring in this case. To simplify subsequent calculations, we will define the origin of our one-dimensional coordinate system to be located at $x_o = 0$, the equilibrium length of our spring:

$$F = ma = m\frac{d^2x}{dt^2} = -\mathrm{K}x \quad \text{or} \quad \frac{d^2x}{dt^2} + \frac{\mathrm{K}}{m}x = 0 \tag{2.1}$$

There is an assumption that is built into Hooke's law that will be examined in the next section. We are assuming that the restoring force produced at the spring's moveable end depends only on the position of that end. That is true for a static extension or compression (e.g., weighing bananas on a spring scale in a market or the Gerber scale in Fig. 2.2). It might not be exact in a dynamical situation since we have assumed that the entire spring is instantaneously aware of the end's location. For that to be true, information about the end's position would have to propagate along the spring with an infinite speed, as discussed in Sect. 2.2.

Equation (2.1) is a second-order, homogeneous, linear differential equation with constant coefficients, and therefore there are two linearly independent solutions. As discussed in Sect. 1.5, we could solve that equation to provide the entire time history of the mass-spring system's response, $x(t)$, using a superposition of sine and cosine functions or complex exponentials, where we have set $\omega_o^2 = \mathrm{K}/m$.

$$x(t) = A\cos(\omega_o t) + B\sin(\omega_o t) \tag{2.2}$$

This solution is periodic and repeats when $\omega_o T = 2\pi$ or integer multiples of 2π, so the *frequency* of oscillation, $f_o = (1/T) = \omega_o/2\pi$, is the reciprocal of the *period*, T.

$$\omega_o = 2\pi f_o = \frac{2\pi}{T} = \sqrt{\frac{\mathrm{K}}{m}} \tag{2.3}$$

We call ω_o the *angular frequency* or the *radian frequency*. The radian frequency simplifies calculations by eliminating the need to explicitly include factors of 2π. It is important to remember that instrumentation and most software typically report frequencies in cycles per second [Hz], not radians per second.

2.1.1 Initial Conditions and the Phasor Representation

It is worthwhile to re-emphasize the fact that the "natural frequency" of oscillation, ω_o, for this simple linear system (see Sect. 1.3) is amplitude independent. As shown in Sect. 1.7.1, simple dimensional arguments require that the frequency be proportional to $\sqrt{\mathrm{K}/m}$. The general solution for the motion of our simple undamped mass-spring oscillator in Eq. (2.2) contains two amplitude-dependent constants: A and B. These constants are determined by the conditions that initiate the system's motion at the instant that defines $t = 0$. We will first consider two ways that could set the system into motion: an initial displacement, x_1, or an initial impulse that propels the mass with an initial velocity, $v_1 \equiv (dx/dt)_o$, where the subscript on the derivative indicates that it is to be evaluated at $t = 0$.

If we hold the mass at position x_1 and then release it at $t = 0$, the sine term vanishes, and the cosine term in Eq. (2.2) will equal one, so $A = x_1$. The instantaneous velocity of the mass is determined by differentiating Eq. (2.2):

$$v(t) = \dot{x}(t) = \frac{dx(t)}{dt} = -\omega_o A \sin(\omega_o t) + \omega_o B \cos(\omega_o t) \tag{2.4}$$

At $t = 0$, the sine term again vanishes and $v(0) = \omega_o B$. Since the mass is at rest at $t = 0$, $v(0) = 0$, so $B = 0$. For this initial displacement, the subsequent motion of the mass is given by $x(t) = x_1 \cos(\omega_o t)$. Initially, the mass moves closer to its equilibrium position; one-quarter period later, it passes through its equilibrium position, $x_o = 0$, reaching its maximum negative position, $-x_1$, a time, $T/2$, after starting, and returns to its initial position, x_1, at $t = T$. In principle, this motion repeats forever with period, T.

If instead the mass is initially at rest at $x(0) = 0$ and the motion is initiated by impulsively striking the mass to impart an initial velocity $v(0) = v_1$, then Eq. (2.4) sets $B = v_1/\omega_o$, while Eq. (2.2) sets $A = 0$. The subsequent motion is given by $x(t) = (v_1/\omega_o) \sin(\omega_o t)$; we have given the mass an initial velocity in the $+x$ direction, and it moves off in that direction until reversing its direction one-quarter cycle later, after coming to rest for an instant at $x_1 = |v_1/\omega_o|$.

Of course, those two exemplary initial conditions are not the only possibilities. In fact, we could impose both of the previous initial conditions simultaneously by displacing the mass by a distance, x_1, and then striking it at $t = 0$ to produce a velocity v_1. The procedure to determine the amplitudes, A and B, is the same and produces the same results.

$$x(t) = x_1 \cos(\omega_o t) + \frac{v_1}{\omega_o} \sin(\omega_o t) \tag{2.5}$$

For an initial displacement in the $+x$ direction, combined with an initial velocity in the same direction, the amplitude of the oscillation will be larger than if either condition was imposed individually, as was done previously. A geometrical interpretation will be useful.

In Fig. 2.1, we imagine the position of the mass being represented by the projection on the x axis of a vector that rotates counterclockwise at an angular velocity, ω_o. The superposition of the two initial conditions produces a vector of length $|\vec{x}| = \sqrt{x_1^2 + (v_1/\omega_o)^2}$. At $t = 0$, the vector makes an angle $\phi = -\tan^{-1}[(v_1/\omega_o)/x_1]$, which is below the x axis in the fourth quadrant. As time increases, the vector rotates, and its projection along the x axis increases until $t = \phi/\omega_o$ when the displacement reaches its maximum positive value.

Another way of expressing the solution provided in Eqs. (2.2) and (2.5) is to express $x(t)$ in terms of a single amplitude $C = \sqrt{x_1^2 + (v_1/\omega_o)^2}$ and the same phase angle, ϕ, as diagrammed in Fig. 2.1.

$$x(t) = C \cos(\omega_o t + \phi) \tag{2.6}$$

The value of ϕ allows a mixture of the appropriate proportion of the sine and cosine terms in Eq. (2.2) to meet the specific initial conditions. It also is ideally suited to the expression of $x(t)$ as the real part of a complex exponential.

$$x(t) = \Re e\left[\widehat{\mathbf{C}} e^{j\omega_o t}\right] = \Re e\left[|\widehat{\mathbf{C}}|\, e^{j\phi} e^{j\omega_o t}\right] = \Re e\left[|\widehat{\mathbf{C}}|\, e^{j(\omega_o t + \phi)}\right] \tag{2.7}$$

Now $\widehat{\mathbf{C}}$ is treated as a complex number (phasor) with magnitude $|\widehat{\mathbf{C}}|$ and phase ϕ. Taking the real part of Eq. (2.7) regenerates Eq. (2.6) and allows the use of the complex exponential to simplify the calculation of integrals and derivatives as shown in Eq. (1.47).

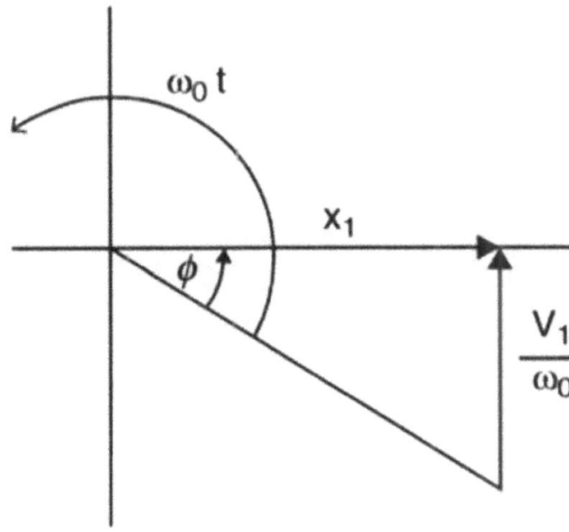

Fig. 2.1 Geometrical representation for the combination of an initial displacement, x_1, and initial velocity, v_1, for a single degree-of-freedom, undamped, simple harmonic oscillator. The vector of length $|\vec{x}| = \sqrt{x_1^2 + (v_1/\omega_o)^2}$ starts at an angle $\phi = -\tan^{-1}[(v_1/\omega_o)/x_1]$ and rotates in the counterclockwise direction at angular velocity, ω_o. The projection of that vector on the horizontal axis represents the position of the mass as a function of time

For example, we can produce the velocity in complex notation, as we did for the trigonometric expression of Eq. (2.4), by simple multiplication of Eq. (2.7) by $j\omega_o$. Similarly, double differentiation of Eq. (2.7) produces the acceleration a (t) by multiplication of Eq. (2.7) by $(j\omega_o)\,(j\omega_o) = -\omega_o^{\,2}$.

$$\begin{aligned} v(t) &= j\omega_o x(t) = \Re\mathrm{e}\left[j\omega_o \widehat{\mathbf{C}} e^{\,j\omega_o t} \right] \\ a(t) &= -\omega_o^2 x(t) = \Re\mathrm{e}\left[-\omega_o^2 \widehat{\mathbf{C}} e^{\,j\omega_o t} \right] \end{aligned} \tag{2.8}$$

These expressions indicate that there is a +90° phase shift (in the counterclockwise sense) between displacement and velocity and 180° phase shift between displacement and acceleration.

The use of complex exponentials is so convenient and ubiquitous in the treatment of problems involving acoustics and vibration that it is customary to simply drop the explicit reference to "taking the real part of the complex expression" and treat the complex expressions as the "result" with the tacit understanding that the behavior of the physical system corresponds to only the projection of the complex vectors onto the real axis.

2.2 The Lumped-Element Approximation

The previous analysis of the undamped simple harmonic oscillator used a spring constant, K, to relate the motion of the end of the spring to the force it imposed on the attached mass. The fact that the force depended only upon the position of the end of the spring implied a *quasi-static approximation* that is equivalent to saying that each location along the spring's entire length had instantaneous "knowledge" of the current location of the end of the spring. Said another way, information about the end position was instantaneously accessible at every location along the entire length of the spring, or the speed at which such information could propagate is infinite. In the quasi-static approximation, the displacement of each "coil" of the spring is linearly proportional to its distance from the fixed end. Figure 2.2 shows an adjustable scale (i.e., ruler) which exploits that proportionality.

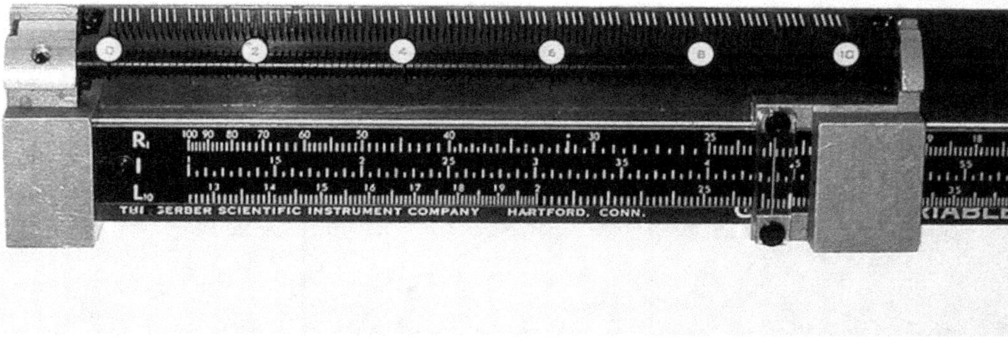

Fig. 2.2 The Gerber scale (H. Joseph Gerber (1924–1996) was a prolific inventor and able businessman who had been granted more than 650 patents over his lifetime) makes use of the quasi-static approximation to produce a ruler with equally spaced markings but adjustable overall length. At the top of the photograph is a triangular spring that has every tenth coil painted red, every fifth coil blue, and the other coils white. Even-numbered labels are mounted on a second spring to keep track of the red marks. Since these are linear springs, which obey Hooke's law, the coils will be equally spaced with respect to any stop position of the slider. Consequently, in the days before computerized plotting and data analysis, this scale could be used for linear interpolation or extrapolation of plotted data points without any computation. The logarithmic scale on the base plate allows it to be used on graphs with logarithmic coordinates

In reality, as the end of the spring is moved, there will be a wave launched from the moving end. Since we will not be addressing wave motion until the next chapter, we can make an estimate of the limitations of the quasi-static assumption using dimensional arguments (i.e., using similitude as discussed in Sect. 1.7). Before doing so, we need to consider the net stiffness of combinations of two or more springs.

2.2.1 Series and Parallel Combinations of Several Springs

If two springs are connected from a fixed mounting to the same load, as illustrated schematically on the left in Fig. 2.3, their individual stiffnesses will add. Springs combined in such an arrangement are said to be "springs in parallel." Since the ends of both springs will be displaced by the same amount, x, the magnitude of the net force, $F_{parallel}$, will be the sum of their forces.

$$-F_{parallel} = -(F_1 + F_2) = \mathrm{K}_1 x + \mathrm{K}_2 x = (\mathrm{K}_1 + \mathrm{K}_2)x = \mathrm{K}_{parallel} x \tag{2.9}$$

The parallel combination of the two springs of stiffness K_1 and K_2 will have an effective spring constant of $\mathrm{K}_{parallel} = \mathrm{K}_1 + \mathrm{K}_2$. The extension to the parallel combination of several springs will result in an effective parallel spring constant that is just the sum of their individual stiffnesses.

If the same two springs are joined end to end, then we say the springs are combined "in series." Their effective series spring constant will be less than that of either spring acting alone. If the position of the free end of the combination is at x and the junction is at x_a, then the forces at x_a produced by either spring must be equal, since that intermediate position is neither accelerating nor decelerating.

$$-F_a = \mathrm{K}_1 x_a = \mathrm{K}_2(x - x_a) \quad \Rightarrow \quad x_a = \frac{\mathrm{K}_2}{\mathrm{K}_1 + \mathrm{K}_2} x \tag{2.10}$$

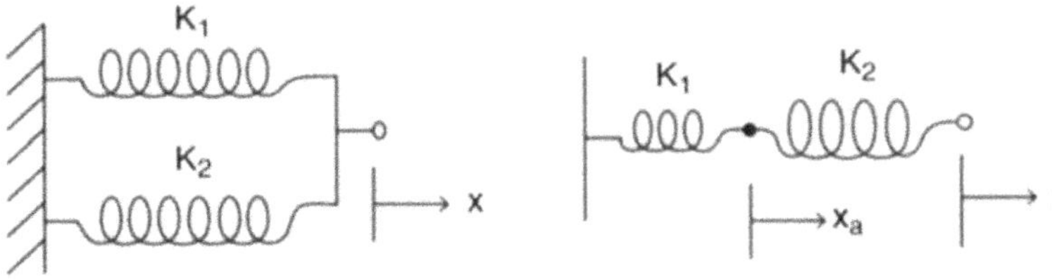

Fig. 2.3 (*Left*) Two springs combined in parallel will exhibit a net effective spring constant that is the sum of the individual spring constants. (*Right*) For the series combination, the forces at the junction between the two springs must be equal and opposite, as dictated by Newton's Third Law of Motion, and the net stiffness will be smaller than the stiffness of either spring

This result makes sense in the limit that $K_2 \gg K_1$, making K_2 effectively rigid, since only K_1 would be extended if x was increased so $x_a \cong x$. Similarly, in the opposite limit, if $K_1 \gg K_2$, then only K_2 would be extended, so $x_a \cong 0$. The other obvious check would be to let $K_1 = K_2$. In that case, we expect each spring to provide half the displacement, i.e., $x_a = x/2$, and the force would be half that of a single spring displaced from its equilibrium position by the same amount.

The force of the series combination can be obtained by substitution of x_a back into the expression for the force.

$$-F_a = -F_{series} = K_1 x_a = \frac{K_1 K_2}{K_1 + K_2} x = \left(\frac{1}{K_1} + \frac{1}{K_2}\right)^{-1} x = K_{series} x$$
$$\frac{1}{K_{series}} = \frac{1}{K_1} + \frac{1}{K_2} \tag{2.11}$$

Again, the extension to a series concatenation of several springs is also simple. Readers with exposure to the basics of electrical circuit theory will recognize that these expressions for the stiffnesses of series and parallel combinations of springs are the exact reverse of those for the series and parallel combinations of electrical resistors or inductors, although the same as for series and parallel combinations of capacitors. This is because for electrical resistors in series, the flow (i.e., electrical current) is continuous and the potentials (i.e., voltages) are additive. For springs in series, the potential (i.e., force) is continuous, and the flow (i.e., displacement) is additive.

2.2.2 A Characteristic Speed

To determine an estimate for the limits of validity of the assumption that we can use the static stiffness of a spring for calculation of the dynamical behavior of a mass-spring oscillator, we can use a dimensional argument to calculate a characteristic speed for a spring of stiffness, K, and length, L. Since such a characteristic speed should be a property of the spring and not any particular application, we need to identify spring properties that are not a function of the spring's length.

The spring's mass, m_s, can be divided by the spring's length, L, to produce a linear mass density, ρ_L, which is a characteristic of the spring and independent of length: $\rho_L = m_s/L$. For a uniform spring, ρ_L will be a constant. Based on our calculation of the overall stiffness of a series combination of springs, the product of the stiffness, K, of a spring with its length, L, is independent of the spring's length and only a property of the spring's construction (i.e., shape and materials). If two identical springs are attached in series, the overall stiffness of the combination is half that of either spring individually, and the overall length is doubled. For a uniform spring material, the product, (LK), is a constant.

We can now use similitude (see Sect. 1.7) to combine ρ_L [kg/m] and (LK) [(N/m)m = N = kg−m/s^2] to produce a speed, c [m/s], based on the units required to specify those two parameters that are characteristics of the spring and independent of its length.

$$\rho_L^a(\mathrm{LK})^b = \frac{M^a}{L^a}\frac{M^bL^b}{T^{2b}} = \frac{L}{T} = c \tag{2.12}$$

Since c does not require any mass units, $a + b = 0$. Since only (KL) involves T, $2b = 1$, so $b = ½$, and therefore $a = -½$. Equating the terms involving L requires $-a + b = 1$, which is also satisfied.

$$c \propto \sqrt{\frac{\mathrm{LK}}{\rho_L}} = L\sqrt{\frac{\mathrm{K}}{m_s}} \tag{2.13}$$

To determine whether it was appropriate to use a static spring stiffness to solve a dynamical problem, we can now simply compare the period, T, of our simple harmonic oscillator given by Eq. (2.3) to the time, t_s, it takes for information to propagate back and forth along a spring of length, L, at speed c: $t_s = 2\,L/c$. If $T \gg t_s$, then there is plenty of time for the entire spring to be influenced by the changing position of its point of attachment to the mass [1].

The dimensional argument that produced Eq. (2.13) cannot provide any multiplicative numerical constant. For now, we will simply express the criterion, $T \gg t_s$, as $m \gg m_s$. Once we are able to calculate the speed of compressional wave propagation along the spring, we will see that our "lumped-element" approximation, used to derive Eq. (2.3), is equivalent to saying that the spring length, L, is much less than the wavelength, λ, of a compressional wave of frequency, f_o. That criterion also leads to the requirement that $m \gg m_s$ to justify neglecting dynamical effects that would alter the effective stiffness of the spring from its static value.

To reiterate, under certain circumstances (usually at lower frequencies), it is possible to treat our mass as having no stiffness and our stiffness as having no mass, allowing us to neglect wavelike behavior associated with the spring's length. Under those assumptions, these components are specified at a point and are called "lumped elements" in the same way that inductors, capacitors, and resistors are considered lumped elements in electrical circuit theory.

2.3 Energy

Just as the definition of work was used to establish the relationship between a restoring force and its associated potential energy, Eq. (1.22) can be used to calculate the work done against the inertia of a particle with mass, m, based on Newton's Second Law of Motion: $\vec{F} = m\left(d\vec{v}/dt\right)$.

$$W = -\int \vec{F}\bullet d\vec{x} = -m\int \frac{dv}{dt}dx = -m\int vdv = -\frac{1}{2}mv^2 + \text{constant} \tag{2.14}$$

A particle of mass, m, has an energy associated with its velocity, v, which is called its kinetic energy.

$$KE = \frac{1}{2}mv^2 \tag{2.15}$$

The relationship between the work done by a conservative force and the change in potential energy created by such a vector force, $\vec{F}$, operating over a vector displacement, $\vec{x}$, was described in Sect. 1.2.1. If the force is produced by a linear spring that obeys Hooke's law, and the force and the displacement are co-linear, then the potential energy takes a particularly simple form.

$$PE = \frac{1}{2}\mathrm{K}x^2 \tag{2.16}$$

In that expression, the displacement, x, is assumed to be the deviation from the spring's equilibrium position, x_o, so Eq. (2.16) also assumes a coordinate system with its origin making $x_o = 0$.

If the origin of our time coordinate is defined so that we can eliminate the cosine term from the position of the mass, $x(t)$, expressed in Eq. (2.2), then we can write $x(t) = B \sin(\omega_o t)$ and $v(t) = \omega_o B \cos(\omega_o t)$. Substitution of these into the expressions for the potential and kinetic energy of a mass-spring system will provide an expression for the total instantaneous energy, $E_{Tot}(t)$.

$$E_{Tot}(t) = KE(t) - PE(t) = \frac{m\omega_o^2 B^2}{2}\cos^2(\omega_o t) + \frac{\mathrm{K}B^2}{2}\sin^2(\omega_o t) \tag{2.17}$$

Since $\omega_o{}^2 = \mathrm{K}/m$, and $\cos^2 x + \sin^2 x = 1$, the total energy is time-independent. The energy oscillates between being entirely kinetic, as the mass moves through $x(t) = 0$, to entirely potential when $x(t) = \pm B$.

$$E_{Tot} = \frac{\mathrm{K}B^2}{2} = \frac{m\omega_o^2 B^2}{2} = (PE)_{\max} = (KE)_{\max} \tag{2.18}$$

Furthermore, the time-averaged values of both the kinetic and potential energies are equal.

$$\begin{aligned} \langle KE \rangle_t &= \frac{m\omega_o^2}{2T}\int_0^T B^2 \cos^2(\omega_o t)dt = \frac{m\omega_o^2 B^2}{4} \\ \langle PE \rangle_t &= \frac{\mathrm{K}}{2T}\int_0^T B^2 \sin^2(\omega_o t)dt = \frac{\mathrm{K}B^2}{4} \end{aligned} \tag{2.19}$$

Conservation of energy can be used to demonstrate the equivalence to the Newtonian formulation of Eq. (2.1).

$$\frac{1}{2}m\left(\frac{dx}{dt}\right)^2 + \frac{1}{2}\mathrm{K}x^2 = \text{constant} \tag{2.20}$$

Differentiation of this expression with respect to time regenerates Eq. (2.1).

$$\left[m\frac{d^2x}{dt^2} + \mathrm{K}x\right]\frac{dx}{dt} = 0 \quad \Rightarrow \quad m\frac{d^2x}{dt^2} + \mathrm{K}x = 0 \tag{2.21}$$

2.3.1 The Virial Theorem

The fact that the sum of the kinetic and potential energies is time-independent is a consequence of energy conservation, since no components have been introduced (yet!) in the system that can dissipate energy. The fact that the maximum and the time-averaged kinetic and potential energies are equal also seems plausible, but it is not universal. This equality is a consequence of the assumed Hooke's law behavior where the power law exponent, $b = 1$: $F \propto x^b$.

For "central forces" (i.e., those forces that act along a line connecting the force with the object acted upon by that force) that obey a power law, the *virial theorem* guarantees that the time-averaged kinetic and potential energies are dependent upon that exponent [2].

$$\langle KE \rangle_t = \frac{b+1}{2} \langle PE \rangle_t \tag{2.22}$$

The equality of the time-averaged potential and kinetic energies is a consequence of our assumption of the linear behavior of our spring. It is useful to remember that this equality does not hold for a nonlinear spring.

A simple counterexample is planetary motion under the influence of gravity. In the simplest case of circular orbits, the centripetal force, F_{cent}, on a planet of mass, m, is provided by Newton's universal law of gravitation[1] due to a larger central body of mass $M \gg m$. Their centers of mass are separated by a distance, R.

$$F_{cent} = m\omega^2 R = \frac{GMm}{R^2} \quad \Rightarrow \quad \omega^2 = \frac{GM}{R^3} \tag{2.23}$$

This is a result known as Kepler's third law of planetary motion: the square of the orbital period is proportional to the cube of the radius of the orbit.[2] The planet's kinetic energy is determined by its tangential velocity, $v_t = \omega R$.

$$KE = \frac{1}{2} m v_t^2 = \frac{m}{2} (\omega R)^2 = \frac{1}{2} \frac{GMm}{R} \tag{2.24}$$

The gravitational force, $F_{grav} = GMm/R^2$, is the negative gradient of the gravitational potential energy.

$$PE = -\frac{GMm}{R} \tag{2.25}$$

For this simple example, both the KE and PE are time-independent, so the average PE is twice the average KE in accordance with Eq. (2.22) for $b = -2$.

2.3.2 Rayleigh's Method

"The object being to approach the truth as nearly as can be done without too great a sacrifice of simplicity."

(J. W. Strutt (Lord Rayleigh) [3])

We can take advantage of the equality of the time-averaged values of the kinetic and potential energy that is unique to the vibration of a system whose restoring force is linear in the displacement from equilibrium. The magnitude of the oscillatory velocity, $|v|$, for a simple harmonic oscillator is equal to the product of the magnitude of the displacement times the oscillatory frequency: $|v| = \omega_o |x|$. Setting KE in Eq. (2.15) equal to PE in Eq. (2.16) leads to the same expression for ω_o that was derived in Eq. (2.3).

This equality facilitates calculation of the correction to the natural frequency of the mass-spring oscillator when $m_s \neq 0$. The displacement of each coil of a spring that is extended from its equilibrium position, x_o, by an amount x_1 is proportional to its distance, x, from the fixed end at $x = 0$, in the quasi-static approximation (as it was for the Gerber scale in Fig. 2.2). The magnitude of the displacement of

[1] The value of $G = 6.6738 \times 10^{-11}$ N/(m/kg)^2 is the most poorly known physical constant, having a relative uncertainty of $\delta G/G = \pm 1.4 \times 10^{-4}$; P. J. Mohr, B. N. Taylor, and D. B. Newell, "The 2010 CODATA Recommended Values of the Fundamental Physical Constants," http://physics.nist.gov/constants

[2] Kepler's third law applies to elliptical orbits. The assumed circular orbit is just a special case for an ellipse without eccentricity. For an elliptical orbit, the square of the period is proportional to the cube of the semi-major axis.

any point on the spring can be expressed as $\xi(x) = x_1(x/L)$. The magnitude of the velocity of the spring at that location is $v(x) = \omega_o \xi(x)$.

Again, defining the linear mass density, $\rho_L = m_s/L$, of the spring in terms of the spring's mass, m_s, and its length, L, as in Eq. (2.13), the kinetic energy of the spring KE_s can be calculated by the mass density of the spring times the square of its maximum local velocity, $v^2(x)$, integrated over the entire length of the spring.

$$(KE_s)_{\max} = \frac{1}{2}\int_0^L \rho_L v_1^2 \left(\frac{x}{L}\right)^2 dx = \frac{\omega_o^2 x_1^2}{6}(L\rho_L) = \frac{1}{3}\left(\frac{1}{2} m_s \omega_o^2 x_1^2\right) \tag{2.26}$$

That result for the maximum kinetic energy of the spring can be added to the maximum kinetic energy of the mass attached to the mobile end of the spring. By Eq. (2.18), the total kinetic energy can be equated to the maximum potential energy of the spring.

$$\begin{gathered} (PE)_{\max} = \frac{1}{2}\mathrm{K}x_1^2 = (KE)_{\max} = (KE_{mass})_{\max} + (KE_s)_{\max} \\ \frac{1}{2}\mathrm{K}x_1^2 = \frac{1}{2}\omega_o^2 x_1^2\left(m + \frac{m_s}{3}\right) \quad \Rightarrow \quad \omega_o = \sqrt{\frac{\mathrm{K}}{m + \frac{m_s}{3}}} \end{gathered} \tag{2.27}$$

In the quasi-static approximation, which we have shown in Sect. 2.2.2 requires that $m \gg m_s$, the effect of the spring's mass on the frequency of oscillation is to add one-third of the spring's mass to the moving mass, m, that is attached to the moving end of the spring and then calculate the resonance frequency using Eq. (2.3). This technique of using the calculation of the kinetic energy of a distribution of masses to determine the natural frequency of a linear vibratory system is known as *Rayleigh's method* [4]. In Sect. 3.4.3 we will use Rayleigh's method to approximate the natural frequency by integrating both the potential and the kinetic energies and then taking the square-root of their ratio. Rayleigh's method is used throughout this textbook because it can "improve computational ease dramatically while only slightly reducing accuracy" [5].

2.3.3 Gravitational Offset

To this point, the effects of gravity on our mass-spring system have been neglected. There has been a tacit assumption that our mass has been constrained to move in only a single direction by some frictionless guide. Mass-spring systems are used as scales for weighing, whether it is a traditional grocer's spring scale, a sport fishing spring scale, or a digital scale which incorporates an elastic load cell and some electronic system for measuring and displaying the deflection caused by the weight of a load.

If our z axis is defined as increasing in the upward direction, then Hooke's law provides the static downward deflection, Δ, from the spring's unloaded equilibrium position, z_o, as shown at the left in Fig. 2.4: $\Delta = -(mg)/\mathrm{K}$.

The right side of Fig. 2.4 is a plot of the gravitational potential energy, $(PE)_{grav} = mg(z - z_o)$, represented as the dotted line, and the elastic potential energy, $PE = (½)\mathrm{K}(z - z_o)^2$, represented as the dashed parabola; and their sum is shown as the solid line. The origin of the z axis in that figure has been translated to place the equilibrium position of the unloaded spring at $z_o = 0$.

The minimum of the sum of those two potential energies is shifted to a negative z value, indicating that the addition of the mass has produced $\Delta < 0$. Using the formalism developed in Sect. 1.2, the net

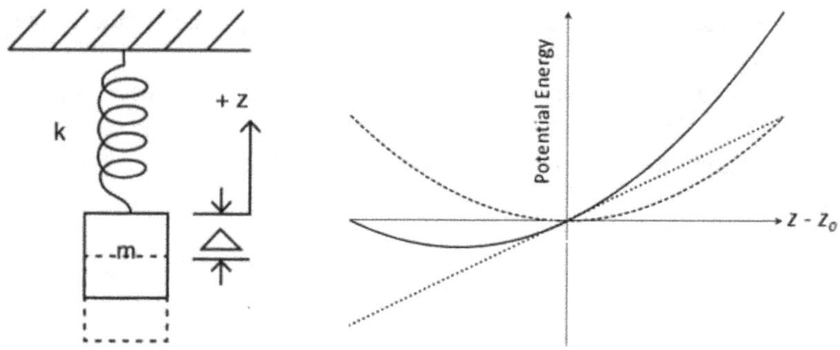

Fig. 2.4 (*Left*) A spring of equilibrium length, L, is suspended from a rigid support, and a mass, m, is hung from the spring causing a static deflection, Δ, from its unloaded equilibrium position, z_o. (*Right*) The *solid line* represents the total potential energy on the vertical axis as a function of position, $(z - z_o)$, plotted on the horizontal axis. The *dotted line* represents the gravitational potential energy, $(PE)_{grav} = mg\,(z - z_o)$. The *dashed parabola* represents the elastic potential energy, $PE = (½)\,K\,(z - z_o)^2$

force at the new position of equilibrium will be given by the first derivative of the potential energy. When the net force is zero, the mass will be at its equilibrium position, $-\Delta$.

$$\frac{d(PE)}{dz} = -Kz - mg = 0 \quad \Rightarrow \quad \Delta = -mg/K \tag{2.28}$$

As shown in Eq. (1.25), the second derivative of the potential energy provides the stiffness, which is unaltered by the displacement produced by the attachment of the weight, mg.

$$\left(\frac{d^2(PE)}{dz^2}\right)_{z=\Delta} = -K \tag{2.29}$$

The fact that the spring stiffness is unaltered by the displacement produced by the attachment of the weight is a consequence of the linearity of Hooke's law. As long as the weight does not take the spring beyond its linear regime (see Fig. 1.4), the oscillation frequency, $\omega_o = (K/m)^{1/2}$, will be the same as in the previous case where we neglected gravity and assumed that the one-dimensional displacements were maintained by some frictionless constraint.

It can be useful to recognize that the natural frequency of the mass-spring system subject to the force of gravity can be expressed in terms of the magnitude of the static displacement, $|\Delta|$, produced when a mass, m, is added to the spring of stiffness, K. From Eq. (2.28), we see that $K = |mg/\Delta|$. Substitution of that result into Eq. (2.3) allows the natural frequency, ω_o, to be specified in terms of the gravitational acceleration, g, and the magnitude of the static displacement, Δ.

$$\omega_o = \sqrt{\frac{g}{\Delta}} \tag{2.30}$$

This form of the result can be convenient for the design and analysis of vibration isolation systems discussed in Sect. 2.5.7.

The pendulum is a simple harmonic oscillator with a natural frequency that is explicitly dependent upon gravity. In this case, there is no "spring" and gravity provides the restoring force. Using the coordinate system of Fig. 1.5, the angular equivalent of Eq. (2.1) is obtained by equating the moment of inertia of the suspended mass, $I = mL^2$, times its angular acceleration, $\ddot{\theta} = d^2\theta/dt^2$, to the angle-dependent torque, $N(\theta)$, derived in Eq. (1.27).

$$\left(mL^2\right)\frac{d^2\theta}{dt^2} + mgL\left[\theta - \frac{\theta^3}{6} + \ldots\right] = 0 \tag{2.31}$$

Restricting the angular displacements to small values, $\theta \ll 1$, the nonlinear term in the restoring force can be neglected. If we restrict the maximum deflection to $\theta_1 \leq 0.10$ rad $= 5.7°$, the ratio of the nonlinear contribution to the linear contribution will be less than $\theta_1^2/6 = 0.17\%$.

After cancellation of common terms, the expression can be cast into the same form as Eq. (2.1) with θ as the independent variable and $\omega_o{}^2 = g/L$.

$$\ddot{\theta} + \frac{g}{L}\theta = \ddot{\theta} + \omega_o^2\theta = 0 \quad \Rightarrow \quad \omega_o = \sqrt{\frac{g}{L}} \tag{2.32}$$

Unlike the natural frequency, f_o, of the mass-spring system, the natural frequency of the pendulum is independent of the mass at the end of the string but is explicitly dependent upon the acceleration due to gravity, g.

This is not because the mass is not important to the motion of the pendulum. The disappearance of mass from Eq. (2.32) is a consequence of the fact that the inertia and the restoring force are both directly proportional to the mass, m. Since the natural frequency depends upon the ratio of the restoring force (or restoring torque, in this case) to the inertia, the mass cancels out of the expression that provides the natural frequency. It is worthwhile to compare Eq. (2.32) for the frequency of the pendulum to Eq. (2.30) for the frequency of a mass-spring system in a gravitational field. In Eq. (2.30), the explicit dependence on the mass has also disappeared, again due to the fact that the stiffness is being expressed in terms of the static displacement, Δ, caused by the gravitational force on the mass.

2.3.4 Adiabatic Invariance

The pendulum provides a simple system that will allow us to demonstrate an additional useful relationship between the energy of a simple harmonic oscillator and its frequency. If we change a constraint on our oscillator, and if we make that change gradually on the time scale set by the period of oscillation, T, then the relative change in the energy is equal to the relative change in the natural frequency.

$$\delta\left(\frac{E}{\omega}\right) = 0 \quad \text{or} \quad \frac{\delta E}{E} = \frac{\delta\omega}{\omega} \tag{2.33}$$

For slow transitions,[3] the ratio of the energy to the frequency is a constant. Although this general relationship can be derived from a Lagrangian mechanical perspective [6], it is much better known by its quantum mechanical manifestation as Planck's law for the emission of electromagnetic radiation, $E = \hbar\omega$, where $\hbar \equiv 1.05457182 \times 10^{-34}$ J - s is Planck's constant, h, divided by 2π. In that application, the ratio of energy to frequency is determined by Planck's constant.

[3] It is the "slowness" of the transition that gives rise to the "adiabatic" designation. If a transition is abrupt, that is, taking place during much less than one period of oscillation, then energy is deposited in a variety of modes other than the original single mode of oscillation. In our pendulum example, a rapid change in the length of the string attached to the pendulum's mass will launch transverse waves of wavelength, $\lambda < L$, along the string, putting the energy into modes of vibration other than the pendulum mode, as illustrated in Fig. 3.8. (If the length is changed at twice the natural frequency, then it is possible to excite a parametric resonance.) If the change is slow compared to T, then the pendulum will continue to swing in the same way but at a frequency that is altered by the slow change in length, δL.

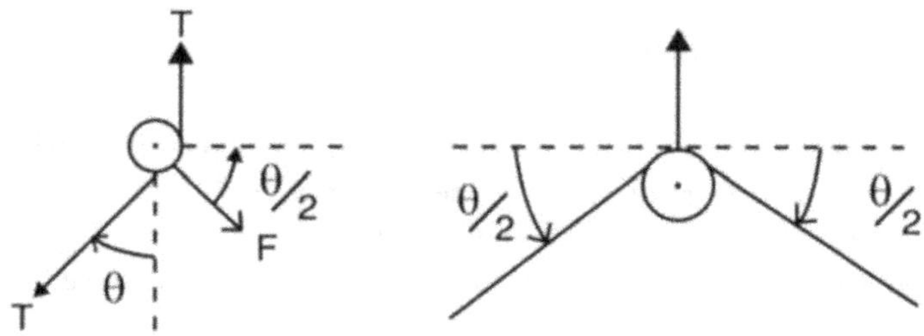

Fig. 2.5 (*Left*) The support point for the pendulum of Fig. 1.5 is replaced by a pair of frictionless pinch-rollers. Only one roller is shown. The string that supports the pendulum's mass has a tension, T. The left pinch-roller must exert a force $F = 2T \sin(\theta/2)$ on the string to keep it from moving the attachment point. The vertical component of that force is $F_z = F \sin(\theta/2) = 2T \sin^2(\theta/2) \cong T\theta^2/2$. (*Right*) This rotated view of the figure on the left simplifies the identification of the angle, $\theta/2$, used in the diagram of the movable pendulum support. The pinch-rollers must do work against this vertical force to move the attachment point lower and shorten the string

The classical relationship can be applied to the simple pendulum. Logarithmic differentiation of Eq. (2.32) relates the relative change in the pendulum's frequency to the relative change in length by assuming that g is constant.

$$\frac{\delta\omega_o}{\omega_o} = -\frac{1}{2}\frac{\delta L}{L} \tag{2.34}$$

To demonstrate that this result is consistent with *adiabatic invariance*, we can imagine that the suspension point for the string that supports the pendulum's mass is squeezed between a pair of frictionless pinch-rollers, as shown in Fig. 2.5. Those rollers can be moved downward to reduce the length of the pendulum by an amount δL. We will calculate the work required to move the pinch-roller along the string. That should be equal to the amount of energy added to the system by the slow variation of this constraint.

Before calculating that force, it is important to recognize that whether the pendulum moves to the right or to the left, the force on the pinch-rollers is always upward. If we let that force move the pinch-rollers, then the pendulum would be doing work on the pinch-rollers, and energy would be removed from the pendulum as the length of the string increases. For the following calculation, we will move the pinch-rollers down, doing work on the pendulum against this force, although either choice would produce a confirmation of adiabatic invariance.

Figure 2.5 shows the force exerted by the pinch-rollers on the string. When the string moves to the left, the pinch-rollers have to exert a force, F, that is downward and to the right: $F = 2T \sin(\theta/2)$. The vertical component of that force in the direction of the motion is $F_z = F \sin(\theta/2) = 2T \sin^2(\theta/2) \cong T\theta^2/2$.

The tension, T, in the string is due to the weight of the mass, mg, and the centripetal acceleration of that mass as it swings to and fro at the natural frequency, ω_o, and with an angular displacement amplitude, θ_1. Selecting the origin of our time coordinate to make $\theta(t) = \theta_1 \sin \omega_o t$ eliminates the need for a phase, ϕ, that appears in Eq. (2.6).

$$T = mg + mL\dot{\theta}^2 = mg + mL\omega_o^2\theta_1^2 \cos^2\omega_o t = mg\left(1 + \theta_1^2 \cos^2\omega_o t\right) \tag{2.35}$$

This expression can be integrated to calculate the work done by slowly sliding the pulley down by a distance, δL.

$$\delta E = W = \int_0^{\delta L} F_z dz = mg\int_0^{\delta L} \left(1+\theta_1^2 \cos^2\omega_o t\right)\frac{\theta_1^2}{2}\sin^2(\omega_o t)dz \tag{2.36}$$

Since we are moving the pulley slowly by a distance, δL, over a time corresponding to many cycles of period, $T = 2\pi/\omega_o$, we can take the time-average of the trigonometric functions: $\langle \sin^2(\omega_o t)\rangle_t = \frac{1}{2}$ and $\langle \cos^2(\omega_o t)\sin^2(\omega_o t)\rangle_t = \frac{1}{8}$.

$$\delta E = \frac{\theta_1^2}{2} mg\left(\frac{1}{2}+\frac{\theta_1^2}{8}\right)\delta L \cong \frac{\theta_1^2}{4} mg(\delta L) \tag{2.37}$$

The energy of the pendulum can be written as the maximum kinetic energy which equals the total energy as the pendulum swings through $\theta = 0$: $E\ (\theta = 0) = ½\ mv^2 = ½\ mL^2\theta_1^{\,2}\omega_o^{\,2}$.

$$\frac{\delta E}{E} = \frac{\frac{\theta_1^2}{4} mg(\delta L)}{\frac{\theta_1^2}{2} mL^2\omega_o^2} = \frac{1}{2}\frac{g(\delta L)}{L^2\omega_o^2} = \frac{1}{2}\frac{\delta L}{L} = -\frac{\delta\omega_o}{\omega_o} \tag{2.38}$$

Since we are shortening the pendulum by doing work on it, $\delta L < 0$, and this result is identical to our expectation based on Eq. (2.34). Although this result only reconfirms our understanding of the relationship between the pendulum's frequency and its length, as we will see in Sect. 6.2.3 and Sect. 6.2.4, and subsequently in Part II as well, adiabatic invariance will allow us to calculate normal mode frequencies of systems that have complicated geometries in Sect. 13.3.5 or will allow us to determine forces when the frequency changes as system constraints are varied in Sect. 15.4.6.

2.4 Damping and Free-Decay

With only a mass and a spring, there is no mechanism by which the vibrational energy can be dissipated. Once set in motion, the mass-spring system has a total energy that is conserved, with the energy being transformed from kinetic to potential forms and vice versa as expressed in Eq. (2.17) throughout each cycle. In principle, the amplitude of oscillation will remain unchanged forever. In this section, we will add a third lumped element that will be able to convert a portion of that energy to heat, thereby reducing the amplitude of the oscillations, until the mass-spring-damper system comes to thermal equilibrium with its environment. As will be demonstrated, all of the energy is initially localized in the kinetic and potential energy of the mass and spring, but the damping element will dissipate this ordered energy, converting it to the disordered form known as heat.

2.4.1 Viscous Damping and Mechanical Resistance

Although damping can be introduced through a variety of physical mechanisms (e.g., friction or sound radiation), the most common and useful dissipative element is the viscous damper, represented in Fig. 2.6 as a *dashpot*. A dashpot is typically envisioned as consisting of a cylinder filled with a viscous fluid in which the motion of a movable vane is resisted by viscous drag, like a spoon being moved slowly through honey. The force of that viscous damping element is proportional to the velocity difference between its ends and is directed opposite to that velocity. Assuming that one end of the dashpot is immobilized, then the force,

$$F_{vis} = -R_m v,$$

where R_m [kg/s] is called the *mechanical resistance*. Adding the viscous force to the previous equation, which included inertia and stiffness, produces the equation of motion for a viscously damped simple harmonic oscillator.

$$m\frac{d^2x}{dt^2} + R_m\frac{dx}{dt} + \mathrm{K}x = 0 \quad \text{or} \quad \ddot{x} + \frac{R_m}{m}\dot{x} + \omega_o^2 x = 0 \tag{2.39}$$

The solution of this second-order differential equation with constant coefficients highlights the extraordinary utility of complex exponentials. We will postulate that the (complex) solution will have the form $\mathbf{x}(t) = \widehat{\mathbf{C}}e^{\eta t}$, substitute into Eq. (2.39), and determine the form of our (as yet) undetermined exponent, η.

$$\left(\eta^2 + \frac{R_m}{m}\eta + \omega_o^2\right)\widehat{\mathbf{C}}e^{\eta t} = 0 \tag{2.40}$$

To satisfy this equation, the term in parentheses must vanish, since the value of $\widehat{\mathbf{C}}e^{\eta t}$ varies with time. This requirement is met by use of the quadratic formula to specify those two roots corresponding to the two values of η.

$$\eta = \frac{-\frac{R}{m} \pm \sqrt{\left(\frac{R_m}{m}\right)^2 - 4\omega_o^2}}{2} = -\frac{1}{\tau} \pm \sqrt{\left(\frac{1}{\tau}\right)^2 - \omega_o^2} \tag{2.41}$$

The choice of $\tau = 2\ m/R_m$ makes sense since τ has the units of time, as required for dimensional homogeneity.

As with any new result, it is always valuable to examine the limiting case that corresponds to a previous solution (i.e., no damping). If we let R_m go to zero, making τ go to infinity, we should recover the earlier result in Eq. (2.7).

$$\lim_{R_m \to 0} \eta = \pm j\omega_o \tag{2.42}$$

After taking the real part of the assumed exponential solution, $\mathbf{x} = \widehat{\mathbf{C}}e^{\eta t}$, once Eq. (2.41) has been substituted for η, we see that τ corresponds to the exponential decay time for the amplitude of the oscillation.

$$x(t) = \Re e\left[\widehat{\mathbf{C}}e^{\eta t}\right] = |\widehat{\mathbf{C}}|\, e^{-t/\tau}\cos(\omega_d t + \phi) \tag{2.43}$$

The damped natural frequency, ω_d, is provided in Eq. (2.45).

Figure 1.18 shows the time evolution of one such damped harmonic oscillator. The motion is no longer strictly periodic. The peak amplitude of each cycle is less than the amplitude of the previous cycle by an amount known as the *logarithmic decrement*. As in the undamped case, the frequency is still amplitude independent. In the damped case, we can define the frequency as the reciprocal of the time between upward or downward zero-crossings, since zero is not amplitude dependent. Equation (2.41) has also led to the introduction of the damped oscillation frequency, ω_d, that will be discussed further in the next section. For the undamped case, where $R_m = 0$, $\tau \to \infty$, $\omega_d = \omega_o$. When $R_m \neq 0$, the period of the oscillatory portion described by Eq. (2.43) becomes longer.

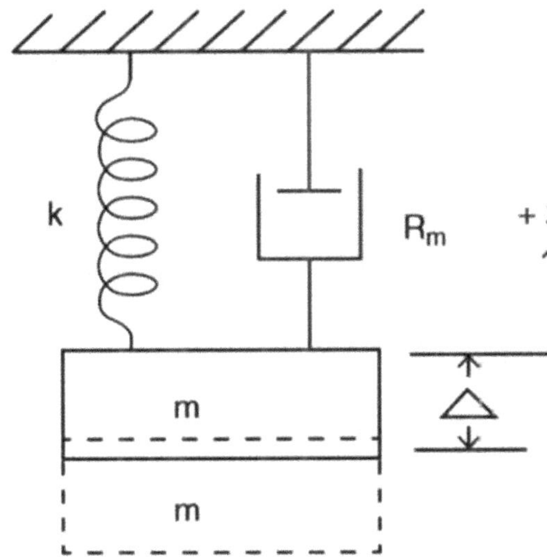

Fig. 2.6 A dashpot representing the viscous resistance, R_m, has been added to the mass and spring system of Fig. 2.4. The dashpot symbol is meant to represent a vane in an oil bath that will provide a viscous force that is proportional and opposite to the velocity of the vane: $F_{vis} = -R_m v$

2.4.2 Free-Decay Frequency and Quality Factor

In the previous chapter's discussion of similitude (see Sect. 1.7.1), the addition of R_m as a third parameter that is required to specify a damped harmonic oscillator introduced a dimensionless group which was labeled Q^2 at that time in Eq. (1.81), without justification.

$$Q^2 = \frac{Km}{R_m^2} = \frac{\omega_o^2 m^2}{R_m^2} = \frac{\omega_o^2 \tau^2}{4} \tag{2.44}$$

Similitude analysis suggests that the square root of this dimensionless group, $Q = (\omega_o \tau)/2$, known as the *quality factor*, must provide the only combination of parameters that can control the system's response, even if it is impossible to specify the exact functional form of that control from only dimensional arguments. To see this, we write $\tau = 2Q/\omega_o = Q/(\pi f_o) = QT/\pi$. This shows that Q corresponds to π times the number of cycles it takes for the amplitude of a damped harmonic oscillator to decay to $e^{-1} \cong 36.8\%$ of its initial amplitude. Similarly, substitution into Eq. (2.41) shows that the damped oscillation frequency, ω_d, is related to the undamped *natural frequency*, ω_o.

$$\omega_d = 2\pi f_d = \sqrt{\omega_o^2 - \left(\frac{1}{\tau}\right)^2} = \omega_o \sqrt{1 - \frac{1}{4Q^2}} \cong \omega_o \left(1 - \frac{1}{8Q^2}\right) \tag{2.45}$$

The expression on the far right is an approximation (using the binomial expansion of Eq. 1.9), which is valid for lightly damped oscillators if $Q^2 \gg 1$. A related expression can be written for the *logarithmic decrement*, δ, where $T_d = 2\pi/\omega_d$ is the period of the damped sinusoid.

$$\begin{aligned} \delta &= \ln\left[\frac{e^{-t/\tau}}{e^{-(t+T_d)/\tau}}\right] = \ln\left[e^{-T_d/\tau}\right] = -\frac{T_d}{\tau} = \frac{2\pi}{\omega_d \tau} \\ \delta &= \frac{\pi}{Q}\left(1 - \frac{1}{4Q^2}\right)^{-1/2} \cong \frac{\pi}{Q}\left(1 + \frac{1}{8Q^2}\right) \end{aligned} \tag{2.46}$$

We can use the parameter values determined by the fit of Eq. (1.116), which is identical in form to Eq. (2.43), to develop some intuition about the results just obtained for a damped harmonic oscillator.

$$V(t_i) = V_o e^{-t_i/\tau} \sin(2\pi f_d t_i + \phi) \tag{1.119}$$

The fit values provided in the caption of Fig. 1.18 are $f_d = 46.141$ Hz and $\tau = 0.0445$ s. Based on Eq. (2.45), let's assume that $\omega_d \cong \omega_o$ and use the measured τ to calculate $Q = (\omega_o \tau)/2 = \pi f_d \tau = 6.45$. Substitution of that result back into Eq. (2.45) suggests that $\omega_d/\omega_o = 0.997$, so treating the natural and damped frequencies as being equal for the initial calculation of the quality factor, Q, was a most reasonable choice.

2.4.3 Critical Damping

The solution to the quadratic formula for η in Eq. (2.41) has three regimes. If $\omega_o \tau = 2Q \gg 1$, then the solutions are oscillatory and exhibit an exponentially decaying amplitude with time, as described in Eq. (1.119) and illustrated in Fig. 1.18. Such solutions are called *underdamped.*

In the opposite limit, the system is *overdamped.* That means that $R_m/\omega_o \gg m$ or equivalently $R_m/\omega_o m = Q^{-1} \gg 1$. In that limit, the behavior of the system is governed by a first-order differential equation, where the inertial term, $m(d^2x/dt^2)$, in Eq. (2.39), can be ignored.

$$R_m \frac{dx}{dt} + \mathrm{K}x = 0 \tag{2.47}$$

Being a first-order differential equation, it possesses only one solution corresponding to exponential relaxation.

$$\int \frac{dx}{x} = -\frac{\mathrm{K}}{R_m} \int dt \quad \Rightarrow \quad x(t) = Ce^{-(\mathrm{K}/R_m)t} \tag{2.48}$$

If the mass is displaced from equilibrium by an amount x_1, it is dragged back to equilibrium by the spring and $x(t) = x_1 e^{-t/\tau}$, where $\tau = R_m/\mathrm{K}$.

An intermediate case can produce *critically damping*. If $\omega_o \tau_{crit} = 1$ (i.e., $Q = ½$), the term under the radical in Eq. (2.41) is zero: $\tau_{crit} = \omega_o^{-1} = (m/\mathrm{K})^{1/2}$. If the system is initially displaced from equilibrium by an amount x_1, then the position decays exponentially to zero, like the solution in Eq. (2.48). Critical damping brings the mass back to its equilibrium position in the least amount of time without crossing zero. For the mass to approach the equilibrium position in the minimum time, the damping should be about 60% of the critical value, although when "critically underdamped," the mass will cross zero before coming to rest.

2.4.4 Thermal Equilibrium and Fluctuations

"The phase (or interferometric) sensor,[4] *whether for magnetic, acoustic, rotation, etc., sensing, theoretically offers orders of magnitude increased sensitivity over existing technologies. In the case of the acoustic sensor*

[4] The fact that these authors are discussing a fiber-optic interferometric sensor is irrelevant, in this context. Such a sensor uses optical interference to detect changes in the difference of the length of two optical fibers. For our purposes, the interferometric sensor only provides a long "optical lever" like the mirrored galvanometer, shown in Fig. 2.6, that will be analyzed later in this section.

> *constructed utilizing optical fiber interferometers, these theoretical predictions have been verified to the limit of state of the art in acoustic measurements… In the case of the magnetic sensor, it appears that fiber sensors operating at room temperature offer detection sensitivities comparable to or exceeding cryogenic SQUID technology, which normally operate between 4 and 10 K."* [7]

The claims in the above quotation are completely wrong (and cost the US taxpayer millions of dollars as those authors chased this "impossible dream"). That quote epitomizes the problem that is encountered if we neglect the fact that the addition of our resistive element to the mass-spring harmonic oscillator opens a two-way street for the exchange of energy with the environment. Before the inclusion of R_m, the total energy of the system was determined by the initial conditions and that total energy remained constant, although the energy shuttled back and forth between its kinetic and potential forms. With a dissipative element, there was a path for energy to leave the oscillator, in this case as heat.[5] That path also connects the oscillator to "the environment" that must share energy with the oscillator by virtue of the fact that the absolute (kelvin) temperature of the environment and R_m is both non-zero. Temperature is a measure of the average kinetic energy of the particles that make up matter. Given sufficient time, every degree of freedom will get its "fair share." This result is known as the *Equipartition Theorem* and can be thought of as a definition of thermal equilibrium, being the condition where, on average, each degree of freedom has the same average energy.

For degrees of freedom that have energies that are quadratic in their position or momentum coordinates,[6] that "fair share" is (½) k_BT, where the Boltzmann's constant[7] is $k_B \equiv 1.380649 \times 10^{-23}$ J/K and T is the absolute (kelvin) temperature.

Before examining the consequences of thermal equilibrium on the limiting behavior of a damped harmonic oscillator due to thermal fluctuations, it may provide some comfort to reflect on the history of science, since it demonstrates that our understanding of heat and temperature took a relatively long time to develop. By comparison, our classical understanding of gravity and dynamics of particles from Newton's falling apple to the orbits of the planets in the solar system, seemingly a far more arcane problem, was completely understood two centuries before we had a fundamental understanding of temperature and the mechanical equivalence of heat.

The year 1642 was an important milestone on the path to our current scientific understanding of the physical world. Galileo died in January of that year and Newton was born on December 25 of the same year, which turned out to be our guarantee that the scientific method would survive the death of Galileo.[8] Kepler (1571–1630) published his first two laws of planetary motion in 1609 and his third (see Sect. 2.3.1) in 1619, based on the observations of Tycho Brahe (1546–1601). Newton's *Philosophiæ Naturalis Principia Mathematica* was first published in 1687. It unified the dynamics of falling bodies on earth with the orbital motion of the planets in our solar system through Newton's laws of motion and his law of universal gravitation.

[5] As will be shown in Chap. 12, energy can also leave by radiation of sound, but the thermodynamic consequences of energy being shed to the environment by any means are identical.

[6] The potential energy of a spring, (½) Kx^2, is quadratic in displacement. Kinetic energy, (½)mv^2 = (½)p^2/m, is quadratic in the momentum, p. For degrees of freedom that have energies that are quartic in the generalized coordinate, each degree of freedom gets (¼) k_BT in thermal equilibrium.

[7] The most accurate value of Boltzmann's constant is determined by a sound speed measurement, e.g., M. de Podesta et al., "A low-uncertainty measurement of the Boltzmann constant," Metrologia **50**, 354–376 (2013). As of 20 May 2019, the value of k_B that is used here is taken to be exact (see Appendix A), and other practical units, such as the kilogram, the second, the meter, and the kelvin, are defined in terms of those "exact" fundamental physical constants.

[8] During the winter holiday season, some godless Humanists (commonly called atheists) celebrate "Continuity Day" to commemorate the continuity of the scientific method.

One would think that the concepts of heat and temperature, which are familiar to us through our own senses, would have been understood much earlier, but exactly the opposite is true. James Prescott Joule (1818–1889) measured the mechanical equivalence of heat (now incorporated into the First Law of Thermodynamics) in 1845,[9] nearly 160 years later than Newton's *Principia.* In fact, the connection between molecular motion and temperature was not established until the work of James Clerk Maxwell (1831–1879) and Ludwig Boltzmann (1844–1906) introduced the kinetic theory of gases and statistical mechanics (see Chaps. 7 and 9). Boltzmann went to his grave defending his theory against opposition within the scientific community at that time, and his gravestone has $S = k \ln W$ at its top, where S is entropy, $k = k_B$ is Boltzmann's constant, and W (for *Wahrscheinlichkeit*) is the number of microscopic states (i.e., combinations of positions and velocities) that correspond to the macroscopic equilibrium state of the system.

The first recorded observation of the mechanical consequences of thermal fluctuations was the microscopic observation in 1827 by botanist Robert Brown (1773–1858) of the motion of pollen particles suspended in water.[10] The theoretical explanation for this random motion was published by Albert Einstein (1879–1955) in 1905 [8] and was the reason, along with the photoelectric effect, that he was awarded the Nobel Prize in 1921.[11] The generalization of the role of thermal fluctuation to mechanical and electrical dissipation did not really start to develop until the 1920s. Most relevant to our immediate interests are thermal fluctuations in simple harmonic oscillators, which started with the observation of random motion in the mirrored galvanometer.

The mirrored galvanometer was patented by William Thompson (Lord Kelvin) in 1858.[12] It consists of a coil of wire, placed in a magnetic field, acting as an electrical meter that uses a mirror to reflect a beam of light to a scale, usually several meters away on a wall. That long "optical lever" provided a sensitive indication of the current passing through the coil. A torsion fiber provided the restoring force for the mirror-coil combination. (A similar arrangement is shown in Fig. 15.33.)

It was noticed that the spot of light reflected from the mirror underwent random motion. At first, it was assumed that the random motions were due to the collision of air molecules with the mirror, and a study was undertaken to measure the fluctuating position of the light spot as air was pumped out of the galvanometer [9]. Even without air, the random motions were still observed. In 1929, Uhlenbeck and Goudsmit were able to show that those fluctuations "depend on the properties of the observed system and on the temperature only" [13] [10]. At the same time, Nyquist was able to demonstrate that electrical

[9] That was the date when he presented his results at a scientific meeting. Those results were not published until 1850: J. P. Joule, "On the mechanical equivalent of heat," Phil. Trans. Roy. Soc. **140**(Part I), 61–82 (1850).

[10] The fact that such motions were not biological in origin was confirmed when similar motion was observed in a drop of water trapped in amber (i.e., fossilized tree sap). No living creature could have survived that long in a water droplet isolated since antiquity.

[11] Einstein's Nobel Prize in Physics did not mention his theory of relativity since it was still too controversial in 1921.

[12] Thompson played a significant role in the laying of the first transatlantic telegraph cable between Ireland and Newfoundland that was successfully completed in 1865. He traveled on the cable-laying ship, the *Great Eastern*, as scientific advisor for all electrical matters. The mirrored galvanometer allowed detection of the weak electrical signals that made it through the cable. He set up two companies to sell these galvanometers and provide engineering consulting services to submarine cable companies. Those endeavors made him a wealthy man. The fascinating story of this first transatlantic cable is told by J. S. Gordon in *A Thread Across the Ocean* (Perennial, 2003): ISBN 978-0060524463.

[13] It is worth mentioning that these two Dutch physicists were also responsible for postulating that the electron had an internal degree of freedom that became known as the "electron spin."

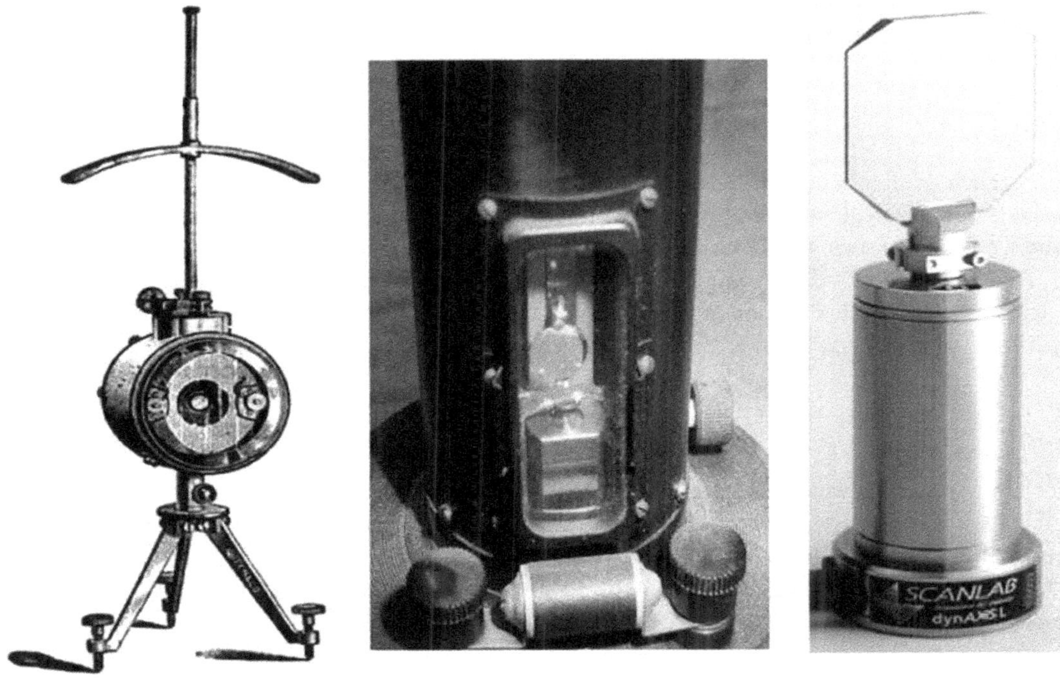

Fig. 2.7 (*Left*) Drawing of Lord Kelvin's patented mirrored galvanometer. (*Center*) Photograph of a mirrored galvanometer from the author's collection. (*Right*) Although such instruments are no longer used to measure electrical currents, the same type of device is used to deflect laser beams for optical scanning purposes.

noise was generated by electrical resistors (Johnson noise [11]) based on their temperature [12] (Fig. 2.7).

Again, this very long historical digression is intended to point out that the connection between temperature and thermal fluctuations took a very long time before it was understood, along with its consequences for setting a minimum detectable signal for sensors [13]. The absurdity of the quote that started this section by claiming that "fiber sensors operating at room temperature offer detection sensitivities comparable to or exceeding cryogenic SQUID technology, which normally operate between 4 and 10 K" [7] is simply a reflection (no pun intended) of those authors' ignorance of the fluctuations that had been observed in the 1920s with mirrored galvanometers [14]. In effect, their "optical lever" was 1–1000 m of optical fiber. The first fiber-optic interferometric sensor that was limited by thermal fluctuations was demonstrated experimentally in 1987 [15].

The Equipartition Theorem contradicts the result in Eq. (2.43), which claims that the amplitude of the motion of the mass connected to the spring and damper will decay to as small a value as you like, if you are willing to wait long enough for the exponential factor, $|\widehat{\mathbf{C}}|\, e^{-t/\tau}$, to reduce the oscillatory amplitude. In thermal equilibrium, the mean value of each independent quadratic term in the energy will contain an average amount of energy that is (½) $k_B T$ [16].

This result can be applied to the kinetic energy of the pollen particle that Brown observed jittering under his microscope in 1827. Today, it is easier to observe the motion of colloidal gold suspensions that are made with particle diameters of 5×10^{-6} cm [17]. The mass of each gold particle is about 10^{-15} g. If we equate their kinetic energy per degree of freedom to (½) $k_B T$, and account for the fact that the particles move in three dimensions (hence, there are three quadratic degrees of freedom corresponding to the velocities v_x, v_y, and v_z), we can solve for the gold particles' root-mean-squared velocity, v_{rms}.

$$v_{rms} = \langle v^2 \rangle^{1/2} = \left\langle v_x^2 + v_y^2 + v_z^2 \right\rangle^{1/2} = \sqrt{\frac{3k_BT}{m}} \tag{2.49}$$

For such a colloidal suspension at $T = 300$ K, $v_{rms} \cong 10$ cm/s, which is what is observed [17]. If we make the same calculation for a 1 g sphere at 300 K, $v_{rms} \cong 3.5 \times 10^{-9}$ m/s $= 11$ cm/year, which is why we don't see such small (but macroscopic) particles moving about randomly.[14]

The same approach can be applied to our single degree-of-freedom mass-spring-damper system to calculate the steady-state motion of the mass by equating the thermal energy per degree of freedom (½) k_BT to the potential energy, (½) Kx^2.

$$x_{rms} = \langle x^2 \rangle^{1/2} = \sqrt{\frac{k_BT}{K}} \tag{2.50}$$

For the mirrored galvanometer, we would get the same result for θ_{rms} except that K would be the torsional stiffness instead of the linear stiffness. Again, for macroscopic masses and ordinary temperatures, the fluctuations in the position of the mass are insignificant, but with the recent ubiquity of micro-electro-mechanical sensors (MEMS) (e.g., accelerometers and microphones), thermal fluctuations can be the dominant noise source [18] (Fig. 2.8).

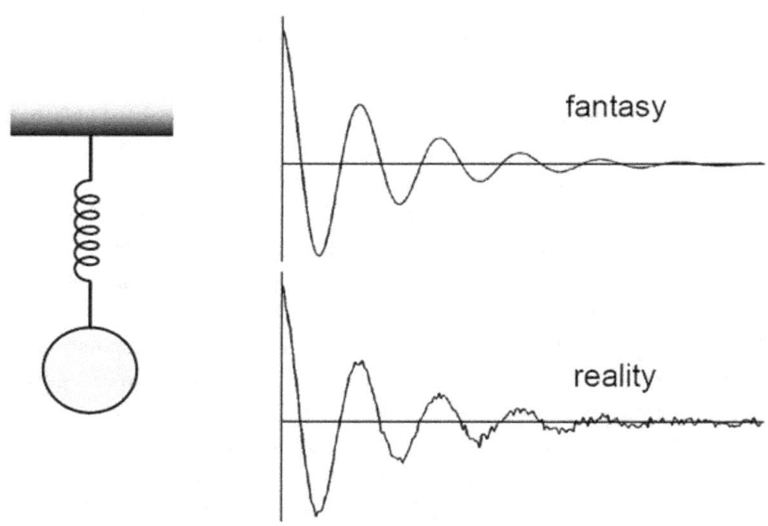

Fig. 2.8 The decay of a mass-spring system due to damping given by the time history in Eq. (2.43) is only true for displacements that are substantially larger than the thermal fluctuation amplitude, x_{rms}, given in Eq. (2.50). In reality, the motion of the mass does not decay to zero but will fluctuate randomly about its equilibrium position indefinitely due to its thermal energy. (Diagram courtesy of T. B. Gabrielson)

[14] The precise measurements of Brownian motion that were used to test Einstein's theory were made by Harvey Fletcher (1884–1981) with oil droplets as part of his Ph.D. thesis under the supervision of R. A. Millikan. His advisor chose to publish the charge quantization demonstrated by that oil drop experiment under his own name and received the Nobel Prize in Physics in 1923. Fletcher published the Brownian motion measurements under his name and went on to a very successful career in acoustics at Bell Labs: M. F. Perry, "Remembering the oil-drop experiment," Phys. Today **60**(5), 56–60 (2007).

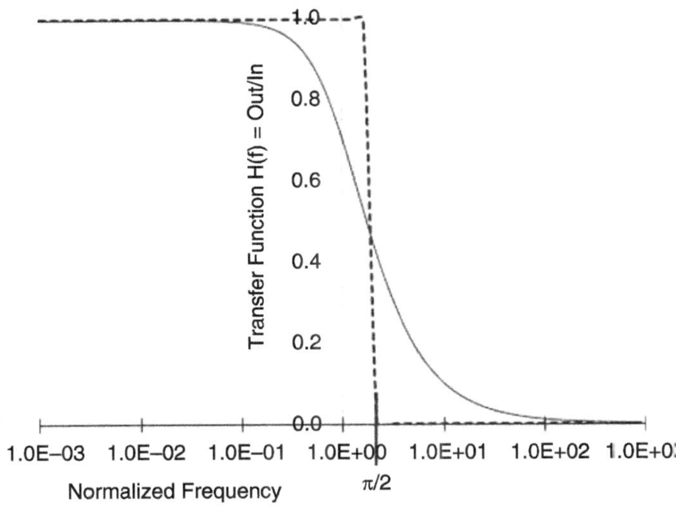

Fig. 2.9 The transfer function, $H(\Omega)$, of a single-stage low-pass filter (*solid line*) is plotted against the normalized frequency, $\Omega = f/f_{-3\ \mathrm{dB}}$, where $f_{-3\ \mathrm{dB}}$ is the frequency at which the amplitude of the transfer function has decreased to $\sqrt{2}$ of its low-frequency limit. The *dashed rectangle* has the value one up to $\Omega = \pi/2$ and zero for $\Omega > \pi/2$, giving it the same area as that under the solid line. For that reason, $(\Delta f)_{EQNB} = \pi f_{-3\mathrm{dB}}/2$

This random motion can be interpreted as the response to a fluctuating force, F_{noise}. Its root-mean-squared value is related to the square root of the product of the absolute (kelvin) temperature, T; the mechanical resistance, R_m; and the *equivalent noise bandwidth*, $(\Delta f)_{EQNB}$, of the system.

$$F_{rms} = \langle F_{noise}^2 \rangle^{1/2} = \sqrt{4k_B T R_m (\Delta f)_{EQNB}} \tag{2.51}$$

The part of this expression which may be unfamiliar at this point is $(\Delta f)_{EQNB}$. At the simplest level, that bandwidth specifies the range of frequencies over which the fluctuating force will preferentially excite the system's response. The "equivalence" is a way to hide the fact that one needs to integrate over the entire system's frequency response function then choose an equivalent "rectangular" filter response function that has a frequency width of $(\Delta f)_{EQNB}$ and unity gain, as shown in Fig. 2.9. We will calculate this *effective bandwidth* for a damped driven harmonic oscillator in Sect. 2.5.

At this point, the attentive reader (hopefully, you!) might become a little anxious. (S)he might reflect thusly: "This section started off claiming that the addition of a mechanical resistance opened a 'two-way street' that let ordered vibrational energy leave the system but also created a path for the random thermal motion to enter the system. I can see in Eq. (2.51) that the fluctuating force is related to that mechanical resistance, but the results for $x_{rms} = \langle x^2 \rangle_t^{\frac{1}{2}} = \sqrt{k_B T/2K}$ and $v_{rms} = \langle v^2 \rangle_t^{\frac{1}{2}} = \sqrt{k_B T/2m}$ do not involve the mechanical resistance at all. What gives?"

The lazy textbook author would provide the following response: "Do you remember our calculation of the frequency of the pendulum? The result, $\omega_o^2 = g/L$, does not involve the mass, yet it is the mass of the pendulum that provided the inertia and the mass of the pendulum that was proportional to the gravitational restoring force. In the final result for frequency, which depends upon the ratio of the restoring force to the inertia, the mass cancelled out. The same is true here, though at this point it is not at all obvious. For a simple harmonic oscillator, $(\Delta f)_{EQNB} \propto R_m/m$, so R_m comes out of the square-root in Eq. (2.51). At the frequency of peak response, ω_o, the velocity of the mass is proportional to F/R_m, thus cancelling R_m out of the final result for v_{rms}. A similar argument holds for x_{rms}."

The thermal noise of any mechanical system that is in thermal equilibrium with its surroundings, no matter how complex, can be analyzed by including this fluctuating force combined with the resistance;

just as in electrical circuits, a fluctuating voltage source, V_{noise}, can be included with the electrical resistance, R_{el} [11].

$$V_{rms} = \langle V_{noise}^2 \rangle^{1/2} = \sqrt{4k_B T R_{el} (\Delta f)_{EQNB}} \tag{2.52}$$

That expression is the "Johnson noise" [11] that is familiar to electrical engineers.

2.4.5 Frictional (Coulomb) Damping*

Most of us are first exposed to friction in high school physics classes where the frictional force, F_d, is the sliding friction between two dry surfaces. That frictional force is characterized by a coefficient of sliding friction, μ, assumed to be velocity independent and proportional to the force (usually gravitational) that is squeezing the two surfaces together.

$$F_d = -\mu m g \tag{2.53}$$

To determine the amplitude decay for frictional damping, we can calculate the work done by the frictional force during one half-cycle and equate that to the change in the potential energy. Figure 2.10 shows the first few cycles that will be used to calculate the amplitude decrease for successive oscillations.

Starting with an initial displacement of $x_1 = x(t = 0)$, the mass will come to rest instantaneously at the other extreme of the motion, $x_{-1} = x(T/2)$, where it changes direction. Since there will be frictional losses, $|x_1| > |x_{-1}|$.

$$\frac{1}{2}K(x_1^2 - x_{-1}^2) = \frac{1}{2}K(x_1 - x_{-1})(x_1 + x_{-1}) = W_d = F_d(x_1 + x_{-1}) \tag{2.54}$$

This results in an expression for the amplitude decrease for the first half-cycle: $(x_1 - x_{-1}) = 2F_d/K$. Repeating the procedure for the next half-cycle, we again see that the amplitude is decreased by the same amount resulting in a loss of $\Delta x(T) = x_1 - x_2 = 4F_d/K$. This will continue, as illustrated in

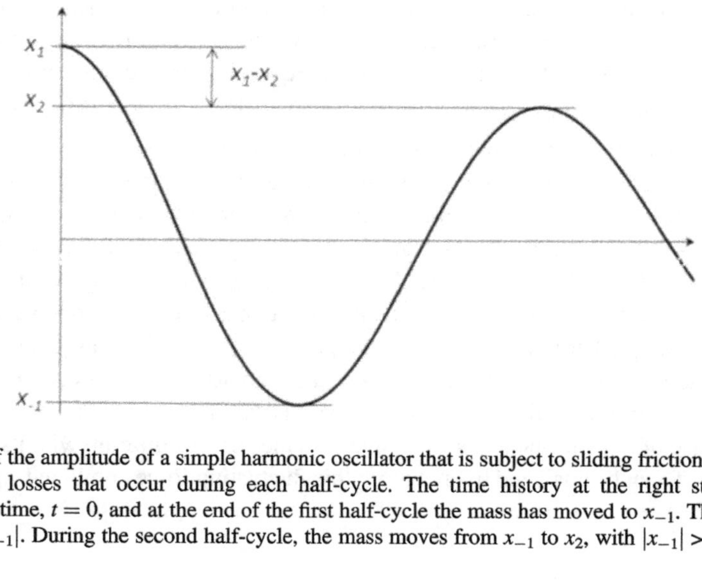

Fig. 2.10 Decay of the amplitude of a simple harmonic oscillator that is subject to sliding friction can be calculated by examination of the losses that occur during each half-cycle. The time history at the right starts with a positive displacement, x_1, at time, $t = 0$, and at the end of the first half-cycle the mass has moved to x_{-1}. The energy dissipation guarantees $|x_1| > |x_{-1}|$. During the second half-cycle, the mass moves from x_{-1} to x_2, with $|x_{-1}| > |x_2|$, etc.

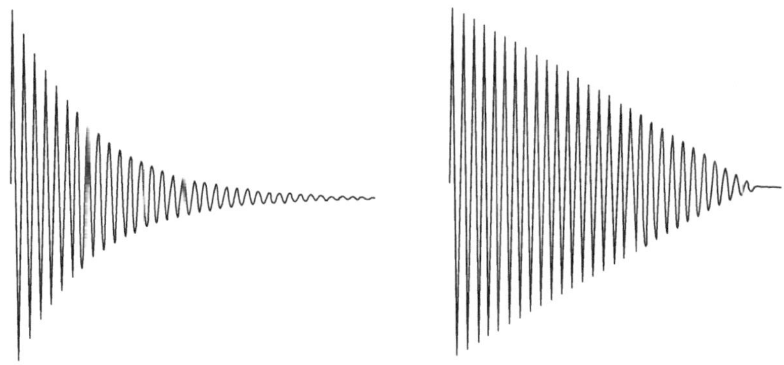

Fig. 2.11 (*Left*) A viscous damping force, $F_{vis} = -R_m v$, produces an exponentially decaying amplitude. (*Right*) A frictional (Coulomb) damping force, $F_d = -\mu\ (mg)$, produces an amplitude that decays linearly with time. (Figures from Pippard [19])

Fig. 2.11, until the mass comes to rest. The decay per cycle is a constant, so the amplitude decreases linearly with time, not exponentially with time [19].

2.5 Driven Systems

Thus far, we have only analyzed the response of a harmonic oscillator that is provided with an initial displacement, an initial velocity, or both. Those solutions, like that shown in Eq. (2.43), are allowed to evolve in time without any further energy input. We have also demonstrated that the imposition of a steady gravitational force will displace the equilibrium position but not influence the oscillator's dynamical response. In this section, we will examine the response of a viscously damped simple harmonic oscillator to a periodic force applied to the mass or to a periodic displacement applied to the end of the spring not attached to the mass. The steady-state solution that supplies the response to both of those driven systems will be provided through the calculation of a *mechanical impedance*. Also, important will be the concept of *resonance*. When the drive occurs at the natural frequency, ω_o, of the undriven system, the steady-state (velocity) response at resonance is only limited by the damping. Resonance is one of the most valuable tools available to the acoustician, and this section will provide our first exposure to its wonders.

2.5.1 Force-Driven SHO

We start by assuming that a sinusoidally varying force of amplitude, F_1, is applied to the mass in a mass-spring-damper system. The force amplitude, F_1, can be chosen to be a real number without any loss of generality since that choice indicates that any phase shifts will be referenced to the phase of the force, which has now been chosen to define the origin of our phase reference (i.e., $\phi = 0$), so $F(t) = F_1 \cos(\omega t)$. When that sinusoidally time-varying force is applied to the mass, then Eq. (2.39) is no longer homogeneous.

$$m\frac{d^2x}{dt^2} + R_m\frac{dx}{dt} + \mathrm{K}x = m\frac{dv}{dt} + R_m v + \mathrm{K}\int v\,dt = F(t) \equiv F_1 e^{j\omega t} \tag{2.55}$$

Instead of writing our differential equation in terms of the displacement of the mass, it is entirely equivalent to write it in terms of the velocity of the mass.

Although the above equation assumes a force is applied to the mass-spring-damper system at a single driving frequency, ω, our discussion of Fourier synthesis (see Sect. 1.4) indicated that, for such a linear system (see Sect. 1.3), we could produce the response to a more complicated periodic driving force by superposition of forces that are applied with other frequencies, amplitudes, and phases.

It is also important to remember that a driven linear system can only respond at the driving frequency, once any transient behavior dies out (see Sect. 2.5.4). Since Eq. (2.39) is homogeneous, solutions to the homogeneous equation, like that in Eq. (2.43), can be added to the solution of Eq. (2.55) in amounts determined by the initial conditions. Because $\tau < \infty$, those transient solutions will eventually die out leaving only the steady-state response.

Again, convenience (complex) numbers (see Sect. 1.5) make the solution to Eq. (2.55) straightforward. Since we know that the linear system's steady-state response occurs only at the driving frequency, ω, we will postulate a solution for the velocity of the mass's response of the form, $v(t) = \Re e\left[\widehat{\mathbf{C}} e^{j\omega t}\right]$, with the understanding that **C** is a phasor (complex number) that can introduce a phase shift between the applied force and the velocity, as well as determine the amplitude of the response. We again revert to phasor notation that allows us to combine both the amplitude and phase into a single complex symbol, $\widehat{\mathbf{v}} e^{j\omega t} = \widehat{\mathbf{C}} e^{j\omega t}$, and substitute that solution into Eq. (2.55) and factor out the common term.

$$\left(j\omega m + R_m + \frac{\mathrm{K}}{j\omega}\right)\widehat{\mathbf{C}} e^{j\omega t} = F_1 e^{j\omega t} \tag{2.56}$$

The amplitude and the phase of the velocity response, as represented by the velocity phasor, **v**, to the applied force of amplitude, F_1, at frequency, ω, can be written directly.

$$\widehat{\mathbf{v}} e^{j\omega t} = \frac{F_1 e^{j\omega t}}{j\left(\omega m - \frac{\mathrm{K}}{\omega}\right) + R_m} \tag{2.57}$$

The entire response of the driven system is characterized by the complex denominator. By analogy with electrical circuit theory, that denominator is identified as the complex mechanical impedance, $\mathbf{Z_m}$.

$$\mathbf{Z_m} \equiv \frac{F}{\widehat{\mathbf{v}}} = j\left(\omega m - \frac{\mathrm{K}}{\omega}\right) + R_m \tag{2.58}$$

The imaginary component of the mechanical impedance is $\Im m[\mathbf{Z_m}] \equiv X_m = \omega m - (\mathrm{K}/\omega)$. It is called the *mechanical reactance* and represents the storage of energy as either elastic stiffness or inertia. It is important to recognize that the reactance vanishes, $X_m(\omega_o) = 0$, when the forcing frequency equals the natural frequency, $\omega = \omega_o = \sqrt{\mathrm{K}/m}$.

We recognized $\Re e\,[\mathbf{Z_m}] = R_m$ as the mechanical resistance that represents the dissipation of energy. At resonance, when $X_m(\omega_o) = 0$, the response is controlled entirely by the resistance; hence it is referred to as the *resistance-controlled* response regime. The steady-state solution for the displacement, $x(t)$; the velocity, $v(t)$; or the acceleration, $a(t)$, in response to the driving force, $F(t)$, can be written compactly in terms of $\mathbf{Z_m}$. The transient response is included in Eq. (2.80).

$$x(t) = \Re e\left[\frac{1}{j\omega}\frac{F_1 e^{j\omega t}}{\mathbf{Z_m}}\right], \quad v(t) = \Re e\left[\frac{F_1 e^{j\omega t}}{\mathbf{Z_m}}\right], \quad \text{and } a(t) = \Re e\left[j\omega\frac{F_1 e^{j\omega t}}{\mathbf{Z_m}}\right] \tag{2.59}$$

As with any new result, it is prudent to interpret the consequences and compare them to our assumptions. The fact that the mechanical impedance has singled out a specific frequency, $\omega_o = \sqrt{\mathrm{K}/m}$, makes it easy to define what is meant by the low- and high-frequency limits. In the low-frequency limit, $\omega \ll \omega_o$, the stiffness is dominant, and substitution of Eq. (2.58) into the above

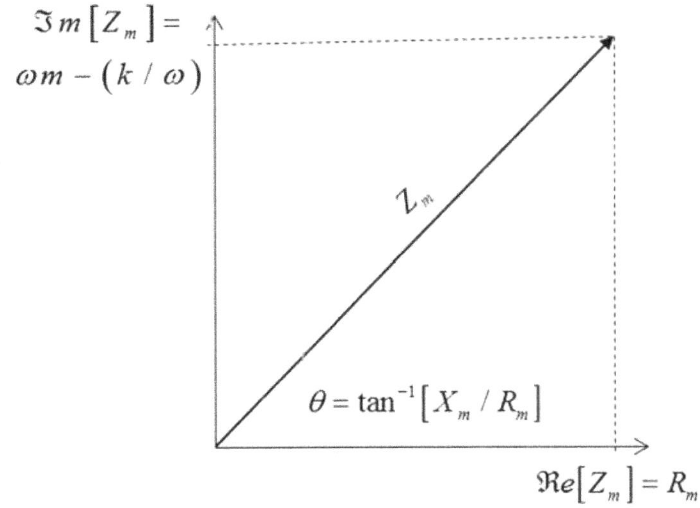

Fig. 2.12 The complex mechanical impedance can be represented in polar form with the frequency-independent real component, $\Re e\ [\mathbf{Z}_m] = R_m$, and the imaginary component, $\Im m\ [\mathbf{Z}_m] = X_m = \omega m - (K/\omega)$. The magnitude of the mechanical impedance, $|\mathbf{Z}_m|$, is given by the length of the hypotenuse, and the phase angle between the force and the velocity is $\Theta = \tan^{-1}[X_m/R_m]$. That phase angle can vary between the "stiffness-controlled limit," $-90°$, well below the resonance frequency (i.e., displacement $x(t) = v(t)/(j\omega)$ out-of-phase with force), and $+90°$, the "mass-controlled limit," well above the resonance frequency (i.e., acceleration, $a(t) = (j\omega)\, v(t)$, is in-phase with the force)

expression, $x(t) \cong -F_1/K$, results in the cancellation of the frequency and the recovery of Hooke's law. For that reason, the low-frequency limit is referred to as the *stiffness-controlled* response regime.

In this high-frequency limit, $\omega \gg \omega_o$, the $j\omega\, m$ term in the mechanical impedance is dominant, so $a(t) \cong F_1/m$ is frequency-independent, and Newton's Second Law is recovered. For that reason, the high-frequency limit is referred to as the *mass-controlled* response regime.

It is useful to have a polar representation of the mechanical impedance magnitude and phase: $|\mathbf{Z_m}| = (\mathbf{Z_m Z_m}*)^{1/2}$ and $\Theta = \arctan(X_m/R_m)$. The geometric interpretation is provided in Fig. 2.12.

$$|\mathbf{Z_m}| = \sqrt{\mathbf{Z_m Z_m^*}} = \left[R_m^2 + \left(\omega m - \frac{K}{\omega}\right)^2\right]^{1/2} = \omega_o m\left[\left(\frac{\omega}{\omega_o} - \frac{\omega_o}{\omega}\right)^2 + \frac{1}{Q^2}\right]^{1/2} \tag{2.60}$$

$$\Theta = \tan^{-1}\left(\frac{X_m}{R_m}\right) = \tan^{-1}\left[\frac{\omega m - \frac{K}{\omega}}{R_m}\right] = \tan^{-1}\left[Q\left(\frac{\omega}{\omega_o} - \frac{\omega_o}{\omega}\right)\right] \tag{2.61}$$

Figure 2.13 shows the magnitude and phase of the velocity response to the sinusoidal force, $F(t) = F_1 e^{j\omega t}$. The complex velocity response can be written in a more symmetric form by using $Q = \omega_o m/R_m$ to substitute for R_m in Eq. (2.57).

$$\mathbf{v}(t) = \frac{F_1 e^{j(\omega t - \Theta)}}{|\mathbf{Z_m}|} = \frac{\frac{F}{R_m Q} e^{j(\omega t - \Theta)}}{\left[\left(\frac{\omega}{\omega_o} - \frac{\omega_o}{\omega}\right)^2 + \frac{1}{Q^2}\right]^{1/2}} = v_1(\omega) e^{j(\omega t - \Theta)} \tag{2.62}$$

Expressions with the form of Eq. (2.62) show up for all driven resonance behavior, not just the driven simple harmonic oscillator. As will become apparent throughout this book, this *Rayleigh line*

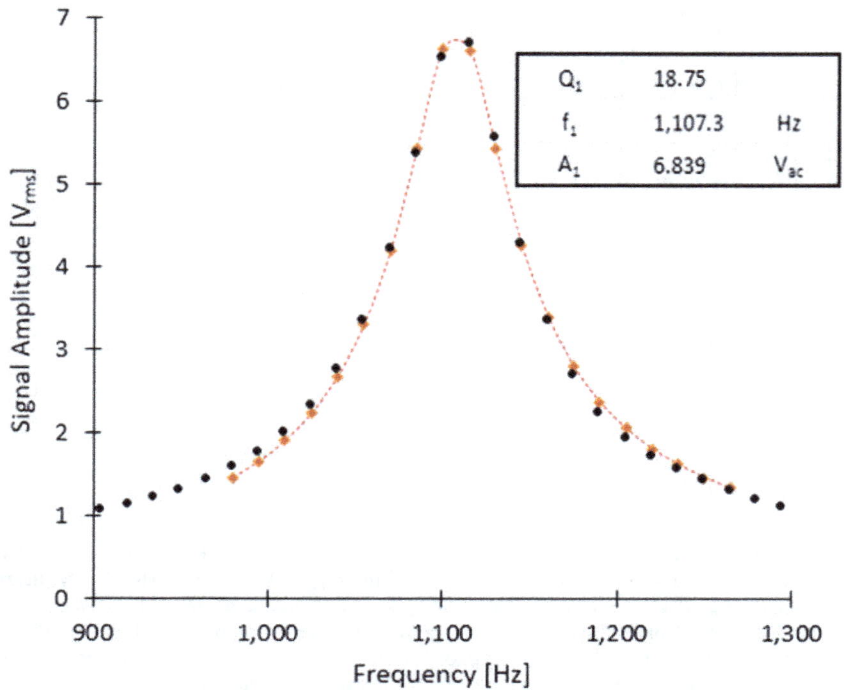

Fig. 2.13 Response of the $n = 1$ torsional resonance of a bar similar to the fundamental mode shown in Fig. 5.16. The measured amplitudes, $A_{exp}(f)$, are shown as the 27 *black dots* plotted for $900 \leq f \leq 1300$ Hz. The *red diamonds* are values of $A_1(f/f_1)$ from Eq. (2.63) that have been joined by a smooth dotted line to illustrate the Rayleigh line shape. The parameter values in the box are those adjusted by a nonlinear curve-fitting algorithm (i.e., "solver") that minimized sum of the squared difference between $A_{exp}\ (f)$ and $A_1(f/f_1)$

shape (aka Lorentz line shape) will describe systems with multiple resonances, like the modes of a long thin cylindrical bar in Fig. 5.16 or the modes of a gas-filled plane wave resonator driven by a loudspeaker in Figs. 10.12 and 10.14, as long as the individual modal frequencies are well-separated.

We can rewrite Eq. (2.62) in a more general form where the subscript, n, refers to the mode number. For the single-degree-of-freedom simple harmonic oscillator that has just been analyzed (or for the Helmholtz resonator discussed in Sect. 8.5), $n = 0$. For the torsional resonance in Fig. 5.16, there are five resonances that would be designated $1 \leq n \leq 5$. At resonance, the amplitude of the response has its maximum value, $A_n(1)$.

$$A_n(\ f/f_n) = \frac{A_n(1)/Q_n}{\sqrt{\left(\frac{f}{f_n} - \frac{f_n}{f}\right)^2 + \frac{1}{Q_n^2}}} \tag{2.63}$$

Here Q_n is the quality factor of the nth mode, and f_n is the resonance frequency of that nth mode. With $A_n(1)$, the frequency response function (i.e., line shape) created by Eq. (2.63) is characterized by three parameters: $A_n(1)$, f_n, and Q_n.

We have already marveled at the ability of modern digital instrumentation and mathematical software to make it possible to acquire an array of data and fit those measurements to a chosen functional form. In Fig. 1.18, 2000 data points were acquired, and a nonlinear curve-fitting algorithm (i.e., "solver") was used to extract the five parameters necessary to characterize the free-decay of a loudspeaker that were specified in Eq. (1.119) by minimizing the sum of the squared difference between those measurements and the fitted function.

In Fig. 2.13, we executed a similar procedure to select the values of the three parameters in Eq. (2.63) that minimize the sum of the squared difference between the measured amplitude values, $A_{exp}(f)$ and $A_n(f/f_n)$, as was done for the least-squares fit to a straight line in Sect. 1.9. In this example, the response amplitudes were generated by the measurement of the fundamental ($n = 1$) resonant torsional modes of a spectrum like that in Fig. 5.15 but for a different bar. The best-fit values are shown in the boxed inset within Fig. 2.13. The slight asymmetry between the fit and the data is due to electromagnetic cross talk (see Sect. 5.4.3) that could be included within the model to improve the fit.

2.5.2 Power Dissipation, the Decibel, and Resonance Bandwidth

The time-averaged power dissipated in the mechanical resistance can be calculated using Eq. (1.73), and the mechanical impedance can be used to provide $v(t)$.

$$\langle \Pi \rangle_t = \frac{1}{2}\Re e[\mathbf{F}^* \hat{\mathbf{v}}] = \frac{F_1^2}{2}\Re e\left[\mathbf{Z}_\mathbf{m}^{-1}\right] = \frac{F_1^2 R_m}{2|\mathbf{Z}_\mathbf{m}^2|} = \frac{1}{2}\frac{F_1^2 R_m}{X_m^2 + R_m^2} \tag{2.64}$$

At resonance, $X_m(\omega_o) = 0$, and the power dissipated in the mechanical resistance has its maximum time-averaged value, $\langle \Pi_{\max} \rangle_t$. The equation is analogous to the power dissipated in an electrical resistor with a potential difference amplitude, V_1, across an electrical resistance, R_{el}.

$$\langle \Pi(\omega_o) \rangle_t = \langle \Pi_{\max} \rangle_t = \frac{1}{2}\frac{F_1^2}{R_m} \tag{2.65}$$

If instead we had used $\mathbf{Z}_m$ to eliminate $\mathbf{F} = \mathbf{Z}_\mathbf{m}\hat{\mathbf{v}}$ in Eq. (2.64), then we would have produced the analog to the electrical power dissipation which is proportional to the square of the current through the resistor.

$$\langle \Pi(\omega_o) \rangle_t = \frac{1}{2}\Re e[\mathbf{F}^* \hat{\mathbf{v}}] = \frac{v_1^2}{2}\Re e[\mathbf{Z}_\mathbf{m}] = \frac{1}{2}R_m v_1^2 \tag{2.66}$$

It is easy to see from Eq. (2.64) that the power is reduced to half of the maximum value when $R_m = \pm X_m$. There are two frequencies at which the power is half the maximum: ω_+ and ω_-. Those two frequencies can be determined by solution of Eq. (2.58).

$$m\omega_+ - (\mathrm{K}/\omega_+) = R_m \quad \text{and} \quad m\omega_- - (\mathrm{K}/\omega_-) = -R_m \tag{2.67}$$

Substituting K from the expression for ω_- into the expression for ω_+ produces an equation that can be factored to determine the required frequency difference.

$$\omega_+^2 - \omega_-^2 = (\omega_+ - \omega_-)(\omega_+ + \omega_-) = \frac{R_m}{m}(\omega_+ + \omega_-) \tag{2.68}$$

The expression for the quality factor from Eq. (2.44) generates another useful relation for Q that emphasizes the fact that Q is a dimensionless measure of the sharpness of a resonance.

$$Q = \frac{\omega_o m}{R_m} = \frac{\omega_o}{\omega_+ - \omega_-} = \frac{f_o}{f_+ - f_-} = \frac{\sqrt{f_+ f_-}}{f_+ - f_-} = \frac{f_o}{(\Delta f)_{-3\ \mathrm{dB}}} \tag{2.69}$$

Due to the symmetry of Eq. (2.62) with respect to ω/ω_o and ω_o/ω, f_o is the geometric mean of f_+ and f_-, not their arithmetic average.

The right-hand expression introduces the "down 3 dB" bandwidth of the resonance, $(\Delta f)_{-3\ \text{dB}} = f_+ - f_-$. This is the full bandwidth of the resonance between the two half-power frequencies and is also known as the *full width at half maximum*, $(\Delta f)_{FWHM}$. The use of the decibel for expression of power ratios will be discussed further in Part II – Fluids, particularly for the specification of sound levels in air in Sect. 10.5. Briefly, the *decibel* (abbreviated "dB") was introduced in the early days of telephony as a means for using addition and subtraction to multiply transducer sensitivities, amplifier gains, and transmission losses in the calculation of the overall response of telephone systems in the century before digital computers became available to engineers and scientists.

$$\text{dB} \equiv 10 \log_{10}\left(\frac{\Pi_1}{\Pi_2}\right) \tag{2.70}$$

In our case, $\Pi_2 = \langle\Pi_{max}\rangle_t$, the time-average power at resonance, and $\Pi_1 = (½)\langle\Pi_{max}\rangle_t$ is the half-power point: $10\log_{10}(½) = -3.01$, hence, the "down 3 dB" designation.

Figure 2.14 plots the velocity amplitude response (above) and phase response (lower right) vs. frequency ratio, f/f_o, for a driven harmonic oscillator with three different values of quality factor, Q. Those amplitude response curves never cross. Driven with the same force, the system with the higher-quality factor will have larger displacement, velocity, and acceleration responses at all drive frequencies. The peak in the velocity response always occurs at f_o. If we were to plot the displacement response, the peak would occur slightly below f_o since $\hat{\mathbf{x}} = \hat{\mathbf{v}}/(j\omega)$. Similarly, the peak in the acceleration response, $\hat{\mathbf{a}} = (j\omega)\hat{\mathbf{v}}$, would occur at a slightly higher frequency than f_o.

The relative power delivered to the mechanical resistance vs. the frequency ratio, f/f_o, is shown at the lower left of Fig. 2.14. The dashed lines, which intersect $(½)\langle\Pi_{max}\rangle_t$, occur at f_- and f_+.

At this point, it is useful to derive still another expression for the quality factor, this time in terms of the energy stored in the spring and the mass at resonance, E_{Stored}, and the energy dissipated in the mechanical resistance during one cycle, $\langle\Pi_{max}\rangle_t T$.

$$\frac{E_{Stored}}{\langle\Pi_{max}\rangle_t T} = \frac{\omega_o E_{Stored}}{2\pi\langle\Pi_{max}\rangle_t} = \frac{\frac{m}{2}v_1^2\omega_o}{2\pi\frac{R_m v_1^2}{2}} = \frac{Q}{2\pi} \tag{2.71}$$

This form can be convenient for calculation of Q since E_{Stored} and $\langle\Pi_{max}\rangle_t$ are both scalars.

Now that we have an expression for the power dissipated in the mechanical resistance of a damped driven simple harmonic oscillator, it is possible to calculate the equivalent noise bandwidth, $(\Delta f)_{EQNB}$, as shown in Eq. (2.73), that was required for our reconciliation of the consequences of the Equipartition Theorem in Eqs. (2.49) and (2.50), and the fluctuating force in Eq. (2.51), by working backward to create an expression for the fluctuating force from the results of the Equipartition Theorem.

$$\frac{1}{2}k_B T = \left\langle\frac{1}{2}mv^2\right\rangle = \frac{m}{2}\left\langle\frac{|F|^2}{|\mathbf{Z_m}|^2}\right\rangle = \frac{mF_{noise}^2}{2}\int_0^\infty \frac{df}{|\mathbf{Z_m}|^2} \tag{2.72}$$

Using the form of mechanical impedance from Eq. (2.60) and defining a dimensionless frequency ratio, $\Omega = f/f_o$, so $df = f_o\, d\Omega$, that definite integral can be evaluated using integral tables [20] or numerically [14].

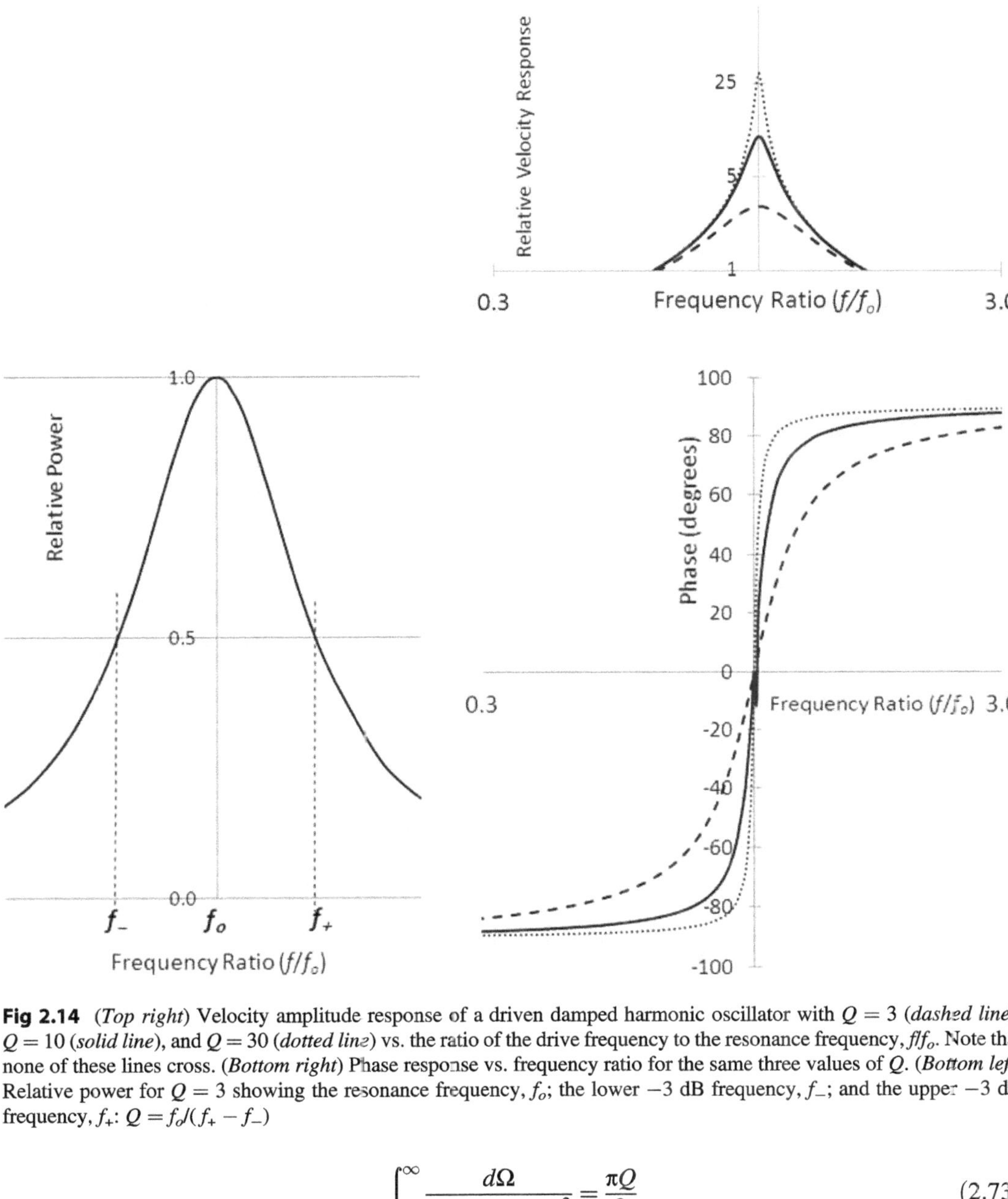

Fig 2.14 (*Top right*) Velocity amplitude response of a driven damped harmonic oscillator with $Q = 3$ (*dashed line*), $Q = 10$ (*solid line*), and $Q = 30$ (*dotted line*) vs. the ratio of the drive frequency to the resonance frequency, f/f_o. Note that none of these lines cross. (*Bottom right*) Phase response vs. frequency ratio for the same three values of Q. (*Bottom left*) Relative power for $Q = 3$ showing the resonance frequency, f_o; the lower -3 dB frequency, f_-; and the upper -3 dB frequency, f_+: $Q = f_o/(f_+ - f_-)$

$$\int_0^{\infty} \frac{d\Omega}{\frac{1}{Q^2} + \left(\Omega - \frac{1}{\Omega}\right)^2} = \frac{\pi Q}{2} \tag{2.73}$$

Substitution of the integral back into Eq. (2.72) results in an expression for the average kinetic energy that is independent of R_m.

$$\left\langle \frac{1}{2} m v_1^2 \right\rangle = \frac{m}{2} \frac{4 k_B T R_m}{\omega_o^2 m^2} f_o \frac{\pi Q}{2} = \frac{1}{2} k_B T \tag{2.74}$$

The equivalent noise bandwidth, $(\Delta f)_{EQNB}$, of the resonance can also be related to the integral in Eq. (2.73): $(\Delta f)_{EQNB} = (\pi/2)(\Delta f)_{-3\ \text{dB}}$. Once again, as we saw in Sect. 2.3.3, we have a final result that is dependent upon a particular parameter (i.e., the noise introduced by the dissipative component, R_m) but which hides the explicit dependence on the parameter that governs that phenomenon.

2.5.3 Resonance Tracking and the Phase-Locked Loop*

Typically, we think of the peak in the amplitude response of a damped driven harmonic oscillator as being the primary characteristic of a resonator driven at its resonance frequency. There is certainly good reason for this prejudice, since resonance can be used to optimize the coupling of an oscillatory force to a useful load, for example, the coupling of an electrodynamic loudspeaker to a thermoacoustic resonator [21]. But we are well aware (see Sect. 1.2) that the slope of the amplitude vs. frequency at that response maximum (or any extremum) is zero. For that reason, if the drive is de-tuned from resonance (e.g., by a change in temperature or some other environmental effect that alters ω_o), the amplitude of the response will decrease whether the de-tuning is either "flat" (i.e., drive frequency below resonance) or "sharp" (i.e., drive frequency above resonance). For that reason, a control system that monitors amplitude cannot determine whether to increase or to decrease the drive frequency to maintain the system on resonance based only on this change in amplitude.

A casual inspection of the resonance response curves in Fig. 2.14 shows that the rate of change of the phase difference with frequency, between the driving force and the velocity response, $d\Theta/d\omega$, is maximum at resonance, where $\Theta = 0$. Because $\Theta \cong 0$ near the resonance frequency, ω_o, it is easy to calculate that derivative, $(d\Theta/d\omega)_{\omega_o}$, using only the first term in the Taylor series for $\tan^{-1} x \cong x$.[15]

$$\left.\frac{d\Theta}{d\omega}\right|_{\omega_o} = \frac{d(\tan^{-1}X_m/R_m)}{d\omega}\Big|_{\omega_o} = \frac{1}{R_m}\frac{d}{d\omega}\left(\omega m - \frac{K}{\omega}\right)\Big|_{\omega_o} = \left[\frac{m + \frac{K}{\omega^2}}{R_m}\right]_{\omega_o} \tag{2.75}$$

Evaluation of that derivative at $\omega = \omega_o$, and substitution of $Q = \omega_o m/R_m$ and $K = \omega_o^2 m$, relates the rate of change of phase with respect to frequency to the quality factor.

$$\left.\frac{d\Theta}{d\omega}\right|_{\omega_o} = \frac{2m}{R_m} = \frac{2Q}{\omega_o} \quad \text{or} \quad \left.\frac{d\phi}{df}\right|_{f_o} = \frac{360^\circ}{\pi}\frac{Q}{f_o} \cong 114.6^\circ\frac{Q}{f_o} \tag{2.76}$$

The right-hand version uses ϕ to indicate the phase difference in degrees, rather than $\Theta = \pi\phi/180^\circ$, which was in radians, since the phase angle is more commonly reported in degrees by instrumentation or in computer models. This also produces another valuable expression for the quality factor in terms of the rate of change of phase difference with frequency at resonance.

$$Q = \left|\left.\frac{\pi f_o}{360^\circ}\frac{d\phi}{df}\right|_{f_o}\right| \cong \left|\left.\frac{f_o}{114.6^\circ}\frac{d\phi}{df}\right|_{f_o}\right| \tag{2.77}$$

The absolute value is specified since the sign of the slope can be either positive or negative, depending upon the parity of the resonances, in systems with multiple resonances that occur for standing waves in continuous vibrators, like strings (see Sect. 3.3) or plane wave resonators (see Sect. 10.6).

In many circumstances, it is much easier to fit a straight line (see Sect. 1.9) to the phase difference vs. frequency, close to the zero-crossing (near resonance), to determine Q than it is to use the -3 dB bandwidth of Eq. (2.69). One such example is provided for the Helmholtz resonator in Sect.

[15] It is easy to see that $\tan^{-1} x \cong x$ by inspection of the first term in the Taylor series expansion of $\tan x \cong x$ in Eq. (1.7) for small values of $x \ll 1$.

8.5.2. The phase vs. frequency for that resonance is plotted near the resonance frequency in Figs. 8.32 and 8.33.

Returning to the control question, we see that phase difference between the drive and the response generates an "error signal" proportional to the de-tuning that provides the information required by a control system to correct deviations between the drive frequency, ω, and the natural frequency, ω_o. It provides a parameter that has one sign if the system is driven sharp and the opposite sign if the system is driven flat. That error signal is nearly linear in its deviation for small frequency shifts away from the resonance frequency.

The extraction of such an error signal, proportional to the phase difference, can be accomplished by multiplication of the drive and response signals followed by a low-pass filtering of the product.[16] The drive signal will be our "reference," having constant amplitude and a frequency, ω. The reference waveform could include a phase shift, ϕ_r, due to the response of electronics, cables, etc., but we will allow ϕ_r to have an adjustable component as well: $F(t) = F_1 \sin(\omega t + \phi_r)$. The steady-state response (signal) will have an amplitude given by Eq. (2.62) and must also have the same frequency, ω, since we assume a linear response, but it will have a different phase, ϕ_s, that will combine Θ of Eq. (2.61) and any electronic phase shifts: $v(t) = v_1 \cos(\omega t - \phi_s)$. The following trigonometric identity will simplify the resulting product.

$$(\sin x)(\cos y) = \frac{1}{2}[\sin(x+y) + \sin(x-y)] \tag{2.78}$$

$$F(t)v(t) = \frac{F_1 v_1}{2}\{\sin[2\omega t + (\phi_r + \phi_s)] + \sin(\phi_r - \phi_s)\} \tag{2.79}$$

Both signals are assumed to be electrical waveforms generated by the diving force and some velocity transducer, and their product is also an electrical waveform.[17] If the product is passed through a low-pass filter with a cut-off frequency, $2\pi f_{-3\text{ dB}} \ll \omega$, then the component at twice the drive frequency will be strongly attenuated, and only a signal proportional to the sine of the phase difference will remain. If the electronically induced phase shifts are adjusted to make the phase difference $(\phi_r - \phi_s) \cong 0$ at resonance, then the output of the low-pass filter will be just the error signal that was sought for control of the drive frequency.[18]

The low-pass filtered signal can be fed back to a voltage-controlled oscillator[19] to drive the system at its resonance frequency. This strategy, known as a *phase-locked loop*, is used in many applications outside of acoustics [22].

---

[16] The "multiplication" can be done either digitally or using analog electronics. Various analog schemes can be implemented ranging from a simple "optional inverter" [11] that synchronizes a square waveform with the reference to multiply the signal by ± 1 to the use of a four-quadrant analog multiplication integrated circuit (e.g., Analog Devices AD633).

[17] Of course, it is possible that those signals were digitized prior to multiplication, and the operation of Eq. (2.79) was executed in a computer. For now, I will continue to treat the signals as analog electrical waveforms ('cause I'm an 'old guy").

[18] An integrator is often used instead of a low-pass filter since it can provide a non-zero output to the voltage-controlled oscillator that forces the phase difference to zero. If a low-pass filter is used, then a non-zero phase difference is required to provide the correction to the voltage-controlled oscillator, so the system will have to be driven slightly off-resonance to provide an adequate error signal. As the gain is increased to minimize the offset, the electronic system can become oscillatory (i.e., unstable).

[19] Analog integrated circuits that are voltage-controlled oscillators are readily available and very inexpensive for a wide variety of frequency ranges. For audio applications, I have used an Exar XR2206 that provides both a square-wave and low-distortion sine wave output and also includes a voltage-controlled amplifier to facilitate automatic gain control.

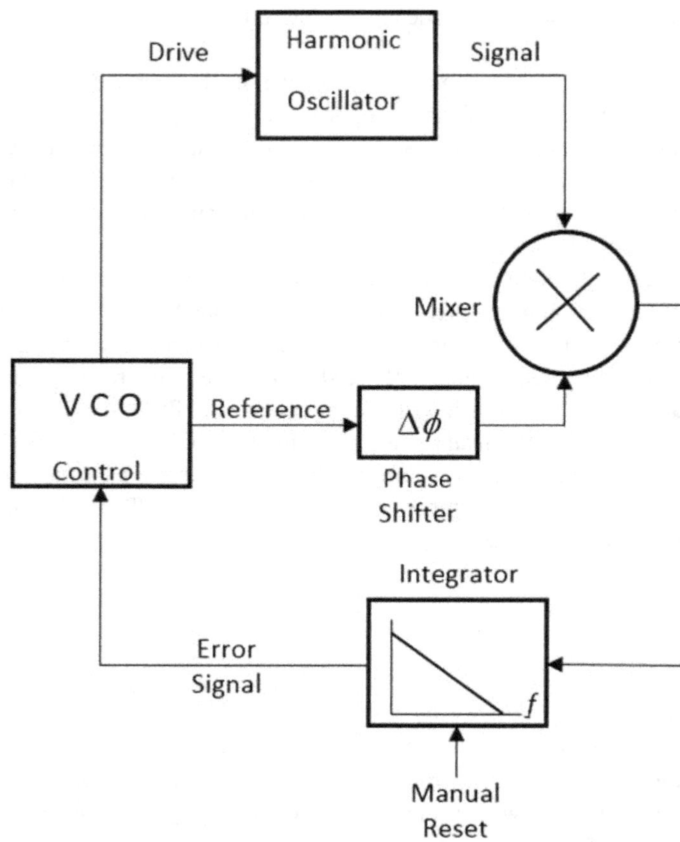

Fig. 2.15 This block diagram illustrates the signal flow in a phase-locked loop frequency tracking circuit meant to maintain a driven harmonic oscillator at its resonance frequency by generation of an error signal proportional to the phase difference between the drive and the response. In this diagram, an integrator is used as the low-pass filter, since it can provide a non-zero output to control the VCO when the phase difference has been driven to zero. The mixer executes the multiplication in Eq. (2.79)

It is not the intention of this textbook to provide detailed electronic strategies for tracking resonances, but it is very important to appreciate the utility of the phase difference between the drive and response of a simple harmonic oscillator, as well as the frequency dependence of the amplitude. Such feedback control systems have to be designed carefully to avoid instabilities, but the combination of digital or analog electronics that can provide phase detection and stability at a level of $\pm 0.01°$ at very modest cost, combined with the high rate of change of phase with frequency shown in Eq. (2.76) for resonators with even modest quality factors, is a very powerful tool in a variety of acoustical control applications. A block diagram of one possible implementation of such a frequency tracking phase-locked loop (PLL) is shown in Fig. 2.15.

2.5.4 Transient Response

Thus far, in our analysis of the damped driven simple harmonic oscillator, we have focused on the steady-state solution that is determined entirely by the oscillator's mechanical impedance, given in Eq. (2.58) or Eqs. (2.60) and (2.61). As mentioned earlier, the equation that describes the dynamics of the undriven system is identical to Eq. (2.55) but homogeneous. The solution to Eq. (2.39) can be added to the steady-state solution to incorporate the initial conditions at the time the harmonic forcing function, $F(t) = \Re e\left[\widehat{\mathbf{F}}e^{j\omega t}\right]$, is applied to the mass through the specification of the undetermined amplitude, $|\widehat{\mathbf{C}}|$, and phase, ϕ, or at the time the excitation is removed.

$$x(t) = |\widehat{\mathbf{C}}|\, e^{-t/\tau} \cos\left(\omega_d t + \phi\right) + \frac{|\widehat{\mathbf{F}}|}{\omega|\mathbf{Z_m}|} \sin\left(\omega t - \Theta\right) \tag{2.80}$$

The contribution of the transient term can initially interfere with the steady-state solution, but eventually the transient solution will diminish after a number of periods that spans any interval longer than τ.

Possibly more important are the consequences of the transient term which follow the removal of the forcing function. The reason is that it is possible that the amplitude of the steady-state solution can become quite large, especially if the system is being driven at a frequency that is close to its natural (resonance) frequency, ω_o. When the forcing ceases, the system will *ring down*, decaying at its damped frequency, ω_d. The acoustic radiation from this ring down is a serious problem for sound reproduction systems at low frequencies producing undesirable "boominess in the bass."

An accomplished musician, playing an electric bass guitar, will frequently damp the motion of a string, abruptly cutting off the force that the instrument (though the pick-up and amplifier) applies to the loudspeaker. If the loudspeaker (in its enclosure) is significantly under-damped, the sound radiated after the electric bass's string has stopped vibrating will be dominated by the decay of the loudspeaker's motion. That sound will decay at ω_d, independent of the note played by the bassist. In certain high-power automotive sound systems, this effect, caused by the ring down, can be particularly unpleasant.

The effects of the transient term in Eq. (2.80) provide insights into the growth and decay of the response of a damped, driven harmonic oscillator that are particularly useful when that system is driven at the resonance (natural) frequency, ω_o. At $t = 0$, the response is zero, so C equals the steady-state amplitude and the transient term is 180° out-of-phase with the steady-state solution. As the transient portion decays with an exponential time constant, τ, the response amplitude grows. For early times, $t < \tau$, after the application of the force, the sum of the transient and the steady-state solution creates a maximum amplitude envelope, $|A(t)|$, that initially grows linearly with time.

$$|A(t)| = A_1\left(1 - e^{-t/\tau}\right) \tag{2.81}$$

In this expression, A_1 is the steady-state amplitude response to the driving force. For times that are less that τ, the exponential can be approximated by the first two terms in the Taylor series expansion of Eq. (1.4), so the initial growth of the amplitude envelope will be approximately a linear function of time after the application of the driving force, as shown in Fig. 2.16.

$$|A(t)| = A_1\left(1 - e^{-t/\tau}\right) \cong A_1\left[1 - \left(1 - \frac{t}{\tau} + \ldots\right)\right] \cong A_1 \frac{t}{\tau} \quad \text{for} \quad t \leq \tau \tag{2.82}$$

Using this linear approximation for the initial growth of the resonance, we can approximate the time, $t_{1/2}$, it takes for the amplitude of the resonance to grow to one-half of its steady-state value, since initially $dA/dt \cong (A/\tau)$.

$$\frac{|A(t_{1/2})|}{A_1} = \frac{1}{2} \cong \frac{t_{1/2}}{\tau} = \frac{Q}{2\pi f_o} = \frac{Q}{2\pi} T \tag{2.83}$$

It takes approximately $(Q/2\pi)$ cycles for the response of a system driven at its resonance frequency to reach half its steady-state amplitude.

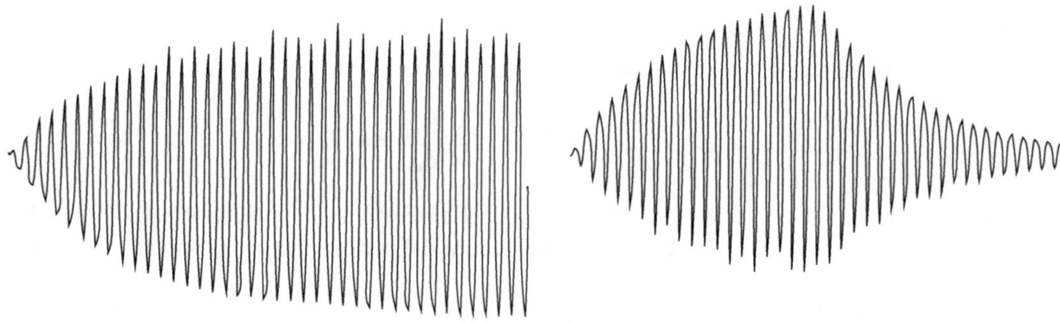

Fig. 2.16 The growth and decay of a damped harmonic oscillator driven at its natural frequency ω_o. In this example, $Q \cong 25$. The response grows to about half of its steady-state value after about five cycles. Once the drive is cut off, the freely decaying sinusoid has a frequency, ω_d, which has a relative difference from ω_o of only 0.02%. (Figures courtesy of J. D. Maynard [23])

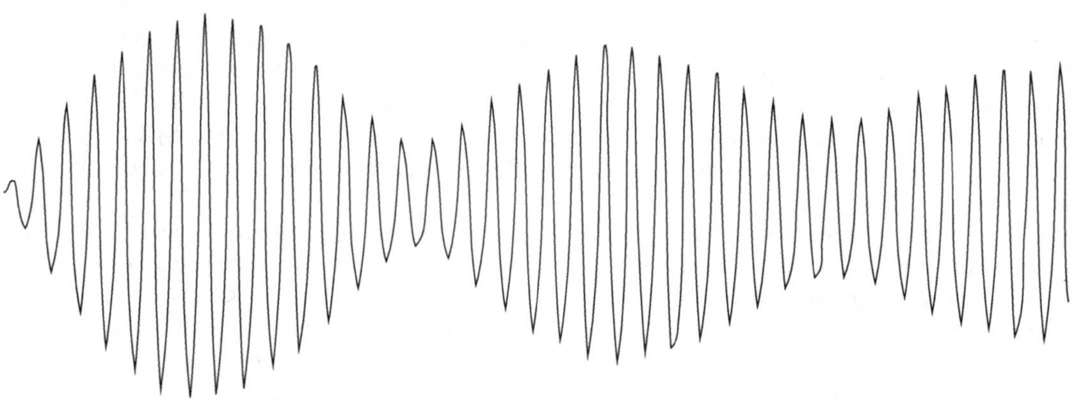

Fig. 2.17 This response is generated by an oscillator with $Q = 80$ driven slightly off resonance. The transient solution beats against the steady-state solution. Forty cycles are shown, but after about 80 cycles, the transient contribution would be negligible. (Figures courtesy of J. D. Maynard [23])

Since the drive frequency is ω_o and the frequency of the transient contribution is $\omega_d \neq \omega_o$, one might ask whether those two solutions might not maintain the phase relation we assumed to fit the initial condition of zero amplitude as time increases. As calculated in Eq. (2.45), the relative differences in those two frequencies are small and are related to the square of the quality factor.

$$\frac{\omega_o - \omega_d}{\omega_o} \cong \frac{1}{8Q^2} \tag{2.84}$$

By the time the resonance has grown to one-half its steady-state amplitude, the phase difference, $\varphi(t_{1/2})$, between the transient and steady-state solutions, will equal $\varphi(t_{1/2}) = \delta\omega \bullet t_{1/2} = (8Q)^{-1} = 45°/Q$, so de-phasing of those two terms of slightly different frequencies is insignificant for high-Q systems.

If the driving frequency is slightly different from the natural frequency, the transient and the steady-state solutions can "beat" against each other until the transient solution has decayed. This behavior is illustrated in Fig. 2.17. When slightly off resonance, the transient and the steady-state solutions can destructively interfere to nearly zero; if the drive is turned off at that time, the ring-down is nearly eliminated.

2.5.5 The Electrodynamic Loudspeaker

The moving-coil electrodynamic loudspeaker is a nearly perfect example of a force-driven, damped, simple harmonic oscillator. It is particularly useful because passing an alternating electrical current through the loudspeaker's voice coil provides an easy method to control the magnitude and frequency of the force applied to the moving mass, m_o (i.e., cone, voice coil, and former). The parts of such a loudspeaker are shown schematically in Fig. 2.18 (right). The voice coil can also be used as a velocity transducer since it will produce an open-circuit voltage that is directly proportional to the velocity of the moving mass.

In this section, we will use data taken from a "typical" electrodynamic loudspeaker[20] to determine the parameters that characterize the behavior of this simple harmonic oscillator: m_o, K, and R_m. The "$B\ell$ – product" is the transduction coefficient that connects the alternating electrical current $I(t) = \Re e\left[\hat{\mathbf{I}}e^{j\omega t}\right]$, to the force on the voice coil: $F(t) = (B\ell)I(t)$. In this case, $(B\ell)$ is the product of the magnitude of the magnetic induction, $|\vec{B}|$ [T],[21] times the length, ℓ [m], of the wire wound around the "former" to create the voice coil. Of course, we want to determine all of these parameters as accurately as possible without causing any damage to the loudspeaker.

The accurate determination of a loudspeaker's parameters is crucial for the design of a speaker system or headphones that are intended to reproduce music and/or speech. It also can provide important insight into the consistency of the manufacturing processes. In this section, one measurement method is described that exploits the description that was developed in this chapter to understand the damped driven harmonic oscillator. Although this approach may not be appropriate for production-line testing, it can provide "ground truth" to certify the validity of a production-line testing procedure.

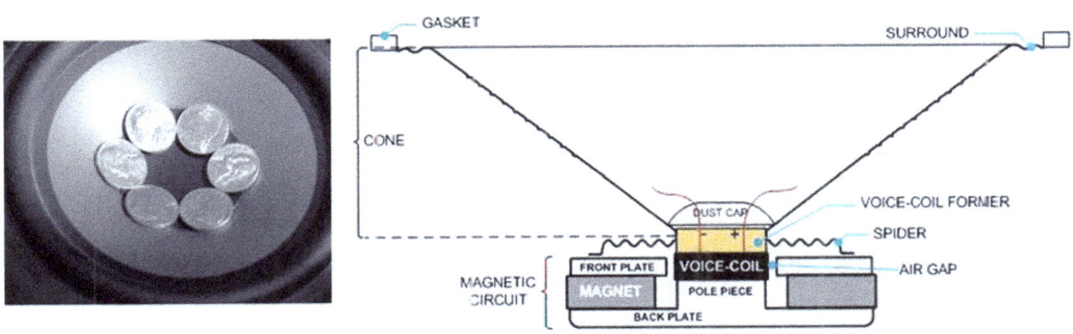

Fig. 2.18 (*Left*) Photograph looking down on a loudspeaker with six nickels resting on the cone surrounding the dust cap. (*Right*) Schematic representation of a moving-coil electrodynamic loudspeaker. The surround and spider center the voice coil in the air gap and provide the restoring stiffness, K, to center the voice coil in the magnet gap. The mass of the voice coil, voice coil former, dust cap, and a portion of the masses of the surround and spider all contribute to the moving mass, m_o. The magnetic circuit, consisting of the permanent magnet material in a high magnetic permeability fixture (back plate, front plate, and pole piece), provides a radial magnetic flux, $|\vec{B}|$, in the air gap that is everywhere perpendicular to the wire of length, ℓ, wound around the voice coil former

[20] Axon model 6S3 Mid-Bass Woofer with a cone diameter of about 6″ (16 cm)

[21] The tesla [T] is the SI unit of magnetic induction. The older c.g.s. unit is the gauss, which still appears in some of the loudspeaker literature: 1.0 T = 10,000 gauss.

The stiffness, K, and moving mass, m_o, can be determined by adding known masses, m_i, to the cone and measuring the loudspeaker's resonance frequencies, f_i, with each added mass. When the voice coil is driven by a constant current source (i.e., the current is independent of the electrical load presented by the voice coil),[22] the voltage across the voice coil has its maximum value at resonance. This is a direct consequence of *Lenz's Law*, which is responsible for the minus sign in *Faraday's law of induction*: $V = -(d\Phi/dt)$. The magnetic flux, Φ, is the product of the magnitude of the magnetic induction, $|\vec{B}|$, and an area with a normal that is perpendicular to the direction of $\vec{B}$. The "back electromotive force (*emf*)," represented by the induced voltage, V, will have its maximum value at resonance since the rate of change of the flux is proportional to the velocity of the voice coil.

Convenient masses to use for such an experiment are the US nickel, which weighs exactly 5.000 g.[23] Care must be taken to keep the electrical current through the voice coil at a low enough level that the peak acceleration of the loudspeaker cone never exceeds the acceleration due to gravity. At greater accelerations, the coins will "bounce," since they cannot fall back fast enough to track the displacement of the cone. The data acquired for measurement of the resonance frequencies, f_i, corresponding to the addition of $m_i = (0.005\ \text{kg})\ i$, is tabulated on the left side of Fig. 2.19 for $i = 0, 1, 2, 3, 4$, and 5.

The natural frequency of the loudspeaker, with the added masses, is given simply by the square root of the ratio of the suspension stiffness, K, and the total moving mass, $m_o + m_i$.

Mass	Freq.	(Period)^2
(Kg)	(Hz)	(sec*sec)
0.000	54.2	3.40E-04
0.005	45.4	4.85E-04
0.010	39.7	6.35E-04
0.015	35.7	7.85E-04
0.020	32.7	9.35E-04
0.025	30.5	1.08E-03
0.005	45.3	4.87E-04

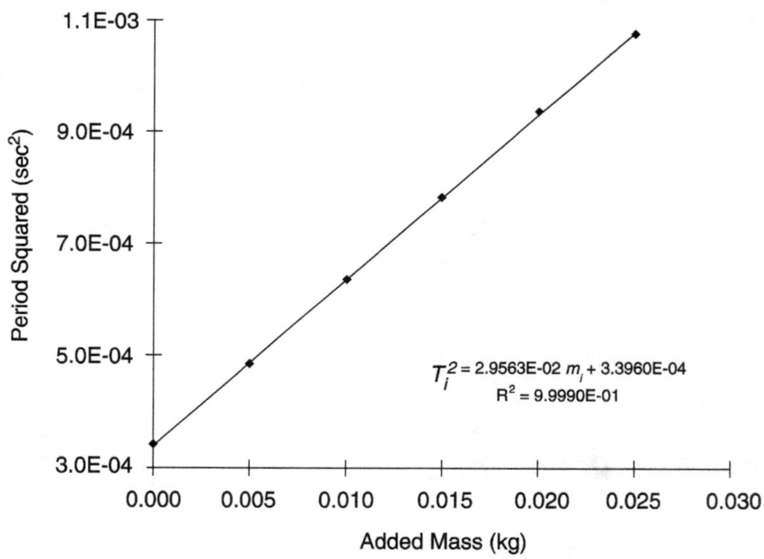

Fig. 2.19 (*Left*) Table of frequencies and squared periods for an electrodynamic loudspeaker's resonance with between zero and five US nickels placed on the cone (see Fig. 2.18). (*Right*) Plot of period squared, T_i^2, vs. added mass, m_i. R^2 is the square of the correlation coefficient (see Sect. 1.9.1)

[22] In practice, it is easy to approximate a constant current source by placing an electrical resistor, R_c (commonly referred to as the "current limiting resistor"), in series with the output of a signal (voltage) generator that is large compared to the largest magnitude of the loudspeaker's input electrical impedance $|\mathbf{Z}_{el}| \ll R_c$. For these measurements, $|\mathbf{Z}_{el}| < 100\ \Omega$, so $R_c \cong 1\ \text{k}\Omega$ was adequate to keep the current sufficiently constant.

[23] The US Mint website provides mass specifications for US coinage: www.usmint.gov. The US penny is 2.500 gm.

$$f_i = \frac{1}{2\pi}\sqrt{\frac{K}{m_o + m_i}} \tag{2.85}$$

That equation has only two unknown parameters: m_o and K. The technique of linearized least-squares fitting, introduced in Sect. 1.9.3, can be used to transform Eq. (2.85) into a linear relationship between the square of the period, $T_i^2 = \left(f_i^2\right)^{-1}$, and the added masses, m_i.

$$\frac{1}{f_i^2} = T_i^2 = \frac{4\pi^2}{K}(m_0 + m_i) = \frac{4\pi^2}{K}m_0 + \frac{4\pi^2}{K}m_i \tag{2.86}$$

In this form, the slope of the line plotted in Fig. 2.19 (Right) is $4\pi^2/K$. The value of the intercept divided by the value of the slope provides the moving mass, m_o.

Visual inspection of the graph in Fig. 2.19 suggests that the tabulated measurements are well-represented by Eq. (2.86). The techniques described in Sect. 1.9.2 relate the square of the correlation coefficient, R^2, to the relative uncertainty in the slope, providing the following value for the measured suspension stiffness and its uncertainty based on those six measured resonance frequencies: $K = 1335 \pm 7$ N/m ($\pm 0.5\%$). Here, the reported uncertainty corresponds to one standard deviation ($\pm 1\sigma$).

The relative uncertainty in the intercept is $\pm 0.6\%$. Since the moving mass involves the ratio of the intercept to the slope, the relative uncertainties in both the slope and the intercept need to be combined in a Pythagorean sum, as discussed in Sect. 1.8.4.

$$\frac{\delta m_o}{|m_o|} = \left[\left(\frac{\delta(slope)}{|slope|}\right)^2 + \left(\frac{\delta(intercept)}{|intercept|}\right)^2\right]^{1/2} = \sqrt{(0.5\%)^2 + (0.6\%)^2} = 0.8\% \tag{2.87}$$

This results in an experimental value for the loudspeaker's moving mass:$m_o = (11.5 \pm 0.1) \times 10^{-3}$ kg. Again, the estimated error reflects an uncertainty of one standard deviation ($\pm 1\sigma$).

The mechanical resistance can be determined from a simple free-decay experiment. I find it convenient to build a simple switching circuit[24] that allows me to drive the speaker near its resonance frequency and then cut the drive while simultaneously triggering a digital oscilloscope and recording the voltage generated by the voice coil that is then connected to the input of that oscilloscope (see Fig. 1.18), which is directly proportional to the velocity of the voice coil.

The data from the free-decay measurement on this loudspeaker were used as the example for fitting an exponentially decaying signal to a straight line by taking the logarithm of the output voltage differences between successive peaks and troughs of the damped sinusoidal waveform. The natural log of these voltage differences is plotted in Fig. 1.15. The linearized least-squares fit produced an exponential decay time of $\tau = (22.18 \pm 0.17) \times 10^{-3}$ s. A fit to the cycle number vs. the time for the voltage zero-crossings, using the same data, provided a measurement of the free-decay frequency and its uncertainty: $f_d = 51.82 \pm 0.17$ Hz.

The mechanical resistance is related to both the decay time, τ, and the moving mass, m_o, $R_m = 2m_o/\tau$, so the relative uncertainties for both quantities have to be combined in a Pythagorean sum, as was done previously in Eq. (2.87). This results in a measured value of $R_m = 1.04 \pm 0.01$ kg/s.

[24] Such a switch can be as simple as a double-pole double-throw toggle switch with one pole controlling a battery that provides a triggering voltage to the oscilloscope and the other pole used to move the voice coil from the output of the signal generator to the input of the oscilloscope that acquires and stores the free-decay time history, like the one shown in Fig. 1.18 and analyzed in Fig. 1.15.

Having determined R_m, it is possible to measure the Bℓ-product if the DC electrical resistance of the voice coil, R_{dc}, is known, and the magnitude of the electrical impedance at resonance, $|\mathbf{Z}_{el}(f_o)| = V_{ac}(f_o)/I_{ac}(f_o)$, is measured. At resonance, the voltage and current are in-phase (thus allowing use of the rms values of V and I), if the voice coil's inductance, L, can be neglected: $\omega_o L \ll R_{dc}$. At resonance, $X_m = 0$, and the expression for the velocity of the voice coil is very simple in the resistance-controlled regime.

$$v(f_o) = \frac{F}{R_m} = \frac{(B\ell)I}{R_m} \tag{2.88}$$

A voltage across the voice coil will be produced, in accordance with Faraday's law, by the back-emf, V_{emf}, generated by the velocity, v, of the voice coil of length, ℓ, moving in the magnetic field of magnetic induction, $|\vec{B}|$.

$$V_{emf}(f_o) = (B\ell)v = \frac{(B\ell)^2 I}{R_m} \tag{2.89}$$

The ohmic voltage across the voice coil, V_Ω, is due to the current through R_{dc}: $V_\Omega = R_{dc} I$. Since both voltages are in-phase with the current, the magnitude of the total electrical impedance measured at resonance, $|\mathbf{Z}_{el}(f_o)|$, is just the sum of those two voltages divided by the electrical current, I.

$$|\mathbf{Z}_{el}(f_o)| = \left|\frac{V}{I}\right|_{f_o} = \frac{(B\ell)^2}{R_m} + R_{dc} \quad \Rightarrow \quad (B\ell) = \sqrt{R_m(|\mathbf{Z}_{el}(f_o)| - R_{dc})} \tag{2.90}$$

This method for measurement of the $B\ell$-product will not have the same accuracy as the mechanical measurements for m_o, K, and R_m, since R_m was measured when the voice coil was an open circuit and there is a non-zero current passing through the voice coil when $|\mathbf{Z}_{el}|$ is being measured. This introduces other losses related to eddy currents induced in the metallic portions of the magnetic circuit and hysteresis in the orientation of the domains of the permanent magnetic materials [24], but with a reasonably large current-limiting resistor, these additional error sources can be minimized.

All of the moving-coil loudspeaker's mechanical parameters have been determined with a relative uncertainty of ±1% or less using a few coins, an electrical resistor to hold current constant, a signal generator, an AC voltmeter, and a digital storage oscilloscope. This approach is so simple it can be used as a teaching laboratory exercise for college freshmen in a first-year seminar [25]. The fact that such simple, inexpensive, and highly accurate determinations of the parameters that characterize the linear behavior of electrodynamic loudspeakers are not widely appreciated, as evidenced by the following quotation [26]:

> The added mass technique, discussed by Beranek [27] and others, requires a known, appreciable mass to be carefully attached to a diaphragm with appropriate distribution and driver orientation. The enclosed volume technique,[25] discussed by Thiele [28] and other, requires the use of an airtight box with known enclosed volume after a driver has been mounted to it. Both approaches can be time consuming and problematic in implementation, with relative bias errors often reaching 10% or higher.

[25] The "enclosed volume technique" is similar to the added mass technique just described except instead of adding masses; the gas stiffness of the enclosure is used to measure the resonance frequency while changing the total stiffness. See Problem 4 at the end of this chapter.

2.5.6 Electrodynamic (Moving-Coil) Microphone

The electrodynamic loudspeaker is a *reversible transducer*. Sound is generated by placing an alternating electrical current through the loudspeaker's voice coil. If the same speaker is placed into a sound field, the oscillating pressure difference (assuming the back of the speaker's cone is isolated from the oscillating pressure) of the sound will cause the speaker's diaphragm to vibrate.

If we start the analysis by assuming that the flow resistance shown in Fig. 2.20 (Left) is infinite, then the air inside the microphone remains at a constant pressure, p_m, and the excess pressure due to one frequency component of a sound wave that impinges on the microphone's diaphragm can be represented as $p(t) = \Re e[\widehat{\mathbf{p}}e^{j\omega t}]$. If the microphone has an area, A_{mic}, then the force on the diaphragm is $F_1(t) = A_{mic}\Re e[|\widehat{\mathbf{p}}|\, e^{j\omega t}]$. The microphone's moving mass (i.e., diaphragm and coil); suspension stiffness, K; and mechanical resistance, R_m, determine the microphone's (complex) mechanical impedance, $\mathbf{Z_m}$, so that the velocity of the coil is given by Eq. (2.62). Again, by Faraday's law, $V_{emf}(t) = (B\ell)v_1(t)$.

Operating as just described, such an electrodynamic microphone would have very uneven frequency response. During the second half of the twentieth century, small loudspeakers were used as both a sound source and a microphone in intercommunication systems for multistory dwellings. A resident could listen to someone at the front door and then press a switch that changed the mode of operation of the loudspeaker into a microphone and hear the voice of the visitor. In such an application, the fidelity was adequate for speech intelligibility.

The success of the electrodynamic microphone in both sound reinforcement and recording applications was due to Benjamin Baumzweiger (who later changed his name to Bauer), in 1937, who added the flow resistance shown in Fig. 2.20 (left) [29]. As just described, if the oscillating acoustic pressure is only applied to the outer surface of the microphone diaphragm, then the response of the microphone will be omnidirectional, if the wavelength of the sound is greater than the size of the diaphragm. If the acoustic pressure could reach both sides of the diaphragm, then the microphone would have a figure-eight (bi-directional) response pattern [30]. Bauer realized that a resistive element, in combination with the internal volume of the microphone, would provide access for the pressure at a small distance from the front of the microphone which could produce an intermediate (cardioid)

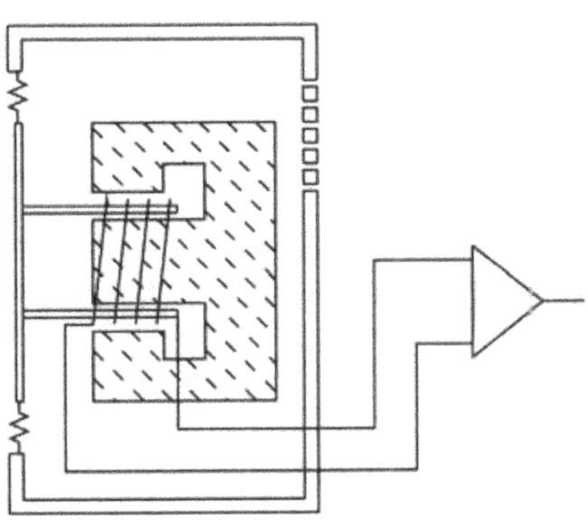

Fig. 2.20 (*Left*) Schematic diagram of an electrodynamic (moving-coil) microphone. An oscillatory pressure difference between the outside and the inside of the microphone's case exerts an oscillating force on the diaphragm causing the coil (shown connected to the input of an amplifier) to vibrate within the radial magnetic field produced by the hatched magnet structure. At the upper right of the microphone's enclosure is a porous medium that provides a flow resistance. Silk cloth is commonly used to provide the required flow resistance (see Sect. 12.5.3) (Diagram courtesy of J. D. Maynard). (*Right*) An electrodynamic microphone is shared by Tina Turner and Janis Joplin (Photo by Amalie R. Rothschild from the collection of Susan Levenstein and Alvin Curran)

Fig. 2.21 The Shure Model 55 "Unidyne®" (*right*) was the first product to incorporate the electrodynamic microphone strategy diagrammed schematically in Fig. 2.19 (left). (*Left*) The later model SM58 used a similar capsule. It is shown in use (with windscreen) in Fig. 2.19 (*right*). (Images used by permission, Shure, Inc.)

directivity pattern and that making the mechanical system resistance-controlled, the frequency response would be fairly uniform over a large range of frequencies.[26]

The first such single-element unidirectional electrodynamic microphone that used a single diaphragm, back-vented through a flow resistance, is the iconic Shure "Unidyne®" Model 55 shown in Fig. 2.21 (right) [31]. Later models, with the same type of microphone capsule, appeared in other packages like the Model SM58, shown in Fig. 2.21 (left), which has been the best-selling vocal microphone for the past 30 years.

2.5.7 Displacement-Driven SHO and Transmissibility

The decision to first analyze a damped simple harmonic oscillator, driven by a single-frequency sinusoidal force of constant amplitude, **F**, that is applied to the mass, was an arbitrary choice. In this section, the response of the same system will be analyzed but driven by a constant amplitude displacement, $x_o(t) = \Re e[\hat{\mathbf{x}} e^{j\omega t}]$, applied at the end of the spring that is not attached to the mass. Since the driving frequency, ω, is determined, the effects of a constant displacement, constant velocity, or constant acceleration that is applied to the end of the spring will be similar, differing only by one or two factors of $j\omega$.

This situation is diagrammed schematically in Fig. 2.22. Such an arrangement is fairly typical of a vibration isolation system that is intended to support a sensitive instrument (phonograph, microphone, disk drive, atomic force microscope, etc.) and isolate it from the vibrations of the foundation on which it is supported.

The mass will be accelerated due to the force applied by the spring and by the dashpot.

$$\mathrm{K}(x_o - x_2) - R_m\dot{x}_2 = m\ddot{x}_2 \quad \text{or} \quad \mathrm{K}x_o = m\ddot{x}_2 + R_m\dot{x}_2 + \mathrm{K}x_2 \tag{2.91}$$

Dividing this equation by m and letting $x_o(t) = \Re e[\hat{\mathbf{x}} e^{j\omega t}]$ transforms the above differential equation into an algebraic equation

[26] A systematic analysis of the directivity pattern of acoustic sensors will be provided in Sect. 12.5.

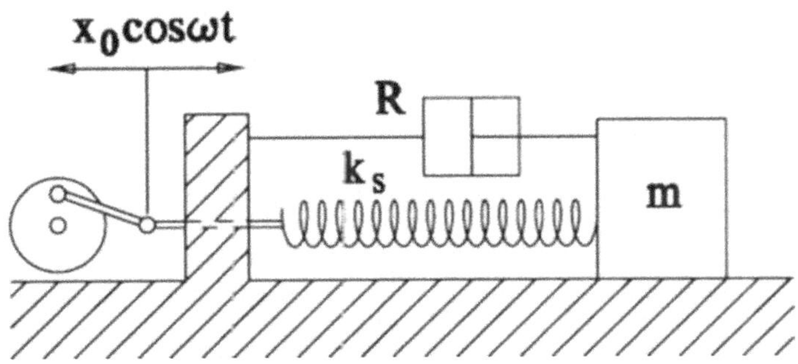

Fig. 2.22 In this configuration, a sinusoidal displacement is applied to the end of the spring located at x_o. A mass is attached to the other end of the spring with the junction between the mass and spring defining the position $x_2(t)$. One terminal of a dashpot, having a mechanical resistance, R_m, is attached to the mass, and the other terminal is rigidly fixed so the dashpot exerts a resistive force, $F = -R_m(dx_2/dt)$. (Diagram courtesy of J. D. Maynard)

$$\omega_o^2 x = -\omega^2 x_2 - j\omega \frac{R_m}{m} x_2 + \omega_o^2 x_2 \tag{2.92}$$

The steady-state solution for the mass's displacement, $x_2(t)$, is then rather reminiscent of our force-driven solution in Eq. (2.57).

$$x_2(t) = \Re\mathrm{e}\left[\frac{\omega_o^2 \widehat{\mathbf{x}} e^{\,j\omega t}}{\left(\omega_o^2 - \omega^2\right) + j\omega \frac{R}{m}}\right] = \Re\mathrm{e}\left[\frac{m\omega_o^2 \widehat{\mathbf{x}} e^{\,j\omega t}}{j\omega\left[R_m + j\left(\omega m - \frac{k}{\omega}\right)\right]}\right] \tag{2.93}$$

If the mass's response is expressed in terms of its complex velocity, $\mathbf{v}_2(t) = \dot{\mathbf{x}}_2(t) = j\omega \mathbf{x}_2(t)$, then the steady-state solution that was expressed in terms of our complex mechanical impedance, $\mathbf{Z_m}$, is recovered, as though $m\omega_o{}^2 x_o$ had been substituted for F_1 in Eq. (2.57).

$$\widehat{\mathbf{v}}_2 = \widehat{\dot{\mathbf{x}}}_2 = \frac{m\omega_o^2 x_o}{\mathbf{Z_m}} \tag{2.94}$$

Many features of this solution should now be familiar. When driven at the resonance frequency, ω_o, the displacement amplitude of the mass, x_2, is "amplified" by the quality factor of the oscillator.

$$\left|\frac{\widehat{\mathbf{x}}_2}{\widehat{\mathbf{x}}}\right|_{\omega_o} = \frac{\omega_o^2 m}{\omega_o R_m} = \frac{\omega_o m}{R_m} = Q \tag{2.95}$$

If the frequency of the drive, ω, is much greater than ω_o, the oscillator operates in its mass-controlled regime, so that the motion of the mass is reduced from that of the foundation.

$$\lim_{\omega \gg \omega_o}\left|\frac{\widehat{\mathbf{x}}_2}{\mathbf{x_o}}\right| = \lim_{\omega \gg \omega_o}\left|\frac{\widehat{\mathbf{v}}_2}{\dot{\mathbf{x}}_\mathbf{o}}\right| = \lim_{\omega \gg \omega_o}\left|\frac{\widehat{\mathbf{a}}_2}{\ddot{\mathbf{x}}_\mathbf{o}}\right| = \frac{\omega_o^2}{\omega^2} \tag{2.96}$$

In this mass-controlled limit, the displacement, velocity, or acceleration decreases by a factor of four for each reduction in ω_o by a factor of two. Of course, it is possible to stack multiple harmonic oscillator stages in series and produce even larger reductions in the transmissibility of a vibration isolator.

The similarity between the force-driven and displacement-driven cases means that the same strategy can be used to reduce the force generated by a vibrating piece of machinery. In the case of a *vibration isolator*, the foundation moves both the spring and the damper. This is not the same as Fig. 2.22, where one end of the resistance is fixed but is mechanically identical to the driven foundation of Fig. 2.23, where the mass is connected to the foundation through both the stiffness and the resistance elements.

As discussed in the following treatment of vibration sensors, our interest is in the relative motion of the mass and its foundation, $\mathbf{z} \equiv \mathbf{x_1} - \mathbf{x_2}$. Equation (2.91) can be rewritten by substituting $\mathbf{z} = \mathbf{x_1} - \mathbf{x_2}$.

$$m\ddot{\mathbf{z}} + R_m\dot{\mathbf{z}} + \mathrm{K}\mathbf{z} = -m\ddot{\mathbf{x}}_1 = m\omega^2\hat{\mathbf{x}}_1 e^{j\omega t} \tag{2.97}$$

The complex *transmissibility*, $\boldsymbol{T}$, is defined as the ratio of the motion of the mass, relative to the foundation, $\mathbf{z}$, divided by the motion of the foundation, $\mathbf{x_1}$, or the ratio of the force on the mass, $\mathbf{F_2}$, to the force on the foundation, $\mathbf{F_1}$ [32]. If the drive frequency, ω, is written in terms of the natural frequency, ω_o, so $\Omega = \omega/\omega_o$, and the mechanical resistance is written in terms of the quality factor, $R_m = \omega_o m/Q$, then the magnitude of the transmissibility, $|\boldsymbol{T}|$, and the phase angle between the displacements or forces, θ_T, based on Eq. (2.93), can be put in a general form.

$$|\mathbf{T}| = \left|\frac{\mathbf{z}}{\mathbf{x_1}}\right| = \left|\frac{\mathbf{F_2}}{\mathbf{F_1}}\right| = \frac{\sqrt{1+(\Omega/Q)^2}}{\sqrt{(1-\Omega^2)^2+(\Omega/Q)^2}} \tag{2.98}$$

$$\theta_T = \tan^{-1}\left[\frac{-\Omega^3/Q}{1-\Omega^2+(\Omega/Q)^2}\right] \tag{2.99}$$

A plot of $|\boldsymbol{T}|$ and θ_T will look identical to those for the driven harmonic oscillator in Fig. 2.14. Similarly, the frequency dependence of the transmissibility of a damped harmonic oscillator (DHO) is compared to the transmissibility for vibration isolators using rubber springs is presented in Sect. 4.5.3 in Figs. 4.30, 4.31, and 4.33.

2.6 Vibration Sensors

The displacement-driven one-dimensional simple harmonic oscillator provides the basis for most common vibration sensors. Typically, such a sensor is housed in a case that will be mounted on a vibrating surface. A schematic diagram of such a sensor is shown in Fig. 2.23. Since the transduction

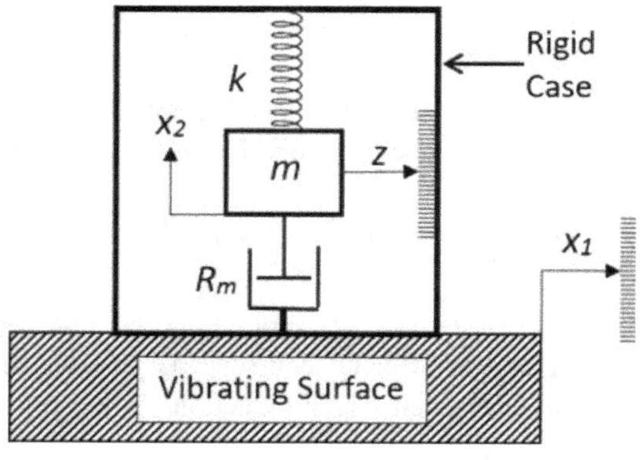

Fig. 2.23 Generic vibration sensor contained within a rigid case that is attached to a vibrating surface. The displacement of the vibrating surface is given by $\mathbf{x_1}(t)$ and the motion of the mass, m, is given by $\mathbf{x_2}(t)$. The displacement of the mass, relative to the case, is given by z, where the scale inside the case represents the sensor's transduction mechanism. Scales are provided for x_1 and for z, where x_1 and x_2 are measured relative to a fixed coordinate system

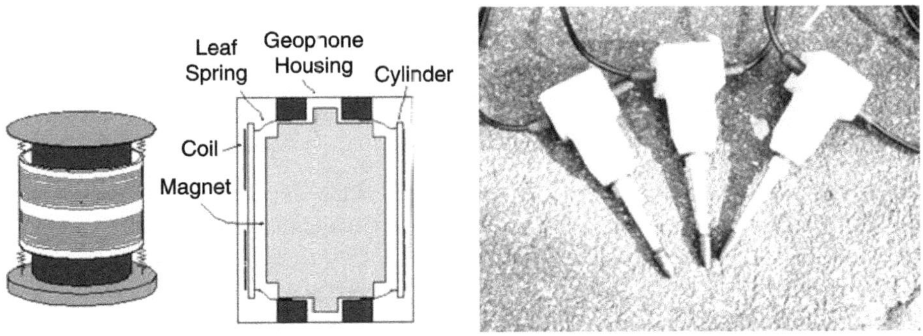

Fig. 2.24 (*Left*) Schematic representation of a geophone containing a permanent magnet surrounded by a coil of wire wound around a cylinder. The leaf spring has a very small stiffness making the natural frequency much lower than the frequencies of interest. The weak suspension means that the coil is effectively stationary and the magnet, attached rigidly to the case, moves with the ground. The relative motion of the coil and magnet generates the electrical signal in the same way as the electrodynamic microphone. (*Right*) Three commercial geophones with spikes that facilitate their insertion into the ground

mechanism (e.g., piezoelectric or electrodynamic) is contained within the sensor, it can only measure the motion of the mass relative to the motion of the case, just as was done for the vibration isolator. The solution is identical to Eqs. (2.98) and (2.99).

Such vibration sensors are called seismic sensors (*seismometers*) or *geophones* if the sensor is operated in its mass-controlled regime where $\Omega \gg 1$. For that circumstance, the mass remains nearly at rest, and the transduction mechanism senses the relative motion of the case. If the transduction mechanism measures the relative displacement, then the sensor is a seismometer. If the transduction mechanism is a coil and magnet (i.e., electrodynamic), then the sensor's output is proportional to the relative velocity of the (nearly stationary) mass and the case, and the sensor is called a geophone.

Seismometers are used to measure earthquake activity, and geophones are usually used for exploration geophysics (e.g., finding sub-surface fossil fuel deposits). Frequently, dozens of geophones will be placed in the ground, and an explosive or vibratory mechanical source will be used to create seismic waves. The outputs of the geophones' response will be processed to create images of sub-surface features. Figure 2.24 shows the typical construction of a geophone.

It is easy to control the damping of the geophone by adjusting the electrical resistance across the output of the coil. The coil's motion, relative to the magnet, will generate the voltage, $V_{emf} = (B\ell)v_{coil}$, in accordance with Faraday's law. As shown in Fig. 1.20, the coil will have some electrical resistance, R_{int}, but if the load resistance, R_{load}, is infinite, no current will flow, $I_{out} = 0$, and the voltage across the coil's terminals will be V_{emf}.

If a load resistance is placed across the geophone's terminals, then the geophone's output voltage, V_{out}, will be reduced, and the current flowing through the coil and the load will be I_{out}, assuming the coil's inductance, L_{coil}, can be ignored (i.e., $\omega L_{coil} \ll [R_{int} + R_{load}]$).

$$V_{out} = \frac{R_{load}}{R_{int} + R_{load}} V_{emf} \quad \text{and} \quad I_{out} = \frac{V_{out}}{R_{load}} \tag{2.100}$$

The time-averaged power dissipated in the electrical circuit, $\langle \Pi_{el} \rangle_t$, will produce an equivalent mechanical dissipation that can be represented as an equivalent mechanical resistance, R_{eq}.

$$\langle \Pi_{el} \rangle_t = \frac{V_{emf} I_{out}}{2} = \frac{1}{2} \frac{V_{emf}^2}{R_{int} + R_{load}} = \frac{1}{2} \frac{(B\ell)^2 v_{coil}^2}{R_{int} + R_{load}} = \frac{1}{2} R_{eq} v_{coil}^2$$
$$R_{eq} = \frac{(B\ell)^2}{R_{int} + R_{load}} \Rightarrow Q_{eq} = \frac{\omega_o m}{R_{eq}} \Rightarrow \frac{1}{Q_{total}} = \frac{R_m}{\omega_o m} + \frac{1}{Q_{eq}} \quad (2.101)$$

A geophone that gets quite a bit of use in our laboratory is the Geo Space GS-14-L3. It is contained within a small cylindrical capsule about 17 mm in both height and diameter. It has a natural frequency of $f_o = \omega_o/2\pi \cong 28$ Hz. The coil has an internal resistance, $R_{int} = 570\ \Omega$, and a moving mass of $m = 2.15$ g. The inductance of the coil, $L_{coil} = 45$ mH, so even at 1.0 kHz, $\omega\, L_{coil}/R_{int} < ½$. Its output voltage as a function of frequency and electrical load resistance, R_{load}, is shown in Fig. 2.25.

If the vibration sensor is operated in its stiffness-controlled regime, $\Omega \ll 1$, then it is called an accelerometer. The motion of the case and the mass are nearly equal, and the transduction mechanism measures the force necessary to maintain their spacing. Two typical piezoelectric accelerometers are shown in Fig. 4.6 and are analyzed in Sect. 4.3.1.

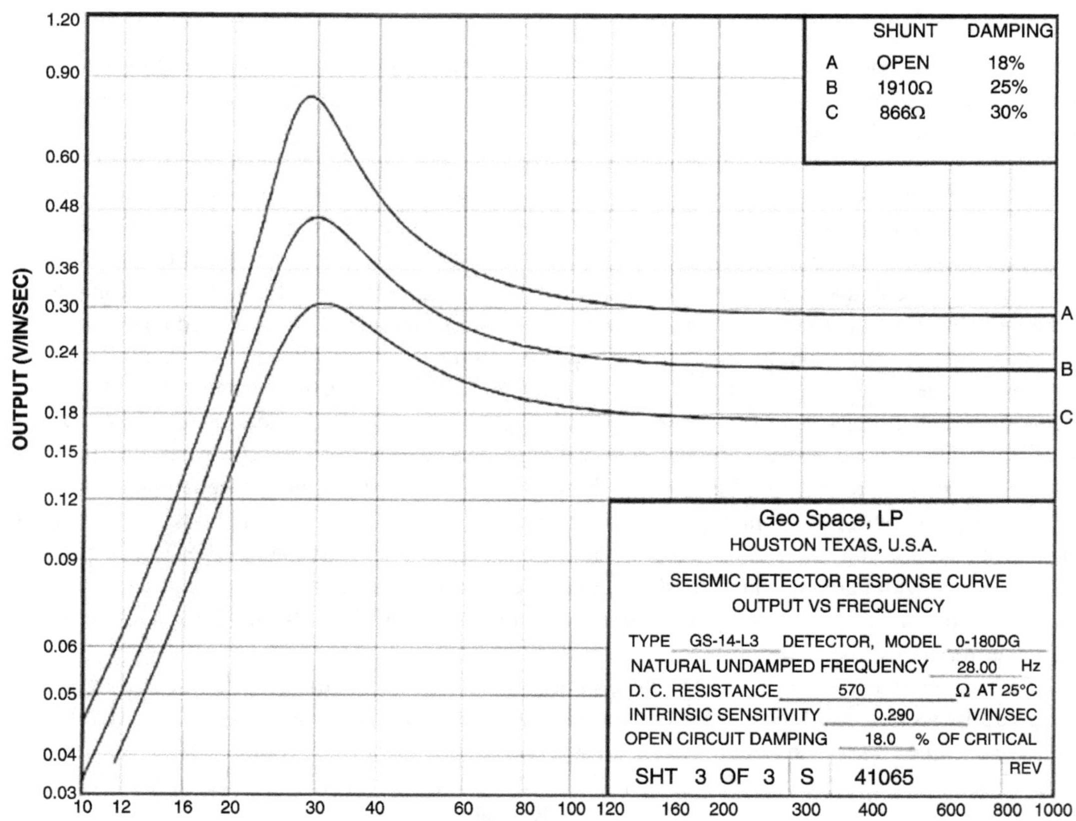

Fig. 2.25 Output sensitivity [V/(in/s)] for a Geo Space GS-14-L3 geophone capsule as a function of frequency for load resistances of $R_{load} = \infty$ (open circuit sensitivity), $R_{load} = 1.91$ kΩ, and $R_{load} = 866\ \Omega$. The external load resistance adds damping, lowering Q, and reducing the sensitivity but making the sensitivity more uniform over a larger range of frequencies. (In the words of Prof. S. J. Putterman, "The flattest response is no response at all.")

2.7 Coupled Oscillators

The techniques developed for our understanding and analysis of the one degree-of-freedom simple harmonic oscillators can be extended to multiple masses joined by multiple springs. This problem is important because we frequently couple one harmonic oscillator to another resonator to drive that system (which may have multiple resonances) as discussed in Sect. 10.7.5 and illustrated in Figs. 10.12 and 10.14. In fact, for closely coupled oscillators, the distinction between the driven system and the driver becomes blurred. Coupled harmonic oscillators are also an important subject because it provides a logical transition to continuous systems as the number of individual masses and springs becomes very large. Since all matter is composed of atoms, it is possible to treat solids as a three-dimensional array of masses joined to their nearest neighbors by springs.

The treatment in this section will again be restricted to motions that are constrained to occur in only one dimension, although we will let the number of degrees of freedom increase to two or more, with the number of coordinates specifying the position of the masses increasing accordingly. In general, for motion restricted to a single dimension, the number of natural frequencies for vibration will also be equal to the number of degrees of freedom, but simple harmonic motion of the coupled system will only be periodic at a single frequency if the system is excited in one of its *normal modes*. If those normal mode frequencies are related by the ratio of integers, then the motion of the coupled system, when disturbed from equilibrium, will be periodic; otherwise the motion will never repeat itself.

Some characteristics of the behavior of coupled systems will be unfamiliar to many readers. For that reason, we will begin our study of coupled oscillators with a very simple pair of coupled undamped oscillators that can be understood intuitively prior to developing the mathematical formalism necessary to write down solutions to the general case.

2.7.1 Two Identical Masses with Three Identical Springs

Thus far we have analyzed the dynamics of a single mass, joined to a spring and damper, that is either oscillating freely or driven by some externally imposed periodic force or displacement. At this point, that development has generated one-hundred numbered equations as well as many others that are contained within the text. Now that we are going to extend this analysis to several such oscillatory systems that are coupled in ways that permit them to exchange their energy of vibration, we might expect the solutions to be even more complex mathematically. That expectation will soon be fulfilled.

There is a simple case that can be understood intuitively. It will give us some basis to test the results of a more rigorous treatment of the general problem. It will also introduce simple behaviors that will be characteristic of the more general case illustrated schematically in Fig. 2.26.

We start by letting $m_1 = m_2 = m$ and $K_1 = K_2 = K_3 = K$. We can then guess the two frequencies that will make the combined motion simple and harmonic and identify the periodic displacements of

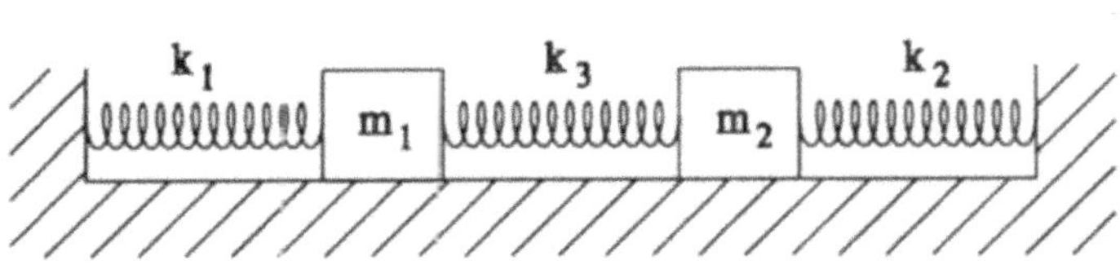

Fig. 2.26 One-dimensional, two degree-of-freedom, coupled harmonic oscillators. The x_1 coordinate measures the displacement of m_1 from its equilibrium position, and x_2 is the measure of m_2's displacement from equilibrium. (Figure courtesy of J. D. Maynard [23])

those two masses that correspond to those two frequencies of oscillation. If $x_1(t) = x_2(t)$, the motion is symmetric, and the middle spring that couples their motion, K_3, is neither compressed nor extended. The frequency of oscillation is just that of either mass oscillating in isolation, $\omega_s = \sqrt{K/m}$, where the subscript "*s*" has been chosen because it corresponds to the *symmetric mode* of oscillation.

In the *antisymmetric mode*, the masses move 180° out-of-phase with each other, $x_1(t) = -x_2(t)$. In that case, the center of the K_3 spring is fixed. From the perspective of either mass, it appears as though the K_3 spring has been cut in half and attached to a rigid boundary. The coupling spring, appearing half as long, is now twice as stiff (see Sect. 2.2.1), so the total effective stiffness experienced by either mass oscillating in the antisymmetric mode is $K_a = K + 2K = 3K$. The frequency of that antisymmetric mode is therefore $\omega_a = \sqrt{3K/m}$.

If the masses are initially displaced by either $x_1(t) = x_2(t)$ or $x_1(t) = -x_2(t)$ and then released, the pair will both oscillate at ω_s or ω_a with the same amplitudes forever. Any other combination of initial displacements will lead to motion that is entirely aperiodic; in fact, the motion, though deterministic, will appear chaotic. The reason for this lack of periodicity is that the ratio of the two normal mode frequencies is an irrational number: $\omega_a/\omega_s = \sqrt{3}$.

2.7.2 Coupled Equations for Identical Masses and Springs

It is not very difficult to write down the equations of motion for each mass using Newton's Second Law of Motion. It is also simple to convert those coupled differential equations to coupled algebraic equations by substituting a time-harmonic solution, just as we have been doing for our single degree-of-freedom oscillators. We will do this first for our simple case of $m_1 = m_2 = m$ and $K_1 = K_2 = K_3 = K$.

$$m\ddot{x}_1 + Kx_1 + K(x_1 - x_2) = m\ddot{x}_1 + K(2x_1 - x_2) = 0 \tag{2.102}$$

By symmetry, the equation of motion for the second mass will interchange the subscripts on the coordinates.

$$m\ddot{x}_2 + K(2x_2 - x_1) = 0 \tag{2.103}$$

Since we seek normal modes, where all parts of the coupled system oscillate at the same frequency, substitution of $x_1 = B_1 \cos \omega\, t$ and $x_2 = B_2 \cos \omega\, t$ converts Eqs. (2.102) and (2.103) into two coupled algebraic equations for the two amplitude coefficients, after dividing through by $\cos \omega t$.

$$\begin{aligned} -m\omega^2 B_1 + K(2B_1 - B_2) = 0 \\ -m\omega^2 B_2 + K(2B_2 - B_1) = 0 \end{aligned} \tag{2.104}$$

That equation can be rearranged to compute a solution for B_1 and B_2.

$$\begin{aligned} \left(2K - \omega^2 m\right)B_1 - KB_2 = 0 \\ -KB_1 + \left(2K - \omega^2 m\right)B_2 = 0 \end{aligned} \tag{2.105}$$

For these coupled linear equations to have a nontrivial solution, the determinant of their coefficients must vanish.

$$\begin{vmatrix} (2K - \omega^2 m) & -K \\ -K & (2K - \omega^2 m) \end{vmatrix} = 0 \tag{2.106}$$

Evaluation of that determinant leads to a *characteristic equation*[27] that can be solved to determine the values of ω that satisfy the original equations.

$$(2K - \omega^2 m)^2 - K^2 = 0 \quad \Rightarrow \quad (2K - \omega^2 m) = \pm K \tag{2.107}$$

The two solutions correspond to the two modes we guessed in Sect. 2.7.1.

$$\omega = \sqrt{\frac{2K \pm K}{m}} \quad \text{so} \quad \omega_s = \sqrt{\frac{K}{m}} \quad \text{and} \quad \omega_a = \sqrt{\frac{3K}{m}} \tag{2.108}$$

Substitution of those frequencies back into either of the algebraic equations for the coefficients of B_1 and B_2 in Eq. (2.104) will provide expressions for $x_1(t)$ and $x_2(t)$ corresponding to those two *normal mode frequencies*, ω_s and ω_a. Starting with the symmetric mode frequency, ω_s,

$$-m\omega_s^2 B_1 + K(2B_1 - B_2) = -KB_1 + K(2B_1 - B_2) = K(B_1 - B_2) = 0. \tag{2.109}$$

Since $K \neq 0$, the result for the symmetric mode has $B_1 = B_2$. Making the similar substitution using the antisymmetric normal mode frequency ω_a,

$$-m\omega_a^2 B_1 + K(2B_1 - B_2) = -3KB_1 + K(2B_1 - B_2) = -K(B_1 + B_2) = 0 \tag{2.110}$$

For the antisymmetric mode, $B_1 = -B_2$.

These results will allow us to write the solution for motions of the individual masses if they are oscillating at one of those two normal mode frequencies.

$$\begin{gathered} x_1(t) = B_s \cos(\omega_s t + \phi_s) \quad \text{and} \quad x_2(t) = B_s \cos(\omega_s t + \phi_s) \\ \text{or} \\ x_1(t) = B_a \cos(\omega_a t + \phi_a) \quad \text{and} \quad x_2(t) = -B_a \cos(\omega_a t + \phi_a) \end{gathered} \tag{2.111}$$

An arbitrary phase has been introduced to accommodate any choice of the time corresponding to $t = 0$.

2.7.3 Normal Modes and Normal Coordinates

Having identified the two frequencies for our simple case and the corresponding motions of both masses at those two frequencies, we will now see that a coordinate transformation yields further simplifications. We can introduce the new coordinates, $\eta_s(t)$ and $\eta_a(t)$, that combine the coordinates of the two masses, $x_1(t)$ and $x_2(t)$, in a way that makes $\eta_a(t) = 0$ for the symmetric mode and $\eta_s(t) = 0$ for the antisymmetric mode.

$$\left.\begin{aligned} \eta_s &= x_1 + x_2 \\ \eta_a &= x_1 - x_2 \end{aligned}\right\} \quad \Rightarrow \quad \left\{\begin{aligned} x_1 &= \frac{1}{2}(\eta_s + \eta_a) \\ x_2 &= \frac{1}{2}(\eta_s - \eta_a) \end{aligned}\right. \tag{2.112}$$

Those transforms allow substitution of $\eta_a(t)$ and $\eta_s(t)$ for $x_1(t)$ and $x_2(t)$ into Eq. (2.103).

[27] In older treatments, the characteristic equation is often called the *secular equation*.

$$\begin{aligned}\frac{1}{2}\left[m(\ddot{\eta}_a + \ddot{\eta}_s) + \mathrm{K}(\eta_s + 3\eta_s)\right] = 0 \\ \frac{1}{2}\left[m(\ddot{\eta}_a - \ddot{\eta}_s) + \mathrm{K}(-\eta_s + 3\eta_s)\right] = 0\end{aligned} \tag{2.113}$$

Adding and subtracting the above equations from each other produces two uncoupled equations for the *normal coordinates*, η_s and η_a.

$$m\ddot{\eta}_s + \mathrm{K}\eta_s = 0 \quad \text{and} \quad m\ddot{\eta}_a + 3\mathrm{K}\eta_a = 0 \tag{2.114}$$

The solutions for the time dependence of these normal coordinates are now trivial, since the above equations are isomorphic to our undamped simple harmonic oscillator equations.

$$\eta_s(t) = C_s \cos(\omega_s t + \phi_s) \quad \text{and} \quad \eta_a(t) = C_a \cos(\omega_a t + \phi_a) \tag{2.115}$$

These normal coordinates provide a complete basis that can be used to express the time dependence of the motion of both masses. The normal coordinates can express initial conditions, for example, those in Eq. (2.116), that do not correspond to initial conditions that produce the simple harmonic motion when the system is displaced from equilibrium into a configuration corresponding to one of those normal modes. Before looking at the time evolution of the positions of the masses for other initial conditions, it will be worthwhile to summarize the results of our simple example that will be present in the solutions to the more general case:

- There will be a number of normal modes that is equal to the number of degrees of freedom.
- A normal mode will have a frequency at which all of the masses will exhibit simple harmonic motion at that frequency.
- Each normal mode will provide a transformation of coordinates that can decouple the equations of motion.
- The normal modes provide a complete set of *basic functions* that can be combined to describe the subsequent motion of the masses for initial conditions that are not those which produce the simple harmonic oscillations characteristic of the normal modes.

2.7.4 Other Initial Conditions

To develop an appreciation for the behavior of our simple system of two identical masses connected by three identical springs, we can construct a solution for a very simple initial condition that does not correspond to one of the normal modes by fixing m_2 at its equilibrium position, $x_2(0) = 0$, and displacing m_1 by an amount $x_1(0) = \alpha$, with both masses initially at rest, $\dot{x}_1(0) = \dot{x}_2(0) = 0$. Those initial conditions can be expressed in terms of the normal modes and their frequencies.

$$x_1(t) = \frac{\alpha}{2}\cos\omega_s t + \frac{\alpha}{2}\cos\omega_a t \quad \text{and} \quad x_2(t) = \frac{\alpha}{2}\cos\omega_s t - \frac{\alpha}{2}\cos\omega_a t \tag{2.116}$$

Upon release of both masses at $t = 0$, the evolution of their positions can be interpreted by use of the following trigonometric identity:

$$\cos x \pm \cos y = \cos\left[\frac{1}{2}(x+y) + \frac{1}{2}(x-y)\right] \pm \cos\left[\frac{1}{2}(x+y) - \frac{1}{2}(x-y)\right]$$
$$= \begin{cases} 2\cos\frac{1}{2}(x+y)\cos\frac{1}{2}(x-y) \\ -2\sin\frac{1}{2}(x+y)\sin\frac{1}{2}(x-y) \end{cases} \tag{2.117}$$

The ratio of the frequencies of our normal modes is an irrational number: $\omega_a/\omega_s = \sqrt{3}$. Their sum and difference can be expressed in terms of the symmetric mode frequency: $\omega_a + \omega_b = (\sqrt{3}+1)\omega_s \cong 2.732\ \omega_s$ and $\omega_a - \omega_b = (\sqrt{3}-1)\omega_s \cong 0.732\ \omega_s$.

$$\begin{aligned} x_1(t) &= \alpha\cos(1.366\omega_s t)\cos(0.366\omega_s t) \\ x_2(t) &= -\alpha\sin(1.366\omega_s t)\sin(0.366\omega_s t) \end{aligned} \tag{2.118}$$

Although this result is deterministic, it is not periodic. As will be demonstrated for the case of weak coupling, the identity in Eq. (2.117) only leads to periodic solutions if the ratio of the frequencies is the ratio of two integers.

2.7.5 General Solutions for Two Masses and Three Springs

Now that the behavior of two coupled harmonic oscillators has been demonstrated for the simplest case, we can progress to the solution of the more general case where the same techniques will allow us to write equations for $m_1 \neq m_2$ and $K_1 \neq K_2 \neq K_3$. For this more general case, Eqs. (2.102) and (2.103) have to be re-written to include the features of Fig. 2.26.

$$m_1\ddot{x}_1 = -(K_1 + K_3)x_1 + K_3 x_2 \quad \text{and} \quad m_2\ddot{x}_2 = -(K_2 + K_3)x_2 + K_3 x_1 \tag{2.119}$$

We expect that the lighter mass will experience greater displacements than the heavier mass. The above equations of motion can be simplified if we re-scale the coordinates by introducing the new primed coordinates.

$$x_1(t) = \frac{x_1(t)'}{\sqrt{m_1}} \quad \text{and} \quad x_2(t) = \frac{x_2(t)'}{\sqrt{m_2}} \tag{2.120}$$

Substitution into Eq. (2.119) provides a pair of equations that resemble those of a driven harmonic oscillator where the coupling spring's stiffness, K_3, is proportional to the driving force due to the motion of the other mass.

$$\frac{d^2x_1'}{dt^2} + \omega_1^2 x_1' = \kappa^2 x_2' \quad \text{and} \quad \frac{d^2x_2'}{dt^2} + \omega_2^2 x_2' = \kappa^2 x_1' \tag{2.121}$$

The new frequencies which appear above are related to the individual masses and springs or to the spring which couples their motion.

$$\omega_1^2 = \frac{(K_1 + K_3)}{m_1}; \quad \omega_2^2 = \frac{(K_2 + K_3)}{m_2}; \quad \kappa^2 = \frac{K_3}{\sqrt{m_1 m_2}} \tag{2.122}$$

If we seek the normal modes for this more general system, as before, those modes must correspond to simple harmonic motion of both masses oscillating harmonically at the same frequency. For that reason, we can substitute harmonic solutions of the same frequency, ω, for $x_1'(t) = \Re e\left[\widehat{\mathbf{x}}_1' e^{j\omega t}\right]$ and

$x_2'(t) = \Re e\left[\widehat{\mathbf{x}}_2' e^{j\omega t}\right]$. This substitution again converts the coupled differential expressions in Eq. (2.121) to coupled algebraic equations.

$$\begin{aligned} \left(\omega_1^2 - \omega^2\right)x_1' - \kappa^2 x_2' &= 0 \\ -\kappa^2 x_1' + \left(\omega_2^2 - \omega^2\right)x_2' &= 0 \end{aligned} \tag{2.123}$$

As before, we impose the requirement that the determinant of the coefficients for the coupled equations vanish so that nontrivial solutions can be identified.

$$\begin{vmatrix} \left(\omega_1^2 - \omega^2\right) & -\kappa^2 \\ -\kappa^2 & \left(\omega_2^2 - \omega^2\right) \end{vmatrix} = 0 \tag{2.124}$$

Evaluation of the determinant results in a secular (characteristic) equation for the frequencies which is biquadratic.

$$\omega^4 - \left(\omega_1^2 + \omega_2^2\right)\omega^2 + \omega_1^2\omega_2^2 - \kappa^4 = 0 \tag{2.125}$$

The solution can be obtained using the quadratic formula to determine ω^2 in terms of the oscillators' parameters as defined in Eq. (2.122).

$$\begin{aligned} \omega &= \left[\tfrac{1}{2}\left(\omega_1^2 + \omega_2^2\right) \pm \tfrac{1}{2}\sqrt{\left(\omega_1^2 - \omega_2^2\right)^2 + 4\kappa^4}\right]^{1/2} \\ &= \frac{1}{\sqrt{m_1 m_2}}\left[\frac{1}{2}[(\mathrm{K}_1 + \mathrm{K}_3)m_2 + (\mathrm{K}_2 + \mathrm{K}_3)m_1] \pm \frac{1}{2}\sqrt{[(\mathrm{K}_1 + \mathrm{K}_3)m_2 - (\mathrm{K}_2 + \mathrm{K}_3)m_1]^2 + 4\mathrm{K}_3^2 m_1 m_2}\right]^{1/2} \end{aligned} \tag{2.126}$$

As before, we have identified the two unique frequencies that allow both masses to oscillate harmonically at the same frequency. If those frequencies are substituted back into the algebraic expressions of Eq. (2.123), the ratios of x_1' to x_2' can be determined.

Since the masses and springs are now different, the normal mode displacements will be more complex than the simple solutions of Sect. 2.7.2. The normal mode descriptions for $x_1(t)$ and $x_2(t)$ will still be classified as symmetric and antisymmetric, but the amplitudes will no longer be identical. The reader is referred to the treatment by Morse for those solutions in their full algebraic ugliness [33].

2.7.6 Driven Oscillators, Level Repulsion, and Beating

In the limit that $\mathrm{K}_3 = 0$, so that there is no coupling between the two masses, it is comforting that Eq. (2.126) regenerates the frequencies of the two isolated oscillators. It is instructive to consider the case of two oscillators of the same frequency, $\omega_1 = \omega_2$, that are weakly coupled so that $\kappa \ll \omega_1$. Under those circumstances, we can use a Taylor series to expand Eq. (2.126) and provide the symmetric and antisymmetric modal frequencies.

$$\begin{aligned} \omega_s &= \sqrt{\omega_1^2 - \kappa^2} = \omega_1\sqrt{1 - \frac{\kappa^2}{\omega_1^2}} \cong \omega_1 - \frac{\kappa^2}{2\omega_1} \\ \omega_a &= \sqrt{\omega_1^2 + \kappa^2} = \omega_1\sqrt{1 + \frac{\kappa^2}{\omega_1^2}} \cong \omega_1 + \frac{\kappa^2}{2\omega_1} \end{aligned} \tag{2.127}$$

The fact that even weak coupling results in a shift of the natural frequencies of the two uncoupled oscillators is known as *level repulsion*. That term is motived by analogy to splitting of energy levels in atomic systems when such levels become coupled, possibly by electromagnetic interactions.[28] At first glance, the symmetric frequency, ω_s, should not be shifted, based on the normal modes of the simpler system of equal masses and springs analyzed in Sects. 2.7.1 and 2.7.2. The fact that ω_s is downshifted is a consequence of the assumption that the motion of our simpler system was initiated by making two identical displacements. If one of the masses is driven, then the coupling spring must be displaced initially to permit the transfer of energy from the driven mass to the other mass.

It is easy to visualize this periodic energy transfer in the weak-coupling limit, with $\omega_s \cong \omega_c$, and $n(\omega_s - \omega_a) = (\omega_s + \omega_a)$, where n is chosen to be an integer. Using the identities of Eq. (2.117), and assuming the oscillations start with $x_1(0) = \alpha$ and $x_2(0) = 0$, the subsequent evolution of $x_1(t)$ and $x_2(t)$ can be written as the product of two trigonometric functions.

$$\begin{aligned} x_1(t) &= \alpha\left[\cos\frac{(\omega_a - \omega_s)}{2}t\right]\left[\cos\frac{(\omega_a + \omega_s)}{2}t\right] \\ x_2(t) &= \alpha\left[\sin\frac{(\omega_a - \omega_s)}{2}t\right]\left[\sin\frac{(\omega_a + \omega_s)}{2}t\right] \end{aligned} \tag{2.128}$$

The second term in both products represents a simple harmonic oscillation at a frequency which is the average of ω_a and ω_s. The first terms define an amplitude modulation envelope that closes (i.e., goes to zero) with a period that is integer multiples of $\pi/(\omega_a - \omega_s)$. Since the peaks in the envelopes for x_1 and x_2 are 90° out-of-phase, it is apparent that all of the energy starts in m_1, but at a time $\pi/2(\omega_a - \omega_s)$, m_1 has come to rest, and all of the energy has been concentrated in m_2, with the situation reversing itself with a period $\pi/(\omega_a - \omega_s)$. This periodic modulation is known as *beating* and is illustrated in Fig. 2.27, where $n = 12$.

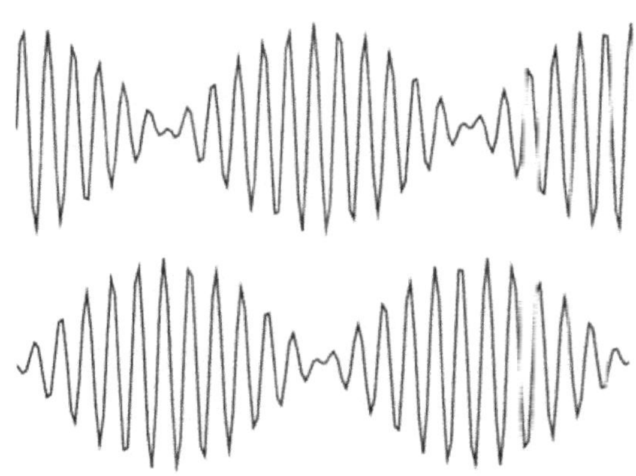

Fig. 2.27 Time evolution of two weakly coupled oscillators, described by Eq. (2.128), is plotted to illustrate the periodic exchange of energy. The upper plot is $x_1(t)$ and the lower is $x_2(t)$. In this figure, the oscillation frequency $(\omega_a + \omega_s)/2$ is 12 times the frequency difference $(\omega_a - \omega_s)$. Initially, all of the motion is in m_1

[28] As mentioned in Sect. 2.3.3, the atomic energy levels are related to the frequencies of light radiated during a level transition through Planck's equation: $E/\omega = \hbar$.

2.7.7 String of Pearls

We can now extend this analysis of coupled oscillators to more than two masses. To make it easier to visualize the motion of several coupled masses, we will first solve the problem of an undamped, single degree-of-freedom oscillator consisting of a single mass suspended at the center of a taught string. The location of the mass on the string will be specified by the x axis, and the displacement of the mass from its position of static equilibrium will be given by its y coordinate. A point mass, m, is shown in Fig. 2.28, attached to the center of a limp string[29] of length $L = 2a$, and is subject to a tension, T. Both ends of the string are immobilized: $y(0) = y(2a) = 0$. If the mass is displaced from equilibrium by a distance, y_1, at the center of the string, there will be a net force on that mass that will be opposite to y; hence, the mass will obey Hooke's law. We assume that the string is massless, $m_s = 0$, and that the tension in the string has a value, T.

It is easy to demonstrate that tension, T, in the string exerts a linear restoring force on the mass if the vertical displacement from equilibrium, y_1, is small enough that $y_1/a \ll 1$. As shown in Fig. 2.28, a displacement, y_1, creates an angle, $\theta = \tan^{-1}(y_1/a)$, between the displaced mass and the former equilibrium position of the string. If the displacement, y_1, is small, the change in the length of the string δL will be second-order in the displacement, y_1, so the increase in tension due to the stretching of the string can be ignored in a first-order (Hooke's law) analysis.

$$\delta L = 2\sqrt{\left(a^2 + y_1^2\right)^{1/2} - a^2} = 2a\sqrt{\left(1 + \frac{y_1^2}{a^2}\right)^{1/2} - 1} \cong a\left(\frac{y_1}{a}\right)^2 \tag{2.129}$$

The restoring force will be linear in the displacement for small enough displacements that $\theta = \tan^{-1}(y_1/a) \cong y_1/a$. From Fig. 2.28, Garrett's First Law of Geometry can be used to calculate the vertical component of the force, F_y, due to the tension, T, of the string.

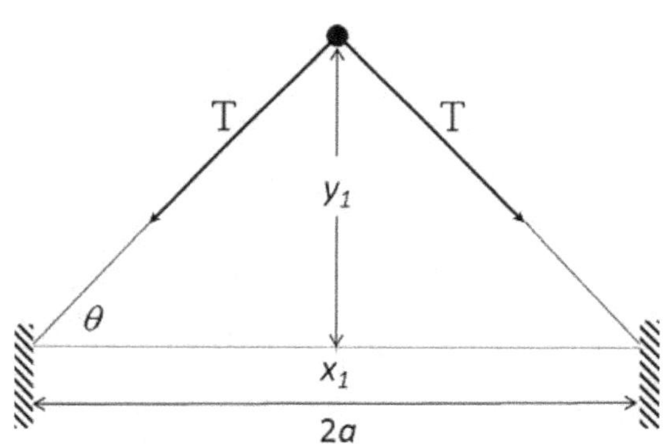

Fig. 2.28 A limp, massless string of length, $L = 2a$, is under tension, T, and has a point mass, m, attached at its center and is fixed at both ends. The mass has been displaced from its equilibrium position by a distance y_1, so the string makes an angle, $\theta = \tan^{-1}(y_1/a)$, with the horizontal. The two vertical components of the tension provide a linear restoring force: $F_y = -2\text{T} \sin\theta \cong -(2\text{T}/a)y_1$

[29] For this case, we will assume that the string has no flexural rigidity and that the restoring force is entirely due to the tension applied to the string. In Sect. 5.5, we will examine the stiff string, which is under tension but also exhibits flexural rigidity.

$$F_y = -2T\sin\theta \cong -\left(\frac{2T}{a}\right)y_1 \tag{2.130}$$

This "mass on a string" is another example of a simple harmonic oscillator that moves in one dimension, along the y axis, and has an effective stiffness constant, $K_{eff} = (2T/a)$. To demonstrate the utility of this configuration for the representation of coupled oscillators, Fig. 2.29 duplicates the normal mode displacements for masses and springs from Eq. (2.111).

In Fig. 2.30, we extend this mass and string model to multiple masses. At equilibrium, each mass is separated from its adjacent masses by a horizontal distance, a. The position of the ith mass is $x_i = ai$, where $1 \le i \le N$. For N masses, the total length of the string is $L = (N + 1)\,a$. Since the string is rigidly fixed at both ends, $y(0) = y(x_{N+1}) = 0$.

The vertical component of the force, F_i, on any mass, m_i, depends only upon the position of that mass with respect to its nearest neighbors.

$$F_i = T(\sin\phi - \sin\theta) \cong T\left(\frac{y_{i+1} - y_i}{a} - \frac{y_i - y_{i-1}}{a}\right) = -\frac{2T}{a}y_i + \frac{T}{a}(y_{i+1} + y_{i-1}) \tag{2.131}$$

Fig. 2.29 The same two normal modes for the simple coupled oscillators with two identical masses and three identical springs that were calculated in Eq. (2.111) are reproduced here as two masses on a tensioned string. The symmetric mode is shown on the left, and the antisymmetric mode, with frequency that is $\sqrt{3}$ greater than the symmetric mode, is shown at the right

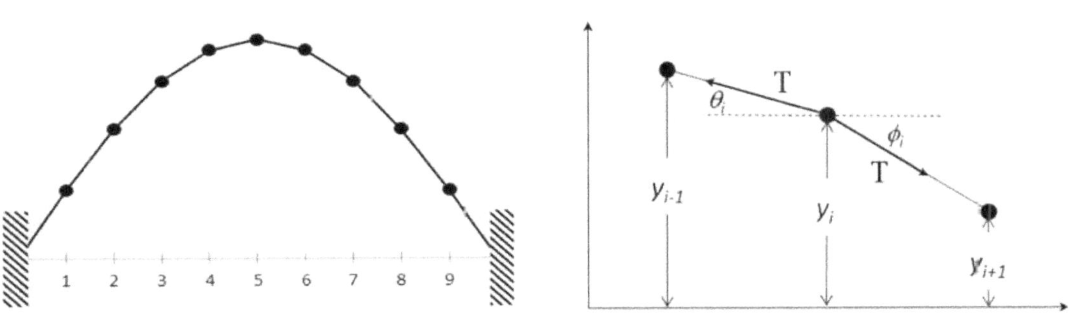

Fig. 2.30 *(Left)* $N = 9$ masses are spaced uniformly on a string of length, $L = (N + 1)\,a$. The coordinate of each mass is specified by its position along the x axis when the string is at its equilibrium position. Each mass is constrained to move only in the vertical direction, and each mass is displaced from its equilibrium position by an amount y_i. *(Right)* The net vertical force on each mass, m_i, is determined by the vertical displacements of its nearest neighbors, y_{i-1} and y_{i+1}, relative to the displacement, y_i. The vertical component of the force on the ith mass is determined by the tension in the string, T, and by the angles $\theta_i = \tan^{-1}[(y_i - y_{i-1})/a]$ and $\phi_i = \tan^{-1}[(y_{i+1} - y_i)/a]$

If we assume all masses are equal, $m_i = m$, and we apply Newton's Second Law of Motion to the ith mass, abbreviating (T/a) as K, the dynamical equations describing the vertical motion of each of the N masses can be written in a compact form.

$$m\ddot{y}_i + 2\mathrm{K}y_i - \mathrm{K}\left(y_{i+1} + y_{i-1}\right) = 0 \quad \text{for} \quad i = 1, 2, \ldots, N \tag{2.132}$$

This set of coupled differential equations could be solved by the techniques used previously in Sects. 2.7.3 and 2.7.5. If we define $\lambda \equiv 2 - (m\omega^2 a/\mathrm{T})$, then the $N \times N$ determinant of the coefficients, D_N, for the algebraic equations, which resulted from the assumption that all masses vibrating in any normal mode must oscillate at the same frequency, ω, has the following form that is characteristic of *nearest-neighbor interactions*:

$$D_N = \begin{vmatrix} \lambda & -1 & 0 & 0 & 0 & \cdots & 0 \\ -1 & \lambda & -1 & 0 & 0 & \cdots & 0 \\ 0 & -1 & \lambda & -1 & 0 & \cdots & 0 \\ \vdots & \vdots & \vdots & \vdots & \vdots & \vdots & \vdots \\ 0 & 0 & 0 & 0 & 0 & \lambda & -1 \\ 0 & 0 & 0 & 0 & 0 & -1 & \lambda \end{vmatrix} \tag{2.133}$$

Fetter and Walecka solve this system of linear equations by development of a recursion relation for the minors [34], but we will take a different approach that anticipates our transition to continuous systems in the next chapter.

Since we have chosen our masses to be equal and to be equally spaced from each other along the string, it is possible to treat the vertical displacements of the masses as a function of the x coordinate: $y_i = y(x_i)$, where $x_i = ai$. Again, we seek normal mode solutions that require that every mass oscillates with the same frequency, ω, but now we will keep track of the amplitudes of vibration of the individual masses by introducing a constant, β, which has the units of inverse length [m^{-1}].

$$y(x_i, t) = Ae^{\,j(\beta x_i - \omega t)} \tag{2.134}$$

Substitution of that expression for displacements into Eq. (2.132) produces the algebraic Eq. (2.135) after cancellation of common exponential time factors.

$$-m\omega^2 + 2\mathrm{K} - \mathrm{K}\left(e^{j\beta a} + e^{-j\beta a}\right) = 0 \tag{2.135}$$

The resulting equation can be solved for the normal mode frequencies and simplified by use of the trigonometric identity, $2\sin^2(x/2) = (1 - \cos x)$, where $2\cos x = (e^{jx} + e^{-jx})$.

$$\omega^2 = \frac{2\mathrm{K}}{m}(1 - \cos\beta a) = \frac{4\mathrm{K}}{m}\sin^2\left(\frac{\beta a}{2}\right) \tag{2.136}$$

Although the frequencies, ω, are a continuous function of β, the normal mode frequencies will become discrete when we impose the end conditions that $y(0) = y(x_{N+1}) = 0$. To satisfy the condition at $x = 0$, we will subtract a solution with the form of Eq. (2.134), but with a negative β, since both the β and $-\beta$ parts will be zero when multiplied by $x = 0$.

$$y(x_i, t) = \Re\mathrm{e}\left[A\left(e^{j\beta x_i - j\omega t} - e^{-j\beta x_i - j\omega t}\right)\right] \tag{2.137}$$

That form of Eq. (2.134) is clearly zero for $x_o = 0$. For x_{N+1}, making $y_{N+1} = y[(N+1)a] = 0$ results in a condition that discretizes the values of β, simultaneously satisfying both boundary conditions.

$$e^{j\beta(N+1)a} - e^{-j\beta(N+1)a} = 2j[\sin\beta(N+1)a] = 0 \tag{2.138}$$

Only discrete values of β_n are now acceptable.

$$\beta_n(N+1)a = n\pi \quad \Rightarrow \quad \beta_n = \frac{n\pi}{a(N+1)}; \quad n = 1, 2, 3, \ldots, N \tag{2.139}$$

Substitution of the quantized values of β_n into Eq. (2.136) provides the normal mode frequencies.

$$\omega_n = 2\sqrt{\frac{k}{m}\sin^2\left(\frac{\beta_n a}{2}\right)} \tag{2.140}$$

Plugging the β_n back into Eq. (2.134) provides the time-dependent displacements for each mass.

$$y(x_i, t) = y_i(t) = A_n \sin\left(\frac{in\pi}{N+1}\right)\sin\omega_n t \tag{2.141}$$

This equation provides N solutions for each mass corresponding to the nth normal mode with normal mode frequency, ω_n. Each normal mode has an amplitude, A_n, that would depend upon the strength of the excitation of the nth mode. The vertical displacements of each mass for each mode are exaggerated for all nine modes in Fig. 2.31, with $N = 9$ and $A_n = 1$ for all n. The ratio of the frequency of each normal mode to the frequency of the $n = 1$ mode, ω_n/ω_1, is tabulated and plotted as a function of mode number, n, in Fig. 2.32.

The lowest-frequency modes correspond to in-phase motion of adjacent masses. The highest-frequency mode, $n = N$, corresponds to out-of-phase motion of the adjacent masses. By increasing the number of masses and decreasing the spacing to hold the overall length constant, $L = (N + 1)a$, the number of modes becomes infinite, and the string with masses attached has a uniform linear mass density, $\rho_L = (m/a)$. The modal frequencies are those given by Eq. (2.142) in this limit.

$$\lim_{N\to\infty}\omega_n^2 = \frac{4\mathrm{T}}{ma}\left[\frac{n\pi}{2(N+1)}\right]^2 = \frac{\mathrm{T}}{\rho_L}\left(\frac{n\pi}{L}\right)^2 \tag{2.142}$$

The quantity $(\mathrm{T}/\rho_L)^{1/2} \equiv c$ (for *celerity*) has the dimensions of speed [m/s]. As we will see when we transition to continuous systems in the next chapter, c is just the speed of transverse wave propagating on a uniform string. The low-frequency normal mode frequencies, $f_n = \omega_n/2\pi$, correspond to integer numbers, n, of half-wavelengths fitting between the fixed ends.

For the discrete case with $N = 9$, it is easy to imagine that the lowest-frequency mode at the top left in Fig. 2.31 approximates a half-sine waveform fit between the two fixed ends, i.e., $\lambda_1 = 2\,L$, and the second mode corresponds to two half-wavelengths (i.e., one entire wavelength) between the fixed ends, $\lambda_2 = L$.

2.8 The Not-So-Simple (?) Harmonic Oscillator

After 60 pages of discussion, 32 figures, and over 140 numbered display equations, the reader would be justified in requesting an explanation for why "simple" is used as a descriptor for the harmonic oscillator. The mass and the spring that are shown to function in its quasi-static limit are simple components. The description of the time history of the motion of a mass connected to a spring involves only simple trigonometric functions with constants that could be fit to specific initial conditions. The expressions for the kinetic energy stored by the mass, $KE = (½)mv^2$, and the potential energy stored by

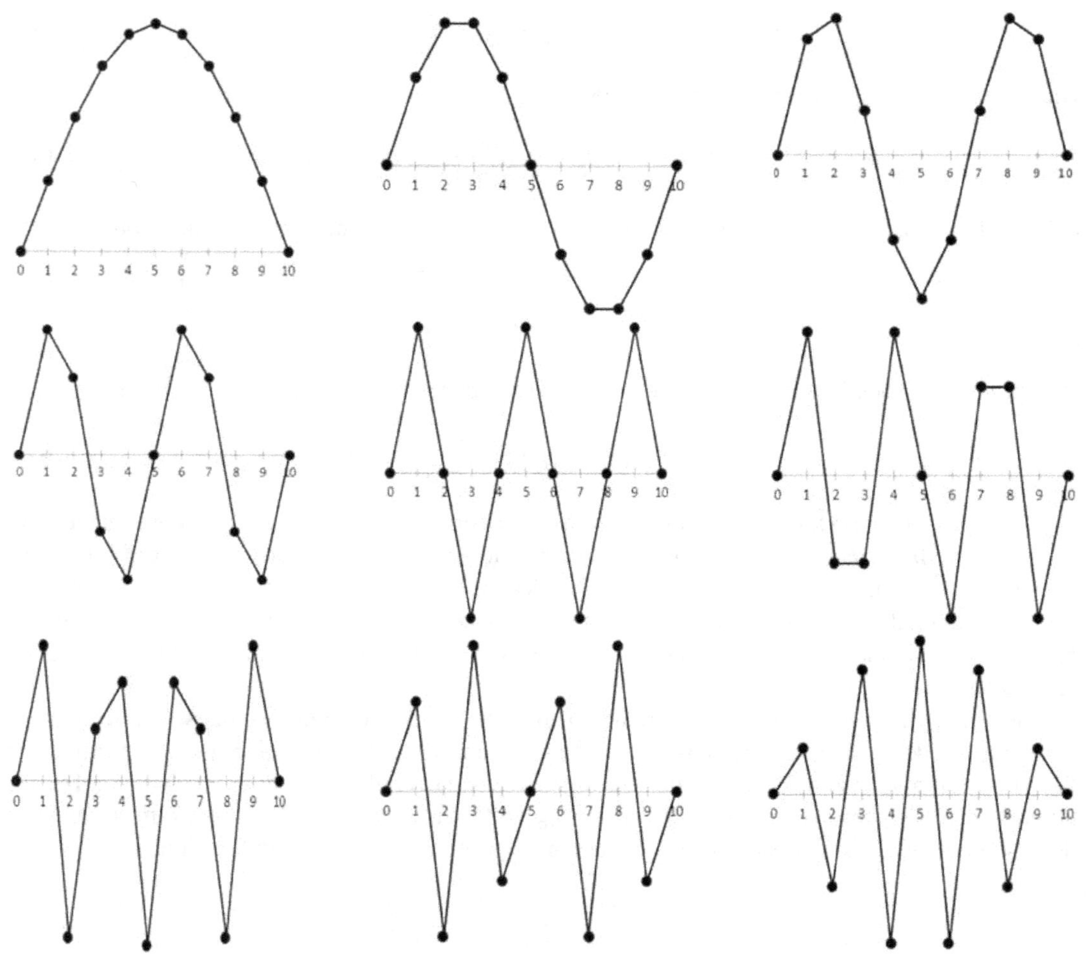

Fig. 2.31 Mode shapes for the normal modes for the nine equal masses (*solid circles*) on a taught string (*straight lines*). The solid circles at $N = 0$ and $N = 10$ represent the fixed ends of the string. The amplitudes are greatly exaggerated for clarity. Please remember these results are only valid for $y \ll a$. The number of zero-crossings increases with increasing mode number. The fundamental ($n = 1$) mode has no zero-crossings, and each subsequent mode has a number of zero-crossings that is one less than the mode number, n. The frequencies of these normal modes are tabulated and plotted in Fig. 2.32

Fig. 2.32 The normalized modal frequencies, ω_n, based on Eq. (2.140), for nine equal masses, are tabulated at the left and plotted vs. mode number, n, at the right

Mode	Freq.
1	1.000
2	1.975
3	2.902
4	3.757
5	4.520
6	5.172
7	5.696
8	6.080
9	6.238

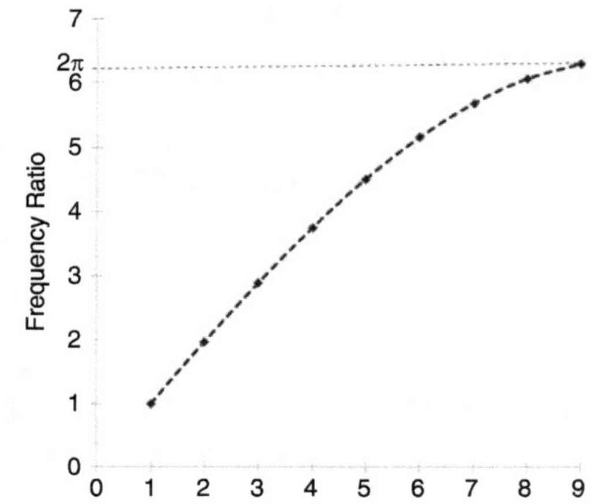

the spring, $PE = (½)Kx^2$, are equally simple, and a gravitational offset had no influence on the dynamical behavior. Even Rayleigh's exploitation of the virial theorem for calculation of oscillatory frequencies would have to be judged "simple," at least in hindsight.

Once the mechanical resistance was added as a means to dissipate energy, the analysis of the oscillator's behavior left the purely mechanical perspective and entered the realm of thermodynamics – the dissipated energy had to leave "the system" and make contact with "the environment." Although such calculations may be unfamiliar, their result incorporated the Equipartition Theorem, which is also quite simple – each quadratic degree of freedom, on average, gets its (½) k_BT share of energy at thermal equilibrium. The steady-state response to systematic injection of energy into the oscillator, by either forcing the mass or jiggling the attachment point of the spring, turned out to be determined entirely by the oscillator's mechanical impedance. The approach to steady state involved addition of solutions to the freely decaying oscillator.

The experimental determination of the stiffness and moving mass for the electrodynamic loudspeaker example was achieved by simply adding some pocket change (i.e., nickels) and recording the frequency shift due to the mass loading. The mechanical resistance was determined by the free-decay time. All three parameters were determined simply to an accuracy of better than $\pm 1\%$ by least-squares fitting, while the force factor (i.e., the $B\ell$-product that determined the force produced by the electrical current flowing in the voice coil) required only a DC electrical resistance measurement and an electrical impedance measurement at the resonance frequency.

When two or more (undamped) oscillators were coupled together, the mathematical description became more complicated, but the simplest case could be analyzed intuitively, based on the previous understanding of individual harmonic oscillators. The identification of the normal modes and their frequencies could be algebraically messy when different masses are joined by springs with different stiffnesses. The time histories of the motion might appear chaotic if the normal mode frequencies were not commensurate (i.e., related by the ratio of integers). One must admit those systems constitute a fairly complicated combinations of otherwise simple harmonic oscillators. Although the coupled solutions were the most complicated results derived in this chapter, the next chapter will simplify the situation once again by treating large numbers of coupled oscillators as a continuum, so an overall "shape" is described instead of the individual displacements of point masses.

Talk Like an Acoustician

Point-mass	Mechanical reactance
Dynamical equation	Resistance-controlled regime
Equation of state	Stiffness-controlled regime
Frequency	Mass-controlled regime
Period	Rayleigh line shape
Angular frequency	Lenz's law
Radian frequency	Faraday's law of induction
Quasi-static approximation	Decibel
Virial theorem	Phase-locked loop
Rayleigh's method	Ring down
Adiabatic invariance	Reversible transducer
Dashpot	Vibration isolator
Mechanical resistance	Transmissibility
Quality factor	Seismometer
Logarithmic decrement	Geophone
Natural frequency	Normal modes

Free-decay frequency	Normal mode frequencies
Underdamped	Symmetric mode
Overdamped	Antisymmetric mode
Critical damping	Characteristic equation
Equipartition Theorem	Secular equation
Half-power bandwidth	Normal coordinates
Full width at half maximum	Basis functions
Equivalent noise bandwidth	Level repulsion
Effective bandwidth	Beating
Mechanical impedance	Nearest-neighbor interactions
Resonance	Celerity

Exercises

Each problem that I solved became a rule which served afterwards to solve other problems. (Rene Descartes)

1. **U-tube oscillations.** The tube shown in Fig. 2.33 is filled with a liquid that has a uniform mass density, ρ. The total length of the fluid column is L, and the cross-sectional area of the tube, S, is constant. When the fluid is displaced from its equilibrium position by a distance, x (as shown), the column of liquid executes simple harmonic motion under the influence of a constant gravitational acceleration, g.

   (a) *Period.* Calculate the period, T, of the oscillations, neglecting damping.

   (b) *Quality factor.* If the oscillations are observed to decrease by 20% during each cycle, calculate Q.

2. **Coefficient of friction oscillator.** Two cylindrical rollers are located a distance $2a$ apart. Their bearings are anchored. The rollers rotate at an angular speed ω_r in opposite directions, as shown in Fig. 2.34. On the top of the rollers rests a bar of length, l, and weight, $W = mg$.

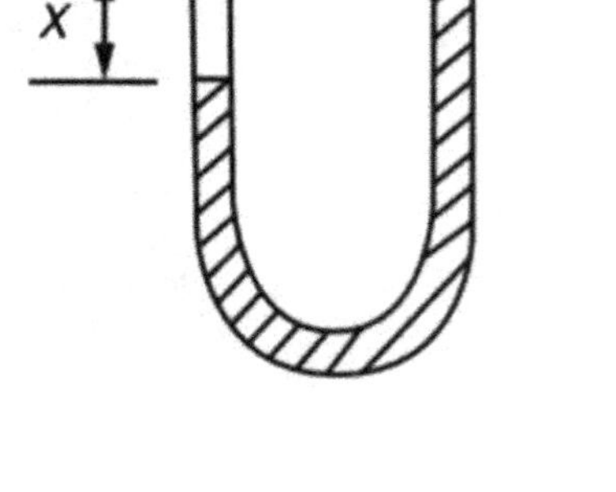

Fig. 2.33 Liquid-filled U-tube

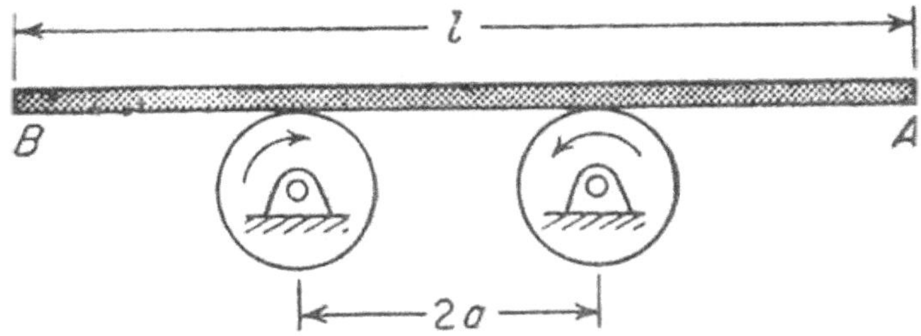

Fig. 2.34 Bar supported on two counter-rotating wheels with centers separated by $2a$

Assuming a dry coefficient-of-friction μ between the rollers and the bar, the bar will oscillate back and forth longitudinally executing simple harmonic motion.[30] The restoring force will be the *difference* between the frictional forces in the x direction, F_y (along the bar), due to the weight (normal force) of the bar in the y direction, $F_y = \mu mg$, of the bar on each roller. (Note that if the end of the bar is closer to one roller, the weight of the bar on the other roller will be larger and will push the bar back toward the center.) Calculate the frequency of the bar's oscillations when the wheels rotate as shown in Fig. 2.34.

3. **Mass loading frequency shift.** A mass-spring system is observed to have a natural frequency of 10.0 Hz. When a mass of 10.0 grams is added to the original mass, m_o, the frequency decreases to 8.0 Hz. Determine the original mass and the spring's stiffness constant, K.
4. **Loudspeaker in an enclosure.** A loudspeaker has a free cone resonance frequency of 55.5 Hz. If a US nickel (mass = 5.00 g) is placed on the speaker's dust cap, the resonance frequency is reduced to 46.3 Hz. After the nickel is removed, the speaker is installed in a sealed enclosure, and the frequency increases to 80.0 Hz. What is the equivalent stiffness (in Newtons/meter) provided by the elasticity of the air contained within the loudspeaker enclosure?
5. **Spar buoy oscillations.** An oceanographic instrument package is placed at the bottom of a spar buoy shown in Fig. 2.35. The mass (in air) of the entire buoy (instruments, batteries, tube, and antenna) is 150 kg. The tubular portion (5 m long and 25 cm diameter) is partially submerged. You may assume the density of the water, $\rho = 1000\ \text{kg/m}^3$, and the acceleration due to gravity, $g = 9.8\ \text{m/s}^2$.
   (a) *Equilibrium depth.* How far from the surface of the water is the **top** of the 5.0 m long tubular section when it is in equilibrium (at rest)?
   (b) *Frequency.* What is the natural frequency of vibration of the buoy if it is displaced (vertically) from its equilibrium position and released?
   (c) *Damping.* When the buoy is displaced from equilibrium, the oscillations decay to *1/e* of their initial value in 30 s. Determine the viscous (mechanical) resistance, R_m, which the water provides to damp the oscillations.
   (d) *Hydrodynamic mass correction.* The water increases the buoy's inertia by adding a mass that is about half that of a sphere, with density ρ (see Sect. 12.5.1) and with the same diameter as the buoy. Recalculate your result for the frequency from part (b) including the effective mass of the water that is driven by the motion of the cylinder.
   (e) *Relative motion.* Assume that there is a "ring" drawn at the static (calm) water level on the tubular section at the normal equilibrium position which you calculated in part (a) of this problem. If a sinusoidal swell with peak-to-trough amplitude of 1.0 m passes the buoy position every 20 s (i.e., a wave with a 20 s period), what is the greatest (peak) distance which the ring

[30] The operation of such an oscillator is provided by Dan Russell in a YouTube video: http://www.youtube.com/watch?v=46lk2FXzwT8

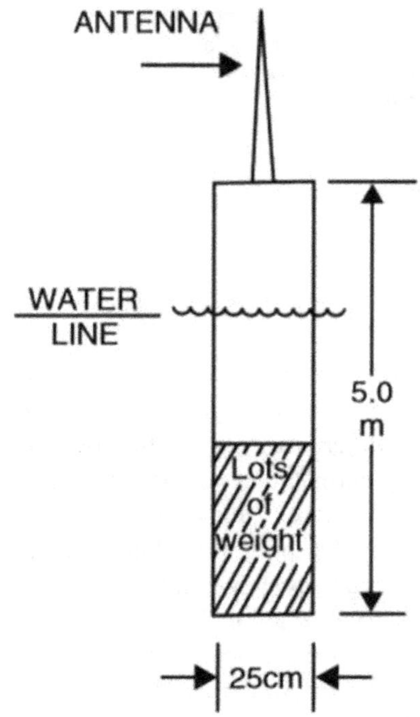

Fig. 2.35 Spar buoy

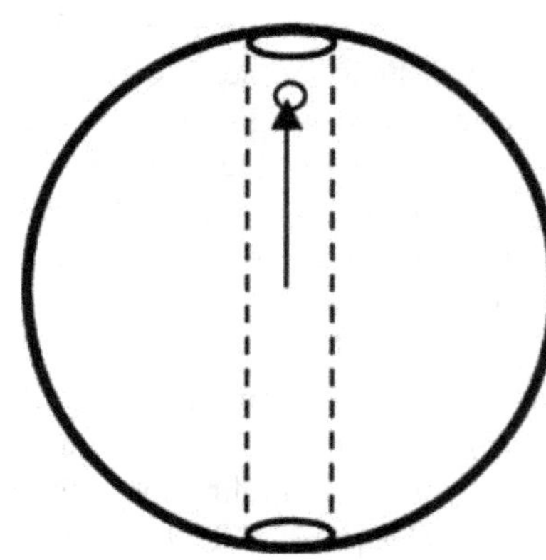

Fig. 2.36 The Earth with a hole passing through its diameter

will be above or below the instantaneous (moving) air-sea interface? Report the amplitude of the motion of the buoy **relative** to the moving water surface, not with respect to the static (calm) water level.

6. **Balls to Bunbury**. Assume the Earth is the large diameter solid sphere of uniform mass density, $\rho_{\oplus}$, shown in Fig. 2.36. A small-diameter cylindrical hole has been bored through the sphere along a diameter. A ball of mass, m, with a diameter smaller than the hole, is dropped from the surface of the sphere down the hole.
   At a distance, r, from the center of the Earth, the magnitude of the gravitational potential energy of the ball, $|U(r)|$, is determined by the mass of the Earth, $M(r) = 4\pi\rho r^3/3$, which is included within a radius less than r. (Note the net force on the small mass due to the mass of the Earth that is at a radius greater than r is completely cancelled. Why?)

(a) *Effective stiffness.* Use Newton's Law of Universal Gravitation, as expressed in either Eq. (2.23) or Eq. (2.25), to determine an effective stiffness constant. Newton's Universal Gravitation Constant, $G = 6.6726 \pm 0.0005 \times 10^{-11}$ m^3/s^2-kg. For the planet Earth, $R_{\oplus} = 6371$ km, the average mass density is $\rho_{\oplus} = 5520$ kg/m^3.

(b) *Oscillation period.* Find the period of oscillation. Express your answer in minutes.

(c) *Orbital period.* Using Eq. (2.23), show that the period you calculated in part (b) is identical to the period of the same mass (or any mass!) in a circular orbit around Earth at a height above the surface $h \ll R_{\oplus}$.

(d) *Superposition.* How is the linear motion of the mass through the Earth related to the orbital motion around the Earth?

7. **An asymmetrically placed mass on a taut string.** A mass, m, is attached on a string of tension, T, and length, L. The mass is located a distance, b, from the left-side rigid support (hence $L-b$ from the right rigid support).
   (a) *Natural frequency.* What is natural frequency if we assume that the string is massless, $m_s = 0$?
   (b) *String mass correction.* Using Rayleigh's method, calculate the natural frequency if $m_s = m/10$, and report the result as a ratio with respect to the natural frequency you calculated in part (a).
8. **Vibration isolator.** A large air conditioning unit is loaded on to a vibration isolation platform. The platform that supports the chiller, which is attached to the isolators, sinks by 2.0 cm when the air conditioner is placed on it.
   (a) *Natural frequency.* What is the natural frequency (in Hz) of the chiller-isolator system? (You may assume that the acceleration due to gravity is 9.8 m/s^2.)
   (b) *Transmissibility.* If the unbalance in the chiller's motor produces an oscillating force in the vertical direction of $F_1 = 1000\ N_{peak}$ at 60 Hz, what is the reduction in the force that the platform applies to the rigid foundation on which it mounted? Report the result as both a decimal ratio and as a dB reduction.
9. **Cantilever spring.** The beam in Fig. 2.37 was drawn by Galileo. When the 50 kg mass "E" was placed on the hook "C," the hook was displaced downward by $\Delta = 4$ cm.

Fig. 2.37 Beam supported by a wall with mass "E" attached at the end

(a) *Frequency*. Neglecting the mass of the beam, what is the frequency of vibration?
(b) *Beam mass correction*. The deflection of the beam, $y(x)$, as measured from the rigid support at $x = 0$ is given by the following curve (from Eq. (4.36):

$$y(x) = \frac{3\Delta}{2L^3}\left(Lx^2 - \frac{x^3}{3}\right)$$

If the mass of the beam is 10 kg, use Rayleigh's method to recalculate the vibration frequency.

10. **Atomic force microscopy.** The beam in Fig. 2.38 is the sensing element of an atomic force microscope.[31] The cantilever is fabricated from silicon nitride (Si_3N_4) and is $w = 30$ μm wide, $L = 100$ μm long, and $t = 3$ μm thick. For this problem, the mass of the tip can be neglected. The stiffness of this cantilever can be related to the Young's modulus of the Si_3N_4, $E = 310$ GPa, by Eq. (2.143).

$$\mathrm{K} = \frac{F}{\Delta} = \frac{Ew}{4}\left(\frac{t}{L}\right)^3 \tag{2.143}$$

(a) *Units*. Show that the units on the right-hand side of Eq. (2.143) correspond to the units of stiffness.
(b) *Stiffness*. What is the value of the stiffness of the Si_3N_4 micromachined cantilever?
(c) *Thermal noise*. What is the root-mean-square displacement, $x_{rms} = \langle x^2\rangle^{1/2}$, of the tip if the cantilever is at a temperature of 20 °C = 293 K?
(d) *Minimum detectible force*. What force on the tip would be required to produce a static deflection, Δ, that is equal to the root-mean-square thermal fluctuation in the position of the tip, x_{rms}, that you calculated in part (c).

Fig. 2.38 Silicon cantilever with spike

[31] Image courtesy of *SecretDisk* under their Attribution-Share Alike 3.0 license.

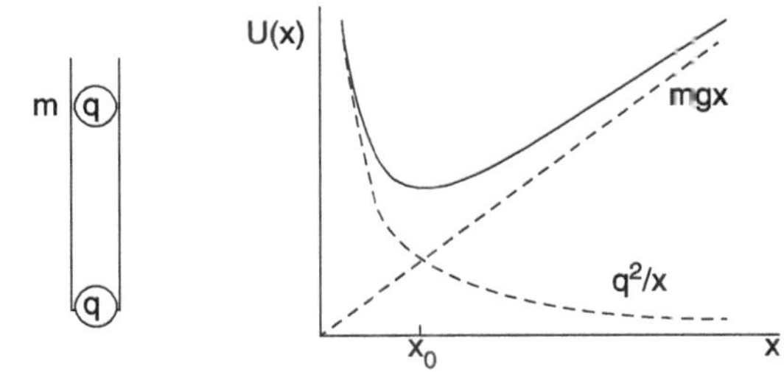

Fig. 2.39 Two electrically charged spheres are constrained to move only in the vertical direction. The potential energy, $U(x)$, of the upper sphere is due to the combination of the force of gravity and the force of electrostatic repulsion. (Figure courtesy of J. D. Maynard [23])

11. **Electrostatic levitation.** Two spheres, made of an electrically insulating material, are contained within a frictionless insulating tube as shown in Fig. 2.39. (You can imagine this as two tiny Styrofoam balls in a drinking straw.) The lower sphere is rigidly attached to the bottom of the tube, and the upper sphere is free to move up or down, without significant damping, only in the vertical direction.

    The mass of each sphere is 0.10 g. The electrostatic force, F, between the spheres is determined by the charges on the spheres, Q_{upper} and Q_{lower}; the separation between their centers, x; and a constant, $\varepsilon_o = 8.85 \times 10^{-12}$ Farads/meter, known as the permittivity of free space. In this case, the force between the spheres is repulsive, since the charge on both spheres is positive.

$$F = \frac{1}{4\pi\varepsilon_o}\frac{Q_{upper}Q_{lower}}{x^2} \tag{2.144}$$

At equilibrium, the centers of the spheres are separated by 1.0 cm. What is the frequency of oscillation of the upper sphere if it is displaced from its equilibrium position by a very small amount? (Let the acceleration due to gravity, $g = 9.8$ m/s^2.)

12. **Lennard-Jones potential.** This potential function, sometimes known as the "6–12 potential" of Eq. (2.145), is plotted in Fig. 2.40. It describes the interaction of two inert gas atoms as a function of their separation and is useful in calculations of scattering or determination of the temperature of condensation from a gas to a liquid. It was first proposed by John Lennard-Jones in 1924 [35].

$$V(r) = 4\varepsilon\left[\left(\frac{\sigma}{r}\right)^{12} - \left(\frac{\sigma}{r}\right)^{6}\right] \tag{2.145}$$

The term proportional to r^{-12} represents an empirical fit to the hard-core repulsion produced by Pauli exclusion at short ranges when the atoms are close enough that the electron orbitals overlap. The r^{-6} term represents the van der Waals attraction due to the mutual interactions of the fluctuating dipole moments.

(a) *Equilibrium separation*. Determine the equilibrium separation of two atoms, r_m, in terms of σ and ε.

(b) *Argon separation*. For argon, $\sigma = 3.42$ Å (1 Å $= 10^{-10}$ m) and $\varepsilon = 124$ K (1 K $= 1.38 \times 10^{-23}$ J). Determine the equilibrium separation of two argon atoms.

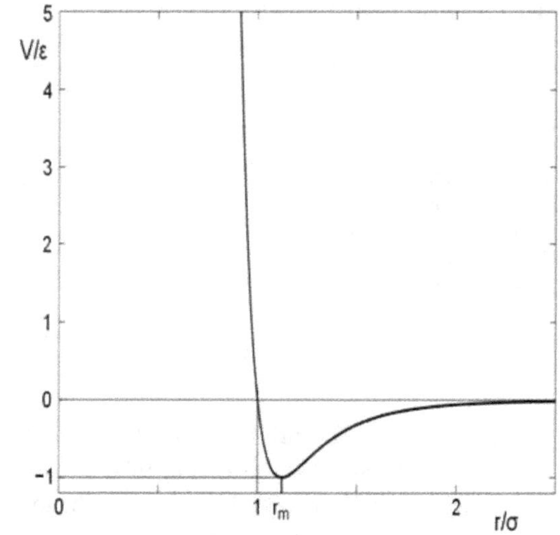

Fig. 2.40 Lennard-Jones interatomic potential as a function of scaled atomic spacing, r/σ

(c) *Antisymmetric vibrational frequency*. The atomic mass of an argon atom is 39.948 amu (1 amu = 1.66×10^{-27} kg). Determine the stiffness constant, $K = d^2V/dr^2$, and use that result to calculate the vibrational frequency of two isolated argon atoms about their equilibrium separation.

13. **Ionic bonding.** Hydrogen chloride (HCl) is a diatomic molecule that forms an ionic bond. The force that bonds the H^+ ion to the Cl^- is the electrostatic attraction between the oppositely charged ions. Of course, Pauli repulsion will separate the two ions if they are drawn too close together. The potential, $V(r)$, for this interaction is the sum of the electrostatic attraction and the Pauli repulsion.

$$V(r) = -\frac{e^2}{4\pi\varepsilon_o r} + \frac{B}{r^9} \tag{2.146}$$

The permittivity of free space is $\varepsilon_o = 8.85 \times 10^{-12}$ Farads/meter [F/m], and the charge on a single electron is $e = 1.602 \times 10^{-19}$ C.

(a) *Repulsion constant*. If the equilibrium separation of centers of the two ions is $r_m = 1.30$ Å (1 Å = 10^{-10} m), determine the value of the constant, B, in Eq. (2.146).

(b) *Effective stiffness*. What is the effective stiffness, K, of the bond about the equilibrium separation, r_m?

(c) *Vibrational frequency*. The mass of the chlorine ion is about 35 times greater than the hydrogen ion. Assume that the position of chlorine atom is fixed, and determine the frequency of oscillation of the hydrogen ion if it is displaced by a small distance, $x_1 \ll r_m$.

(d) *Reduced mass*. Since the mass of the chlorine ion is not infinite, the oscillation actually takes place about the center of mass, which remains at rest. This can be accounted for by using the reduced mass, μ, given in Eq. (2.147), rather than using the mass of the hydrogen ion, so $f_o = (K/\mu)^{1/2}/2\pi$.

$$\mu = \frac{M_{Cl^-} M_{H^+}}{M_{Cl^-} + M_{H^+}} \tag{2.147}$$

Fig. 2.41 Damped, driven harmonic oscillator

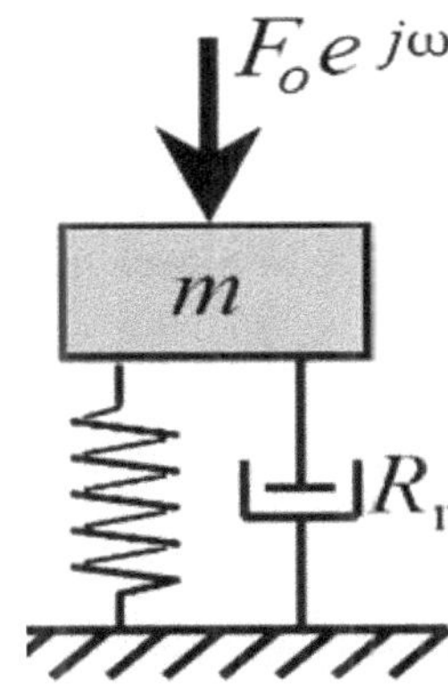

By what percentage is this frequency shifted from the value that assumed the chlorine atom had infinite mass?

14. **Damped driven SHO.** The system in Fig. 2.41 has a mass, $m = 0.100$ kg, that is attached to a rigid foundation with a spring of stiffness, K = 1000 N/m, and a dashpot with mechanical resistance, R_m, that is to be determined in part (b) of this problem.
    (a) *Natural frequency.* What is the natural frequency of the mass-spring system, in hertz, if the damper is not present ($R_m = 0$)?
    (b) *Mechanical resistance.* After an impulsive excitation, it is found that amplitude of free vibration of the mass decays to e^{-1} of its value in 2.0 s. Determine the value of R_m and report your results in kg/s.
    (c) *Driven response.* If the mass is driven with a time-harmonic force of amplitude 10 N, at a frequency of 20 Hz, $F(t) = 10e^{j40\pi t}$, what is:
        (i) The magnitude of the time-harmonic displacement of the mass?
        (ii) The phase of the mass's velocity with respect to that of the driving force?
        (iii) The time-averaged power dissipated in R_m?
15. **Damped harmonic oscillator.** A 1.00 kg mass is suspended from a spring. A 0.40 kg mass is hung by a string below the 1.00 kg mass. The additional mass causes the spring to be extended by an additional 4.0 cm. The string is cut at $t = 0$ and the additional mass drops to the floor. The resulting harmonic oscillations of the 1.00 kg mass are observed to decay by e^{-1} in 1 s.
    (a) *Spring stiffness.* What is the spring's stiffness?
    (b) *Mechanical resistance.* What is the oscillator's mechanical resistance, R_m?
    (c) *Displacement.* Equation (2.43) provides an expression for the motion of the mass, $x(t)$. Determine the values of the constants C, τ, ω_d, and ϕ that describe the motion for $t \geq 0$.
16. **Five identical masses.** Determine the five modal frequencies, and sketch the corresponding five mode shapes (like Fig. 2.31) for five masses, each with mass, m, that is equally spaced along a massless string of total length, $L = 6a$, that is fixed at both ends (i.e., $y(0) = y(6a) = 0$) and is stretched to a tension, T. You may assume that the vertical displacements of all of the masses are small compared to their separation (i.e., $y \ll a$). The normal mode frequencies can be reported in terms of T, a, and m.
17. **Coupled oscillators.** Nine objects of equal mass are joined by ten identical springs to two rigid boundaries as shown in Fig. 2.42. The masses are constrained to move along the line joining them. In Fig. 2.42, the equilibrium positions of the numbered masses are indicated by the equally spaced tick marks. In that diagram, the nine masses are shown at a particular phase of their collective

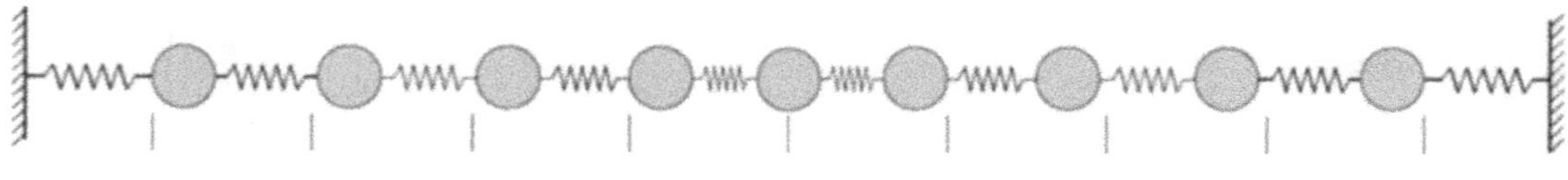

Fig. 2.42 Nine identical masses connected to rigid boundaries by nine identical springs

Fig. 2.43 Two pendula coupled by a spring

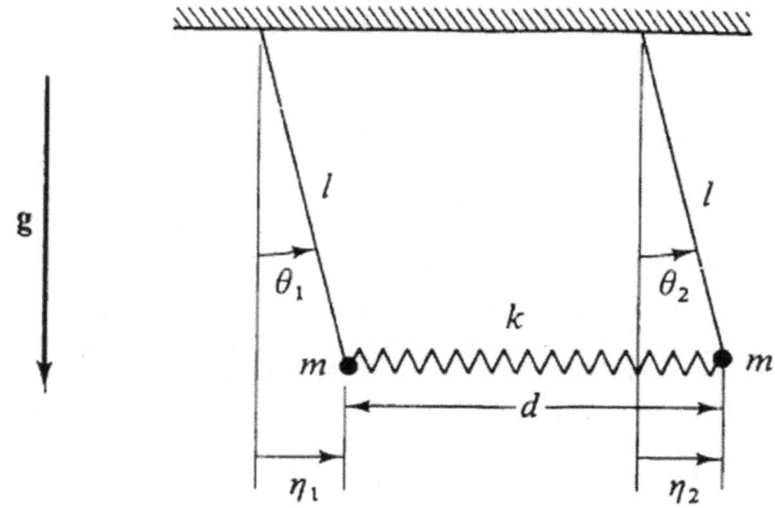

oscillation corresponding to a normal mode frequency of 20 Hz. What is the frequency of the fundamental (lowest-frequency) mode of oscillation?

18. **Two coupled pendula.** Consider the two identical pendula with masses, *m*, and lengths, ℓ, which are joined by a spring of stiffness, K, diagrammed in Fig. 2.43.
    (a) *Coupled equations.* Assume that the angular displacements, θ_1 and θ_2, are both small enough that $\sin\theta \cong \theta$. Write the coupled differential equations that relate $d^2\theta/dt^2$ to the angular displacements.
    (b) *Modal frequencies.* After making the harmonic substitution, calculate the determinate-of-coefficients of the resulting coupled algebraic equations to determine the symmetric and antisymmetric normal mode frequencies in terms of ℓ, *m*, K, and gravitational acceleration, *g*.
19. **Electrodynamic loudspeaker.** In this exercise, you will characterize a 5″ electrodynamic loudspeaker (Morel MW 142, S/N 0496). You may neglect all acoustic loading and assume that the speaker measurements in Table 2.1 are made in a vacuum. Figure 2.44 contains the catalog description of the speaker (Right) and a diagram of its parts (Left) taken from Hunt's *Electroacoustics* [36].
    You are to determine the values of the loudspeaker's parameters and their uncertainty using least-squares curve fitting and error propagation using the measurements summarized in Table 2.1.
    (a) *Stiffness and moving mass.* Determine the original moving mass, m_o, and dynamic stiffness, K, of the loudspeaker and the uncertainties in those values based on the data in the first two columns of data at the left end of Table 2.1.
    (b) *Mechanical resistance.* If the driver is driven at its resonance frequency and then the power is cut and the voice coil ($R_{dc} = 5.1\ \Omega$) is connected to a high impedance load, the exponential free-decay time, τ, can be related to the mechanical resistance, $R_m = 2m_o/\tau$, of the loudspeaker. The columns labeled "Time" and "*V*(peak)" in Table 2.1 are the results of

Table 2.1 (Left to right) Measured frequency vs. added mass (shaded), free-decay amplitude vs. time, LVDT output vs. position (shaded), and LVDT output vs. DC electrical current through the voicecoil

Mass (gm)	Freq. (Hz)	Time (ms)	*V(peak)* (mV$_{dc}$)	*Position* (in.)	V_{out} (mV$_{dc}$)	I_{dc} (mA$_{dc}$)	V_{out} (mV$_{dc}$)
0	55.5	29.25	–823	0.700	947.7	–253.3	388
0.877	53.6	35.50	0	0.650	728.3	–202.1	343
2.877	49.9	39.55	+318	0.600	512.4	–153.6	301
5.877	45.6	45.32	0	0.550	296.3	–101.4	256
10.877	40.2	49.40	–144	0.500	81.4	–54.6	216
20.877	33.4	54.97	0	0.450	–133.5	0	169.1
		59.00	+69	0.400	–247.3	+49.8	136.0
		64.47	0			+102.6	91.3
		68.75	–33			+151.4	46.1
						+202.5	1.6

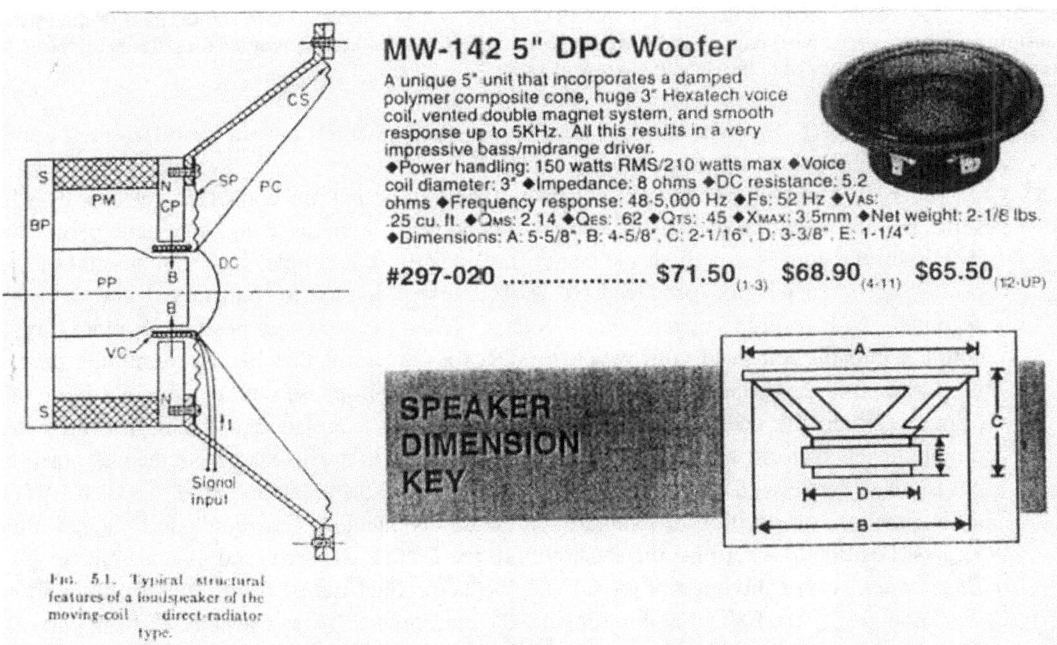

Fig. 2.44 Electrodynamic loudspeaker used to produce the results summarized in Table 2.1

such a free-decay measurement. V(peak) is the value of the voltage at the peak (or trough) of each cycle. The peaks occur at integer multiples of the period T, $n = 0, 1, 2, \ldots$ The troughs occur at half-odd integer values of $n = {}^1/_2, {}^3/_2, {}^5/_2, \ldots$

$$V(peak) = V_1 e^{-nT/\tau} \tag{2.148}$$

Take the natural log of that expression to convert to a linear relation, and find R_m and its uncertainty by creating a plot like Fig. 1.15.

(c) *LVDT calibration.* To determine the transduction coefficient, $B\ell$, the displacement of the loudspeaker cone has been measured as a function of DC current, I_{dc}, through the voice coil:

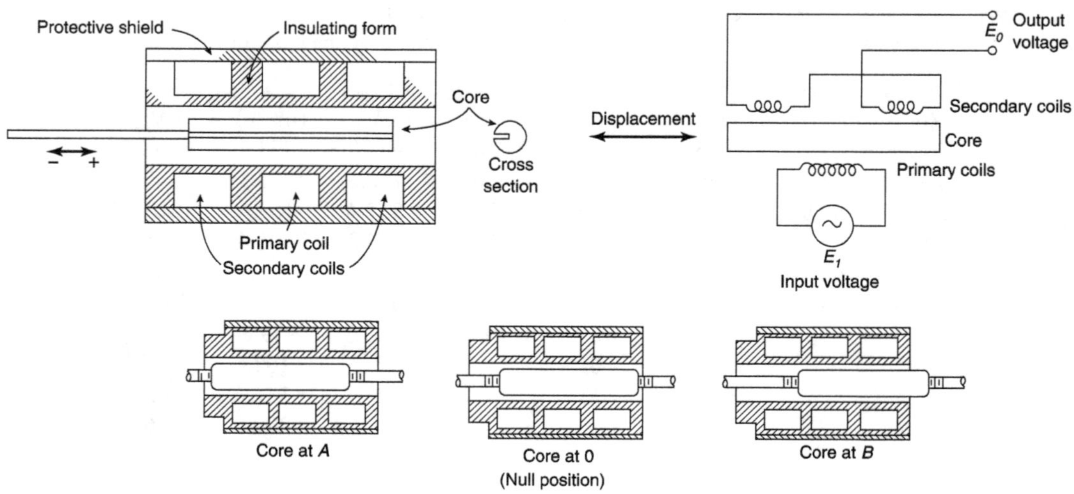

Fig. 2.45 (*Above*) Cross-sectional diagrams of an LVDT at the left and electrical schematic diagram of the astatic transformer showing the (driven) primary and (output) secondary coils. (*Below*) Position of the LVDT core displaced to the left, centered, and displaced to the right. (Courtesy of Lucas-Schaevitz)

$F = -\mathrm{K}x = -(B\ell)I_{dc}$. The displacement is measured by a linear variable differential transformer (LVDT) shown in Fig. 2.45.

The LVDT output (secondary) consists of two coils that are connected (electrically) in series but are wound in opposite directions. The movable core has a high magnetic permittivity. When the core is exactly at the center (null position), it couples an equal amount of the oscillating magnetic flux (produced by an AC current applied to the primary coil) to both secondary coils resulting in zero output voltage. When the core is in position *A*, more flux is coupled into the left-hand core which produces a waveform that has an amplitude that is proportional to the distance from the null position and is in-phase with the signal driving the primary. When the core is in position *B*, more flux is coupled into the right-hand core, producing a waveform amplitude that is also proportional to the distance from the null position but is out-of-phase with the signal driving the primary. The voltage output of one such LVDT as a function of position is tabulated in columns labeled "Position" and "V_{out}." Plot V_{out} vs. Position to determine the sensitivity of the LVDT, $\mathrm{d}V_{out}/\mathrm{d}x$, and its uncertainty.

(d) *Bℓ-product.* After calibration of the LVDT, the $B\ell$-product can be determined from the fit to "I_{dc}" and "V_{out}," the LVDT calibration, and the measured stiffness constant, K, from part (a). Find the value of $B\ell$ and its uncertainty.[32]

(e) *Constant current response.* Assume the loudspeaker is driven by a constant sinusoidal current of 100 $\mathrm{mA_{rms}}$ at all the frequencies of interest in this exercise.[33] Plot the magnitude and phase of the voice coil's velocity, $\boldsymbol{v}$, as a function of frequency between 10 Hz and 1 kHz. Use a $\log_{10}$ axis for frequency.

(f) *Power dissipation.* Determine the power dissipated in the speaker's mechanical resistance, R_m, at the resonance frequency, ω_o.

---

[32] This calculation assumes that the dynamic stiffness constant measured in the added mass experiment is the same as the static stiffness constant that is used in this part of the exercise to convert the displacement of the cone to the force produced by the current in the voice coil. Due to the viscoelastic behavior of the cone's surround, those two stiffnesses can differ due to relaxation of the elastomeric surround (see Fig. 1.16). For this exercise, that difference can be neglected.

[33] In normal operation, speakers are usually driven at constant voltage, but the variation in the speaker's electrical impedance with frequency makes that calculation more difficult.

20. **Jug hustlers.** One of the most common tools in the oil exploration business is the geophone. It is used for seismic surveys, and the surveyors who plant and recover this ground vibration velocity sensors are known as "jug hustlers," since the geophones look like little jugs (see Fig. 2.24). In this problem, you will calculate the response of a GeoSpace® Corp. Model GS-30 CT geophone. The moving mass (mostly the coil and bobbin), $m_o = 11.2$ gm ($\pm 3.5\%$), and the natural frequency, $f_o = \omega_o/2\pi = 10$ Hz.
    (a) *Damping*. The open-circuit damping factor, $\zeta_o = 0.316 = (2Q)^{-1}$. Find the mechanical resistance, R_m, of the moving coil and its relative uncertainty.
    (b) *Sensitivity*. The electrodynamically generated, open-circuit voltage, $V_{emf} = (B\ell)v$, where v is the relative velocity of the coil with respect to the case. If the transduction coefficient, $(B\ell) = 70$ N/A, calculate the root-mean-squared (rms) open-circuit voltage, V_{rms}, generated by the geophone if the peak-to-peak ground motion is 10.0 μm at 20, 50, 100, and 200 Hz.
    (c) *Thermal noise*. Use the Equipartition Theorem to determine the value of V_{rms} due to the thermal motion of the coil if the temperature, $T = 20$ °C.
21. **Rayleigh line-shape fit.** The data in Table 2.2 provides the measured amplitude at 22 frequencies around resonance for the third (i.e., $n = 3$) torsional mode of a cylindrical rod made from Eastman Tritan® Copolyester TX1001. Using the Rayleigh line-shape of Eq. (2.63), determine the best values for f_3, $A_3(1)$, and Q_3 by minimizing the squared difference between the measurements and the values produced by Eq. (2.63) by using your choice of "solver" software.
    (a) *Parameter estimates*. The convergence of any solver's results will depend upon the quality of the initial guess for the parameters that the solver is allowed to vary. Estimate f_3 and $A_3(1)$ by choosing the largest amplitude value in Table 2.2 and its corresponding frequency. Estimate $Q_3 \cong f_3/(f_+ - f_-)$ where f_+ and f_- are the "down 3 dB" frequencies, as in Eq. (B.3).

Table 2.2 Resonance amplitude vs. frequency

Frequency [Hz]	Amplitude [mV$_{ac}$]
3245	682.5
3260	714.9
3275	749.7
3290	787.1
3305	827.4
3320	870.1
3335	915.3
3350	957.1
3365	997.2
3380	1034.1
3395	1059.0
3410	1073.0
3425	1074.7
3440	1063.6
3455	1036.4
3470	998.0
3485	953.3
3500	905.1
3515	855.9
3530	806.1
3545	758.7
3560	713.8

(b) *Plot.* On a single graph with properly labeled axes (including units), plot the data in Table 2.2 as "points" and the Rayleigh line shape as a smooth line using the parameter estimates of part (a).

(c) *Optimize fit.* Have your solver software optimized the values of f_3, $A_3(1)$, and Q_3 by minimizing the squared difference between the measurements in Table 2.2 and the values produced by Eq. (2.63). Report the optimized values, and update your plot in part (b) using the best-fit parameter values.

(d) *Error estimate.* How could you try to estimate the uncertainty in the fit parameters? Use your scheme to report your estimate in the relative uncertainty ($\pm 1\sigma$) for your optimized values of f_3, $A_3(1)$, and Q_3.

References

1. R.C. Cross, M.S. Wheatland, Modeling a falling slinky. Am. J. Phys. **80**(12), 1051–1060 (2012)
2. H. Goldstein, *Classical Mechanics* (Addison-Wesley, Reading, 1950), pp. 69–71
3. J.W. Strutt (Lord Rayleigh), *The Theory of Sound*, vol. I, 2nd edn. (Macmillan, London, 1894), pp. 112; reprinted (Dover, 1945): ISBN: 486-60292-3
4. J.W. Strutt (Lord Rayleigh), *The Theory of Sound*, vol. I, 2nd edn. (Macmillan, London, 1894), pp. 109–110; reprinted (Dover, 1945): ISBN: 486-60292-3
5. G.W. Swift, *Thermoacoustics: A Unifying Perspective for Some Engines and Refrigerators*, 2nd edn. (ASA Press/ Springer, Cham, 2017); ISBN: 978-3-319-66932-8
6. M. Greenspan, Simple derivation of the Boltzmann-Ehrenfest adiabatic principle. J. Acoust. Soc. Am. **27**(1), 34–35 (1955)
7. T.G. Giallorenzi, J.A. Bucaro, A. Dandridge, G.H. Sigel Jr., J.H. Cole, S.C. Rashleigh, P.G. Priest, Optical fiber sensor technology. IEEE J. Quantum Electron. **QE-18**(4), 626–665 (1982)
8. A. Einstein, Über die von der molekularkinetischen Theorie der Wärme geforderte Bewegung von in ruhenden Flüssigkeiten suspendierten Teilchen (On the motion of small particles suspended in a stationary liquid, as required by the molecular kinetic theory of heat). Ann. Phys. **17**(8), 549–560 (1905)
9. W. Gerlach, E. Lehrer, Uber die Messung der rotatorishen Brownschen Bewegung mit Hilfe einer Drehwage. Naturwiss **15**(1), 15 (1927)
10. G.E. Uhlenbeck, S. Goudsmit, A problem in Brownian motion. Phys. Rev. **34**, 145–151 (1929)
11. P. Horowitz, W. Hill, *The Art of Electronics*, 2nd edn. (Cambridge University Press, Cambridge, 1989). §7.11; ISBN: 0-521-37095-7
12. H. Nyquist, Thermal agitation of electric charge in conductors. Phys. Rev. **32**, 110–113 (1928)
13. C.W. McCombie, Fluctuation theory in physical measurements. Rep. Prog. Phys. **16**, 266–319 (1953)
14. T.J. Hofler, S.L. Garrett, Thermal noise in fiber optic sensors. J. Acoust. Soc. Am. **84**(2), 471–475 (1988)
15. D.L. Gardner, T. Hofler, S.R. Baker, R.K. Yarber, S.L. Garrett, A fiber optic interferometric seismometer. J. Lightwave Technol. **LT-5**(7), 953–960 (1987); S.L. Garrett, D.L. Gardner, Multiple axis fiber optic interferometric seismic sensor. U.S. Patent No. 4,893,930 (January 16, 1990)
16. F. Reif, *Fundamentals of Statistical Mechanics and Thermal Physics* (McGraw-Hill, New York, 1965), §7.5
17. W.P. Allis, M.A. Herlin, *Thermodynamics and Statistical Mechanics* (McGraw-Hill, New York, 1952)
18. T.B. Gabrielson, Mechanical-thermal noise in micromachined acoustic and vibration sensors. IEEE Trans. Electron Devices **40**(3), 903–909 (1993)
19. A.B. Pippard, *The Physics of Vibration* (Cambridge University Press, Cambridge, 1989); ISBN: 0-521-37200-3
20. I.S. Gradshteyn, I.M. Ryzhik, *Tables of Integrals, Series, and Products* (Academic, London, 1980). isbn:0-12-294760-6. See equation 3.264(2)
21. R.S. Wakeland, Use of electrodynamic drivers in thermoacoustic refrigerators. J. Acoust. Soc. Am. **107**(2), 827–832 (2000)
22. R.E. Best, *Phase-Locked Loop: Design, Simulation, and Applications*, 6th edn. (McGraw-Hill, New York, 2007): ISBN: 9780079375-8
23. J.D. Maynard, Topics in physics, Part IV-Acoustics, Penn State Lecture Notes, 2012
24. P. Campbell, *Permanent Magnet Materials and Their Application* (Cambridge University Press, Cambridge, 1994); ISBN: 0 521 56688 6
25. S.L. Garrett, J.F. Heake, Hey kid! Wanna build a loudspeaker? The first one's free. Audio Engineering Society Convention Paper 5882 (2003 October 10–13, New York, NY); S.L. Garrett, Two-way loudspeaker enclosure

assembly and testing as a freshman seminar. Paper #526,17th International Congress on Sound and Vibration, 18–22 July 2010, Cairo, Egypt
26. T.W. Leishman, B.E. Anderson, Evaluation of moving-coil loudspeaker and passive radiator parameters using normal-incidence sound transmission measurements: Theoretical developments. J. Acoust. Soc. Am. **134**(1), 223–236 (2013)
27. L.L. Beranek, *Acoustics* (Acoustical Society of America, Woodbury, 1996). pp. 185–188, 211–230: ISBN: 0-88318-494-X
28. A.N. Thiele, Loudspeakers in vented boxes: Parts I and II. Proc. Inst. Radio Eng. (Australia) **22**, 487–502 (1961); reprinted in J. Audio Eng. Soc. **19**, 382–392, 471–483 (1971)
29. B. Baumzweiger (a.k.a., Bauer), Microphone apparatus, U.S. Patent No. 2,184,247 (December 19, 1939)
30. W.L. Dooley, Ribbon microphones. J. Acoust. Soc. Am. **136**, 2130 (2014)
31. M.S. Pettersen, 75th anniversary of the Shure Unidyne® microphone. J. Acoust. Soc. Am. **136**, 2129 (2014)
32. J.C. Snowdon, *Vibration and Shock in Damped Mechanical Systems* (Wiley, London, 1968)
33. P.M. Morse, *Vibration and Sound*, 2nd edn. (McGraw-Hill, New York, 1948). Reprinted (Acoustical Society of America, 1976). isbn:0-88318-287-4. The normal mode displacement are given in Eqs. (7.3) though (7.5), pp. 55–58
34. A.L. Fetter, J.D. Walecka, *Theoretical Mechanics of Particles and Continua* (Dover, Mineola, 2003); ISBN: 0-486-43261-0
35. J.E. Lennard-Jones, On the determination of molecular fields. Proc. R. Soc. Lond. A **106**, 463–477 (1924)
36. F.V. Hunt, *Electroacoustics: The Analysis of Transduction and Its Historical Background* (Wiley, New York, 1954); reprinted (Acoust. Soc. Am., 1982). ISBN: 0-88318-401-X

3

String Theory and Vibration

The vibrating string has been employed by nearly every human culture to create musical instruments. Although the musical application has attracted the attention of mathematical and scientific analysts since the time of Pythagoras (570 BC–495 BC), we will study the string because its vibrations are easy to visualize and it introduces concepts and techniques that will recur throughout our study of vibration and the acoustics of continua. A retrospective of these concepts and techniques is provided in Sect. 3.9, near the end of this chapter. You might want to skip ahead to read that section to understand the plot before you meet the characters.

In the analyses of the previous chapter, we sought equations that would specify the time histories, $x(t)$, of the displacement of discrete masses constrained to move only along one dimension, focusing particularly on single-frequency time-harmonic motion. At the end of that chapter, we examined the limit as the number of masses became infinite and the inter-mass spacing decreased in a way that held constant the linear mass density, ρ_L, of our "string of pearls" in Sect. 2.7.7. In this chapter, we will develop continuous mathematical functions of position and time that describe the shape of the entire string. The amplitude of such functions will describe the transverse displacement from equilibrium, $y(x, t)$, at all positions along the string.[1]

3.1 Waves on a Flexible String

We begin this exploration, as we did with the simple harmonic oscillator, by seeking an equation-of-state (i.e., a relationship between forces and displacements), but now we will examine an infinitesimal segment of string, with length, dx, acted upon by the tension applied at both ends. As we did with the "string of pearls" in Sect. 2.7.7, we will assume that the string has a constant linear mass density, ρ_L, and that the string has no flexural rigidity (i.e., no bending stiffness); thus the string can only apply forces produced by the tension, T, to influence the string's motion. We will also assume that the string's displacements from equilibrium are small enough that the displacements create length changes that only contribute modifications to the tension that are second-order in the ratio (y/L), as shown in Eq. (2.129), and can be neglected in a linear (i.e., first-order) analysis.

Figure 3.1 shows the infinitesimal length of string as just described. It is acted upon by the vertical components of the tension at either end, $F_y(x)$ and $F_y(x + dx)$. As will be demonstrated, if the string does not have any curvature, there will be no net transverse force. We can apply Newton's Second Law to that segment of length, dx, once we calculate the net transverse force, dF_{net}.

Since x, y, and t are all independent variables, partial derivatives are used to remind us to hold the other two variables constant when we take a derivative with respect to the third variable.

$$dF_{net} = (\mathrm{T}\sin\theta)_{x+dx} - (\mathrm{T}\sin\theta)_x \cong \left[(\mathrm{T}\sin\theta)_x + \left.\frac{\partial(T\sin\theta)}{\partial x}\right|_x dx + \ldots\right] - (\mathrm{T}\sin\theta)_x \qquad (3.1)$$

[1] Since we are discussing transverse waves, it is necessary to recognize that we have two orthogonal options for the specification of the plane of the polarization for those displacements. We will let the x axis designate the direction of propagation, determined by the undisturbed stretched string. The string's displacements from equilibrium could be in the y direction, the z direction, or some superposition of both. Since we are starting with the assumption of perfect azimuthal symmetry around the axis of the string, the choice is irrelevant. When we apply periodic external forces (see Sect. 3.8) and/or displacements, the polarization plane will be specified. If the string is "twisted," that symmetry might become broken and the plane of polarization could change with time.

This approximate expression utilizes the first two terms in the Taylor series of Eq. (1.2). After cancellation of the two leading $(\mathrm{T}\sin\theta)_x$ terms of opposite sign, we are left with an expression for the net transverse force that can be expanded by the product rule of Eq. (1.10).

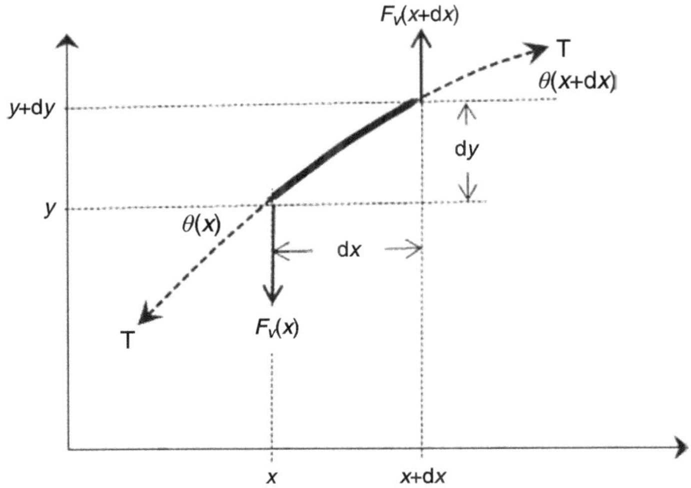

Fig. 3.1 An infinitesimal length, $ds = \sqrt{dx^2 + dy^2} \cong dx$, of a string with uniform linear mass density, ρ_L, that is acted upon by the tension, T, exerting a force along the direction of the string. The angle that the string makes with the horizontal at x is $\theta(x)$, and the angle at $x + dx$ is $\theta(x + dx)$. If these angles are different, there will be a net transverse force on that segment, causing it to accelerate

$$dF_{net} = \left.\frac{\partial(\mathrm{T}\sin\theta)}{\partial x}\right|_x dx \cong \left.\frac{\partial\left(\mathrm{T}\frac{\partial y}{\partial x}\right)}{\partial x}\right|_x dx = \left(\left.\frac{\partial \mathrm{T}}{\partial x}\right|_x \left.\frac{\partial y}{\partial x}\right|_x + \mathrm{T}\left.\frac{\partial^2 y}{\partial x^2}\right|_x\right) dx \tag{3.2}$$

We have again imposed the assumption that $\partial y/\partial x \ll 1$, so $\sin\theta \cong \tan\theta \cong dy/dx$, where these two differential lengths, dx and dy, are shown in Fig. 3.1.[2] For now, we will assume that the tension along the string is independent of position, so $\partial\mathrm{T}/\partial x = 0$, although we will reconsider this restriction later in Sect. 3.4.3, when we consider the oscillations of a heavy chain that has a tension that decreases as we recede from its point of suspension (since the lower links are supporting less mass than the upper links).

For our simplified case, our equivalent of an equation-of-state for the string shows that the net transverse force, dF_v, on an infinitesimal string segment is the product of the constant tension, T, and the curvature of the string.[3] This is consistent with our intuition. Strings that are displaced but remain straight, like those strings attached to the discrete masses shown in Figs. (2.23) through (2.26), only apply forces at the "kinks" produced at the positions of the discrete masses or at the fixed boundaries.

Once again, we will use Newton's Second Law to provide the dynamical equation that describes the acceleration of that infinitesimal string segment with mass, $\rho_L\, dx$, acted upon by dF_{net}.

$$(\rho_L dx)\frac{\partial^2 y(x,t)}{\partial t^2} = \mathrm{T}\frac{\partial^2 y(x,t)}{\partial x^2}dx \tag{3.3}$$

After cancellation of the common differential, this solution is written in a form known as the *wave equation* where we identify the *speed of transverses waves*, $c = \sqrt{\mathrm{T}/\rho_L}$.

---

[2] In Fig. 3.1, dy has been exaggerated to make the drawing easier to read.

[3] The exact mathematical definition of "curvature," κ, involves both the second and first derivatives: $\kappa = \left(\frac{\partial^2 y}{\partial x^2}\right)/\left[1+\left(\frac{\partial y}{\partial x}\right)^2\right]^{3/2}$. In our first-order analysis, we can neglect the square of the first-order term $(\partial y/\partial x)^2$, since it is second-order, and let $\kappa \cong \partial^2 y/\partial x^2$.

$$\frac{\partial^2 y(x,t)}{\partial t^2} - \left(\frac{\mathrm{T}}{\rho_L}\right)\frac{\partial^2 y(x,t)}{\partial x^2} = \frac{\partial^2 y(x,t)}{\partial t^2} - c^2\frac{\partial^2 y(x,t)}{\partial x^2} = 0 \tag{3.4}$$

Such a second-order homogeneous partial differential equation must have two independent solutions. As demonstrated below, those solutions can have any functional form so long as the arguments of those functions only involve two particular combinations of the space and time coordinates.

$$y(x,t) = f_-(x-ct) + f_+(x+ct) \tag{3.5}$$

At this point, the form of the two functions, f_- and f_+, is entirely arbitrary, but the requirement imposed by the wave equation suggests that after some time t, f_- $(x - ct)$ will be identical to its shape at $t = 0$, except at a distance, $x = ct$, farther to the right. Similarly, f_+ $(x + ct)$ will be identical to its shape at $t = 0$, except at distance, $x = -ct$, farther to the left. Because these functions propagate with speed, c, f_- $(x - ct)$ and f_+ $(x + ct)$ are called *traveling wave* solutions.

The wave equation can be written in a "linear operator" form (see Sect. 1.3) represented by the *d'Alembertian operator.*

$$\begin{aligned}\frac{1}{c^2}\frac{\partial^2 y(x,t)}{\partial t^2} - \frac{\partial^2 y(x,t)}{\partial x^2} &\equiv \Box \mathrm{y(x,t)} \\ = \left(\frac{1}{c}\frac{\partial y(x,t)}{\partial t} - \frac{\partial y(x,t)}{\partial x}\right)&\left(\frac{1}{c}\frac{\partial y(x,t)}{\partial t} + \frac{\partial y(x,t)}{\partial x}\right) = 0\end{aligned} \tag{3.6}$$

In the lower expression, the operator has been factored to make it easy to see that the solutions shown in Eq. (3.5) satisfy the wave equation one or the other term in the rightmost version of Eq. (3.6) vanish.

The proof that Eq. (3.5) is a solution to the wave equation of Eq. (3.4) or Eq. (3.6) is simplified by the introduction of a new variable, $w_\pm = x \pm ct$, and the use of the chain rule.

$$\frac{\partial y}{\partial x} = \frac{\partial f_\pm}{\partial w_\pm}\frac{\partial w_\pm}{\partial x} = \frac{\partial f_\pm}{\partial w_\pm} \tag{3.7}$$

The same transformation can be applied to the partial derivative with respect to time.

$$\frac{\partial y}{\partial t} = \frac{\partial f_\pm}{\partial w_\pm}\frac{\partial w_\pm}{\partial t} = \pm c\frac{\partial f_\pm}{\partial w_\pm} \tag{3.8}$$

Substituting Eqs. (3.7) and (3.8) into the factored version of Eq. (3.6) shows that f_- makes the operator, W, with the sum identically zero.

$$\Box = \left(\frac{1}{\mathrm{c}}\frac{\partial \mathrm{y}}{\partial \mathrm{t}} - \frac{\partial \mathrm{y}}{\partial \mathrm{x}}\right)\left(\frac{1}{\mathrm{c}}\frac{\partial \mathrm{f}_-}{\partial \mathrm{t}} + \frac{\partial \mathrm{f}_-}{\partial \mathrm{x}}\right) = \left(\frac{1}{\mathrm{c}}\frac{\partial \mathrm{y}}{\partial \mathrm{t}} - \frac{\partial \mathrm{y}}{\partial \mathrm{x}}\right)\left(\frac{-\mathrm{c}}{\mathrm{c}}\frac{\partial \mathrm{f}_-}{\partial \mathrm{w}_-} + \frac{\partial \mathrm{f}_-}{\partial \mathrm{w}_-}\right) = 0 \tag{3.9}$$

Since the second operator is now identically zero, the first operator has no effect. Since the order of application of the factored linear (differential) operators is irrelevant, we could apply the operator with the minus sign to f_+ to obtain the same result. The factorization of the second-order partial differential equation we called the wave equation in Eq. (3.4) into the product of two one-way first-order linear differential operators can be useful in problems where the propagation is only in one direction (i.e., for situations where there is no reflected wave).

The validity of the solutions in Eq. (3.5) to the wave equation in Eq. (3.4) can be established directly by calculation of the second partial derivative with respect to position, x, or with respect to time, t, produced by a second application of the chain rule.

$$\frac{\partial^2 y}{\partial x^2} = \frac{\partial}{\partial x}\left(\frac{\partial y}{\partial x}\right) = \frac{\partial}{\partial x}\left(\frac{\partial f_\pm}{\partial w_\pm}\right) = \frac{\partial^2 f_\pm}{\partial w_\pm^2}\frac{\partial w_\pm}{\partial x} = \frac{\partial^2 f_\pm}{\partial w_\pm^2} \tag{3.10}$$

$$\frac{\partial^2 y}{\partial t^2} = \frac{\partial}{\partial t}\left(\frac{\partial y}{\partial t}\right) = \frac{\partial}{\partial t}\left(\pm c\frac{\partial f_\pm}{\partial w_\pm}\right) = \pm c\frac{\partial^2 f_\pm}{\partial w_\pm^2}\frac{\partial w_\pm}{\partial t} = c^2\frac{\partial^2 f_\pm}{\partial w_\pm^2} \tag{3.11}$$

In the limit of small displacements, it is important to appreciate that any disturbance will propagate along the string at a constant speed forever without reducing its amplitude or changing shape. That would not be true if our analysis had incorporated dissipation or if nonlinear (i.e., second-order) contributions had not been suppressed.

The result would have been identical if we had chosen to exchange the order of the space and time coordinates. The reader should be convinced that replacing the solution in Eq. (3.5) by the form in Eq. (3.12) still satisfies Eq. (3.4) or Eq. (3.6).

$$y(x,t) = f_-(ct - x) + f_+(ct + x) \tag{3.12}$$

The function with the minus sign in its argument still corresponds to a right-going wave, and the plus sign corresponds to the left-going wave. As will be shown later in this chapter, the choice of Eq. (3.12) or of Eq. (3.5) might simplify a particular calculation.

Figure 3.2 illustrates the propagation of a Gaussian pulse with the form of Eq. (1.87) but with an argument which is proportional to $x \pm ct$.

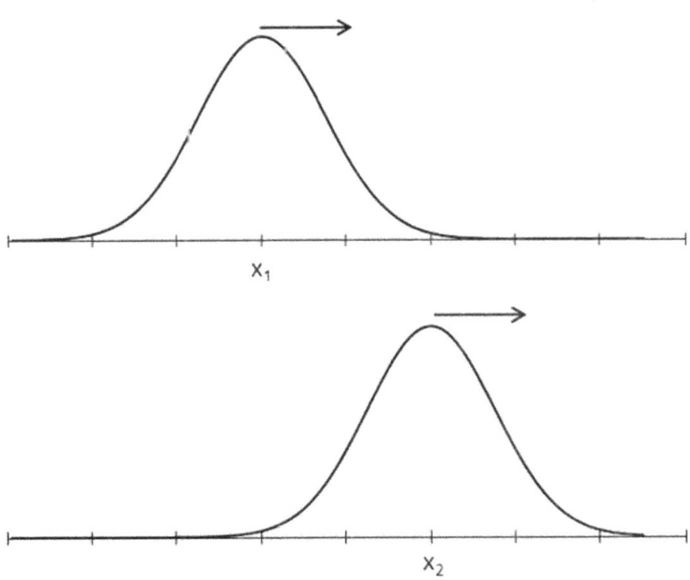

Fig. 3.2 (*Above*) A Gaussian pulse with displacement, $y(x, t)$, given by Eq. (3.13), is propagating to the right. The pulse is shown at an instant, t_1, where it has its maximum displacement centered at position, x_1. (*Below*) At a later time, $t_2 > t_1$, the center of the pulse has moved to a new position, x_2. The change in position is related to the change in time by the speed of transverse waves on a string of uniform linear mass density, ρ_L, and uniform tension, T, $c = \sqrt{T/\rho_L} \Rightarrow (x_2 - x_1) = c(t_2 - t_1)$

$$y(x,t) = \frac{1}{\sigma\sqrt{2\pi}} e^{\left[-\frac{1}{2}\left(\frac{x \pm ct}{\sigma}\right)^2\right]} \tag{3.13}$$

For Fig. 3.2, I have chosen the right-going argument, $x - ct$. The upper plot of Fig. 3.2 captures the pulse at an instant, t_1, with the pulse centered at x_1. The lower plot shows the same pulse at a later instant, t_2, after the center of the pulse had advanced to x_2. Over that time interval, $\Delta t = t_2 - t_1$, all portions of the pulse have advanced by a distance, $\Delta x = x_2 - x_1$. The speed of such a transverse disturbance, whatever the shape, is defined by the usual kinematic relation: $c = \Delta x/\Delta t$.

3.2 Pulse Reflections at a Boundary and the Utility of Phantoms

Infinite strings are hard to come by, and a method for applying a uniform tension to such a beast is even harder to imagine. A string of finite length will be terminated by two boundaries. To initiate our analysis of the reflection of a pulse from such a boundary, we will examine two limiting cases: A rigid ("*fixed*") boundary will suppress the transverse motion of the string by providing whatever vertical force, $F_v = -T \sin(\theta) \cong -T(\partial y/\partial x)_{boundary}$, that would be required to keep the string's attachment point immobile.[4] The opposite limit is a *free boundary*. Such a termination will maintain the tension in the string but will allow the end of the string to move up or down without constraint. [5] Since no vertical forces are produced by the boundary, the pulse must arrive at such a free boundary and leave without any slope, $F_y = T(\partial y/\partial x)_{boundary} = 0$. The common "textbook" way to envision such a boundary is to picture the string attached to a massless ring that slides along a frictionless pole.

Of course, these are only idealized limits that simplify calculations and help to develop our intuition. (If the bridge on a guitar or violin did not move, then the body would not radiate sound.) In this chapter, we will be able to analyze the consequences of a broad range of termination conditions that may be massive (e.g., a mass at the end of a pendulum), elastic (e.g., a string attached to a flexible cantilevered beam), resistive (e.g., a string tensioned by our massless ring but with the ring connected to a dashpot), or a combination of all three.

We will begin by analyzing a pulse that is impinging upon a rigid boundary. This situation is shown schematically in Fig. 3.3 for a Gaussian pulse of the same kind that was displayed in Fig. 3.2. The point of rigid attachment is indicated by the hatched barrier located at $x = 0$. At that boundary, $y(0, t) = 0$ for all times. We also know that the linearity of our general solution to the homogeneous wave equation, given in Eq. (3.5) or Eq. (3.12), allows superposition of any number of such right- and left-going pulses that may be required to satisfy the initial conditions or the boundary constraints.

One fact that is not intuitively obvious is that we can choose to imagine the string extending beyond the termination, as long as the superposition of the pulses coming from that nonexistent extension can be combined with pulses on the physical portion of the string to satisfy the requirements imposed at the boundary. In Fig. 3.3, the pulse at the extreme left, with the arrow above pointing to the right, indicates the real pulse on the real string that is approaching the boundary at $x = 0$. We will satisfy the boundary

[4] It is useful to consider that two expressions have been introduced for the vertical force on a string. At a "point," like that where the string is attached to a boundary or where an harmonic force or displacement is applied to the string (see Sect. 3.7), that vertical force is proportional to the slope of the string at that location: $F_v = -T \sin(\theta) \cong -T(\partial y/\partial x)_{boundary}$. For an infinitesimal segment of string with length, dx, where the vertical forces on two nearby ends are due to the string's tension, the net vertical force is proportional to the string's curvature: $dF_{net} \cong T(\partial^2 y/\partial x^2)dx$.

[5] A "free" boundary condition for a string under uniform tension rarely occurs in realizable physical systems, although a string whose tension is provided by centripetal acceleration can have such a boundary condition (see Problem 9) as does the "heavy chain" described in Sect. 3.4.3. The purpose for introduction of the free condition on a string at this point is only to preview the appearance of free boundary conditions at the ends of vibrating bars that will be examined in Chap. 5.

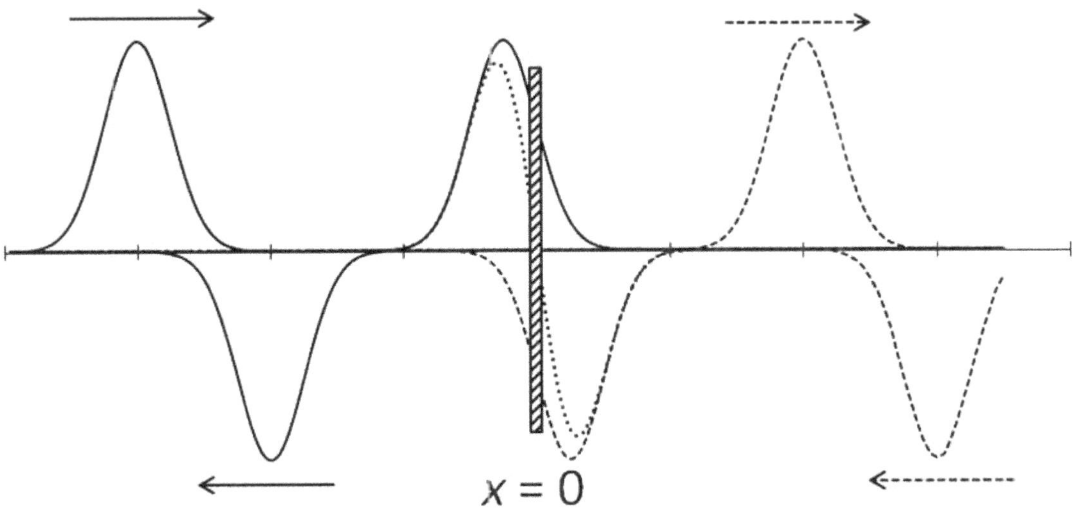

Fig. 3.3 The positive Gaussian pulse (*solid line*) at the far left is shown approaching a rigid boundary at $x = 0$, represented by the *hatched rectangle*. To satisfy the fixed boundary condition, $y\ (0, t) = 0$, an inverted phantom (*dashed line*), shown at the far right, is launched along an imaginary extension of the string. The phantom is approaching the boundary from an equal distance, but propagating in the opposite direction, as indicated by the dashed arrow below the phantom pulse. The superposition of those two pulses (*dotted line*) makes $y\ (0, t) = 0$. As the two pulses leave the boundary, the phantom propagates along the actual string, and the original pulse becomes the phantom

condition, $y\ (0, t) = 0$, by imagining a pulse that is identical to that initial pulse but inverted, approaching from an equal distance behind the boundary. That phantom is shown at the extreme right with a dashed line and a dashed arrow below it. Both pulses are propagating at the transverse wave speed, c, although in opposite directions.

Since the real pulse and the phantom are always equidistant from the boundary, they reach it at the same time and superimpose linearly within their regions of overlap. One instant during that overlapping time interval is illustrated in the vicinity of the boundary in Fig. 3.3 by the dotted line. At that instant, the "real" pulse has started to excite the nonexistent portion of the string, and the phantom has started to appear on the real string. Because of the equality of their amplitude and the congruency of their shape, at $x = 0$ their sum, indicated by the dotted line, will be zero. At the instant during overlap, shown in Fig. 3.3, the sum exhibits a greater slope and decreased amplitude. We have assumed an infinitely rigid termination, so the vertical force, $F_y(x, t) = -T(\partial y/\partial x)_{x=0}$, required to immobilize the string at the boundary, is available to enforce the complete immobilization.

At the instant the two pulses completely overlap, their superposition will mean that the real string (and the fictitious extension) will be flat everywhere. Subsequently, the vertical force which immobilized the string at the boundary will launch the inverted pulse traveling in the opposite direction in the real portion of the string. At a later time, the inverted pulse is shown propagating away from the barrier along the real string, and the original pulse has crossed the barrier and has become the phantom.

We should pause momentarily to reflect (no pun intended) on the process that was used to satisfy the conditions at the real string's rigid termination. The invention of the phantom provided a simple method to satisfy the boundary condition by trading a semi-infinite string for an infinitely long one. We should remember that there are only real strings and real forces. On the other hand, we would be remiss if we did not also marvel at the human imagination and intelligence that produced such a simple way to understand such a complex combination of wave–barrier interactions.

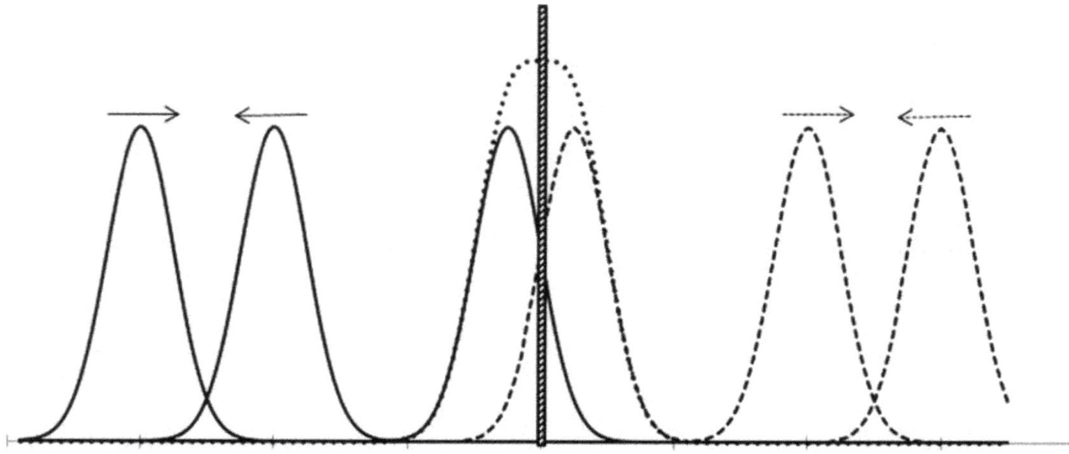

Fig. 3.4 The positive Gaussian pulse (*solid line*) at the far left is shown approaching a force-free boundary at $x = 0$, represented by the *hatched rectangle*. To satisfy the boundary condition, $(\partial y/\partial x)_{x=0} = 0$, a phantom (*dashed line*), shown at the far right, is launched along an imaginary extension of the string, approaching the boundary from an equal distance, but propagating in the opposite direction. The superposition of those two pulses (*dotted line*) maintains the slope of their sum to be zero at $x = 0$ and doubles their amplitude at the instant of complete overlap. As the two pulses leave the boundary, the phantom propagates along the actual string, and the original pulse becomes the phantom

It is easy to repeat this analysis for the other limiting case, a boundary that applies no vertical forces on the string. In that case, we require that the slope of the pulse vanishes for all times at $x = 0$ (i.e., $-T(\partial y/\partial x)_{x=0} = 0$). Figure 3.4 is identical to Fig. 3.3 except that the phantom is not inverted. Near the region of the boundary, where the pulses overlap, their superposition always maintains zero slope. At the exact instant of overlap, the pulse has twice its original amplitude.

This approach to the solution of boundary-value problems is common in acoustics and other fields of physics (e.g., electrostatics, optics). It is known as the *method of images*. When we study sound sources in near boundaries (see Sect. 12.4.1 and Figs. 12.13 and 12.14) or in rooms (see Sect. 13.1.1), or underwater in proximity to an air–water interface, we will sprinkle image sources about in regions that are outside those spaces that are physically accessible to the waves to satisfy boundary conditions.

The same results, for both the fixed and free boundaries, can be expressed mathematically by the appropriate choice of functions, with $w_{\pm}$ in their arguments. If we again choose a Gaussian pulse and abbreviate the waveform of Eq. (3.13) as $y(x, t) = G_{\pm}(x \pm ct)$, then the rigid boundary condition at $x = 0$ is satisfied by a solution $y(x, t) = G_{-}(x - ct) - G_{+}(x + ct)$. With the additional restriction that when $x \leq 0$, we need not concern ourselves with the nonexistent portions of the string.

To satisfy the free boundary condition, we require that the slope vanishes at the boundary $x = 0$.

$$\left(\frac{\partial y}{\partial x}\right)_{x=0} = \left(\frac{\partial G_{-}}{\partial x}\right)_{x=0} + \left(\frac{\partial G_{+}}{\partial x}\right)_{x=0} = \left(\frac{\partial G_{-}}{\partial w_{-}}\right)_{x=0} + \left(\frac{\partial G_{+}}{\partial w_{+}}\right)_{x=0} = 0 \tag{3.14}$$

Evaluation of those expressions at $x = 0$, with $w_{-} = -ct$ and $w_{+} = ct$, leads to the requirement that both G_{-} and G_{+} be positive pulses.

3.3 Normal Modes and Standing Waves

We will now examine the free vibrations of a uniform string of length, L, and will start by rigidly fixing both ends of the string so that $y(0, t) = y(L, t) = 0$. Those boundary conditions would provide a close approximation to a guitar, violin, *erhu* (二胡), *sitar* (सितार), harp, ukulele, Balkan tambura, or piano string. Recalling that a normal mode is a particular displacement that results in all parts of the system oscillating at a single frequency, we will seek the normal mode vibration frequencies for a uniform string by again assuming a right- and left-going solution to the wave equation and then combining those solutions in a way that satisfies both boundary conditions simultaneously. This will produce a theoretically infinite set of normal mode frequencies. It should come as no surprise that the mode shapes for the lower-frequency modes will resemble the lower-frequency modes of our "string of pearls" that were diagrammed in Fig. 2.30.

The solutions to the wave equation provided in Eq. (3.5) or Eq. (3.12) involve arguments that have the dimensions of length. The arguments of any mathematical function (e.g., sines, cosines, or exponentials) must be dimensionless. To provide dimensionless arguments that simplify our mathematics, we will scale the argument of the wave functions by a scalar constant, k, which has the dimensions of an inverse length [m^{-1}] that is known as the *wavenumber*.

$$k(ct \pm x) = kct \pm kx = \omega t \pm kx \tag{3.15}$$

As before, the minus sign indicates propagation in the direction of increasing x (to the right), and the plus sign corresponds to propagation in the direction of decreasing x (to the left). The rightmost version of Eq. (3.15) imposes the requirement that $kc = \omega$. Since we seek time-harmonic normal modes at radian frequency, $\omega = 2\pi f$, we can choose either a complex exponential or a sinusoidal expression as our function of the (now dimensionless) argument required of solutions to the linear wave equation.

$$\begin{gathered} y(x,t) = \Re e\left[\widehat{\mathbf{y}} e^{j(\omega t \pm kx)}\right] = |\widehat{\mathbf{y}}| \Re e\left[e^{j\phi} e^{j(\omega t \pm kx)}\right] \\ \text{or} \quad y(x,t) = |\widehat{\mathbf{y}}| \cos(\omega t \pm kx + \phi) \end{gathered} \tag{3.16}$$

These solutions are periodic in both space and time. We can define the wavelength, λ, of the disturbance that propagates to the right along the string as $y(x, t) = y(x + n\lambda, t)$, where n can be any positive or negative integer or zero. Based on that definition, the wavelength can be related to the wavenumber.

$$\lambda = \frac{2\pi}{k} = \frac{2\pi}{(\omega/c)} = \frac{c}{f} \quad \Rightarrow \quad c = \lambda f = \frac{\omega}{k} \tag{3.17}$$

Similarly, $y(x, t) = y(x, t + nT)$, where $T = (1/f) = (2\pi/\omega)$ is the period of the oscillation. Fig. 3.5 illustrates the propagation of the harmonic (i.e., single frequency) disturbance along the string as a function of both position and time.

3.3.1 Idealized Boundary Conditions

Our experience with the reflection of pulses from fixed and free boundaries in Sect. 3.2 suggests that we will be able to satisfy the boundary conditions on this string by superposition of left- and right-going waves and letting the displacement amplitude, $\mathbf{y}(x, t)$, be a complex function of x and t.

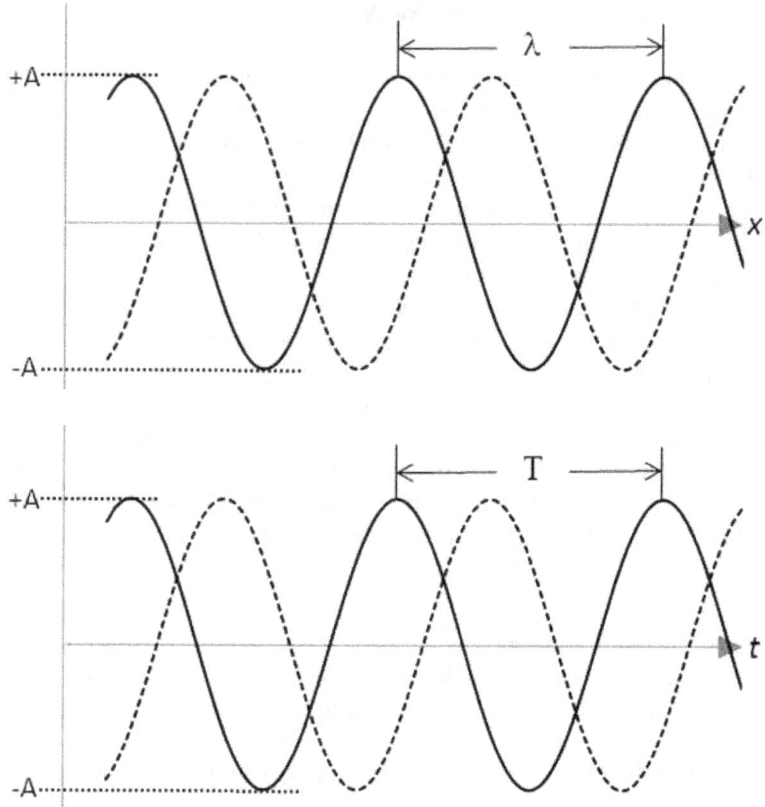

Fig. 3.5 A time-harmonic wave that propagates to the right along a uniform string. (*Above*) The solid line is the wave at time, t_1, and the dashed line is the same wave at slightly later time, $t_2 > t_1$. The distance between successive peaks or troughs is the wavelength, $\lambda = (2\pi/k)$, where k is the wavenumber. (*Below*) The solid line is the time history of a wave passing a point, x_1, and the dashed line is the same wave passing a point, $x_2 > x_1$, that is farther to the right. The wave repeats itself after a time, $T = (1/f) = (2\pi/\omega)$, that is, the period of the harmonic disturbance. The unit of frequency, f, is given in hertz [Hz], and the unit of angular frequency, ω, is in radian/second [rad/s] or just [s^{-1}]

$$\mathbf{y}(x,t) = A\mathrm{e}^{\,j(\omega t-kx)} + Be^{\,j(\omega t+kx)} \tag{3.18}$$

The boundary condition at $x = 0$ is $|\mathbf{y}(0,\ t)| = 0$. Evaluation of this restriction on Eq. (3.18) requires that $A = -B$.

$$\mathbf{y}(x,t) = B\left(e^{\,j(\omega t+kx)} - \mathrm{e}^{\,j(\omega t-kx)}\right) = Be^{j\omega t}\left(e^{jkx} - e^{-jkx}\right) = \widehat{\mathbf{C}}e^{j\omega t}\sin kx \tag{3.19}$$

In this expression, we have used the fact that $2j\sin x = (e^{jx} - e^{-jx})$ and have absorbed $2jB$ into the new complex amplitude constant (phasor), $\widehat{\mathbf{C}}$.[6] The imposition of the boundary condition at $x = 0$ has provided a rather satisfying form for our solution since $\sin(0) = 0$, as required.

[6] At this point, the specification of the amplitude constant, C, is irrelevant. As with the simple harmonic oscillator, or any other linear system (see Sect. 1.3), that constant will be determined by the initial conditions which determine the values of C_n for each of the n normal modes.

The imposition of the boundary condition at $x = L$ now produces the quantization of the frequencies that represent the normal modes of the fixed-fixed string. To make $|\mathbf{y}(L, t)| = 0$, only those discrete values of k_n that make $\sin k_n L = 0$ will satisfy this second boundary condition.

$$\begin{aligned} &\sin k_n L = 0 \quad \Rightarrow \quad k_n L = n\pi \; ; \quad n = 1, 2, 3, \ldots \\ &\Rightarrow \quad 2\pi f_n = \omega_n = k_n c = \frac{n\pi c}{L} \quad \Rightarrow \quad \lambda_n = \frac{2L}{n} \quad \text{or} \quad L = n\frac{\lambda_n}{2} \end{aligned} \tag{3.20}$$

The physical interpretation of this result is simple: the normal mode shapes of a uniform fixed-fixed string correspond to placing n sinusoidal half-wavelengths within the overall length, L, of the string. Substitution of these normal mode frequencies, ω_n, and wavenumbers, k_n, into the functional form of Eq. (3.19) provides the description of the mode shapes, $y_n(x, t)$, for each of the normal modes.

$$y_n(x,t) = \Re e\left[\widehat{\mathbf{C}}_{\mathbf{n}} e^{\,j\omega_r t} \sin\left(n\frac{\pi x}{L}\right)\right]; \quad n = 1, 2, 3, \ldots \tag{3.21}$$

These solutions are called *standing waves*. It is worthwhile to remember that these standing waves are equivalent to the superposition of two counter-propagating traveling waves of equal amplitude that were our starting point in Eq. (3.18). The term "standing wave" refers to the fact that the amplitude envelope is fixed in space (i.e., standing), although the transverse displacement varies sinusoidally in time. The locations where the $\sin k_n x$ term vanishes are called the displacement *nodes* (or just nodes) of the standing wave. In this case, there are always two nodes located at the rigid boundaries. As the mode number, n, increases, $(n - 1)$ additional nodes appear along the string and are equally spaced from each other for a string of uniform mass density and constant tension.

The locations where $\sin k_n x = 1$ are designated *anti-nodes* and are the location of the largest amplitudes of harmonic motion. Those anti-nodes are equidistant from the nodes for a string of uniform density and tension. The separation distance of adjacent nodes and anti-nodes for a uniform string is $\lambda_n/4$.

This result also accounts for the ubiquity of musical instruments that are based on the tones produced by the vibrations of a fixed-fixed string.

$$f_n = \frac{c}{\lambda_n} = n\left(\frac{c}{2L}\right); \quad n = 1, 2, 3, \ldots \tag{3.22}$$

The lowest-frequency mode, $n = 1$, is known as the *fundamental frequency*, $f_1 = c/2L$. The higher normal mode frequencies are known as the *overtones*. The second mode, $f_2 = c/L$, is called the "first overtone." I find this "overtone" terminology to be needlessly confusing and do not use it in this textbook.

The frequencies in Eq. (3.22) form a *harmonic series*. The frequency of each normal mode is an integer multiple of the fundamental mode: $f_n = nf_1$. Pythagoras recognized the intervals produced by a string that was shortened by ratios of integers produced tonal combinations that were pleasing to the ear. A string whose length was halved produced the same note as the full-length string but one octave higher in frequency. A string that is one-third as long as the original produces a note that is a "perfect fifth" above the octave. (See Table 3.1 for the frequency ratios of musical frequency intervals.) Keeping the string length constant but fitting integer numbers of half-wavelengths into the string produces the same frequency as the fundamental of the shorter string.

Although I know of no physical manifestation for the normal modes of a fixed-free string (unless the mass of the string provides its tension, as for the "heavy chain" in Sect. 3.4.3),[5] it is easy to calculate those normal mode frequencies. Those frequencies will provide some useful insights that are

Table 3.1 The ratios (intervals) between the frequencies of two tones that are judged to be "consonant" are written as a fraction and as a decimal

	Frequency ratio		Equal temperament	
Interval	Fractional	Decimal	Semitones	Ratio
Octave	2:1	2.0000	12	2.0000
Perfect fifth	3:2	1.5000	7	1.4983
Perfect fourth	4:3	1.3333	5	1.3348
Major sixth	5:3	1.6667	9	1.6818
Major third	5:4	1.2500	4	1.2599
Minor sixth	8:5	1.6000	8	1.5874
Minor third	6:5	1.2000	3	1.1892

To avoid difficulties that arise in the transposition of musical keys, an "equal temperament" scale was created (see Sect. 3.3.3) so that the two tones, which are separated by an integer number of "semitone" intervals, have a frequency ratio that is a multiple of $2^{1/12} = 1.05946$. In terms of the equal temperament scale, those intervals can be represented by $2^{1/12}$ raised to the number of semitones between the intervals

applicable to acoustic resonators or vibrating bars.[7] Leaving the boundary condition at $x = 0$ fixed, so $y(0, t) = 0$, we can start with Eq. (3.19). Since the end at $x = L$ is free, $(\partial y/\partial x)_{x=L} = 0$. This second condition again quantizes the values of k_n that simultaneously satisfy both boundary conditions.

$$\mathbf{F_y}(L,t) = -T\left(\frac{\partial \mathbf{y}}{\partial x}\right)_{x=L} = -T\widehat{\mathbf{C}}_{\mathbf{n}}e^{\,j\omega_n t}k_n \cos k_n L = 0 \tag{3.23}$$

This can only be satisfied for all times if $\cos k_n L = 0$.

$$\begin{aligned} &\cos k_n L = 0 \quad \Rightarrow \quad k_n L = \left(\frac{2n-1}{2}\right)\pi \quad \Rightarrow \quad \omega_n = \left(\frac{2n-1}{2}\right)\frac{\pi c}{L} \\ &\Rightarrow \quad \lambda_n = \frac{4L}{2n-1} \quad \text{or} \quad L = (2n-1)\frac{\lambda_n}{4} \quad \text{for} \quad n = 1, 2, 3, \ldots \end{aligned} \tag{3.24}$$

Again, the simple interpretation is that odd-integer multiples of one-quarter wavelengths are fit within the overall length of the fixed-free string of length L. This produces a harmonic series that includes only odd-integer multiples of the fundamental, $f_1 = c/4L$, $f_2 = 3c/4L$, $f_3 = 5c/4L$, etc.

3.3.2 Consonance and Dissonance*

"Agreeable consonances are pairs of tones which strike the ear with a certain regularity; this regularity consists in the fact that the pulses delivered by the two tones, in the same time, shall be commensurable in number, so as not to keep the eardrum in perpetual torment." Galileo Galilei, 1638 [1]

The perception of musical (and nonmusical) tones has been an interesting topic for experimental psychologists long before the term "experimental psychologist" came into existence. A perceptual feature of hearing that seems to span many cultures is that of the *consonance* or *dissonance* when two notes with different fundamental frequencies are presented simultaneously to a listener. The ratio of the frequency of the higher-frequency tone to the frequency of the lower-frequency tone is known as the *interval* between the two notes. The intervals that are judged to be "harmonious" are called consonant, while other ratios are judged to be "annoying" (producing "perpetual torment") or dissonant. Table 3.1

[7] For example, the modes of a clarinet behave like those of a fixed-free string because the reed behaves as a closed (fixed) end and the bell behaves as an open (free) end.

Table 3.2 This is the harmonic sequence produced by a uniform string that has its fundamental normal mode frequency ($n = 1$) set to "Concert A," $A_4 = f_1 = 440$ Hz

Mode	1	2	3	4	5	6	7	8
Freq. [Hz]	440	880	1320	1760	2200	2640	3080	3520
Note	A_4	A_5	E_5	A_6	$C_7^\#/D_7^b$	E_7	$F_7^\#/G_7^b$	A_7
Intervals	Root	Octave	Fifth	2nd Oct.		Fifth		3rd Oct.
	Octave (2:1)			Major Third (5:4)				
		Fifth (3:2)			Minor Third (6:5)			
			Fourth (4:3)					
				Major Triad (4:5:6)				

The subscripted musical designation of the notes are based on middle-C $\equiv C_4 = 263.61$ Hz. The first four harmonics consist only of octaves and a perfect fifth. The fifth and seventh harmonics are not exact musical intervals, but the sixth and eighth are again a perfect fifth and an octave. The lower intervals shown in this table are between adjacent harmonics. With the exception of the seventh mode frequency, $f_7 = 3080$ Hz, they are all intervals deemed "consonant," as listed in Table 3.1

lists several consonant frequency intervals and the "musical" designation of that interval (e.g., octave, perfect fourth, minor third, etc.).

A harmonic series is not produced for a nonuniform string or for a uniform string subject to nonuniform tension (e.g., a hanging chain), or a uniform string that has some stiffness (in addition to the restoring force provided by the tension, as discussed in Sect. 5.5), or a uniform string subject to boundary conditions that are not perfectly rigid or perfectly free (see Sect. 3.6).

There are various physical explanations for humans finding some intervals pleasing (consonant) and others unpleasant (dissonant), but a few simple arguments can be based upon the relationships between the harmonics of the two tones either being identical, or creating a consonant interval, or being separated by a sufficient frequency difference that the superposition of the overtones do not produce perceptible beating (i.e., a high rate of amplitude modulation due to the interference between the two tones).

Helmholtz concluded that dissonance occurred when the difference between any overtones of the two tones produced between 30 and 40 beats per second [2]. Later investigations have shown that the dissonance caused by the beat rate will depend upon frequency [3]. Table 3.2 lists a series of harmonics that start with "Concert A_4," a frequency defined as 440 Hz. $A_4 = 440$ Hz is used to set the tuning of pitch for most Western musical instruments. As can be seen, the overtones for a single string produce mostly octaves and perfect fifths.

3.3.3 Consonant Triads and Musical Scales*

As we have just seen, the normal modes of a perfect string generate a harmonic series with relationships that are mostly judged to be pleasing to the ear (i.e., consonant). That is a good start, but music is made by a succession of notes that produce a melody within some rhythmic (i.e., temporal) structure. The construction of a series of musical notes (i.e., a musical scale) that is also pleasant is rather complicated. It is possible to construct a musical scale, suitable for Western music (as well as country music), which is based on the consonant intervals in Table 3.1. As will be illustrated briefly [4], it is not possible to create such scales for which all intervals are consonant and which can be readily transposed so that a melody which is written in a particular key signature (i.e., based on a scale

Table 3.3 Frequency ratios for notes and intervals for the just intonation musical scale constructed from three major triads based on the root (in this case C), the perfect fourth (F), and the perfect fifth (G)

Note	C		D		E		F		G		A		B		C
Root ratio	1:1		9:8		5:4		4:3		3:2		5:3		15:8		2:1
Interval ratio		9:8		10:9		16:15		9:8		10:9		9:8		16:15	

The intervals between adjacent notes are only major whole tones (9:8), a minor whole tone (10:9), or semitones (16:15)

that starts on a specific note) can be played in some other key. A compromise that has been found acceptable, called the scale of *equal temperament* (also known as the tempered scale), places 12 semitones within an octave, all of which that have a constant frequency ratio of $2^{1/12} = 1.05946$.

A musical scale that has eight notes within an octave can be constructed based on three major triads. Such a scale is called the *scale of just intonation*. As seen in Table 3.2, a three-note major chord is built from a minor third (6:5) above a major third (5:4), producing a frequency ratio of 4:5:6 that is perceived to have a particularly pleasant sound. Those of you familiar with American blues and rock music will recognize that many melodies are based on a succession of three such major chords based on a first (root) note, the perfect fourth, and the perfect fifth. These are known as the tonic (I chord), the subdominant (IV chord), and the dominant (V chord).

If we choose C as our root, then the tonic is {C, E, G}. The dominant (V) major chord is built on the perfect fifth G to produce {G, B, D}, thus introducing B as a major third above the perfect fifth, (5/4)(3/2) = 15:8, and D which is a perfect fifth above a perfect fifth, (3/2)(3/2) = 9/4. Reducing the frequency of D by an octave places it within the original octave where we are building the scale, thus producing the interval 9:8, known as a major whole tone. The process is repeated one last time for the subdominant (IV) chord rooted on the perfect fourth, {F, A, C}. A is a major third above a perfect fourth (5/4)(4/3) = 5:3, and C is the octave. The intervals between the notes and the root, as well as the intervals between adjacent notes, are shown in Table 3.3.

The scale of just intonation introduces two other triads with frequency ratios of 10:12:15 that are known as minor triads. Like the major triad, they are constructed from a major third (5:4) and a minor third (6:5), but with the major third above the minor third.

It all sounds good until we examine two of the new fifths that were generated in this process. Five of the fifths are perfect: G:C = B:E = C:F = D:G = E:A = 3:2, but the ratio of B:F is not actually a fifth, B:F = (8/3)/(15/8) = (64/45) = $1.422 \neq 3{:}2 = 1.500$. Similarly D:A which should be a fifth is D:A = (10/3)/(18/8) = (40/27) = $1.481 \neq 3{:}2 = 1.500$. Examination of the perfect fourth reveals two more problematic intervals. Again, five of the fourths are perfect: F:C = G:D = A:E = C:G = E:B = 4:3 = 1.333, but two are not. F:B = (15/8)/(4/3) = 45:32 = $1.406 \neq 1.333$ and D:A = (18/8)/(5/3) = 27:20 = $1.350 \neq 1.333$.

If we try to construct a major triad other than I, IV, and V, used to produce the just scale, we see that sharps and flats must be added that correspond to the "black keys" on a piano. Those sharps and flats do not necessarily have the same frequencies. For example, if we construct the major triad with D as its root, then to produce the major third, we need to introduce $F^{\#}$ that has a ratio of 5:4, so $F^{\#}$ = (9/8)(5/4) = (45/32) = 1.406:C. If we construct the minor third below A, we introduce G^{b} = (5/3)/(6/5) = (25/18) = 1.389:C. On the piano, $F^{\#}$ and G^{b} are the same note.

Instead of building a scale on major triads, it is possible to construct a scale that attempts to preserve the perfect fifth and perfect fourth by noticing that the octave is produced by the combination of a perfect fifth and a perfect fourth: (4/3)(3/2) = 2. All 12 notes of the *chromatic scale* can be produced by increasing frequency by a perfect fifth or decreasing it by a perfect fourth and then bringing the note

Table 3.4 Comparison of the frequency ratios of intervals produced by a just intonation scale, a Pythagorean scale, and an equally tempered scale

	Just		Pythagorean		Equal temperament	
Interval	Ratio	Cents	Ratio	Cents	Ratio	Cents
Octave	2:1	1200	2:1	1200	2:1	1200
Perfect fifth	3:2	702	3:2	702	1.498	700
Perfect fourth	4:3	498	4:3	498	1.335	500
Major sixth	5:3	884	1.687	906	1.682	900
Major third	5:4	386	1.265	408	1.260	400
Minor sixth	8:5	814	1.580	792	1.587	800
Minor third	6:5	316	1.184	294	1.189	300

The frequency ratios are also given in "cents," corresponding to a frequency ratio of $2^{1/1200} = 1.0005778$, to simplify the intercomparison of the intervals among the three scales

back within a single octave to create the *Pythagorean scale*. This approach unfortunately detunes the major and minor thirds from their just intervals, as shown in Table 3.4. Also problematic is the fact that 12 multiplications should return to an octave above the root, except that $(3/2)^{12} = 129.75$, but $2^7 = 128$. That means the Pythagorean process (also known as the "circle of fifths") misses the octave by just under one-quarter semitone.

A workable compromise is achieved by constructing a scale based on a definition of a semitone that is a frequency ratio of $2^{1/12} \cong 1.0595$. There is an octave between each note that is separated by 12 semitones, and a whole step is two semitones: $2^{1/6} = 1.1225 \cong 9/8 = 1.125$. The major scale consists of five notes that are separated by one whole step and two separated by one semitone. The equal temperament now allows musical instruments with fixed pitch to be played in any key (e.g., keyboard instruments like the piano or organ and fretted string instruments like the guitar, banjo, and mandolin). This also eliminates the problems we encountered when major triads generated sharps and flats that had different frequencies but corresponded to the same black keys on a piano keyboard. Of course, the price is that none of the just intervals other than the octave retain their exact frequency ratios.

Table 3.4 provides a comparison between the just, Pythagorean, and equally tempered scales. To make that comparison easier, a frequency ratio that is called a "cent" is introduced. One cent is one one-hundredth of an equal-temperament semitone (in the logarithmic sense), or $2^{1/1200} = 1.000578$.

Although the focus of this textbook is not musical acoustics, the harmonicity of the fixed-fixed modes of an idealized string expressed in Eq. (3.22) is culturally significant, as is the psychological perception of consonance and dissonance. Also, an understanding of the difficulties introduced by the scales used by musicians to specify frequencies, and the compromise provided by equal temperament, should be appreciated by anyone who would call themselves an acoustician.

Moreover, this digression provides the opportunity to marvel at the skill of the professional musicians. They are able to exploit the strong feedback between their sense of pitch and their playing technique in a way that allows them to "tune" their notes to enhance the melody if their instrument permits the "bending" of a note by breath control and/or fingering, thus escaping the limitations otherwise imposed by the convenience of equal temperament.

3.4 Modal Energy

Just as we were able to calculate the kinetic and potential energy stored in undamped harmonic oscillators (consisting of discrete masses and springs) in terms of their amplitudes of excitation, we can calculate the energy stored in a normal mode of a continuous system like the string. Instead of summing the kinetic energies of the discrete masses and potential energies of springs, we will integrate over the length of the string to calculate these energies. As before, Rayleigh's method (Sect. 2.3.2) can be used to approximate the natural frequencies of the system, even if we perturb the modes by adding discrete masses, or analyze the modes of a string having a nonuniform mass density distribution, or a variable tension.

Before doing so, we need to consider the process we used to calculate our results by linearizing our equation-of-state relating forces to displacements (or in fluids, pressure changes to volume changes). We imposed an assumption that $\partial y/\partial x \ll 1$, so $\sin\theta \cong \tan\theta \cong dy/dx$, as illustrated in Fig. 3.1. This is equivalent to neglecting the nonlinear terms in the Taylor series expansions of the sine and tangent functions, given in Eqs. (1.5) and (1.7). We did this by eliminating the terms proportional to θ^3 and higher powers while retaining the term that is linear.

Because the lowest-order term in both the kinetic and potential energies is quadratic in the displacements from equilibrium, there is no first-order term that will dominate the result and which would let us neglect second-order terms. As was the case for the simple harmonic oscillator, where the expressions for the kinetic energy stored by the mass, $KE = (½)mv^2$, and the potential energy stored by the spring, $PE = (½)Kx^2$, were both quadratic in the amplitudes, we need to produce similar expressions for the vibration of strings.

The maximum kinetic energy of the nth standing wave normal mode is easy to calculate from Eq. (3.21) since it provides an expression for the transverse displacement of a fixed-fixed string. We also know that all parts of the string must oscillate at the same frequency, ω_n, when excited in the nth normal mode. To simplify our notation, we'll let $C_n^2 = \left|\widehat{\mathbf{C}}_\mathbf{n}\right|^2$.

$$(KE)_n = \frac{1}{2}\int_0^L \rho_L \left(\frac{\partial y}{\partial t}\right)^2 dx = \frac{\rho_L C_n^2 \omega_n^2}{2}\int_0^L \sin^2\left(n\frac{\pi x}{L}\right)dx = \frac{1}{4}\rho_L L C_n^2 \omega_n^2 = \frac{m_s}{4}v_1^2 \qquad (3.25)$$

The integral of $\sin^2 x\, dx$ or $\cos^2 x\, dx$ over an integer number of half-wavelengths is $L/2$, since $\sin^2 x + \cos^2 x = 1$. This is shown explicitly also for the orthogonality condition expressed in Eq. (3.54). The result in Eq. (3.25) is one-half the value of the maximum kinetic energy that the string would have if all of its mass, $m_s = \rho_L L$, were oscillating at the maximum velocity amplitude, $v_n = \omega_n C_n$.

The maximum potential energy can be calculated from the work done against the tension due to the change in the length of the string caused by the vertical displacements produced along the string by the wave. In our earlier linearization process for masses on strings, we neglected the time-dependent changes in the total length of the string due to the transverse displacement, because those length changes, δL, approximated in Eq. (2.129), were second-order in the displacement from equilibrium, $\delta L \cong a(y/a)^2$. Since we now want to calculate the intrinsically second-order potential energy, that second-order length change is no longer negligible. The Pythagorean theorem can be used to approximate the arc length, ds, of each differential element, as pictured in Fig. 3.1. To determine the local extension, $\delta L(x)$, that is, the difference between the arc length, ds, and the length of the undisturbed string, dx, the Pythagorean sum can be simplified using a binomial expansion.

$$\delta L(x) = ds - dx = \sqrt{dx^2 + dy^2} - dx = dx\left[\sqrt{1 + \left(\frac{\partial y}{\partial x}\right)^2} - 1\right] \cong \frac{1}{2}\left(\frac{\partial y}{\partial x}\right)^2 dx \qquad (3.26)$$

The result is again second-order in the displacement from equilibrium, but now it will not be ignored, as it was previously in the calculation of the string's effective tension. That length change becomes the dominant contribution to the potential energy, since there are no first-order contributions.

The potential energy will be the integral of this length change times the tension.[8] Equation (3.21) can be used to calculate the string's slope.

$$\frac{\partial y_n(x,t)}{\partial x} = C_n k_n \cos(k_n x) e^{j\omega_n t}; \quad n = 1, 2, 3, \ldots \tag{3.27}$$

Substitution into the expression for $\delta L(x)$ in Eq. (3.26) produces a result that is analogous to Eq. (3.25) for the potential energy of the nth standing wave mode.

$$(PE)_n = \frac{\mathrm{T}}{2}\int_0^L \left(\frac{\partial y}{\partial x}\right)^2 dx = \frac{\mathrm{T} C_n^2 k_n^2}{2}\int_0^L \cos^2(k_n x)\,dx = \frac{C_n^2 c^2 \rho_L L}{4}\frac{\omega_n^2}{c^2} = \frac{m_s}{4} v_1^2 \tag{3.28}$$

We take comfort in the equality of the maximum kinetic and potential energies (and hence their time-averaged values), since it is consistent with the virial theorem (Sect. 2.3.1), under the assumption that our restoring force is linear in the displacement from equilibrium (i.e., obeying Hooke's law).

To calculate the total energy, the time phasing of the kinetic and potential energy contributions must be taken into account. If we take the real part of the normal mode solution in Eq. (3.21), then $y(x, t)$ is proportional to $\cos(\omega_n t)$. Since the kinetic energy includes the square of the transverse velocity, $(\partial y/\partial t)^2$, the kinetic energy is proportional to $\sin^2(\omega_n t)$. The potential energy involved $(\partial y/\partial x)^2$, so it will be proportional to $\cos^2(\omega_n t)$. As shown below, the total energy, $E_{tot}(t)$, in the nth normal mode, with maximum velocity, v_1, is a constant.

$$E_{tot}(t) = (KE)_n + (PE)_n = \frac{1}{4} m_s v_1^2 \left[\sin^2(\omega_n t) + \cos^2(\omega_n t)\right] = \frac{1}{4} m_s v_1^2 \tag{3.29}$$

When the string is passing through its equilibrium position, $y(x, t) = 0$, all of the energy is kinetic. When it is at its extreme displacement and momentarily at rest, all of the energy is potential. Since no loss mechanisms have been included in the model thus far and the attachment points are rigid, the total energy must be conserved.

3.4.1 Nature Is Efficient

"Nature uses as little as possible of anything." Johannes Kepler [5]

As before (see Sect. 2.3.1), we can still use the energy to approximate the normal mode frequencies, even for continuous systems, if we equate the maximum kinetic and potential energies. To illustrate this approach and to develop confidence that will empower us to apply Rayleigh's method to attack problems that might be difficult to solve exactly, we will apply his method to the calculation of the fundamental mode frequency of the fixed-fixed string. Since we have already found the exact solution, provided in Eq. (3.22), we can easily determine the error introduced by making such an approximation.

We know that the exact shape of the uniform fixed-fixed string vibrating in its nth normal mode consists of n half-sine functions between the two rigid supports: $\lambda_n = 2L/n$. Using Eq. (3.25), we can write the kinetic energy of the nth mode.

---

[8] We can ignore the product of the length change and the tension change since that would be the product of two second-order quantities making it fourth-order in the relative displacement from equilibrium.

$$(KE)_n = \frac{\rho_L L}{4}(\omega_n C_n)^2 \tag{3.30}$$

Using Eq. (3.28), and writing $k_n = 2\pi/\lambda_n$,

$$(PE)_n = \frac{\pi^2 \mathrm{T} L}{\lambda_n^2} C_n^2 = \frac{n^2\pi^2 \mathrm{T} L}{4L^2} C_n^2. \tag{3.31}$$

I like to call the coefficient of C_n^2 in Eq. (3.31) the *stability coefficient* (see Sect. 1.2) and the coefficient of $(\omega_n C_n)^2$ in Eq. (3.30) the *inertia coefficient*, although those designations are somewhat archaic [6]. The radian frequency of the nth normal mode, ω_n, is then given by the square root of the ratio of the stability coefficient to the inertia coefficient.

$$\omega_n = \sqrt{\frac{\text{stability}}{\text{inertia}}} = \left[\frac{n^2\pi^2 \mathrm{T} L}{4L^2}\frac{4}{\rho_L L}\right]^{1/2} = \left[\frac{n^2\pi^2 c^2}{L^2}\right]^{1/2} = \frac{n\pi c}{L} \tag{3.32}$$

This regenerates the result of Eq. (3.20) for the normal mode frequencies of a uniform fixed-fixed string.

Now let us imagine that we cannot solve Eq. (3.4) exactly for a fixed-fixed string. To use Rayleigh's method to approximate the fundamental frequency, f_1, we need to postulate a mode shape that satisfies the boundary conditions, has no zero-crossing (since we want the fundamental mode frequency), and can be easily used to evaluate the kinetic and potential energies of vibration. If we define the center of the string at $x = 0$, then the following function will clearly make $y(-L/2) = y(+L/2) = 0$.[9]

$$y_1(x,t) = C_1\left[1 - \left(\frac{2x}{L}\right)^m\right]\cos\omega_1 t \tag{3.33}$$

If the exponent $m = 1$, then our trial function is just a triangle with its peak in the center. If $m = 2$, then the trial function becomes a parabola. These trial functions are plotted in Fig. 3.6. Since the fundamental mode is symmetric about $x = 0$, we can calculate the energies for only half of the string.

$$(KE)_1 = \frac{\rho_L}{2}\int_0^{L/2}\left(\frac{\partial y}{\partial t}\right)^2 dx = \frac{\rho_L m^2 L}{2(m+1)(2m+1)}(C_1\omega_1)^2 \tag{3.34}$$

$$(PE)_1 = \frac{\mathrm{T}}{2}\int_0^{L/2}\left(\frac{\partial y}{\partial x}\right)^2 dx = \frac{\mathrm{T} m^2}{(2m-1)L} C_1^2 \tag{3.35}$$

Application of Eq. (3.32), using the inertia and stability coefficients calculated above, determines ω_1 for any value of the exponent, m.

$$\omega_1^2 = \frac{2(m+1)(2m+1)}{2m-1}\frac{\mathrm{T}}{\rho_L L^2} \tag{3.36}$$

For $m = 1$, $\omega_1^2 = 12(T/\rho_L L^2)$. The exact answer is $\pi^2(T/\rho_L L^2) = 9.87(T/\rho_L L^2)$, so this approximation, using a "triangular" trial function, overestimates the frequency by about 11%. For the parabolic

[9] We will be generating non-integer values for the exponent m, so we should have two expressions for Eq. (3.33), one for positive values of x and one for negative values. Since the problem is symmetric about $x = 0$, we avoid writing the second expression by only integrating over $0 < x \le L/2$ when we calculate the stability and inertial coefficients.

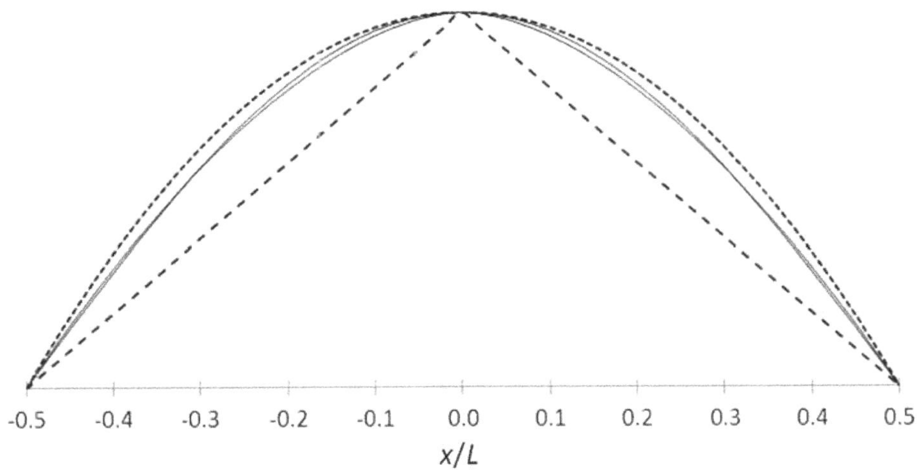

Fig. 3.6 The four trial functions used to approximate the frequency, f_1, of the fundamental normal mode of a uniform fixed-fixed string using Rayleigh's method are shown. The *inner* and *outer dashed curves* represent Eq. (3.33) for $m = 1$ (*triangle*) and $m = 2$ (*parabola*). The two fine solid lines between them are nearly indistinguishable. They represent the exact sinusoidal solution and Eq. (3.33) with the optimized value of the exponent from (3.37) with $m \cong 1.725$

trial function, with $m = 2$, $\omega_1^2 = 10(T/\rho_L L^2)$, which provides an overestimate of the frequency by less than 0.7%.

As Lord Rayleigh pointed out, "When therefore the object is to estimate the longest proper period of a system by means of calculations founded on an assumed type [the 'type' is what we have called the trial function], we know a priori that the result will come out too small." [7] Said another way, Rayleigh is acknowledging the fact that nature minimizes the total energy by using the exact "trial function," hence minimizing the normal mode frequency (or equivalently, maximizing the period).

Unless we select nature's shape function as our trial function, the frequency obtained from Eq. (3.32) will always be an overestimate, as we saw with our $m = 1$ and $m = 2$ solutions for the fundamental mode of a fixed-fixed string. The parabolic trial function was closer to nature's sinusoidal mode shape, so its frequency error was smaller, but the parabolic trial function still predicted a frequency that was slightly greater than the exact solution.

Rayleigh was able to exploit this fact by realizing that he could improve the trial function by minimizing the frequency (or the square of the frequency, if that is more convenient) by adjusting a trial function's parameter. Taking the derivative of Eq. (3.36) with respect to the exponent, m, and setting that derivative to zero,[10] we can find the value of m that provides the lowest frequency that can be obtained with an assumed trial function of the form in Eq. (3.33).[11]

[10] When the fraction is written as the ratio of two quadratic functions in the form $u = (A + 2Hm + Bm^2)/(a + 2hm + bm^2)$, then the minimum or maximum value is given by the solution of the following quadratic equation: $(ab - h^2)u^2 - (aB + bA - 2hH)u + (AB - H^2) = 0$.

[11] If you are too lazy to work out the derivative (or want an independent check), it is possible to use an equation solver, like that available in many mathematical software packages, to determine the value of m which minimizes Eq. (3.36). The result from such a solver is $m = 1.724745$, producing a coefficient of 9.89898.

$$\frac{d}{dm}\left[\frac{2(m+1)(2m+1)}{(2m-1)}\right]=0 \quad \Rightarrow \quad m=\frac{1}{2}\left(\sqrt{6}+1\right)\cong 1.72474 \tag{3.37}$$

Substitution of that optimized exponent into Eq. (3.36) gives $\omega_1{}^2 = 9.899\ (T/\rho_L L^2)$ which only overestimates the normal mode frequency by less than 0.15%.

Figure 3.6 shows the four trial functions used for this exercise. It is reassuring that the triangular shape produced a modal frequency that was only slightly greater than 10% higher than the exact result. The parabolic trial function is quite close to the exact solution, so an error of 0.7% seems reasonable. In Fig. 3.6, the difference in the shapes between the exact sinusoidal function and Eq. (3.33) with $m = ½\left(\sqrt{6}+1\right)$ is nearly imperceptible and produces an approximate frequency that is less than 0.15% greater than the exact result.

3.4.2 Point Mass Perturbation

Let us calculate the shift in the normal mode frequencies produced by the addition of a small mass, m_o, to the string at some location, $0 < x < L$, between the rigid supports using Rayleigh's method. We know the exact string shapes for the unloaded uniform string, so if we assume that $m_o \ll \rho_L L = m_s$, the unloaded normal mode shapes should provide excellent trial functions. The result will provide a useful check when we calculate the exact solution for a mass-loaded string.

The stability coefficient is unchanged by the addition of a point mass. The inertia coefficient must now include the sum of the kinetic energy contribution made by the added mass, m_o, located at x, and the kinetic energy of the uniform string as given by Eq. (3.25). Since we are using the unperturbed normal mode shapes, the velocity of the mass will be $v(x) = C_n\,\omega_n \sin(n\pi x/L)$.

$$(KE)_n=\frac{1}{4}\rho_L L C_n^2\omega_n^2+\frac{m_o}{2}C_n^2\omega_n^2\sin^2\left(\frac{n\pi x}{L}\right)=\frac{C_n^2\omega_n^2}{4}\left[m_s+2m_o\sin^2\left(\frac{n\pi x}{L}\right)\right] \tag{3.38}$$

Since the "interesting" change produced by the added mass occurs in the denominator of Eq. (3.32), we will write the expression for the square of the period, $T_n{}^2$, instead of the square of the radian frequency, $\omega_n{}^2$.

$$T_n^2=\frac{4\pi^2}{\omega_n^2}=4\pi^2\frac{\text{inertia}}{\text{stability}}=\frac{4\pi^2 m_s}{TLk_n^2}\left[1+2\frac{m_o}{m_s}\sin^2\left(\frac{n\pi x}{L}\right)\right] \tag{3.39}$$

Recognizing that $c^2 = (TL/m_s)$, the combination, $(ck_n)^2 = \omega_n{}^2$, is just the square of the unperturbed radian frequency.

$$f_n=\frac{1}{T_n}=\frac{\omega_n}{2\pi}\cong\frac{nc}{2L}\left[1-\frac{m_o}{m_s}\sin^2\left(\frac{n\pi x}{L}\right)\right] \quad \text{for} \quad m_o << m_s \tag{3.40}$$

As expected, the addition of the mass has decreased the normal mode frequencies. That decrease depends both on the location of the added mass and the mode number, as well as the magnitude of the perturbation, m_o.

If we designate the perturbed modal frequencies as f_n', and the unperturbed frequencies given in Eq. (3.22) as f_n, application of the binomial expansion to Eq. (3.39) allows us to invert the expression in square brackets and take its square root under the assumption that $m_o \ll m_s$.

Table 3.5 The effect on the modal frequency shift of Eq. (3.41) caused by a small mass, m_o, placed at $x = L/4$ on a string of mass, $m_s \gg m_o$, is determined by $\sin^2(n\pi x/L)$. Since $x = L/4$ is a location of a node for the $n = 4$ mode, the added mass has no effect on that normal mode frequency. That position is the location of maximum displacement for the $n = 2$ and $n = 6$ modes, producing the largest frequency shifts

Mode	$\sin^2(n\pi x/L)$
1	0.5
2	1.0
3	0.5
4	0.0
5	0.5
6	1.0

$$\frac{f_n - f'_n}{f_n} = \frac{\delta f_n}{f_n} \cong \frac{m_o}{m_s} \sin^2\left(\frac{n\pi x}{L}\right) \quad \text{if} \quad m_o << m_s \tag{3.41}$$

Table 3.5 lists the values of $\sin^2(n\pi x/L)$ for a small mass located one-quarter of the way from one rigid support of a uniform string, $x = L/4$. Because the $n = 4$ normal mode shape has a node at the location of the added mass, the mass has no effect on the normal mode frequency, f_4, or on any other mode that has a mode number which is an integer multiple of four, f_j, $j = 4, 8, 12$, etc.

3.4.3 Heavy Chain Pendulum (Nonuniform Tension)*

As one final example of Rayleigh's method applied to a string, we will obtain an approximate solution to a problem which actually was the first problem to necessitate the use of what are now known as Bessel functions to produce the exact solution [8]: a fixed-free string with a tension that is a linear function of position along the string. We will consider a chain (or string) with a constant linear mass density, ρ_L, but hung under the influence of gravity, so that the tension decreases linearly with the distance from the fixed attachment point.

To remind ourselves that the string's vertical orientation is central to the definition of this problem, we will choose the z axis to specify position along the string in its equilibrium (straight) state If the (fixed) top end of the string is defined as $z = 0$ and we let z increase as we go downward, then the tension in the string, as a function of position, T(z), is a linear function of distance from the support point.

$$\mathrm{T}(z) = g\rho_L(L - z) \tag{3.42}$$

At the top, the entire weight, $\rho_L gL$, must be supported, causing the tension to be greatest at the attachment point, $z = 0$. We will still let the transverse displacement of the string from its equilibrium state be y (z, t).

Returning to our definition of the net vertical force on an infinitesimal length of string, given in Eq. (3.2), we see that we can no longer treat T as a constant, although we will still make the small-amplitude (linear) approximation.

$$dF_y = \left.\frac{\partial(\mathrm{T}\sin\theta)}{\partial z}\right|_y dz \cong \left.\frac{\partial\left(\mathrm{T}\frac{\partial y}{\partial z}\right)}{\partial z}\right|_y dz \tag{3.43}$$

Application of Newton's Second Law produces an equation of motion that is not the wave equation of Eq. (3.4).

$$\rho_L \frac{\partial^2 y}{\partial t^2} = \frac{\partial}{\partial z}\left[T(z)\frac{\partial y}{\partial z}\right] = \rho_L g \frac{\partial}{\partial z}\left[(L-z)\frac{\partial y}{\partial z}\right] \tag{3.44}$$

Having no desire to attempt a solution to this new equation, we will use Rayleigh's method to approximate the fundamental normal mode frequency, f_o.

The first step will be to choose a trial function that satisfies the boundary condition for the fundamental mode, $y_o\,(0, t) = 0$, while providing sufficient flexibility that we are able to minimize the resulting frequency. Of course, a polynomial trial function is preferred to simplify the necessary differentiations and integrations. An adjustable mix of a linear and quadratic dependence upon z is a reasonable choice. The adjustable parameter, β, will be used to optimize the trial function.

$$y_o(z,t) = C_o \cos(\omega_o t)\left(\frac{z}{L} + \beta\frac{z^2}{L^2}\right) \tag{3.45}$$

Calculation of the kinetic energy will involve the square of the velocity.

$$\dot{y}_o^2 = \omega_o^2 C_o^2 \sin^2(\omega_o t)\left(\frac{z}{L} + \beta\frac{z^2}{L^2}\right)^2 \tag{3.46}$$

Substitution into the equation for kinetic energy is straightforward but a little messy.

$$(KE)_o = \frac{\rho_L \omega_o^2 C_o^2}{2}\int_0^L \left(\frac{z^2}{L^2} + 2\beta\frac{z^3}{L^3} + \beta^2\frac{z^4}{L^4}\right) dz \tag{3.47}$$

Having chosen a polynomial trial function, integration is simple.

$$(KE)_o = \frac{\rho_L \omega_o^2 C_o^2 L}{2}\left(\frac{1}{3} + \frac{\beta}{2} + \frac{\beta^2}{5}\right) = \frac{\rho_L \omega_o^2 C_o^2 L}{60}\left(10 + 15\beta + 6\beta^2\right) \tag{3.48}$$

The calculation of the potential energy requires the square of the derivative of y with respect to z.

$$\left(\frac{\partial y}{\partial z}\right)^2 = C_o^2 \cos^2(\omega_o t)\left(\frac{1}{L} + 2\beta\frac{z}{L^2}\right)^2 = C_o^2 \cos^2(\omega_o t)\left(\frac{1}{L^2} + 4\beta\frac{z}{L^3} + 4\beta^2\frac{z^2}{L^4}\right) \tag{3.49}$$

Since the potential energy must include the dependence of tension on position, the integral contains more terms but is just as easy to evaluate.

$$(PE)_o = \frac{\rho_L g C_o^2}{2}\int_0^L (L-z)\left(\frac{1}{L^2} + 4\frac{\beta z}{L^3} + 4\frac{\beta^2 z^2}{L^4}\right) dz = \frac{\rho_L g C_o^2}{12}\left(3 + 4\beta + 2\beta^2\right) \tag{3.50}$$

As before, the ratio of the stability coefficient to the inertia coefficient provides an estimate of the upper limit to the square of the normal mode frequency, ω_o^2, for the fundamental mode of an oscillating chain with a free end as a function of our adjustable parameter, β.

$$\omega_o^2 \le \frac{5\left(3 + 4\beta + 2\beta^2\right)}{10 + 15\beta + 6\beta^2}\frac{g}{L} \tag{3.51}$$

The use of the "≤" sign in Eq. (3.51) is an acknowledgment of Rayleigh's recognition that this method never underestimates the modal frequency. If we let $\beta = 1$, thereby making the linear and quadratic combinations in Eq. (3.45) be equal, $\omega_o^2 = (45/31)\,(g/L) = 1.4516\,(g/L)$. We should be able to reduce this result, bringing it closer to the exact value, if we minimize the fractional coefficient of (g/L) in Eq. (3.51). That can be done analytically[11] or by use of an equation solver. I chose the latter and found $\beta = 0.6385$, making $\omega_o^2 = 1.4460\,(g/L)$.

The exact solution is given by the first zero-crossing of the zeroth-order Bessel function of the first kind, J_o, and produces a result which is identical to Rayleigh's method (after optimization) to within four decimal places: $\omega_o^2 = 1.4460\,(g/L)$.

Since the fundamental mode corresponds to the motion of the chain staying nearly straight and moving almost like a rigid pendulum, the frequency could have been approximated from the moment of inertia of the chain, treated as a rigid rod: $I = (m_s L^2)/3$. The linear restoring torque, given by Eq. (1.25), acting as though all of the mass were concentrated at the chain's midpoint produces a torsional stiffness: $K_{torq} = m_s g L/2$, resulting in $\omega_o^2 = 1.500\,(g/L)$. The actual fundamental frequency is 1.8% lower than this "rigid pendulum" approximation (i.e., $\beta = 0$).

3.5 Initial Conditions

Thus far, the amplitude coefficients, C_n, for the normal modes have remained undetermined. As was the case with the vibrational amplitudes of the simple harmonic oscillator, the values of these coefficients are determined by the initial conditions. For the excitation of the normal modes of a string, the two limiting cases are a plucked or a struck string. A plucked string (e.g., guitar or harp) is pulled away from its equilibrium position so that the string forms two legs of a triangle with the separation between the fixed ends as its base. It is then released at $t = 0$. For a struck string (e.g., piano or dulcimer), a hammer is usually "bounced" off the string at $t = 0$, imparting an initial velocity at the instant of contact.[12] The bowed instruments of the violin family or the Chinese *erhu* (二胡) are better described as driven strings that include a complex feedback system of the slip-stick friction produced by the bow that is mode-locked to the modal frequency of the string [9].

The imposition of the initial conditions on a string is facilitated by the use of the Fourier series (see Sect. 1.4). The harmonic modes of the fixed-fixed string in Eq. (3.21) form a complete set of orthogonal basis functions that allow any initial displacement or velocity to be expressed as the superposition of normal modes.[13]

$$y(x,t) = \sum_{n=1}^{\infty} y_n(x,t) = \sum_{n=1}^{\infty} C_n e^{j\omega_n t} \sin\left(n\frac{\pi x}{L}\right) \tag{3.52}$$

An initial displacement, $y_o\,(x, 0)$, or an initial velocity, $v_o\,(x, 0)$, can then be written in terms of that series.

[12] This is the physicists' definition of the striking process. In reality, there are many other factors that influence the excitation of a struck string including the curvature of the hammer's face as well as the compliance (e.g., hard wood or soft felt) of the material on the hammer's face.

[13] It is worth pointing out that the orthogonality of these basis functions for standing waves on a string is a special case that is only true for idealized (rigid) Dirichlet boundary conditions (after P. G. L. Dirichlet, 1805–1859) or (free) Neumann boundary conditions (after Carl G. Neumann, 1832–1925). For other boundary conditions, the normal modes are not orthogonal. See, for example, Fig. 3.8 where the integral of the product of the lowest (pendulum) mode (dashed line) and the first nearly half-wavelength mode (solid line), when integrated over the length of the string, is clearly non-zero.

$$\begin{aligned} y(x,0) &= y_o(x) = \sum_{n=1}^{\infty} C_n \sin\left(n\frac{\pi x}{L}\right) \\ \dot{y}(x,0) &= v_o(x) = j\sum_{n=1}^{\infty} \omega_n C_n \sin\left(n\frac{\pi x}{L}\right) \end{aligned} \tag{3.53}$$

As in Sect. 1.4, we exploit the orthogonality of the basis functions (i.e., normal modes) to determine the values of the coefficients, C_n.

$$\int_0^L \sin\left(n\frac{\pi x}{L}\right) \sin\left(m\frac{\pi x}{L}\right) dx = \begin{cases} L/2 & \text{if n = m} \\ 0 & \text{if n} \neq \text{m} \end{cases} \tag{3.54}$$

Multiplication of Eq. (3.53) by sin $(m\pi x/L)$ and integration over the length of the string extracts the amplitude coefficients, C_n.

$$\begin{aligned} \int_0^L y_o(x) \sin\left(m\frac{\pi x}{L}\right) dx &= \sum_{n=1}^{\infty} C_m \int_0^L \sin\left(m\frac{\pi x}{L}\right) \sin\left(n\frac{\pi x}{L}\right) dx = \left(\frac{L}{2}\right) C_m \\ C_n &= \frac{2}{L}\int_0^L y_o(x) \sin\left(n\frac{\pi x}{L}\right) dx \end{aligned} \tag{3.55}$$

The same procedure will also produce the coefficients for an initial velocity distribution.

$$C_m = \frac{2}{\omega_m L}\int_0^L v_o(x) \sin\left(m\frac{\pi x}{L}\right) dx \tag{3.56}$$

For the plucked case, the solution for the C_n coefficients that describe the subsequent (i.e., $t > 0$) motion of a string of length, L, initially displaced from its straight condition by a transverse distance, $y_o(x_o, 0) = h$, at a distance, x_o, from one attachment point can be determined if the initial shape of the string is expressed by the function in Eq. (3.57). Values of the C_n coefficients can be calculated by breaking the integral of Eq. (3.55) into two parts. Following the terminology used for most Western stringed instruments, we will designate $x = 0$ as the location of the "bridge" and $x = L$ as the location of the "nut," which is typically the end closest to the tension adjusters (e.g., tuning pegs, machine heads).

$$y_o(x,0) = \begin{cases} h\left(\dfrac{x}{x_o}\right) & \text{for} \quad 0 \leq x < x_o \\ h\left(\dfrac{L-x}{L-x_o}\right) & \text{for} \quad x_o \leq x < L \end{cases} \tag{3.57}$$

The initial shape will look much like the dashed triangle in Fig. 3.6, though that figure would pertain to $x_o = L/2$, with the amplitude being greatly exaggerated, since $h \ll L$ is required to ensure the assumed subsequent linear behavior.

$$C_n = \frac{2h}{L}\int_0^{x_o} \frac{x}{x_o} \sin\left(\frac{n\pi x}{L}\right) dx + \frac{2h}{L}\int_{x_o}^L \frac{(L-x)}{(L-x_o)} \sin\left(\frac{n\pi x}{L}\right) dx \tag{3.58}$$

The usual formulas for the integration of sine functions are provided below:

$$\int \sin ax\, dx = -\frac{1}{a}\cos ax + C; \qquad \int x \sin ax\, dx = \frac{1}{a^2}\sin ax - \frac{1}{a} x\cos ax + C \tag{3.59}$$

After repeated applications of Eq. (3.59) and collection of like terms, the amplitude coefficients for a string plucked at x_o are obtained.

Table 3.6 The amplitude coefficients, C_n, for a plucked string are given in Eq. (3.60)

x_o/L	2:1	3:1	4:1	5:1	6:1	7:1	8:1
0.500	0	−0.111	0	0.040	0	−0.020	0
0.400	0.155	−0.069	−0.063	0	0.028	0.013	−0.010
0.333	0.250	0	−0.062	−0.040	0	0.020	0.016
0.200	0.405	0.180	0.063	0	−0.028	−0.033	−0.025
0.143	0.450	0.250	0.140	0.072	0.028	0	−0.016
0.125	0.462	0.268	0.163	0.097	0.051	0.020	0
0.100	0.476	0.291	0.192	0.129	0.085	0.053	0.030
0.050	0.494	0.322	0.235	0.181	0.144	0.116	0.095

In this table, the plucking position is given by x_o/L. The ratios of the subsequent harmonic amplitudes, C_n/C_1, are provided in the other columns. A negative ratio indicates that the harmonic has a phase that is opposite that of the fundamental. It can be seen that any harmonic which has a node at the plucking position has zero amplitude (e.g., every other column for $x_o/L = ½$ is zero). For the central plucking position, $x_o/L = 0.50$, the non-zero ratios decrease in proportion to n^{-2} as expected for the (isosceles) triangular wave in Eq. (1.45). For the plucking position closest to the bridge, $x_o/L = 0.50$, the relative amplitudes decrease nearly in proportion to n^{-1} as expected from Eq. (3.61)

$$C_n = \frac{2}{\pi^2}\frac{h}{n^2}\frac{L^2}{x_o(L-x_o)}\sin\left(\frac{n\pi x_o}{L}\right) \tag{3.60}$$

For any given plucking displacement, h, no modes will be excited that would have had nodes at the plucking location, since $\sin(n\pi x_o/L) = 0$ if $n = L/x_o$, or any integer multiple of that ratio, as shown in Table 3.6. For example, if the string is plucked at its center, then none of the even harmonics will be excited, since all such modes have nodes at the string's center, $x_o = L/2$.

The result of Eq. (3.60) for the amplitude coefficients of a plucked string has a strong influence on the tonality of the resulting vibrations. This is illustrated in Table 3.6 where it is obvious that the relative strength of the higher harmonics becomes greater as the plucking position approaches the bridge. In the limit that x_o/L is very small, the relative amplitude of the harmonics falls off as n^{-1}.

$$\lim_{x_o/L\to 0}[C_n] = \frac{2h}{\pi n} \tag{3.61}$$

This limiting behavior is approached in the last row of Table 3.6, where $x_o/L = 0.05$.

The sound is more "mellow" if the string is plucked near the center and is "twangy" when plucked near the bridge. This is also the reason that many electric guitars have multiple pickup locations (e.g., the Fender Stratocaster has three pickups located between the bridge and the place where the neck joins the body). An electromagnetic guitar pickup will be less sensitive to any mode that has a node that is close to the pickup location. The closer the pickup is to the bridge, the smaller the amplitude of any mode will be, but also the less likely that any mode will have a node near the pickup location. Again, the pickup closer to the position where the neck joins the body will produce a more "mellow" sound and the one near the bridge will sound more "twangy."

3.5.1 Total Modal Energy

Having calculated the behavior of a plucked string that is released at $t = 0$, we can now calculate the work it took to displace the string by a transverse distance, h, at a position, x_o, before it was released (i.e., at times $t < 0$). Energy conservation guarantees that the work done to displace the string before release must be equal to the sum of the energies of the individual normal modes of vibration that were excited. Since we have shown that the total energy of any mode is the sum of the kinetic and potential energies and that the maximum value of the kinetic and potential energies is equal, we can simplify this calculation by summing the maximum kinetic energies, $(KE)_n$, expressed in Eq. (3.30).

This can be demonstrated easily for a string plucked at its center $x_o = L/2$, although it must also be true for any plucking position. Substitution of $x_o = L/2$ into the general result of Eq. (3.60) provides the amplitude coefficients for the fixed-fixed string plucked at its center.

$$C_n(x_o = \tfrac{1}{2}L) = \frac{1}{n^2}\frac{8h}{\pi^2}\sin\left(\frac{n\pi x_o}{L}\right); \quad n = 1, 3, 5, \ldots \tag{3.62}$$

Since the string is plucked at the center, excitation of all even harmonics is suppressed. The maximum kinetic energy stored in each mode is proportional to the square of the maximum velocity, ${v_n}^2 = {\omega_n}^2 {C_n}^2$, of that mode, where we have used Eq. (3.20) to relate ω_n to the mode number, n.

$$v_n^2 = \omega_n^2 C_n^2 = \left(\frac{n\pi c}{L}\right)^2 \left(\frac{1}{n^2}\frac{8h}{\pi^2}\right)^2 = \left(\frac{8hc}{\pi n L}\right)^2 \tag{3.63}$$

The total energy of all of the modes is then expressed as a summation over the odd values of mode number, n.

$$E_{tot} = \frac{m_s}{4}\sum_{n=odd}^{\infty}\left(\frac{8hc}{\pi n L}\right)^2 = m_s c^2\left(\frac{4h}{\pi L}\right)^2 \sum_{n=0}^{\infty}\frac{1}{(2n+1)^2} = \frac{2T}{L}h^2 \tag{3.64}$$

The final expression on the right-hand side of Eq. (3.64) uses the fact that the sum of that infinite series is $\pi^2/8$ [10] and $m_s c^2 = \rho_L L c^2 = LT$.

We have already calculated the total vertical component of the force that the tension of a fixed-fixed string exerts when its center is displaced from equilibrium by a distance, y, in Eq. (2.129). The total work done to displace the center by a distance, h, is therefore given by the integral of the vertical component of that force, $F_y\,(y)$, times the displacement, y, in the vertical direction.

$$W = -\int_0^h F_y y\,dy = \int_0^h \left(\frac{4T}{L}\right) y\,dy = \frac{2T}{L}h^2 \tag{3.65}$$

The equality of the results of Eqs. (3.64) and (3.65) can be viewed as a confirmation of our faith in the conservation of energy or as an interesting way to find the sum of the infinite series in Eq. (3.64).

3.6 "Imperfect" Boundary Conditions

The harmonic series produced by a string that is rigidly immobilized at both ends is a special case.[14] If the boundary conditions are more general, we will see that the standing wave frequencies no longer form a harmonic series and the mode shapes are no longer orthogonal. Also, for the case of a mass-loaded string, a new "lumped-element" mode (i.e., the pendulum mode) appears. That additional normal mode frequency is not harmonically related to the standing wave modes. The wavelength of that pendulum mode may be many times greater than the length of the string joining the rigid attachment point to the mass that provides the termination at the other end.

The analytical "tools" we've developed thus far are entirely adequate to treat a string that is terminated by a mass that is not infinite (i.e., a nonrigid termination). If the end of the string at $x = 0$ is rigidly fixed, then we already know that the form of the solution imposed by that end condition must be given by Eq. (3.19). We can use Newton's Second Law to impose the boundary condition at the point of attachment of the mass, M, located at $x = L$, which will result in the quantization of the allowable normal mode frequencies.

$$F_y(L,t) = -\mathrm{T}\left(\frac{\partial y}{\partial x}\right)_{x=L} = M\left(\frac{\partial^2 y}{\partial t^2}\right)_{x=L} = j\omega M\left(\frac{\partial y}{\partial t}\right)_{x=L} \tag{3.66}$$

The right-hand version expresses the force which the string applies to the mass (and by Newton's Third Law of Motion that must be the negative of the force that the mass exerts on the string) in terms of the mechanical impedance of the terminating mass, $j\omega M$. Substitution of Eq. (3.19) into the above expressions is straightforward.

$$\left(\frac{\partial y}{\partial x}\right)_L = C_n k_n \cos(k_n L) e^{j\omega_n t} \quad \text{and} \quad \left(\frac{\partial y}{\partial t}\right)_L = j\omega_n C_n \sin(k_n L) e^{j\omega_n t} \tag{3.67}$$

Plugging these expressions into Eq. (3.66) provides an equation that will yield the quantized values of the normal mode frequencies, ω_n, and wavenumbers, k_n.

$$\frac{\cos k_n L}{\sin k_n L} = \cot(k_n L) = \frac{M}{\mathrm{T}}\left(\frac{\omega_n}{k_n}\right)^2 k_n = \frac{M}{\mathrm{T}} c^2 k_n = \frac{M}{m_s}(k_n L) \tag{3.68}$$

This is a *transcendental equation* for $(k_n L) \equiv x$. It does not have a closed-form algebraic solution. Figure 3.7 provides the quantizing solutions for $(k_n L) \equiv x$, which occur at the intersections of $\cot x$ and the straight line of slope (M/m_s) passing through the origin. Thus far, we have assumed that the tension, T, is not a function of the string's mass, m_s, as it was for the "heavy chain" with variable tension analyzed in Sect. 3.4.3.

Before examining the solutions in some important limits, it is worthwhile to consider the general type of normal modes that arise for this fixed mass-loaded string. If $M = 0$, then our current analysis should regenerate the modes of the fixed-free string, given in Eq. (3.24), which correspond to odd-integer multiples of a quarter wavelength fitting within the overall length of the string, L. If

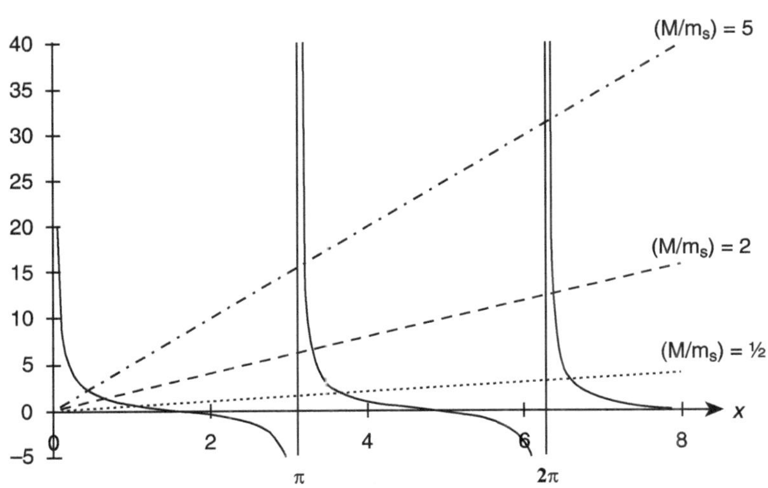

Fig. 3.7 A graphical solution of Eq. (3.68): $\cot x = (M/m_s)\,x$, hence both axes are dimensionless. If $M = 0$, the intersection of $\cot x$ with the x axis defines the modes of a fixed-free string with $(k_n L) = (2n-1)(\pi/2)$. The *dotted line* corresponds to $M/m_s = ½$, the *dashed line* $M/m_s = 2$, and the *dash-dot line* $M/m_s = 5$. As M/m_s becomes very large, the intersections approach $(k_n L) = n\pi$. In that limit, the modes become those of a fixed-fixed string. Intersections of the lines with $x < \pi/2$ correspond to the lumped-element "pendulum" mode

$M = 0$, those modes are defined by the intersections of cot $(k_n L)$ with the x axis of the graph in Fig. 3.7. Those intersections occur at $(k_n L) = (2n - 1)(\pi/2)$, so $\omega_n = ck_n = c(k_n L/L) = (2n - 1)(\pi c/2L)$, or $f_n = (2n - 1)(c/4L)$, in agreement with Eq. (3.24), where $n = 1, 2, 3$, etc.

As the terminating mass, M, increases, the slope of the straight line increases, and the values of $(k_n L)$ for all normal modes decrease. If the terminal mass is much greater than the mass of the string, $M \gg m_s$, the intersection of the straight line and the cotangent start to approach the cotangent's asymptotes, so that the standing wave modes approach $(k_n L) = n\pi$ for $n = 1, 2, 3$, etc. This limit produces the (nearly) harmonic series we expect for a fixed-fixed string, since a very large mass approximates a second rigidly fixed end condition.

The one-quarter wavelength mode for the $n = 0$ case becomes the pendulum mode when $M > 0$. The pendulum frequency is given by the intersection of the straight line with the cotangent function and occurs for $(k_n L) < \pi/2$. Since we seek solutions for small values of x when $M \gg m_s$, those solutions can be approximated if the cotangent function is expanded in a Taylor series.

$$\lim_{x\to 0}[\cot x] = \frac{1}{x} - \frac{x}{3} - \frac{x^3}{45} - \frac{2x^5}{945} - \cdots \tag{3.69}$$

If we retain only the first term, substitute into Eq. (3.68), and let $T = Mg$, then

$$\frac{1}{x} \cong \frac{M}{m_s}x \quad \Rightarrow \quad x^2 = (k_o L)^2 = \frac{\omega_o^2 L^2}{c^2} \cong \frac{m_s}{M} \quad \Rightarrow \quad \omega_o^2 \cong \frac{g}{L} \tag{3.70}$$

The frequency of that pendulum mode is subscripted as $n = 0$, a convention we imposed without justification when specifying the natural frequency of a simple (mass-spring) harmonic oscillator in Chap. 2. We will use this designation again for the normal mode frequency of a Helmholtz resonator in Chap. 8 (e.g., the resonance frequency of an ocarina or the frequency generated when blowing over beverage bottles like those shown in Fig. 8.1).

Equation (3.70) has recovered the pendulum frequency calculated for the first time in this textbook using similitude, resulting in Eq. (1.80), and then again from the solution of the differential equation, yielding Eq. (2.32). In both cases, we assumed that "information" could pass along the entire string in time intervals that are short compared to the oscillation period. This is equivalent to saying that the wavelength along the string, when the system oscillates in the pendulum mode, is much greater than the physical length of the string, $\lambda_o = c/f_o \gg L$. That allowed us to treat the pendulum as a "lumped-element" system in which all of the mass was concentrated at the pendulum bob.

If we include the second term in the Taylor series expansion of cot x, then we see the effects of the string's mass, m_s, coming into play in just the same way as when we used energy considerations to incorporate the mass of a spring in our calculation of the mass-spring natural frequency given in Eq. (2.27).

$$\frac{1}{x} \cong \left(\frac{M}{m_s} + \frac{1}{3}\right)x \quad \Rightarrow \quad x^2 = \frac{\omega_o^2 L^2}{c^2} \cong \frac{1}{\left(\frac{M}{m_s} + \frac{1}{3}\right)} \cong \frac{m_s}{M}\left(1 - \frac{m_s}{3M}\right) \tag{3.71}$$

The second term in the expansion of cot x tells us that we can improve the accuracy of our solution by adding one-third the mass of the string to the mass of the bob.

$$\omega_o^2 = \left(\frac{g}{L}\right)\left(1 - \frac{1}{3}\frac{m_s}{M}\right) \tag{3.72}$$

3.6.1 Example: Standing Wave Modes for $M/m_s = 5$

Let's examine the modes corresponding to the wavelength being a significant fraction of the string's length. As an example, let's do the calculation for the case of $M = 0.10$ kg and $m_s = 0.020$ kg. We will neglect the fact that the tension is a function of the position along the string, but we will use the average tension, $T = (M + \frac{1}{2}m_s)g = 1.078$ N, to calculate the speed of transverse wave motion. The string's linear mass density is $\rho_L = m_s/L = 0.020$ kg/m, if we let $L = 1.00$ m. The speed of transverse waves on that string will be $c = (T/\rho_L)^{1/2} = 7.34$ m/s.

For this example, we need to solve Eq. (3.68) for $M/m_s = 5$. I employed an equation solver to find $k_oL = 0.4328$, $k_1L = 3.204$ (2% greater than π), and $k_2L = 6.315$ (0.5% greater than 2π). With the value of the wave speed, and $L = 1.00$ m, those solutions correspond to $\omega_n = ck_n$, so $\omega_o = 3.18$ rad/s, $\omega_1 = 23.52$ rad/s, and $\omega_2 = 46.36$ rad/s. The displacement amplitudes for the three lowest-frequency normal modes, y_o (x), y_1 (x), and y_2 (x), are exaggerated in Fig. 3.8.

3.6.2 An Algebraic Approximation for the Mass-Loaded String

"One measure of our understanding is the number of independent ways we are able to get to the same result."
R. P. Feynman

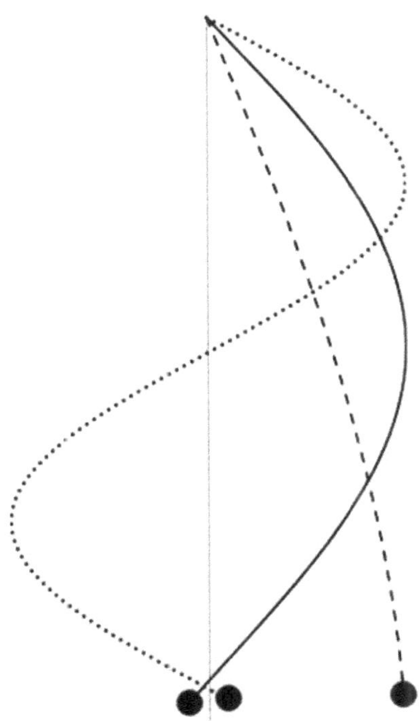

Fig. 3.8 Greatly exaggerated displacement amplitudes of the three lowest-frequency normal modes of a mass-loaded string with $M/m_s = 5$. The (lumped element) "pendulum" mode is shown by the *dashed line*. The corresponding wavelength, $\lambda_o \gg L$, and the normal mode frequency, $\omega_o = 3.18$ rad/s, based on the example in the text (i.e., $c = 7.34$ m/s and $L = 1.00$ m). The fundamental mode ($n = 1$) is shown by the *solid line* which puts just over one full wavelength on the string and produces a frequency $\omega_1 = 23.52$ rad/s. The second standing wave mode ($n = 2$) is shown by the *dotted line* which puts just over one wavelength on the string and produces a frequency, $\omega_2 = 46.36$ rad/s. The ratio, $\omega_2/\omega_1 \cong 2$, as it would be for a fixed-fixed string, but the frequency of the pendulum mode has no simple integer relation to the standing wave modes

Equation (3.72) provides a good approximation for the frequency of the pendulum mode obtained by using the first two terms in the Taylor series approximation to cot x. We can approximate the standing wave normal modes, $n \geq 1$, if we remember that the normal mode is a solution where all parts of the system oscillate at the same frequency.

Careful inspection of Fig. 3.8 shows that a short section of the string connected to the mass is moving out-of-phase with the adjacent section of the string that oscillates with an integer multiple of half-wavelengths. There are nodes very close to the mass for the $n \geq 1$ modes. We can think of those modes as having exactly an integer number of half-wavelengths above the node connected to a very short pendulum below the node, since the node is equivalent to a rigid termination. Since the normal mode frequency of the pendulum part must be the same as the frequency of the standing wave part, we can use that understanding to approximate the normal mode frequencies without resorting to the solution of a transcendental equation.

Figure 3.9 is an exaggeration of the maximum amplitude for the $n = 1$ mode that divides the length of the string into two very unequal parts. The longer segment of the string, which contains a fixed-fixed standing wave mode, is designated L_s. The short pendulum portion is designated L_p. Based on Eq. (3.70), $L_p \cong g/\omega_n^2$. Fitting n half-wavelengths of a fixed-fixed normal mode between the support and the node makes $L_s = nc/2f_n = n\pi c/\omega_n$. The sum of those two segments must equal the total equilibrium length of the string (since we are assuming small displacements), L, resulting in a quadratic equation for ω_n.

$$L = \frac{n\pi c}{\omega_1} + \frac{g}{\omega_1^2} \quad \Rightarrow \quad L\omega_1^2 - n\pi c\omega_1 - g = 0 \tag{3.73}$$

This equation has a well-known algebraic solution.

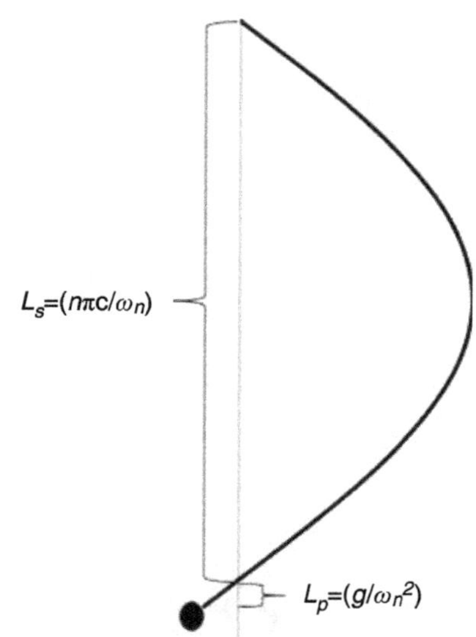

Fig. 3.9 The exaggerated maximum transverse displacement of a mass-loaded string oscillating in its $n = 1$ mode shows that the longer segment of the string, above the displacement node (the node close to the mass), is oscillating in the $n = 1$ fixed-fixed standing wave mode. The length of that segment is $L_s = \lambda/2 = \pi c/\omega_n$. The distance between the node and the pendulum bob, based on Eq. (3.70), is $L_p = g/\omega_n^2$. The sum of those lengths must equal the equilibrium length of the string, $L = L_s + L_p$, leading to a quadratic equation (3.73) for ω_n. The frequencies for this example from Eq. (3.68) are ω_{exact}, and the frequencies from the quadratic approximation are ω_{quad}. Comparison of these frequencies for the first four standing wave modes, $1 \leq n \leq 4$, are provided in Table 3.7

Table 3.7 The radian frequencies for the first four standing wave modes, $1 \leq n \leq 4$, of the example in Fig. 3.9 using Eq. (3.68) are ω_{exact}. The frequencies from the quadratic approximation of Eq. (3.74) are ω_{quad}

n	1	2	3	4
ω_{exact}	23.5169	46.3510	69.3333	92.3538
ω_{quad}	23.4843	46.3311	69.3195	92.3434
$\omega_{\text{exact}} - \omega_{\text{quad}}$	0.0326	0.0199	0.0137	0.0104
Diff/ω_{exact}	0.14%	0.04%	0.02%	0.01%

$$\omega_n = \frac{n\pi c \pm \sqrt{n^2\pi^2c^2 + 4Lg}}{2L} \cong \frac{n\pi c}{L} + \frac{g}{n\pi c} \tag{3.74}$$

The right-hand expression exploits the fact that $(4Lg/\pi^2c^2) = (4/\pi^2)\,(m_s/M) \ll 1$ in our example, so the radical can be approximated using the binomial expansion.

The results for the first four standing wave modes are provided in Table 3.7. The frequency, ω_{exact}, is obtained from the numerical solutions of the transcendental equation, Eq. (3.68), and ω_{quad} comes from the solutions to the quadratic approximation of Eq. (3.74). The ratio of their difference, $\omega_{exact} - \omega_{quad}$, to the exact frequency, ω_{exact}, is reported as a percentage.

3.6.3 The Resistance-Loaded String*

The calculation of the normal modes of a stiffness-loaded string is left to the exercises (see Problem 11), since it can be treated in a way that is very similar to the mass-loaded string. If we want to consider a pure resistance attached to the string at $x = L$, our previous strategy runs into an interesting complication. Since the boundary at $x = 0$ is still rigidly fixed, Eq. (3.19) must still be valid. The vertical component of the force that a pure resistance would apply to the string is proportional to the velocity of the string at the point of attachment.

$$F_v(L) = -R_m\dot{y}(L) = -j\omega R_m C\sin(kL)e^{j\omega t} \tag{3.75}$$

By Newton's Third Law, that force must be equal and opposite to the force the string applies to the resistance.

$$F_v(L) = \mathrm{T}\left(\frac{\partial y}{\partial x}\right)_L = k_n \mathrm{T} C_n\cos(k_nL)e^{j\omega t} = j\omega_n R_m C_n\sin(k_nL)e^{j\omega t} \tag{3.76}$$

This equation can be re-written in the form of Eq. (3.68), but with an interesting difference.

$$\cot k_nL = \frac{j\omega_n R_m}{k_n\mathrm{T}} = j\frac{cR_m}{\mathrm{T}} = j\frac{c^2R_m}{c\mathrm{T}} = j\frac{R_m}{\rho_L c} \tag{3.77}$$

If k_n is a real number, then $\cot k_n L$ must be a real number. In that case, Eq. (3.77) cannot be satisfied since the right-hand side of the equation is purely imaginary. On the other hand, if k_n were a complex number, say $\boldsymbol{k_n} = k_n + j\alpha_n$, then the following identity would produce the necessary imaginary pre-factor.

$$\cot \mathbf{z} = j\coth(j\mathbf{z}) \tag{3.78}$$

Returning to the form of Eq. (3.77), but allowing $\boldsymbol{k_n}$ to be a complex number, we can use the following identities to separate that equation into two individual equations for its real and imaginary components [10].

$$\begin{aligned} \sin(x \pm jy) &= \sin x \cosh y \pm j \cos x \sinh y \\ \cos(x \pm jy) &= \cos x \cosh y \mp j \sin x \sinh y \end{aligned} \tag{3.79}$$

Equating the real parts of Eq. (3.76) produces one equation.

$$-\cos(k_n L)\sinh(\alpha_n L) = \frac{\rho_L c}{R_m}\cos(k_n L)\cosh(\alpha_n L) \tag{3.80}$$

Equating the imaginary parts produces another.

$$-\sin(k_n L)\cosh(\alpha_n L) = \frac{\rho_L c}{R_m}\sin(k_n L)\sinh(\alpha_n L) \tag{3.81}$$

One solution would be to let $\sin k_n L = 0$, automatically satisfying Eq. (3.81), producing the real component of the wavenumber, k_n, that yields fixed-fixed standing wave solutions (if $\alpha_n L$ could be neglected), while simultaneously requiring that $\cos k_n L = 1$. With $\cos k_n L = 1$, the real part, given by Eq. (3.80), becomes the quantizing condition on $\alpha_n L$.

$$\frac{\sinh(\alpha_n L)}{\cosh(\alpha_n L)} = \tanh(\alpha_n L) = \frac{\rho_L c}{R_m} \le 1 \tag{3.82}$$

For all values of x, $|\tanh x| \le 1$. Our use of this expression requires that the terminal resistance be large compared to the string's characteristic resistance, $R_m/\rho_L c > 1$. This is certainly true in the limit where R_m becomes nearly infinite, since that would approximate an immobile boundary resulting in fixed-fixed normal modes, as required by $\sin k_n L = 0$, so $k_n L = n\pi$, for $n = 1, 2, 3$, etc. Also, with $R_m/\rho_L c \gg 1$, we can make a Taylor series expansion of $\tanh(\alpha_n L) \cong (\rho_L c/R_m) \ll k_n L = n\pi$. This makes the real portion of the wavenumber dominate the imaginary component, and we have solutions that are now familiar.

If $(R_m/\rho_L c) = 1$, corresponding to the "matched" impedance case, discussed in the subsequent section describing a driven semi-infinite string (Sect. 3.7), then $\alpha_n L$ must be infinite; all of the energy will be absorbed. There will be no reflected wave, hence, no normal mode standing wave solution.

The real part of the transcendental equation, given in Eq. (3.80), can be satisfied if $\cos k_n L = 0$, producing values of $k_n L$ corresponding to the odd-integer half-wavelengths that we expect for a fixed-free string, and forcing $\sin k_n L = 1$. This leads to the quantizing condition on $\alpha_n L$ for values of mechanical resistance that are small compared to the string's characteristic impedance, $R_m < \rho_L c$.

$$\frac{\sinh(\alpha_n L)}{\cosh(\alpha_n L)} = \tanh(\alpha_n L) = \frac{R_m}{\rho_L c} < 1 \tag{3.83}$$

If the damping produced by the mechanical resistance is very small relative to the string's characteristic impedance, then $(R_m/\rho_L c) \ll 1$, again prompting a Taylor series expansion of $\tanh \alpha_n L \cong \alpha_n L = (R_m/\rho_L c)$ and making the real component of the wavenumber times the length, $k_n L$, again be dominant. Substituting these values back into our assumed solution for a fixed condition at $x = 0$, given in Eq. (3.19), describes the resulting normal mode shapes for small damping.

The transition between the fixed-free limit and the fixed-fixed limit is not abrupt. Even when $\alpha_n L \cong k_n L \cong n\pi$, which is approximately true in either limit, the amplitude of the function decays to $e^{-\pi} \cong 0.04$ within one wavelength. Effectively, any standing wave behavior disappears between the fixed-free and fixed-fixed transition.

To complete the substitution back into the mode shape equation, we also need the complex frequency. Since the resistance only dictates a boundary condition, the entire string is still characterized by a uniform value of the linear mass density and a constant tension. The vibrations of the string must still be described by the wave equation as written in Eq. (3.4), so c is a real scalar constant. This demonstrates that $\mathbf{k_n}c = ck_n + jc\alpha_n = \boldsymbol{\omega_n} \equiv \omega_n + j/\tau$, where we have chosen the imaginary part of the angular frequency to be consistent with our choice for the exponential decay time used with the damped simple harmonic oscillator in Eq. (2.43). Equating the imaginary parts, $1/\tau = c\alpha_n$.

$$\begin{aligned} y_n(x,t) &= \Re e\left[\widehat{\mathbf{C}}\sin(k_n x)e^{j\omega_n t}\right] \cong \Re e\left[\widehat{\mathbf{C}}e^{-t/\tau_n}e^{j\omega_n t}\right]\sin(k_n x) \\ &= \left|\widehat{\mathbf{C}}\right|e^{-t/\tau}\sin(k_n x)\cos(\omega_n t + \varphi) \end{aligned} \tag{3.84}$$

3.7 Forced Motion of a Semi-Infinite String

Our desire to examine the behavior of a string in response to a time-harmonic excitation should come as no surprise. As before, the steady-state response will be characterized by the string's mechanical impedance. We start that analysis by driving a semi-infinite string that is excited by the time-harmonic transverse vibration of the end of the string at fixed amplitude, $y(0, t) = y_o \cos(\omega t)$.

As before, for any linear system (see Sect. 1.3), the steady-state response of the string can only occur at the driving frequency, ω. We also assume that our displacement source is sufficiently robust that it can provide whatever force is required to maintain the specified amplitude of oscillation, y_o. Although semi-infinite strings under uniform tension are rare commodities, this analysis will demonstrate that a string of finite extent can behave as though it were semi-infinite if the undriven end is terminated by a mechanical resistance equal to the *radiation resistance* that was derived for the resistance-loaded uniform string in Sect. 3.6.3.

The behavior of the entire string can be specified by matching the imposed time-harmonic end displacement to a right-going solution to the wave equation. Since the string is assumed to be infinite in length as we go away from the driver, there will be no need for a left-going solution that would represent reflections moving back toward the source. To simplify our mathematics, we will multiply the argument of the wave function by a constant, k, that has the dimensions of an inverse length [m^{-1}] and is known as the *wavenumber*.

$$k(x \pm ct) = kx \pm kct = kx \pm \omega t \tag{3.85}$$

As before, the –sign indicates propagation in the direction of increasing x (i.e., to the right), and the +sign corresponds to propagation in the direction of decreasing x (i.e., to the left). The rightmost version of Eq. (3.85) imposes the requirement that $kc = \omega$, since we know that a time-harmonic driving force at radian frequency, $\omega = 2\pi f$, must create a steady-state response only at ω. We chose either a complex exponential or a sinusoidal expression as our function of $x - ct$ or $ct - x$.

$$\begin{aligned} y(x,t) &= \Re e\left[\widehat{\mathbf{y}}_{\mathbf{o}}e^{j(kx-\omega t)}\right] = \Re e\left[|\widehat{\mathbf{y}}_{\mathbf{o}}|e^{j\phi}e^{j(\omega t - kx)}\right] \\ \text{or } y(x,t) &= |\widehat{\mathbf{y}}_{\mathbf{o}}|\cos(\omega t - kx + \phi) \end{aligned} \tag{3.86}$$

Matching the drive at $x = 0$ to either right-going solution in Eq. (3.86) requires that $\phi = 0$. These solutions are periodic in both space and time.

The vertical component of the force that must be provided by our constant displacement drive to the point of attachment can be calculated from Eq. (3.2) where we again assume a constant tension, T.

$$\mathbf{F_y}(0,t) = -\mathrm{T}\left(\frac{\partial y}{\partial x}\right)_{x=0} = jk\mathrm{T}|\widehat{\mathbf{y}}_\mathbf{o}|e^{j\omega t} = F_o e^{j\omega t} \tag{3.87}$$

As with the case of the driven simple harmonic oscillator, it is useful to define the steady-state response of a driven system in terms of its mechanical impedance, as was done in Eq. (2.58), that is, the ratio of the force, $F_o e^{j\omega t}$, to the velocity, $\dot{\mathbf{y}}(0,t) = (\partial y/\partial t)_{x=0} = j\omega\mathbf{y}(0,t) = j\omega|\widehat{\mathbf{y}}_\mathbf{o}|e^{j\omega t}$. In this case, we will designate this impedance as the string's *input mechanical impedance*, $\mathbf{Z_{m,0}}$, since it is the impedance of the string where the displacement is imposed by the driver at $x = 0$.

$$\mathbf{Z_{m,0}} = \frac{\mathbf{F_y}(0,t)}{\dot{\mathbf{y}}(0,t)} = \frac{jk\mathrm{T}|\widehat{\mathbf{y}}_\mathbf{o}|e^{j\omega t}}{j\omega|\widehat{\mathbf{y}}_\mathbf{o}|e^{j\omega t}} = \frac{\mathrm{T}}{c} = \frac{\rho_L c^2}{c} = \rho_L c \tag{3.88}$$

It is useful to think about the fact that $\mathbf{Z_{m,0}} = \rho_L c$ is a real number. In our analysis of the driven simple harmonic oscillator, the only real part of the mechanical impedance was the mechanical resistance, R_m, which was also our only means for absorbing power. As yet, no "resistance" has appeared in our model of the flexible string, so we have included no mechanism that would dissipate energy.

For our driven string, this input mechanical resistance is a *radiation resistance*. In terms familiar to the microwave engineer, it would be known as the *characteristic impedance* of the transmission line (i.e., our string). What appears as a "dissipation mechanism" is merely an accounting anomaly—the instantaneous power, $\Pi(t)$, delivered to the string, propagates along the string. (Warning: Do not attempt to use this "accounting" argument with the Internal Revenue Service.)

$$\langle\Pi(t)\rangle_t = \frac{1}{T}\int_0^T F_y \bullet \dot{y}\, dt = \frac{1}{T}\int_0^T \frac{F_o^2}{\rho_L c}\cos^2\omega t\, dt = \frac{F_o^2}{2\rho_L c} = \frac{\rho_L c}{2}\dot{y}_o^2 \tag{3.89}$$

Again, the similarity of the above result to the expressions for the Joule heating, Π_{el}, of an electrical resistor, R_{dc}, is useful: $\langle\Pi_{el}\rangle_t = (VI/2) = (V^2/2R_{dc}) = (I^2 R_{dc}/2)$.

It is possible to avoid the issues introduced by a string that must be infinitely long in the $+x$ direction. If the string is terminated at a free boundary, but the "massless ring" to which the string is attached is joined to a mechanical resistance (e.g., a dashpot) with $R_m = \rho_L c$, then all of the energy that is propagating along that string will be dissipated in R_m, and there will be no reflections, even though the string is not infinitely long.

This "matched load" strategy is very common in electrical engineering. To avoid reflections along a cable attached to a measuring instrument (e.g., oscilloscope, spectrum analyzer), an electrical resistance with a value equal to the cable's characteristic impedance is placed across the input terminals of the instrument. For common co-axial cables, this resistance is typically 50 Ω. This *nonreflecting termination* technique is so common that many instruments have a built-in, switch-selectable input resistance with the option of providing a 1 MΩ or 50 Ω input electrical resistance.

3.8 Forced Motion of a Finite String

If the termination of our uniform string is anything besides the characteristic impedance, $\rho_L c$, then there will be reflections. Those reflections will influence the driver and can make the amplitude of the input mechanical impedance take values from zero to $\pm\infty$. If the termination provides no dissipation and there are no mechanisms to create propagation losses along the string, the input mechanical impedance will be entirely imaginary.

For that lossless case, with either a fixed or free termination, if we were to launch a pulse of duration, Δt, from the driven end of a string (like those shown in Fig. 3.2 through Fig. 3.5) of total length, L, the pulse would be reflected and return to the driver after a time, $t = 2\,L/c$.[14] Being reflected from the driven end, the process would repeat indefinitely with a period, $T = 2\,L/c$, and there would be no superposition if $\Delta t < L/c$. Such multiple reflections would be periodic, but not harmonic.

3.8.1 Displacement-Driven Finite String

We can calculate the steady-state behavior for a string of length, L, which is terminated by a fixed boundary and excited by a single-frequency time-harmonic displacement. Instead of placing our drive at $x = 0$, as we did for the semi-infinite case, the calculations are greatly simplified if we let the fixed end be located at $x = 0$ and drive from the end at $x = L$, in this case with a forced displacement amplitude, $y(L, t) = y_o \cos(\omega t)$. By letting y_o be a real scalar, we are only requiring that the phase of the string's response will be referenced to the phase of the displacement drive. This simplification arises from the fact that we already have an exact expression for the string's displacement if $y(0, t) = 0$, given in Eq. (3.19), so we can just equate the displacement of the string at $x = L$ to the imposed harmonic displacement, $y(L, t) = y_o \cos(\omega t)$.

$$y(L,t) = y_o \cos(\omega t) = \Re e\left[\widehat{\mathbf{C}} \sin(kL) e^{j\omega t}\right]$$
$$\Rightarrow \quad \left|\widehat{\mathbf{C}}\right| \equiv C = \frac{y_o \cos(\omega t)}{\sin(kL)} \tag{3.90}$$

Substitution of C into Eq. (3.19) produces the steady-state response of the entire string to the displacement drive at $x = L$.

$$y(x,t) = y_o \frac{\sin(kx)}{\sin(kL)} \cos(\omega t) \tag{3.91}$$

The displacement of the string is infinite everywhere whenever $kL = n\pi$. This corresponds to placing an integer number of half-wavelengths between the driver and the rigid termination. We call this condition *resonance*. It will be easy to understand what is happening in this critically important circumstance if we calculate the mechanical input impedance that the string presents to the displacement drive located at $x = L$.

At any point along the string, the vertical component of the force can be calculated from Eq. (3.91). By Newton's Second Law, the vertical force, F_y, on the string will be equal and opposite to the force applied at $x = L$.

$$F_y(x,t) = -\mathrm{T}\left(\frac{\partial y}{\partial x}\right) = -y_o \mathrm{T} k \frac{\cos(kx)}{\sin(kL)} e^{j\omega t} \tag{3.92}$$

The velocity of the string at any location is just as simple to calculate.

[14] The termination could be lossless but possibly be mass-like or stiffness-like. In that case, the apparent distance, L_{eff}, between the driver and termination, might be slightly different than the physical length, L.

$$\dot{\mathbf{y}}(x,t) = j\omega y_o \frac{\sin(kx)}{\sin(kL)} e^{j\omega t} \tag{3.93}$$

Their ratio gives the mechanical impedance of the string at any location.

$$\mathbf{Z_m}(x) = \frac{F_y(x)}{\dot{\mathbf{y}}(x)} = -\frac{\mathrm{T}k}{j\omega}\frac{\cos kx}{\sin kx} = j\frac{\mathrm{T}}{c}\cot kx = j\rho_L c \cot kx \tag{3.94}$$

In this case, the complex mechanical impedance, $\mathbf{Z_m}$ (x), anywhere along the string, is a purely imaginary quantity since force and velocity are 90° out-of-phase with respect to each other. The fixed boundary condition and the absence of any loss mechanisms means that none of the input energy can be dissipated.

Before using this result to explain the resonance, it is always prudent to examine a new result in a limit that we have already analyzed by other means (after checking that the units make sense, of course). Back in Chap. 2, we calculated the vertical component of the force exerted on a mass by a string with tension, T, that was displaced from equilibrium by a distance, y. In the small-amplitude limit, $y \ll a$, the result was given in Eq. (2.124). That expression provided the vertical component of the force due to two taut strings of length, $a = L/2$. In this case there is only one string of length, L, so F_y $(x = L) = -(\mathrm{T}/L)\, y$, producing an equivalent stiffness constant, (T/L). At very low frequency (i.e., $L \ll \lambda$), Eq. (3.69) guarantees that $\cot(kx) = (kx)^{-1}$ in the limit that $kx \to 0$. Also, since the motion is simple harmonic in time, $\mathbf{y} = \dot{\mathbf{y}}/j\omega = -j\dot{\mathbf{y}}/\omega$. Using all of these assumptions to evaluate Eq. (3.94) shows that the input impedance at $x = L$ is consistent with the static limit producing a stiffness, $\mathrm{K} = \mathrm{T}/L$.

$$\lim_{\omega \to 0}[-j\rho_L c \cot(kL)] = \frac{-j\rho_L c}{kL} = \frac{-j}{\omega}\frac{\mathrm{T}}{L} = \frac{1}{j\omega}\frac{\mathrm{T}}{L} \tag{3.95}$$

Returning now to the question of the resonance of a finite string driven by a constant amplitude harmonic displacement, $y(L,\, \mathrm{t}) = y_o \cos(\omega\, t)$, we see from the mechanical impedance in Eq. (3.94) that the string applies a force to the displacement driver that is infinite when $kL = n\pi$. Therefore, the required magnitude of the vertical force is $|\mathbf{F_y}| = |\mathbf{Z_m}(L)|(dx/dt) = \infty$; an infinite force is required to produce infinite displacements at all locations.

Of course, a real string, driven by a real displacement drive, will not produce infinite displacements. Long before the displacements become infinite, we will have exceeded the limitations of our linear approximation, and the average tension in the string will start to increase with increasing amplitude, causing the speed of the transverse vibrations to increase, thus de-tuning the resonance condition [11]. Also, practical considerations, including the effective output mechanical impedance of the driver and dissipation in the string, radiation of sound from the string, and losses introduced by motion at the "fixed" end, will all limit the transverse displacement amplitudes.

Another useful perspective on this resonance behavior is provided if we consider the string as an *impedance transformer*. This perspective is diagrammed schematically in Fig. 3.10. Movement away from the fixed end at $x = 0$ by an integer number of half-wavelengths transforms the impedance by 1:±1. Odd-integer numbers of half-wavelengths provide a 1:−1 transformation and displacements by full wavelengths correspond to a 1:1 transformation. Since resonance occurs when an integer number of half-wavelengths is interspersed between the fixed end at $x = 0$ and the driven end at $x = L$, the resonance corresponds to the transformation of the infinite impedance of the fixed end to an infinite impedance at the driving location.

At the nodal locations, both the displacement and the velocity vanish, but the slope of the line at that location (and hence the vertical component of force at that location) has its maximum value. The mechanical impedance of the string at those nodal points is therefore infinite. Said another way, a

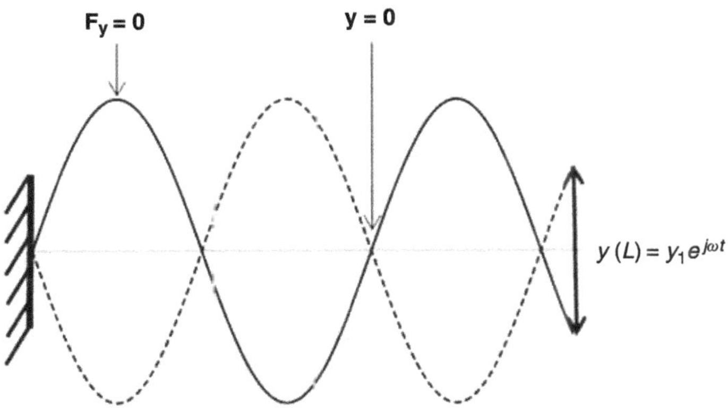

Fig. 3.10 Displacement amplitude envelope of a uniform string, fixed at $x = 0$, driven by a constant displacement drive at $x = L$ indicated by the *solid double-headed arrow*. At the locations of maximum displacement, the vertical component of the force, F_y, is zero, as is the string's mechanical impedance. At the nodal locations, both the string's displacement and its velocity are zero. At such nodal locations, the string's mechanical impedance is infinite. In this figure, the constant displacement drive is not located at a node so the string's displacements are finite, though grossly exaggerated in this figure

quarter wavelength displacement corresponds to a $1{:}\pm\infty$ transformer. At an anti-node, the impedance is zero, but a quarter wavelength away in either direction is infinite if the string is lossless.[15]

3.8.2 Mass-Loaded String in the Impedance Model

The expression for the mechanical impedance of the string in Eq. (3.94) is not a function of the drive but only of the string's properties (i.e., tension and linear mass density) and the fact that its end at $x = 0$ is rigidly fixed. For that reason, we should be able to use Eq. (3.94) to calculate the normal modes of a string that is fixed at one end and has a mass attached to the other end. An impedance-matching analysis must yield the same results as did our Newtonian analysis in Sect. 3.6.

The mechanical impedance of a mass, M, is $Z_m = j\omega M$. By Newton's Third Law, the force that the string applies to that mass must be equal and opposite to the force that the inertia of the mass applies to the string. Setting the impedance of the mass to the negative of the impedance of the string produces the same transcendental equation as derived before in Eq. (3.68).

$$
\begin{gathered}
j\omega_n M = j\rho_L c \cot(k_n L) \\
\Rightarrow \quad M = \frac{m_s}{k_n L}\cot(k_n L) \quad \text{or} \quad \frac{M}{m_s}\,(k_n L) = \cot(k_n L)
\end{gathered} \tag{3.96}
$$

Each successive value of n corresponds to the mass being transformed another half-wavelength away from the rigid termination at $x = 0$. The same approach can be used to solve for the normal modes of a string fixed at $x = 0$ and terminated by a spring at $x = L$.

[15] This results again highlights the difference between the vertical force applied to a point, F_y, vs. the net force, dF_{net}, on an infinitesimal string segment of length, dx, provided in Eq. (3.2) and used to derive the wave equation in Eq. (3.4).4 The imposition of a fixed displacement or fixed force drive is equivalent to specification of a boundary condition. When this imposed force or displacement is applied to the string at a location other than one of the string's boundaries, it can be thought of as making the string into two independent segments with the drive being applied at a point to two loads corresponding to the two impedances that represent those two separate string segments.

3.8.3 Force-Driven Finite String

Having an expression for the mechanical impedance of a string that is attached rigidly at $x = 0$, we can easily calculate the behavior of the same string driven by a time-harmonic vertical force of fixed amplitude applied at $x = L$: $F_y(L) = F_o \cos(\omega t)$. By letting F_o be a real scalar, we are only requiring that the phase of the string's response will be referenced to the phase of the driving force. In that case, whenever the constant force drive encounters an impedance that is zero, the vertical force produced by the string will be zero. Imposition of a non-zero force at that location implies that the vertical force everywhere else on the string must be infinite.

The mechanical impedance of the string goes to zero whenever $k_n L = (2n - 1)(\pi/2)$, where $n = 1$, 2, 3, etc. This resonance condition corresponds to an odd-integer number of quarter-wavelengths between the rigid point of attachment at $x = 0$ and the constant force driver location at $x = L$. It is exactly the opposite of the resonance condition for a constant displacement (or constant velocity) drive.

Although a constant displacement drive is easy to visualize, it may be helpful to provide an example of how a constant amplitude force drive can be applied to a string. With a constant displacement drive, we are assuming that the driving mechanism can provide whatever force is necessary to keep the amplitude of the displacement constant. With a constant force drive, we assume that the force can produce whatever displacement is required to hold the amplitude of the vertical component of the force constant.

Such a fixed amplitude time-harmonic force drive can be created by placing a magnetic field of magnetic induction, $\vec{B}$, perpendicular to the direction of the string and running an alternating electrical current, $I(t) = \Re e\left[\hat{\mathbf{I}} e^{j\omega t}\right]$, through the string, assuming that the string is made of a material that conducts electricity. The vertical force on the string is given by the Lorentz equation: $F_y(t) = (B\ell)I(t)$, where it is assumed that the magnetic induction, $\vec{B}$, is constant over a length, $\vec{\ell}$, of the string and that $\vec{F}_v$, $\vec{B}$, and $\vec{\ell}$ are all mutually perpendicular [11].

If the electrically conducting string is fixed at both ends and driven by the current, $I(t)$, applied at a frequency that would excite one of the fixed-fixed string's normal modes (i.e., $f_n = nc/2\,L$), then if $\left|\vec{\ell}\right| \ll LL$, and if the magnetic field is centered over a displacement anti-node, where $|\mathbf{Z}_m| = 0$, the string would be excited at resonance. Being a displacement maximum, the driven anti-node point has zero slope, hence zero force, but the electromagnetic force created by the magnetic field's interaction with the current is non-zero, so the force everywhere else on the string must become infinite. If the magnetic field were placed over a displacement node, then $|\mathbf{Z}_m| = \infty$, and the string would remain at rest.

3.8.4 An Efficient Driver/Load Interaction

There are no displacement drivers that can produce infinite forces, and there are no force drivers that can produce the required unlimited displacements. As with our "perfect" boundary conditions, the idealized force and displacement drives are useful because they inform our intuition and because there are circumstances where the loads which the string (or any "driven" system) places on the driver are far less demanding than the infinite or zero impedances that led to infinite responses. Although conceptually convenient, the cases where the driver overpowers the load are cases where the transfers of energy from the driver to the load are extraordinarily inefficient. In fact, it is this inefficiency that allows the load and driver to be treated independently—the effects of one upon the other are minimal.

The efficient coupling of a driver to an acoustic load requires the matching of the driver's impedance to the load's impedance. Since only resistive loads can absorb power, optimization of this matching process requires first that the reactive components (i.e., the stiffness and moving mass) of the combined system cancel. With that cancellation, the power of the driving mechanism (e.g., electrodynamic or piezoelectric forces) is not consumed by the acceleration and deceleration of masses nor the compression or extension of stiffness elements. Here we must recognize that the load and the driver can also provide stiffness or mass (e.g., see Sect. 2.5.5 or Sect. 10.7.5), depending upon the driving frequency.

Matching of the resistive components of the driver and load is also required for high-efficiency power transfer. Since power is proportional to the product of force and velocity, a load resistance, R_{load}, that is much greater than the drive's internal mechanical resistance, R_m, will require greater forces to generate time-averaged power transfer since $\langle \Pi \rangle_t = F^2/2R_{load}$. Since no drive mechanism is 100% efficient, the required force will result in some energy loss within the transduction mechanism. Similarly, if R_{load} is too small relative to the mechanical resistance of the driver, R_m, then larger velocities, v, will be required to provide the required time-averaged power to the load, $\langle \Pi \rangle_t = R_{load} v^2/2$. If the velocity of the drive is equal to that of the load, then the power dissipated internally within the driver will be proportional to $R_m v^2/2$. One solution is provided by the use of some nearly lossless "lever" system that allows the velocity of the driver to differ from the velocity of the load[16] [12].

The cancellation of the reactive components suggests operation at resonance; hence the highest efficiency can only be achieved over a limited range of frequencies. The matching of the driver's resistance and the resistance of the load may be limited by the available "leverage." A very large "piston" (e.g., speaker cone) that improves the matching of the resistive components might also be so massive that the piston's mass requires too much of the motor's force to accelerate and decelerate it. An exponential horn is a common method used to increase the radiation loading on a smaller piston, but that solution may require the horn to be large (i.e., expensive), particularly for matching at low frequencies (see Sect. 10.9).

This discussion of efficient driving strategies is only meant to raise awareness for subsequent applications that do not involve strings and to emphasize that the idealized concepts introduce practical limitations. Also, there are situations where efficiency is not the primary objective. In sound reproduction systems, sometimes the primary objective is uniform output power (i.e., "flatness") over the broadest possible bandwidth (i.e., frequency range). In those cases, Professor Putterman's perspective is worthy of consideration:[17] "The flattest response is no response at all." One can trade efficiency for bandwidth just as the sensitivity of a geophone was traded for increased bandwidth, as shown in Fig. 2.24. The "art" is in knowing how to make an optimal trade-off based on the specified design goals.

3.9 "I've Got the World on a String . . .": Chapter Summary

The physical system that has been the focus of this chapter is an idealized string that is "limp" in the sense that any displacement of the string from its equilibrium condition is restored only by tension and that the string itself has no stiffness (i.e., the string has no flexural rigidity). Although the musical

[16] In acoustical systems, this "lever" might be the area of the piston driving the acoustic load. The piston transforms the velocity of the driver (e.g., the voice coil velocity) to the volume velocity (i.e., oscillatory mass flow rate) of the fluid driven by the piston. Calculations to determine the piston area that optimizes the efficiency of the coupling between an electrodynamic linear motor (e.g., moving-coil loudspeaker) and a standing wave resonator are provided in Sect. 10.7.5.

[17] S. J. Putterman is a Professor of Physics at the University of California, Los Angeles.

consequences of the string's normal modes of vibration have been briefly noted, the goal of this chapter has been to introduce many of the perspectives and mathematical techniques that will appear repeatedly throughout this textbook to analyze the vibrations of continua (i.e., fluids and solids). It is much easier to illustrate the transverse motion of a string in a textbook than it is to visualize the local concentrations of air molecules that are excited by the passage of a sound wave consisting of longitudinal expansions and compressions. Because these ideas are so important, we will close this chapter by reviewing some of the most significant results.

As stated in Sect. 1.1, two of the most useful techniques in mathematics, substitution and Taylor series, have been employed throughout. As with our analysis of the simple harmonic oscillator, we began by identifying the relationship between forces and displacements (i.e., an equation-of-state), in this case, a transverse force and a slope, $F_y = -\mathrm{T}\ (\partial y/\partial x)$. Newton's Second Law of Motion then provided the connection between the acceleration of an infinitesimal segment of the string and the net force produced by the tension acting on both ends of that segment, determined by a Taylor series expansion of the force at one end of the segment relative to the force at the other end. We assumed small displacements from equilibrium so that the Taylor series expansion could be truncated beyond the linear term.

This process led us to combine those tensile and inertial effects into a homogeneous second-order partial differential equation that related transverse displacements to position along the string and to time. That equation is called "the wave equation." The first thing we used that wave equation to reveal was that it could be satisfied by any two functions, $f_+(w_+)$ and $f_-(w_-)$, as long as the argument of those functions had the form $w_\pm = x \pm ct$, where c was the speed of propagation for the transverse displacements. It was also the last thing that the wave equation revealed.

Although claiming that "the wave equation is the least useful equation in acoustics" is considered a heresy by many acousticians, the wave equation did not make any further contribution to the investigations within this chapter, nor will it play a more significant role in our other analyses. In this chapter, when we considered a string that had a variable tension, $\mathrm{T}(z)$, in Sect. 3.4.3, Eq. (3.44) set up the wave equation for that case. We did not bother to solve it.[18] Instead, we employed a plausible solution that satisfied the boundary conditions and calculated the corresponding potential and kinetic energies of vibration to determine the normal mode frequencies from their ratio. When the "trial function" contained an "adjustable parameter," we improved the accuracy of our approximation by minimizing the frequency with respect to variation of the adjustable parameter, an approach known as "Rayleigh's method."[19]

It might seem strange to use an approximate solution when the exact solution is known, but an approximate solution can be a better choice in many circumstances. For example, the use of a polynomial approximation can be much more computationally efficient than the use of exact solutions when simultaneous solution of many coupled equations is performed by a computer [13]. In our case, I had no interest in slowing the development of our investigation of string vibrations by having to introduce Bessel's equation, although it will be introduced in Chap. 6, when Bessel functions will be required to analyze two-dimensional vibrating systems that possess circular symmetry.

With the functional dependence on $w_\pm$ established, the concept of "ideal" boundary conditions was introduced, and the reflection of an arbitrary pulse from such boundaries was explained in terms of the superposition of that pulse and an "image" pulse that approached the boundary traveling in the opposite direction, along a string that did not exist in the real world. We then went on to impose

[18] The solution to Eq. (3.44) can be obtained by a clever change of variables resulting in the spatial dependence being described by a Bessel function (see Fig. 6.8) of the transformed coordinates rather than a trigonometric or exponential function. Execution of this approach is a standard problem in many textbooks on mathematical physics.

[19] In quantum mechanical applications, this is commonly called the Rayleigh-Ritz method.

two such boundary conditions to define a string of finite length. The imposition of one boundary condition restricted the form of the solutions, and the imposition of the second boundary condition quantized the frequencies and wavenumbers of the normal modes of a finite string's vibrations.

Although no physical boundary condition is an exact match to the idealized cases (e.g., fixed-fixed or fixed-free), it is a good enough approximation in many cases that we could use the normal mode frequencies, and their harmonic relationships, to develop a rudimentary appreciation for the sounds produced by stringed musical instruments and combine those results with our understanding of the psychological concepts of consonance and dissonance. This led to a brief consideration of the difficulties that are encountered when we attempt to preserve consonance while creating the notes of a musical scale that will be both pleasing and practical.

Again, the fundamental equations relating stiffness and inertia to displacements and accelerations (and not the wave equation!) allowed us to develop an understanding of the energy content of the string's vibration. Those energies were exploited to calculate normal mode frequencies in cases where the exact solution was known, as well as cases where the linear mass density of the string was either perturbed at a point or was a continuous function of position. The use of energy methods to calculate the normal mode frequencies for the case where tension was a function of position has already been mentioned in this summary.

As for any linear system, the solutions for the string's motion always involved arbitrary amplitudes that could not be determined until the initial conditions were specified. We found that the Fourier series was an ideal method to project such initial conditions (at $t = 0$) onto a linear superposition of normal mode shapes that formed a complete orthogonal basis describing the spatial and temporal evolution of the system for all $t > 0$. The relative amplitudes of the normal modes provided some insight into the tonality produced by the location where a string was plucked, as well as explaining the difference in the sound produced by an electrodynamic pickup as a function of its distance from the electric guitar's bridge.

Although this provides an important piece of our understanding of the physics involved in the production of musical sounds by plucked or hammered strings, it is by no means a complete understanding. The attack and decay of the sound of a plucked string, though related to the harmonic structure, are dependent upon other effects (e.g., damping and motion of the bridge that couples the string's energy to the radiating surface of the instrument's body). The attack and the decay are equally important to our perception of musical sounds and our ability to differentiate the sounds produced by specific musical instruments or to identify the individual musicians playing those instruments by the sounds they produce.

Techniques for the calculation of normal mode solutions (i.e., mode shapes and frequencies) for "imperfect" boundary conditions demonstrated that such imperfections (e.g., an end that was "fixed" by a finite mass or a "free" end that was attached to a stiffness or resistance) used the same matching of the wave's motion to the boundary's mechanical impedance but produced equations for quantized modal frequencies and wavenumbers that had no closed-form algebraic solutions and produced normal mode frequencies that were not exact integer multiples of the fundamental ($n = 1$) mode's frequency, f_1. Again, a small amount of physical insight was leveraged into approximate solutions that did have simple algebraic forms and impressive accuracy. In addition, for the case of a fixed, mass-loaded string, an additional "lumped-element mode" was identified that had a frequency, $f_o < f_1$, and did not exhibit any harmonic relationship to the fundamental frequency, f_1. That normal mode is the pendulum mode.

A brief investigation of the resistance-loaded string introduced behavior that was more complicated to compute, but that approached the expected fixed-free behavior when the resistance was very small (i.e., $R_m/\rho_L\, c \ll 1$) and fixed-fixed behavior in the opposite limit. That problem also introduced the need to treat both the wavenumber and the frequency as complex numbers to incorporate dissipative behaviors associated with the connection to a mechanically resistive element.

The last topic was the driven string. For a semi-infinite string, the characteristic impedance, $\rho_L c$, was derived as the solution to the steady-state problem. For the string of finite length, the steady-state response was shown to be critically dependent upon both the frequency of the drive and the nature of the driver's excitation. A constant displacement drive produced a very different steady-state response than a constant force drive when applied at the same location and at the same frequency. In the absence of any dissipation, there were frequencies for either driver type that resulted in an infinite response at all points along the string. We identified this behavior as "resonance."

In this chapter and the previous one, we have accumulated a rather impressive arsenal of concepts and mathematical techniques. Although others will be forthcoming, those accumulated thus far should serve us well throughout this textbook.

Talk Like an Acoustician

Wave equation	Interval
d'Alembertian operator	Equal temperament
Transverse wave speed	Scale of just intonation
Traveling wave	Pythagorean scale
Fixed boundary	Chromatic scale
Free boundary	Stability coefficient
Method of images	Inertia coefficient
Wavenumber	Transcendental equation
Fundamental frequency	Nonreflecting termination
Overtones	Input mechanical impedance
Harmonic series	Radiation resistance
Standing waves	Characteristic impedance
Nodes	Resonance
Anti-nodes	Impedance transformer
Consonance and dissonance	

Exercises

1. **Mersenne's Laws**. Marin Mersenne wrote the following three laws that govern the frequency of a vibrating string in his 1637 work, *Traité de l'harmonie universelle* [14]:
   (i) For a given string and a given tension, the time [period] varies as the length.
   (ii) When the length of the string is given, the time varies inversely as the square root of the tension.
   (iii) Strings of the same length and tension vibrate in times which are proportional to the square root of the linear density.

   Show that all three of Mersenne's Laws are a direct consequence of the fact that there is only one combination of tension, length, and linear mass density that have the units of time [15].
2. **Traveling wave**. The transverse displacement of a wave propagating to the right on an infinite string of linear mass density, $\rho_L = 1.00 \times 10^{-4}$ kg/m, is given by $y(x, t) = (0.05)\cos(4t - 3x)$, where y and x are in meters and t is in seconds.
   (a) *Amplitude*. What is the peak-to-peak amplitude of the wave?
   (b) *Transverse wave speed*. Determine the propagation speed of the wave.
   (c) *Tension*. What is the tension in the string?
   (d) *Wavelength*. What is the wavelength of the wave?
   (e) *Transverse string speed*. What is the maximum speed of the string's motion in the transverse direction?

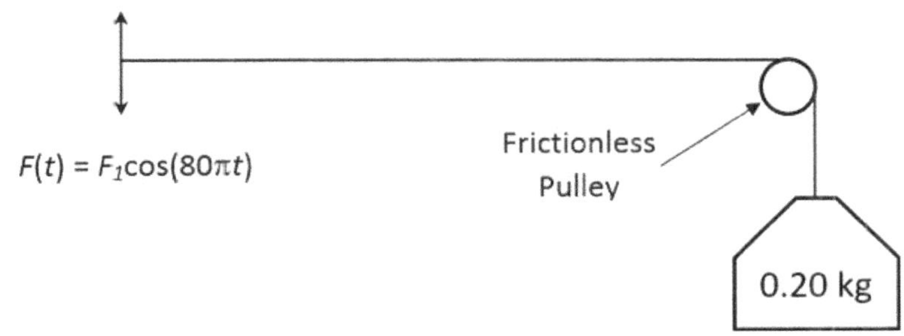

Fig. 3.11 Force-driven string tensioned by a mass hung from a frictionless pulley

3. **Driven semi-infinite string**. A string of constant linear mass density, $\rho_L = 0.10$ kg/m, is stretched to a tension of T = 10.0 N. It is excited by a constant displacement generator which produces a transverse displacement of $y(0, t) = 0.020 \cos(20t)$, where the amplitude is expressed in meters and the time in seconds.
   (a) *Wavelength*. What is the wavelength of the wave that is being generated?
   (b) *Wave function*. Write a wave function which describes the transverse displacement of the string caused by the wave for all positions, $x > 0$ and all times $t > 0$.
   (c) *Power*. How much time-averaged power does the generator have to deliver to the string to maintain the wave?
4. **Force-driven fixed string**. A uniform string is driven at one end at 40 Hz, and the other end runs over a pulley and is attached to a mass of 0.20 kg as shown in Fig. 3.11. You may consider the pulley to provide a rigid termination and let $g = 9.8$ m/s^2.
   (a) *String mass*. In a distance 1.20 m from the pulley, there are three loops (i.e., half-wavelengths). If the distance between the drive and the pulley is 2.00 m, what is the mass of that 2.00 m length of string?
   (b) *Water absorption*. If the string should absorb some water and become heavier but is still driven at 40 Hz, for each of the following quantities, indicate whether they would increase, decrease, or remain the same.

(i)	Linear mass density, ρ_L.	__________
(ii)	Tension, T.	__________
(iii)	Propagation speed, c.	__________
(iv)	Wavenumber, k.	__________
(v)	Period, $T = f^{-1}$.	__________
(vi)	Wavelength, λ.	__________
(vii)	Radian frequency, ω.	__________

5. **Plucked string**. A string of length, L, rigidly constrained at both ends, is displaced by a transverse distance, h, at a distance, $(1/3)L$, from one end and released at $t = 0$.
   (a) *Harmonic amplitudes*. Determine the amplitude of the first four standing wave modes in terms of the initial displacement, h.
   (b) *Frequency of vibration*. If $L = 0.25$ m, T = 10.0 N, and $\rho_L = 1.0 \times 10^{-4}$ kg/m, determine the frequency of the fundamental ($n = 1$) mode of vibration.
   (c) *Energy of vibration*. What is the total energy of all the modes excited by plucking the string with an initial displacement of $h = 1.0$ cm?
6. **Initial velocity harmonic amplitude coefficients**. Calculate the values of C_n for a string that is excited by the following triangular velocity distribution applied to one-half of the string's length.

$$v_o(x,0)=\begin{cases}\dfrac{v_1 x}{L} & \text{if } \ 0<x<\dfrac{L}{4}\\ \dfrac{v_1}{L}\left(\dfrac{L}{2}-x\right) & \text{if } \ \dfrac{L}{4}<x<\dfrac{L}{2}\\ 0 & \text{if } \ x>\dfrac{L}{2}\end{cases}$$

7. **Nonuniform mass distribution**. Use Rayleigh's method to calculate the normal mode frequencies of a string which has the following parabolic variation linear mass density, $\rho_L(x)$, where m_s is the mass of the string before the parabolic contribution is added. The expression below places $x = 0$ at the center of the string with length, L, so $-{}^{L}/_{2} \le x \le +{}^{L}/_{2}$. Assume $m_o = m_s/5$ and report your frequencies in terms of the exact frequencies for a uniform string of the same length with constant linear mass density, $\rho_L(x) = m_s/L$.

$$\rho_L(x)=\frac{m_s}{L}\left[1-\frac{m_o}{L^2 m_s}\left(x^2-\frac{L^2}{4}\right)\right]$$

   (a) *Uniform density trial functions*. Calculate the approximate normal mode frequencies using the exact shapes for the uniform string given in Eq. (3.21).
   (b) *Quadratic polynomial trial function*. Calculate the frequency of the fundamental ($n = 1$) normal mode using the power law shape function of Eq. (3.33) with $m = 2$.
   (c) *Optimized power law trial function*. Write the frequency of the fundamental mode in terms of m, as was done in Eq. (3.36), and minimize the frequency with respect to m, to determine the best value for m. Use that result to calculate a new approximation to the frequency of the fundamental mode.

8. **An "inverse" problem**. The data summarized in Table 3.8 provides the location of the nodes on a string that is 1 m long and tensioned with a force of 3.948 N. For both parts, the string is driven at its tenth standing wave mode.
   (a) *Uniform density*. A string of uniform density is driven at $f_{10} = 31.4159$ Hz. Determine the linear mass density, ρ_L, of that string for the node spacing above.
   (b) *Nonuniform density*. The uniform string is replaced with one of nonuniform density, $\rho_L(x)$, and the frequency of the 10th mode is found to be $f_{10} = 31.7565$ Hz. Use the "half-wavelength" of each segment between adjacent nodes (length $x_i - x_{i-1}$), to approximate the mean linear mass density at the mean location, $(x_i + x_{i-1})/2$. Write an expression for the linear mass density as a function of position along the string [16].

Table 3.8 The position of the nodes of the tenth mode of a fixed-fixed string with total length, $L = 1.00$ m, is provided. For the uniform string, the nodes are equally spaced. For the nonuniform string, the spacing between the nodes decreases monotonically from the end at $x = 0$ to the end at $x = L$

1	Uniform (m)	Variable (m)
End	0	0
1	0.1	0.1315
2	0.2	0.2505
3	0.3	0.3607
4	0.4	0.4642
5	0.5	03625
6	0.6	0.6564
7	0.7	0.7467
8	0.8	0.8338
9	0.9	0.9181
End	1	1

9. **Whirling string**. The tension in a string that is rotated at a constant angular frequency, ω_a, is produced by the centripetal acceleration of the string's mass. For a string with a uniform linear mass density, ρ_L, the tension, $T(x)$, is given by the integral of the centripetal force from x to the end of the string at $x = L$.

$$T(x) = \int_x^L \rho_L \omega_a^2 x\, dx = \left(\frac{\rho_L \omega_a^2}{2}\right)(L^2 - x^2)$$

Use the parabolic trial function of Eq. (3.45) to approximate the lowest-frequency mode of the whirling string by Rayleigh's method. Compare your result to the pendulum frequency for a pendulum bob of mass, $M = \rho_L L$, on a massless string and an equivalent acceleration, $g = \omega_a^{\,2} L$.

10. **Mass-loaded string**. A mass, $M = 0.20$ kg, is hung from a string that has a mass, $m_s = 0.05$ kg, and length (measured from the fixed end to the center of mass of the end mass), $L = 1.00$ m. Neglect the fact that the tension in the string is a function of position and approximate the tension as being constant, $T = g(M + m_s/2)$. Let the acceleration due to gravity be $g = 9.8\ \text{m/s}^2$.
    (a) *Pendulum frequency*. Assuming $m_s = 0$, what is the pendulum frequency, f_o, for small oscillations of the terminal mass, M.
    (b) *Transverse wave speed*. Assuming a constant tension, $T = g\,(M + ½m_s)$, what is the speed of transverse waves on the string?
    (c) *Pendulum mode*. Determine the frequency of the mode which has only one node located at the fixed attachment point of the string.
    (d) *(Nearly) Half-wavelength mode*. Determine the frequency of the mode which has one node located at the fixed attachment point of the string and one other node close to the mass.
    (e) *Transverse displacements*. For the (nearly) half-wavelength mode, determine the ratio between the largest transverse displacement of the string and the transverse displacement of the terminal mass, M.

11. **Stiffness-loaded string**. A string of length, L, is rigidly fixed at one end and terminated at the other end by a spring with spring constant, K_{sp}, so that $F_y(L) = -K_{sp}y$.
    (a) *Normal mode frequencies*. Write a transcendental equation similar to Eq. (3.68) for the mass-loaded string that can be solved for the normal mode frequencies of the stiffness-loaded string.
    (b) *Stiff spring limit*. Show that your equation for the normal modes in part (a) reduces to those of a fixed-fixed string if $K_{sp} = \infty$ and for those of a fixed-free string if $K_{sp} = 0$.
    (c) *Normal mode frequencies*. Calculate the lowest three normal mode frequencies, f_n, where $n = 1, 2,$ and 3, in terms of the lowest-frequency "idealized" fixed-free normal mode frequency, $f_{fix\text{-}free} = c/4L$, if $K_{sp} = T/L$.

References

1. G. Galilei, *Discorsi e dimostrazioni matematiche intorno à due nuove scienze,* 1638; translated by H. Crew and A. de Salvio, *Dialogs Concerning Two New Sciences* (McMillan, 1914), reprinted (Dover, 1954); ISBN 486-60099-8.
2. H.L.F. von Helmholtz, *On the Sensations of Tone as a Philosophical Basis for the Theory of Music* (Longsman, 1885), reprinted (Dover, 1954); ISBN 0-486-60753-4.
3. R. Plomp, W.J.M. Levelt, Tonal consonance and critical bandwidth. J. Acoust. Soc. Am. **38**, 548–560 (1965)
4. A more complete discussion of scales, as well as other topics related to musical acoustics, is available in T. D. Rossing, *The Science of Sound* (Addison-Wesley, 1982); ISBN 0-201-06505-3.
5. M.J. Kepler, *De Fundamentis Astrologiae Certioribus* [On the more Certain Fundamentals of Astrology or On Giving Astrology Sounder Foundations], 1601

6. H. Lamb, *The Dynamical Theory of Sound* (E. Arnold, London, 1931), pg. 87. Reprinted (Dover, 1960).
7. J.W. Strutt (Lord Rayleigh), *The Theory of Sound*, 2nd ed. (Macmillan, 1894), Vol. I, pg. 112; reprinted by Dover, 1945; ISBN 486-60292-3
8. D. Bernoulli, 1732.
9. L. Cremer, *The Physics of the Violin*, translated by J. S. Allen (MIT Press, 1984); ISBN 0-262-03102-7.
10. I.S. Gradshteyn, I.M. Ryzhik, *Tables of Integrals, Series, and Products* (Academic, 1980); ISBN 0-12-294760-6.
11. G. Penelet, S. Garrett, Periodic self-modulation of an electrodynamically-driven heated wire near resonance. J. Acoust. Soc. Am. **145**(2), 998–1017 (2019)
12. E. Mitchell, *Development of a thermoacoustic refrigerator inline model to assist in design*, Master of Science in Acoustics, Penn State (Dec 2012).
13. A. Migliori and J.L. Sarrao, *Resonant Ultrasound Spectroscopy* (Wiley, 1997); ISBN 978-0471123606.
14. M. Mersenne, *Traité de l'harmonie universelle; où est contenu la musique theorique & pratique des anciens & modernes, avec les causes de ses effets* (Guillaume Baudry, Paris, 1627). For a complete discussion of Mersenne's contributions see F. V. Hunt, *Origins in Acoustics* (Yale, 1978); ISBN 0-300-02220-4, reprinted (Acoust. Soc. Am., 1992).
15. J.W. Strutt (Lord Rayleigh), *The Theory of Sound*, 2nd ed. (Macmillian, 1894), Vol. I, pg. 182; reprinted (Dover, 1945); ISBN 486-60292-3.
16. J.R. McLaughlin, Inverse spectral theory using nodal points as data - a uniqueness result. J. Differ. Equ. **73**, 354–363 (1988). Good vibrations, Am. Sci. 86, 342–349 (1998)

4

Elastic Properties of Solids

If we take a piece of solid matter that is initially at rest and apply equal and opposing forces to the sample, Newton's Second Law of Motion guarantees that the sample will remain at rest because the net force on the sample is zero. If the sample is an elastic solid, then those forces will cause the solid to deform by an amount that is directly proportional to the applied forces. When the forces are removed, the sample will return to its original shape and size. These are the characteristics that are required if we say the behavior of the solid is "elastic."

Most solids will behave elastically if the forces and their resulting deformations are sufficiently small. If the forces are larger, it is possible to fracture the solid or to cause it to yield so that it does not return to its original shape after the forces are removed. Most metals will fracture if the forces that are applied to them cause deformations that change the length relative to the original length by a few percent. Glass will fracture at less than 1% change in relative length, and rubber can behave elastically for relative deformations that exceed 100%. Different solid materials exhibit a very wide range of behaviors, but most exhibit elastic behavior over some range of deformations.

This chapter will quantify the elastic behavior of solids by introducing the concepts of stress and strain and expressing their linear relationship through the definition of elastic moduli that depend only upon the material and not the shape of the sample. Those concepts will allow us to generalize Hooke's law. As before, the combination of a linear equation of state, like Hooke's law, with Newton's Second Law of Motion, will allow us to describe the wave motion in solids in the next chapter. Further insight will be gained by analyzing vibrational modes of bars, both through the understanding of those modes and by the measurement of those modal frequencies to accurately determine the elastic moduli of the materials.

4.1 Hooke, Young, Poisson, and Fourier

We begin our investigation into the elasticity of solids by considering the rectangular block of material shown in Fig. 4.1. Since the material is elastic, the force is proportional to the extension, $F \propto \Delta l$; this is the relationship we have called Hooke's law in previous chapters. If two such pieces are joined end to end to create a piece twice as long, but with the same cross-sectional area, $S = wh$, then the same force would create twice the extension, $2\Delta l$. We have analyzed this behavior before with springs joined in series, as shown in Fig. 2.3 (right), and also observed twice the extension for the same force. We can therefore write the proportionality of Hooke's law to exploit this dimensionless measure of deformation as $F \propto (\Delta l/l)$. That dimensionless ratio is the *strain* and it is commonly represented by ε.

Following our analysis of springs in Chap. 2, we can also combine the two solid blocks in parallel so that the overall length of the combination is still l but the cross-sectional area is doubled. To create the same stretch, Δl, the force would have to be doubled. We can combine the dependence on length, l, and on cross-sectional area, S, into a single expression, which is a generalization of Hooke's law, that

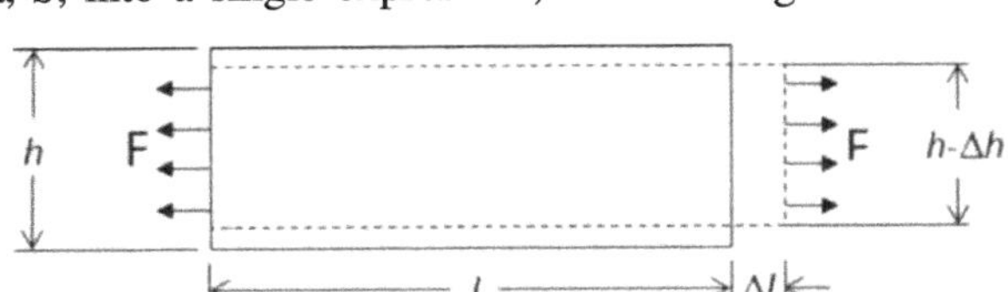

Fig. 4.1 In the absence of any applied forces, the volume of this rectangular bar is $V = lhw$. The bar is shown being stretched by equal and opposite forces, F, applied to the ends having cross-sectional area, $S = wh$. The bar responds elastically to the applied force by increasing its length an amount, Δl. In this two-dimensional representation, we see that the height of the bar is reduced by Δh. The width is also reduced by an amount, Δw

introduces a new constant, E, called *Young's modulus*. Young's modulus depends only on the material used to make the block (and its temperature). We will ignore the small difference between the adiabatic and isothermal moduli [1].[1] The ratio of the force, F, to the area, S, over which it is applied, $\sigma = F/S$, is called the *stress* and has the units of pressure [Pa].

$$\sigma = \frac{F}{S} = E\frac{\Delta l}{l} = E\varepsilon \tag{4.1}$$

We read Eq. (4.1) as "stress is proportional to strain." In this case, the constant of proportionality is Young's modulus, E. Our introduction of this dimensional constant, which is a property of the material, preserves dimensional homogeneity in a way that would please Mr. Fourier (see Sect. 1.6). Since strain, ε, is dimensionless, the units of Young's modulus are also pressure [Pa].

In Fig. 4.1, we see that in addition to the elongation of l, the height, h, of the rectangle is simultaneously reduced, as is the width, w (not shown). We can introduce another constant to relate the lateral contraction to the longitudinal extension.

$$\frac{\Delta h}{h} = \frac{\Delta w}{w} = -\nu\frac{\Delta l}{l} \tag{4.2}$$

That constant of proportionality, ν, is called *Poisson's ratio*[2] and is clearly dimensionless.

If the volume of the sample is conserved, when it is strained along its length, as shown in Fig. 4.1, then logarithmic differentiation (see Sect. 1.1.3) can be used to determine the volume-conserving value of Poisson's ratio.

$$\frac{\Delta V}{V} = \frac{\Delta l}{l} + \frac{\Delta h}{h} + \frac{\Delta w}{w} = (1 - 2\nu)\frac{\Delta l}{l} = 0 \Rightarrow \nu = \frac{1}{2} \quad \text{if} \quad \Delta V = 0 \tag{4.3}$$

The deformation of rubber is nearly volume conserving (see Sect. 4.5.1), so its value of Poisson's ratio is very close to ½, but for most common solid construction materials, like metals and plastics, $1/3 \gtrsim \nu \gtrsim 1/4$. Poisson's ratio for cork is nearly zero. This makes it a convenient material for sealing wine bottles since it does not expand as force is being applied to push the cork into the neck of a bottle. As we continue our investigation of elasticity, we will determine the fundamental limits on the values of Poisson's ratio that are imposed by material stability and energy conservation.

4.2 Isotropic Elasticity

If the elastic material is homogeneous and isotropic, its elastic behavior is completely specified by E and ν. Our introduction of E and ν was convenient, since our sample had a rectangular shape and was unconstrained along the sides and the forces were normal to the ends where they were applied. The choice of E and ν is not our only option for the two independent constants required to specify the elastic behavior of an isotropic solid. As we are about to demonstrate by the use of the principle of

[1] For gases, this is an important distinction (e.g., the adiabatic bulk modulus of air is 40% larger than the isothermal bulk modulus). The difference between the adiabatic and isothermal Young's modulus can be expressed in terms of the material's absolute (kelvin) temperature, T; the (volumetric) coefficient of thermal expansion, $\alpha = (1/V)(\partial V/\partial T)_p$; mass density, ρ; and specific heat (per unit mass) at constant pressure, c_p.

$$E_{ad} = \frac{E_{iso}}{1 - E_{iso}T\alpha^2/9\rho c_p}$$

In most solids, this is a small effect. At room temperature, $E_{iso}T\alpha^2/9\rho\, c_p$ is about 0.44% for aluminum and 340 ppm for copper.

[2] Simeon Denis Poisson (1781–1840).

superposition (see Sect. 1.4), we can relate E and ν to other moduli that are convenient for the specification of an isotropic material's response to different combinations of stresses.

4.2.1 Bulk Modulus

We can use Young's modulus, E, and Poisson's ratio, ν, to calculate the deformation of a solid that is subject to a hydrostatic compression. If we imagine our solid being submerged in a fluid, as the fluid pressure increases, the volume of the solid will decrease, but its shape will remain the same. Pascal's law guarantees that the hydrostatic pressure will be the same on all faces if the sample is at rest and the size of the sample is small enough that any gradients in the gravitational field can be ignored. This situation is shown schematically in Fig. 4.2. We can combine Pascal's law with the principle of superposition to calculate the change in length of any side and combine those changes using Eq. (4.3) to compute the relative change in volume of the sample, $\Delta V/V$.

We start by calculating the change in length, Δl_l, due to the force, F_l, on the ends normal to the l-direction. The pressure, P, provides the stress, $\sigma = P$.

$$\frac{\Delta l_l}{l} = -\frac{\sigma}{E} \tag{4.4}$$

The minus sign reminds us that the increased pressure decreases the length of the sample. At the same time, the pressure, P, should produce the same strain in the height, $(\Delta h_h/h) = -(\sigma/E)$. Poisson's ratio determines the influence this change in height will have on the change in length.

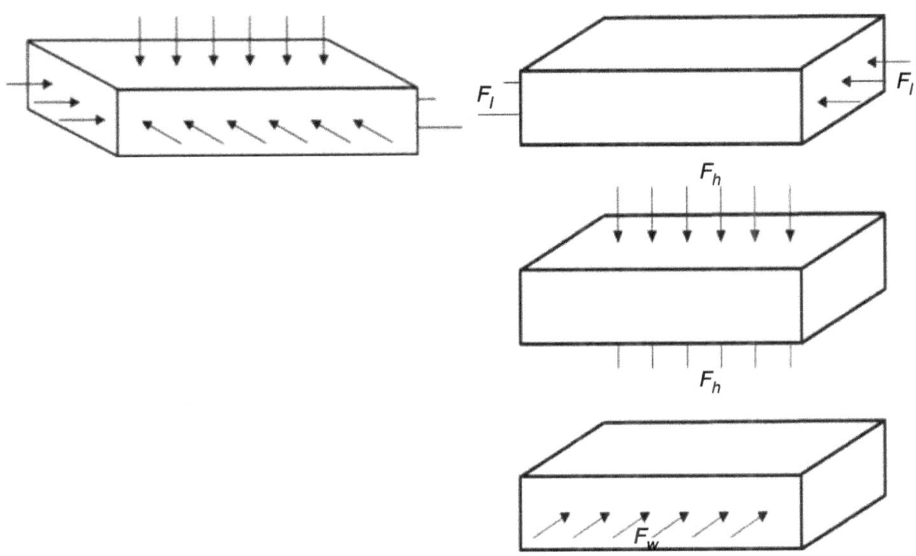

Fig. 4.2 The deformation of a rectangular object subject to hydrostatic pressure (*above*) can be analyzed by the superposition of the pressures applied to the orthogonal faces shown at the right. Assuming unit area for all forces, the pressure applied along the l-direction is F_l. It causes l to become shorter while h and w become longer, as dictated by Poisson's ratio. At the same time, F_h reduces h and increases l and w, and F_w decreases w and increases l and h. The overall change in volume can be calculated by the superposition of the three individual deformations

$$\frac{\Delta l_h}{l} = -\nu \frac{\Delta h_h}{h} = \nu \frac{\sigma}{E} \tag{4.5}$$

Since both the pressure and the material are isotropic, there is an equal change in length, $\Delta l_w/l$, caused by $\Delta w/w$ and Poisson's ratio, ν. The total strain in length, $\Delta l/l$, must be the sum (i.e., superposition) of these three contributions.

$$\frac{\Delta l}{l} = \frac{\Delta l_l}{l} + \frac{\Delta l_h}{l} + \frac{\Delta l_w}{l} = -\frac{\sigma}{E}(1 - 2\nu) \tag{4.6}$$

This result now allows us to relate the *volumetric strain*, $\Delta V/V$, to the hydrostatic pressure, $P = \sigma$, again using logarithmic differentiation, as in Eq. (4.3), and recognizing that the strain along all three of the sample's axes must be the same.

$$\frac{\Delta V}{V} = \frac{\Delta l}{l} + \frac{\Delta h}{h} + \frac{\Delta w}{w} = -3(1 - 2\nu)\frac{\sigma}{E} \tag{4.7}$$

We can now define a third elastic modulus, known as the *bulk modulus*, *B*, to relate changes in hydrostatic pressure to volumetric strain.

$$P = \sigma = -B\frac{\Delta V}{V} = -\frac{E}{3(1 - 2\nu)}\frac{\Delta V}{V} \tag{4.8}$$

In addition to providing the relationship between the bulk modulus, Young's modulus, and Poisson's ratio, Eq. (4.8) restricts Poisson's ratio to being less than one-half. To guarantee the stability of matter, the bulk modulus can never be negative. If the bulk modulus were negative, then the application of a small amount of pressure would cause the volume of the material to increase rather than decrease. The product of that change in pressure and the change in volume would do work on the surroundings. This would allow us to get useful work out of any material with $B < 0$ and would violate energy conservation. Similarly, if the solid were immersed in a fixed volume of liquid and the pressure of the liquid decreased (due to a change in temperature?), then the solid would become smaller, decreasing the fluid pressure further, and ultimately the solid would disappear [2].

4.2.2 Modulus of Unilateral Compression

The same use of superposition that allowed us to relate the bulk modulus, Young's modulus, and Poisson's ratio can be exploited again to calculate the *modulus of unilateral compression*, sometimes also known as the *dilatational modulus*. We will use *D* to designate this modulus. The modulus of unilateral compression is similar to Young's modulus except that the cross-section is not allowed to change, that is, $\Delta h_l/h = \Delta w_l/w = 0$, when $\Delta l_l/l \neq 0$. An example of a deformation due to a unilateral compression is shown in Fig. 4.3. This combination of pressure and strain is important for the propagation of plane longitudinal waves in solids.[3]

We would expect the modulus of unilateral compression to be greater than Young's modulus, since it should be more difficult to compress a sample if the walls are not permitted to bulge out. As before, we will superimpose three contributions to the strain in the length.

[3] If the diameter of a sound beam is much greater than its wavelength, the associated compressions and expansions do not allow the material to "squeeze out" because there is a compression both above and below that is trying equally hard to squeeze the material in the transverse direction.

Fig. 4.3 A plane wave of sound in a solid will cause a square element of the solid to be deformed into a rectangle. The height (and also width) of the element remains constant, while the length is changed. The undeformed element is shown as a *solid line*, and the unilaterally compressed element is shown by the *dashed lines*

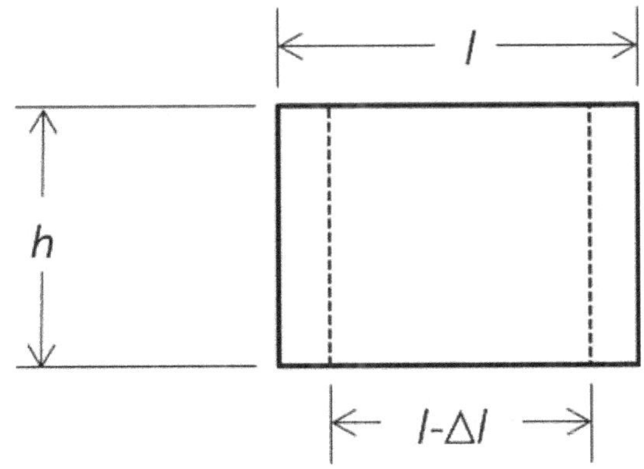

$$\frac{\Delta l_l}{l} = \frac{1}{E}\frac{F_l}{S_l} - \frac{\nu}{E}\frac{F_h}{S_h} - \frac{\nu}{E}\frac{F_w}{S_w} = \frac{1}{E}\left[\frac{F_l}{S_l} - \nu\left(\frac{F_h}{S_h} + \frac{F_w}{S_w}\right)\right] \tag{4.9}$$

$$\frac{\Delta l_h}{h} = \frac{1}{E}\left[\frac{F_h}{S_h} - \nu\left(\frac{F_l}{S_l} + \frac{F_w}{S_w}\right)\right] \tag{4.10}$$

$$\frac{\Delta l_w}{w} = \frac{1}{E}\left[\frac{F_w}{S_w} - \nu\left(\frac{F_l}{S_l} + \frac{F_h}{S_h}\right)\right] \tag{4.11}$$

Imposition of the constraint that $\Delta h/h = \Delta w/w = 0$ dictates the required stress needed to keep the cross-section constant. Those stresses can be determined by simultaneous solution of Eqs. (4.10) and (4.11).

$$\frac{F_h}{S_h} = \frac{F_w}{S_w} = \frac{\nu}{1-\nu}\frac{F_l}{S_l} = \left(\frac{\nu}{1-\nu}\right)\sigma \tag{4.12}$$

Substitution of this result back into Eq. (4.9) relates the strain in l to the stress in the direction normal to l.

$$\frac{\Delta l_l}{l} = \frac{1}{E}\left(1 - \frac{2\nu^2}{1-\nu}\right)\frac{F_l}{S_l} = \left(\frac{(1+\nu)(1-2\nu)}{1-\nu}\right)\frac{\sigma}{E} \tag{4.13}$$

By inverting this expression, we can write the modulus of unilateral compression, D, in terms of Young's modulus and Poisson's ratio.

$$\sigma = D\frac{\Delta l}{l} = \frac{(1-\nu)}{(1+\nu)(1-2\nu)}E\frac{\Delta l}{l} \tag{4.14}$$

Figure 4.4 shows that $D/E > 1$, if $\nu > 0$, as expected, since the sample is "stiffened" if the cross-section is constrained to remain unchanged.

The appearance of $(1 + \nu)$ in the denominator of Eq. (4.14) implies that if Poisson's ratio is negative, it cannot be more negative than -1 without violating the stability requirement, so $-1 < \nu \leq ½$.

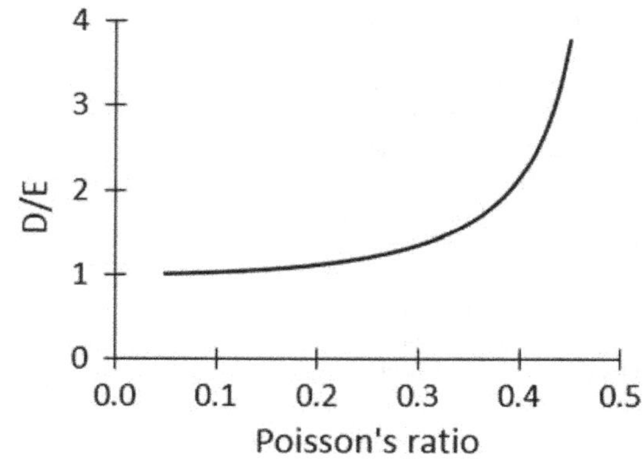

Fig. 4.4 The modulus of unilateral compression, D (also known as the dilatational modulus), is always greater than Young's modulus, E, if Poisson's ratio, ν, is greater than zero. The ratio, D/E, is plotted for $0 < \nu < ½$

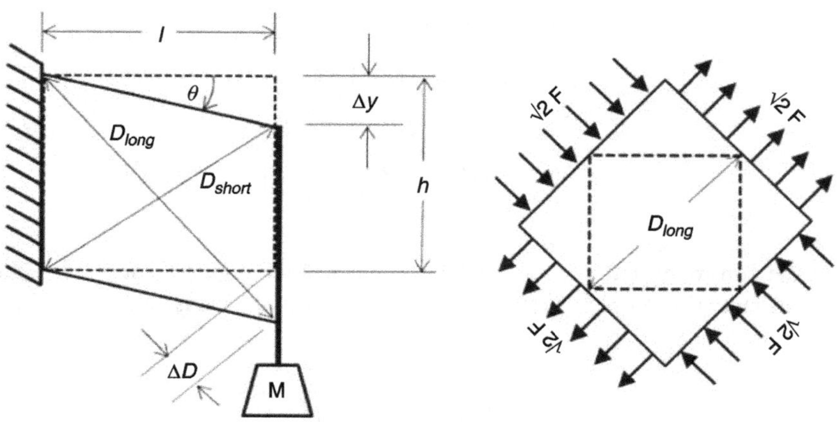

Fig. 4.5 (*Left*) A cube of initially square cross-section, $l = h = w$, is rigidly attached along one face and is loaded by a shear force, $F = Mg$, on the opposite face. This shear stress causes the square to deform into a rhombus with no changes to the cube's width, $\Delta w = 0$. The area is conserved by the shearing, but the shape has changed. The two diagonals that were initially of equal length are now unequal. (*Right*) D_{long} has been rotated by 90° in this figure. The same deformation can be achieved by applying a force $F' = F\sqrt{2}$ to a cube with sides $l' = h' = l\sqrt{2}$ that circumscribes the original cube and is rotated by 45°. The normal force compresses the larger square along two parallel faces and expands it when applied in the opposite direction along the orthogonal faces

4.2.3 Shear Modulus

A hydrostatic compression changes the volume of a sample but not its shape. A shear deformation, shown in Fig. 4.5, changes the shape of a sample but not its volume. Figure 4.5 shows one face of an initially cubical sample, $l = h$, which is rigidly supported along its left face and has a shearing force, $F = Mg$, applied along the opposite face by a suspended mass, M. The resulting shear strain will deform the square into a rhombus. Since h has not changed, the areas of the square and rhombus are equal, as are the volumes of their corresponding solids, the cube and rectangular parallelepiped (sometimes referred to as a rhomboid).

We can again use superposition to express the shear modulus in terms of Young's modulus and Poisson's ratio. This process is simplified if we examine Fig. 4.5 and realize that the shear deformation has taken the two diagonals of the square, which were originally of equal length, and made one diagonal shorter and the other longer, $D_{long} > D_{short}$. The cube is represented in Fig. 4.5 by a two-dimensional cross-section that ignores the width, w, since it is not changed by the shearing. The compression and expansion of the two diagonals can be reproduced if the original square is circumscribed by a larger square that is rotated by 45 ° with respect to the original. The larger square has edge lengths, $l' = h' = l\sqrt{2}$.

If we apply compressive forces, $F' = F\sqrt{2}$, to parallel faces, then the applied stress would be unchanged, since the surface area of the larger cube is $\sqrt{2}$ greater than that of the original cube. Those compressive forces will make one of the diagonals shorter. Similarly, if an equal tensile stress is applied to the cube's orthogonal faces, then the other diagonal is lengthened. The net forces and the net torque are zero, so the cube remains at rest during deformation. Also, we see that a pure shear is equivalent to the superposition of equal compressive and tensile stresses applied at right angles to each other and at 45 ° to the original faces of the cube.

We again superimpose the two orthogonal stresses to calculate the change in the length and the height of the deformed cube.

$$\frac{\Delta l'}{l'} = \frac{\Delta D_{long}}{D} = -\frac{\Delta h'}{h'} = -\frac{\Delta D_{short}}{D} = \frac{1}{E}\frac{F}{S} + \nu\frac{1}{E}\frac{F}{S} = \left(\frac{1+\nu}{E}\right)\frac{F}{S} \tag{4.15}$$

In Fig. 4.5, the strain is a dimensionless quantity, as is the angle, θ , by which the cube is sheared. (Keep in mind that in these calculations, D is the unstrained diagonal, not the modulus of unilateral compression.)

$$\tan\theta \cong \theta = \frac{\Delta y}{l} = \sqrt{2}\frac{\Delta D}{l} = 2\frac{\Delta D}{D} = 2\left(\frac{1+\nu}{E}\right) \tag{4.16}$$

The far-right term takes advantage of the fact that we are assuming linear elastic behavior, so $\Delta y << l$. Inverting this result leads to the definition of the shear modulus, G, that again allows us to assert Hooke's law as "stress is proportional to strain," this time for shearing stresses.

$$\sigma_{yx} = \frac{F_y}{S_x} = G\theta = \frac{E}{2(1+\nu)}\theta \tag{4.17}$$

In this version of Hooke's law, a new nomenclature has been introduced that is convenient for the description of stress. σ_{yx} represents a force, F_y, that is applied in the y direction on a surface of area, S_x, which has its normal in the x direction. In this notation, the normal hydrostatic stresses would be $\sigma_{xx} = \sigma_{yy} = \sigma_{zz} = P$ and $\sigma_{xy} = \sigma_{yz} = \sigma_{zx} = 0$, since fluids cannot sustain static shearing forces.[4]

Again, for stability, we must insist that $G > 0$, so $\nu > -1$. Combined with the similar restriction imposed by the bulk modulus, the values of Poisson's ratio are therefore restricted: $-1 < \nu < ½$. Most materials have Poisson's ratios that are between zero (e.g., cork) and ½ (e.g., rubber). *Auxetic materials*, which have negative values of Poisson's ratio, are uncommon. Love reported $\nu = -0.14$ for single-crystal pyrite [3]. Most auxetic materials are anisotropic, have very low mass densities, and usually involve some complicated internal structure, like solid foams. A Poisson's ratio as small as $\nu = -0.7$ has been reported for solid foam with a tetrakaidecahedral (14-sided) unit cell [4].

[4] In fluids, shearing forces diffuse; they are not restored elastically. See Sect. 9.4.

Table 4.1 For any isotropic solid, two elastic constants are sufficient to completely specify the solid's elastic behavior. Given any two known elastic moduli, listed in the left-hand column, equations are provided for the calculation of the other three isotropic elastic moduli

Given	Young's (E)	Poisson's (ν)	Bulk (B)	Dilatational (D)	Shear (G)
E,ν	E	ν	$\frac{E}{3(1-2\nu)}$	$\frac{(1-\nu)E}{(1+\nu)(1-2\nu)}$	$\frac{E}{2(1+\nu)}$
B,E	E	$\frac{1}{2}-\frac{E}{6B}$	B	$3B\frac{(3B+E)}{(9B-E)}$	$\frac{3BE}{9B-E}$
B,ν	$3B(1-2\nu)$	ν	B	$\frac{3B(1-\nu)}{1+\nu}$	$\frac{3B(1-2\nu)}{2(1+\nu)}$
G,E	E	$\frac{E}{2G}-1$	$\frac{EG}{3(3G-E)}$	$G\frac{(4G-E)}{(3G-E)}$	G
G,B	$\frac{9BG}{3B+G}$	$\frac{3B-2G}{6B+2G}$	B	$B+\frac{4G}{3}$	G
G,ν	$2G(1+\nu)$	ν	$\frac{2G(1+\nu)}{3(1-2\nu)}$	$\frac{2G(1-\nu)}{1-2\nu}$	G

4.2.4 Two Moduli Provide a Complete (Isotropic) Description

If a solid is isotropic and homogeneous, then two moduli provide a complete description of the solid's elastic behavior. The results of the calculations that relate Young's modulus, Poisson's ratio, the bulk modulus, the modulus of unilateral compression (or dilatational modulus), and the shear modulus are summarized in Table 4.1.[5]

4.3 Real Springs

In Chap. 2, we defined a spring constant that provided a parameter, K, to relate forces and displacements governed by Hooke's law. That definition allowed us to explore the behavior of simple harmonic oscillators consisting of masses, springs, and dashpots. Having developed a system for understanding the elastic behavior of solids, we can now explore a small amount of the vast territory that is dedicated to the design of a highly engineered product that has a very significant impact on the design or isolation of vibrating systems – the spring.

Real springs are much more complicated than Hooke's law may lead one to believe. Our analysis of the simple harmonic oscillator, in Chap. 2, represented the spring through a single parameter, the spring stiffness, K, which concealed the multiple design trade-offs that must be made to approach optimum performance for any spring. In addition to stiffness, attention must also be paid to the spring's moving mass; to limitations on the tensile stresses that might cause the material to yield or fracture, thus exceeding the elastic limit (see Fig. 1.4); or to application of compressive stresses that would cause buckling (see Sect. 4.3.4), as well as issues related to fatigue life in a material that is subjected to fully reversing cyclic loading for millions or billions of cycles [5].

In this section, we will explore only a few strategies for shaping materials in ways that make them suitable for applications where they are intended to store (and in some cases, absorb) elastic energy using the spring materials efficiently. We will start with simple use of the bulk material to provide stiffness, but then look at how shapes such as cantilevered beams, tubes in torsion, and helical coils

[5] The Lamé constants, λ and μ, also are taken as the two isotropic moduli in some elasticity calculations, although they are not listed in Table 4.1. The shear modulus, G, and μ are identical and $\lambda = \frac{E\nu}{(1+\nu)(1-2\nu)} = \frac{G(E-2G)}{3G-E} = \frac{2G\nu}{1-2\nu}$.

produce acceptable and convenient trade-offs in a few applications and how rubberlike materials are used as springs for vibration isolation where they simultaneously provide both elasticity and damping.

4.3.1 Solids as Springs

Although the majority of spring applications require forming and shaping of the elastic material, as well as surface treatments (e.g., shot peening, tumble deburring) and heat treatments (e.g., precipitation hardening, quick quenching) [6], there are a few circumstances where the material is used "in bulk," as a block, rod, or tube to provide compressional or shear stiffness. This is most common with rubber springs used as vibration isolators. We will postpone discussion of rubber springs until later in this chapter because the viscoelastic behavior of those elastomeric materials can simultaneously contribute damping as well as stiffness. Here we will examine two piezoelectric accelerometer designs and a resonant piezoelectric underwater sound source that use the elasticity of the bulk materials to provide stiffness.[6]

An accelerometer is a vibration measurement sensor that converts mechanical accelerations into an electrical signal that can be displayed and/or recorded by some electronic measurement instrument. The base of an accelerometer is usually attached to some vibrating structure using a threaded stud, a magnet, an adhesive or wax, etc., that attempts to ensure that the motions of the sensor and of the vibrating structure are identical. As we have seen from our study of the displacement-driven simple harmonic oscillators in Sect. 2.5.6, if the driving frequency, ω, is less than the natural frequency, $\omega < \omega_o = \sqrt{K/m}$, then the displacement of the spring is directly proportional to the force produced by the acceleration, a, of the mass, m: $xe^{j\omega t} = -mae^{j\omega t}/K$.

Figure 4.6 shows cross-sectional diagrams of two types of piezoelectric accelerometers. We start by calculating the stiffness of the version (Fig. 4.6, right) that compresses a *test mass* (sometimes also called a *seismic mass* or a *proof mass*) against a hollow cylinder made of a *ferroelectric ceramic* material[7] using a steel "preload stud." Since ceramic materials are much stronger in compression than under tension, static compression provided by the preload stud guarantees that the piezoelectric spring never goes into tension, as well as holding the mass-spring "sandwich" together. The total stiffness of the cylinder and stud will be the sum of their stiffnesses: $K_{total} = K_{stud} + K_{piezo}$.

If we treat the pre-load stud as a thin rod, then its stiffness, K_{stud}, will be related to its geometry and Young's modulus of the material that is used to make the stud. We will let the length of the stud (between the mass and the base) be l, let its diameter be D_{stud}, and assume the stud is made of some generic steel. The stiffness of the piezoelectric tube will depend upon its Young's modulus and its inner and outer radii, a_{in} and a_{out}. Its elastically active length, l, is the same as the stud's length. Using the appropriate cross-sectional areas, the stiffnesses are given by

[6] I will apologize for indulging in a bit of a fraud at this point by introducing piezoelectric solids. So far, we have addressed the elastic behavior of isotropic solids that require only two independent moduli to completely specify their elastic behavior. Piezoelectric solids are intrinsically anisotropic crystalline materials that not only require more than two elastic constants but also require specification of the electrical impedance that provides the load across their electrodes, $\mathbf{Z}_{load}$. For example, a piezoelectric material's stiffness will be different if the electrodes are electrically shorted together (i.e., $\mathbf{Z}_{load} = 0$) or left as an "open circuit" (i.e., $\mathbf{Z}_{load} = \infty$). Anisotropic elasticity will be addressed later in this chapter (see Sect. 4.6), but for more complex crystalline substances, there are necessarily more than two independent elastic constants. In practice, once the relevant constants have been identified for a specific deformation, they are incorporated into Hooke's law in the same way as the elastic constants of an isotropic solid.

[7] A ferroelectric ceramic behaves like a piezoelectric crystal. The difference is the piezoelectric behavior of the crystal is intrinsic, and a ferroelectric material (usually a ceramic or polymer) only exhibits piezoelectric behavior after the material has been "polarized," usually by application of a large electric field at elevated temperatures [7].

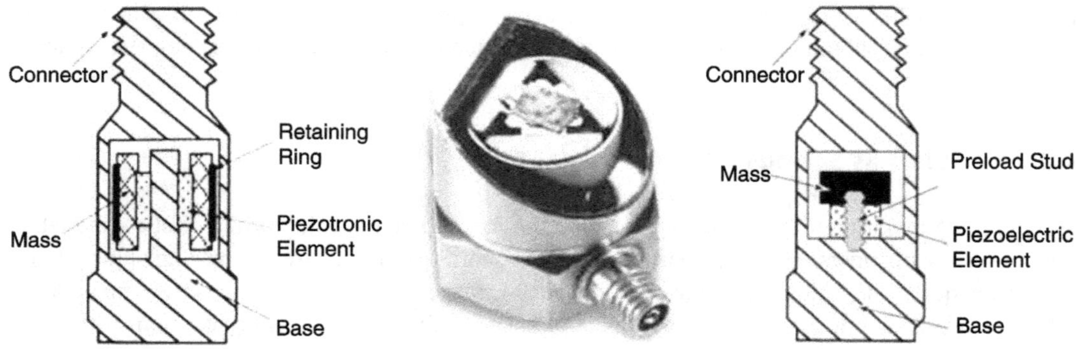

Fig. 4.6 (*Left*) Cross-sectional schematic diagram of an accelerometer that uses the shear stiffnesses of multiple piezoelectric elements (called "piezotronic" in this drawing) to both determine the natural frequency, $\omega_o = (K/m)^{1/2}$, of the mass-spring combination and to generate an electrical signal that can be used to display and/or record those accelerations (*Center*) Photograph of an accelerometer that uses three blocks of piezoelectric material in shear and three masses. All of the masses are joined by a retaining ring so that the sensor behaves as a single mass-spring oscillator. At the top center of the photograph, on the post supporting the three shear elastic elements, there is a small integrated electronic circuit that acts as an impedance transforming amplifier to electrically buffer the output of the piezoelectric elements. (*Right*) Cross-sectional schematic diagram of an accelerometer that uses a tubular piezoelectric cylinder as the spring to support a "test mass," also known as the "seismic mass." That spring is compressed and relieved by the forces produced by the accelerations of the supported (seismic) mass. In that design, the "preload stud" is a threaded rod that squeezes the mass and piezoelectric spring together so that the piezoelectric element never experiences tensile strains. (Figures courtesy of PCB Piezotronics, Inc.)

$$K_{stud} = \frac{E_{steel}\left(\frac{\pi}{4} D_{stud}^2\right)}{l} \quad \text{and} \quad K_{piezo} = \frac{\pi E_{piezo}\left(a_{out}^2 - a_{in}^2\right)}{l} \tag{4.18}$$

Let's assign some nominal values to those one dimensions and to the elastic moduli. For the stud, let $D_{stud} = 2.0$ mm (about the correct value for a 4–40 machine screw) and let $l = 6.0$ mm. Representative values for steel[8] are $E_{steel} = 195$ GPa and $\rho_{steel} = 7700$ kg/m^3. We'll choose lead zirconium titanate (PZT) as our piezoelectric (ferroelectric) ceramic, with representative values of $E_{PZT} = 50$ GPa and $\rho_{PZT} = 7500$ kg/m^3. The ceramic tube will have the same length as the stud with an outer radius $a_{out} = 4.0$ mm and an inner radius $a_{in} = 2.0$ mm. To be relatively consistent with the dimensions in Fig. 4.6, the test mass will be a steel cylinder with $D_{mass} = 12.0$ mm and a height $h = 4.0$ mm. Plugging these values into Eq. (4.18) makes $K_{stud} = 1.02 \times 10^8$ N/m and $K_{piezo} = 3.14 \times 10^8$ N/m for a total stiffness of $K_{total} = 4.16 \times 10^8$ N/m.

For the test (seismic) mass, $m_{test} = \rho_{steel}\,(\pi/4)\,D_{mass}{}^2\,h = 3.5 \times 10^{-3}$ kg. Based on Eq. (2.27), we should add one-third of the mass of the springs to the test mass to approximate the dynamic mass, m_o.

$$m_o = m_{test} + \left(\frac{m_{PZT} + m_{stud}}{3}\right) = m_{test} + \frac{\pi}{12}\left[4\rho_{PZT}\left(a_{out}^2 - a_{in}^2\right)l + \rho_{steel} D_{stud}^2 l\right] \tag{4.19}$$

Using the same values for the dimensions and the mass densities, $m_o = 4.10 \times 10^{-3}$ kg, so $f_o = (\omega_o/2\pi) = \sqrt{K_{total}/m_o}/2\pi = 50.7$ kHz.

To demonstrate that these numbers are not unrealistic, Fig. 4.7 shows the calibration card for a typical accelerometer of this type that has been in use in my laboratory for decades. The relative

[8] It is important to remember that in a textbook example, it is convenient to define "some generic steel," but properties of steel (e.g., modulus, yield strength, endurance limit, heat capacity, thermal conductivity, electrical conductivity, etc.) vary with alloy composition and temper. In a commercial design, the choice of the material is very important.

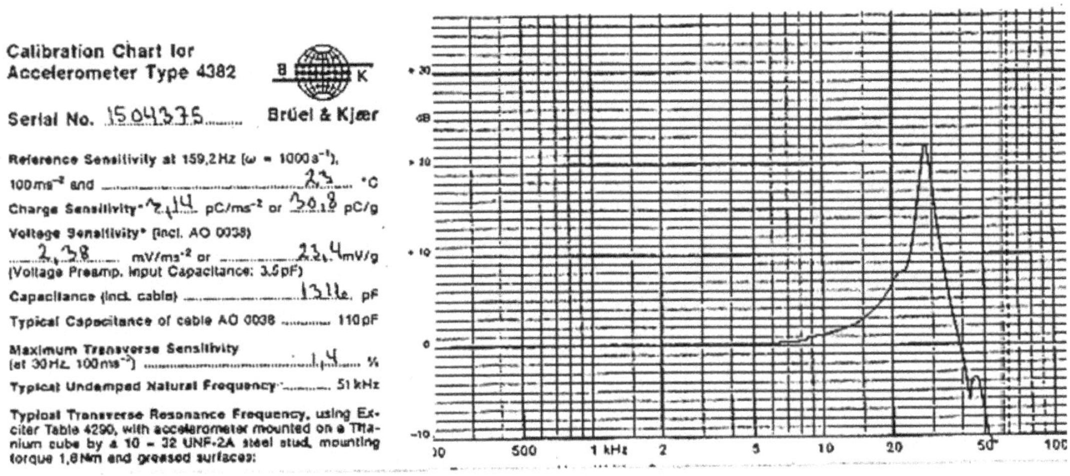

Fig. 4.7 Calibration card for a Brüel and Kjær Type 4382 piezoelectric accelerometer. The graph provides the response relative to its low-frequency sensitivity, which is "flat" (i.e., frequency independent) to about 6 kHz. The mass-spring resonance frequency occurs just below 30 kHz

response (in dB) as a function of frequency shows a peak at just below 30 kHz. The response is constant with an increase of 1.0 dB in the sensitivity at 10 kHz but is "flat" (frequency independent) below that frequency where the sensor is operating well within its stiffness-controlled frequency regime, $\omega < \omega_o$.

Let's repeat the analysis for the "shear mode" accelerometer shown in Fig. 4.6 (left and center). It consists of three flat quartz (SiO_2) plates supported along their inside surfaces by a post attached to the accelerometer's case and along their outside surfaces by three masses that are attached to each other by a retaining ring. As the case moves upward, the masses exert a downward shear force on the quartz plates in much the same way as the suspended mass exerts a static shear force on the sample diagrammed in Fig. 4.5 (left). We will use $G_{quartz} = 2.23 \times 10^{10}$ Pa as the shear modulus of quartz and $\rho_{quartz} = 2650$ kg/m^3. We'll let $m_{test} = 5$ grams for the test mass (0.005 kg) and choose plausible dimensions for the quartz plates: $l = 2.0$ mm, $h = 4.0$ mm, and $w = 5.0$ mm. Using the appropriate cross-sectional areas, the stiffnesses are given by

$$K_{quartz} = -\frac{F_y}{\Delta y} = G_{quartz}\frac{wh}{l} \tag{4.20}$$

Remembering that the stiffnesses, K_{quartz}, of each of the three plates are additive, $K_{total} = 3K_{quartz} = 6.7 \times 10^8$ N/m. The total mass for the three quartz plates is $3m_{quartz} = 0.318$ gram, one-third of which is added to the test mass to make $m_o = 5.11 \times 10^{-3}$ kg. Again, $f_o = (\omega_o/2\pi) = \sqrt{K_{total}/m_o}/2\pi = 57.6$ kHz.

The piezoelectric effect will produce an alternating electrical output based on the alternating stresses in the piezoelectric elements in either accelerometer design. This is not the time to belabor the differences between the two designs or their electrical transduction mechanisms, since these accelerometers were introduced primarily as vibration sensors and simple examples of the direct use of the elastic properties of solids to provide stiffness in real (i.e., useful) physical systems.

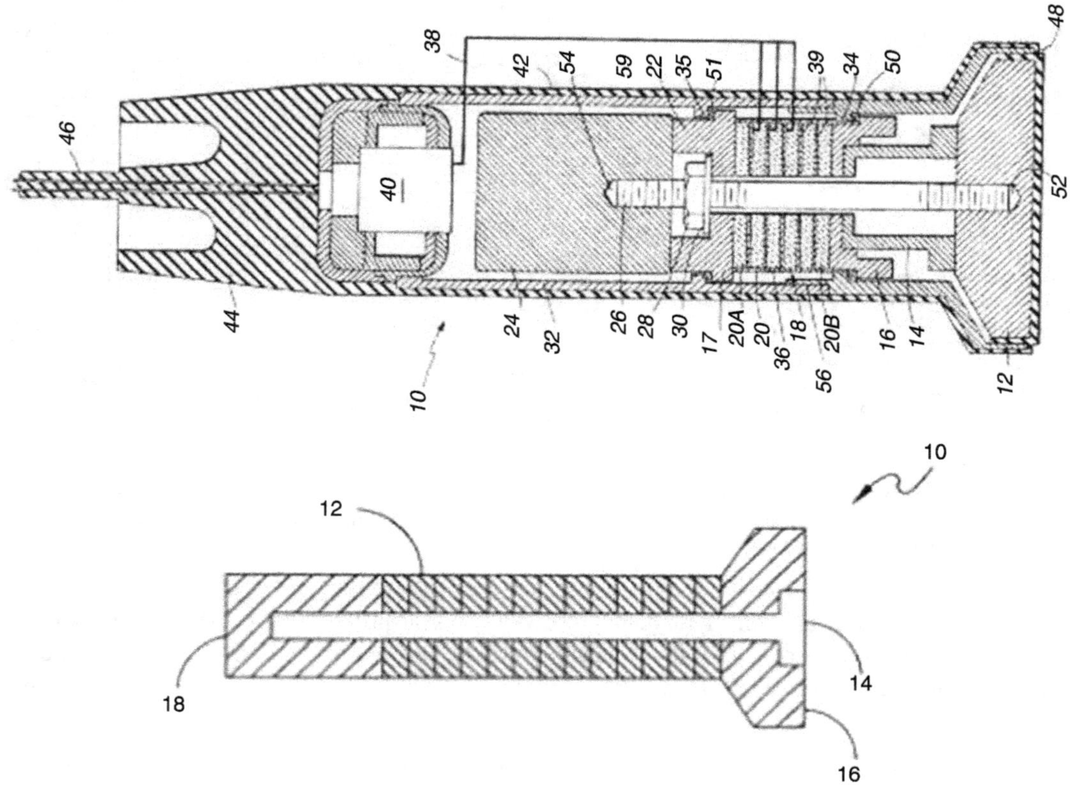

Fig. 4.8 These cross-sectional diagrams represent a *Tonpilz* transducer [7]. (*Above*) This figure was taken from a patent [8] that shows many details of a projector intended for efficient production of sound under water. A tensioning rod (52) and nut (28) compress a stack of piezoelectric rings (20, 20A, 20B) between the radiating surface, known as the "head mass" (12), and a "tail mass" (24) that acts as an inertial counter balance. The mechanically active components (i.e., head mass, tail mass, and stack) are contained within the waterproof housing (44) and are resiliently supported (17 and 34) near a node in the oscillatory motion. That node occurs at a position within the piezoelectric stack. Other parts, such as an electrical impedance matching transformer (40), are not related to our calculations. (*Below*) A simplified cross-section (10) that focuses on only the critical vibrating components [9]: tail mass (18), head mass (16), tensioning rod (14), and the stack of piezoelectric elements (12) that provide both the stiffness and the "motor mechanism" to drive the simple harmonic oscillator at its resonance frequency in the asymmetric mode

Since the piezoelectric effect is both linear and reversible, it is also possible to drive the piezoelectric elements as "motors." The next example is a *Tonpilz*[9] *transducer* that is common in naval SONAR systems and torpedo guidance systems [7].

Figure 4.8 shows a detailed drawing of such an underwater projector on the top [8] and a simplified cross-sectional diagram on the bottom [9]. We will further simplify our analysis by treating this transducer as two masses joined by a spring. As with the coupled harmonic oscillators in Sect. 2.7, this transducer is a one-dimensional oscillator that has two degrees of freedom corresponding to the positions of the head and tail masses. It will therefore possess two normal mode frequencies: a lower-frequency symmetric mode (both masses moving in-phase) and a higher-frequency antisymmetric mode. In this case, the symmetric mode corresponds to a uniform translation with both masses

[9] *Tonpilz* is from the German *ton* (tone) and *pilz* (mushroom). Apparently, the piezoelectric stack is the stem and the head mass is the "singing mushroom" cap.

moving at the same velocity in the direction joining their centers. The frequency of that mode is zero because there is no restoring force. Our interest is in the antisymmetric mode, where the head and tail masses move in opposite directions, thus compressing or expanding the spring that joins them.

The modal frequency for such a combination is well-known (see Sects. 2.7.1, 2.7.2, and 2.7.3) because it is analogous to the vibration of a diatomic molecule [10]. If the spring were massless, then the frequency of the antisymmetric mode is $\omega_a = \sqrt{K/\mu}$, where $\mu = (m_1 m_2)/(m_1 + m_2)$ is known as the *reduced mass.*

This result is easy to visualize for $m_1 = m_2$. In that case, we know that the amplitude of the motions of both masses must be equal and opposite, so the midpoint of the spring will be a node. The normal mode frequency corresponds to all parts of the system oscillating at the same frequency. We can exploit that fact to calculate ω_a by calculating the motion of one mass and one spring of twice the spring constant that has half the total length: $\omega_c = \sqrt{2K/m_1} = \sqrt{K/\mu}$. We will use the same approach to calculate the resonance frequency of our *Tonpilz* transducer that has different values for masses, $m_1 \neq m_2$, and a spring, being the piezoelectric stack and tensioning rod, which will also contribute non-negligible mass.

First, let's choose some reasonable values for the components in our Tonpilz example. Since the objective of this design is the generation of sound in water, we would like the tail mass to be larger than the head mass so that the motion of the end in contact with the water will be greater than the motion of the "counter weight" provided by the tail mass. As will be demonstrated in Sect. 12.8.3, a circular piston whose circumference is less than the wavelength of the radiated sound, $2\pi a < \lambda_{fluid} = c_{water}/f$, has to accelerate a mass of fluid, m_{rad}, that is equivalent to the mass of fluid contained in a cylindrical volume with the same area as the piston, πa^2, and a height, $h = (8/3\pi)a$. If we let the radius of the head mass be 5.0 cm, then the *radiation mass* is $m_{rad} = (8/3)\, a^3 \rho_{water} = 0.333$ kg.

Since the design goal is to make the head mass as light as possible, the radiation mass places a natural limit on how a natural limit on how advantageous it might be to make the head mass small. For this example, let's make the head mass (including the radiation mass of the fluid) about twice the radiation mass, $m_h = 0.70$ kg. The head mass should be as stiff as possible so that it acts as a piston and does not flex. That combination of lightness and stiffness can be improved by making the head mass "ribbed" to increase rigidity and reduce its moving mass. Such a "ribbed piston" is shown in Fig. 4.9. We will let the tail mass be five times the head mass, $m_t = 5m_h = 3.5$ kg.

The spring's stiffness will again be provided by a stack of piezoelectric ceramic rings that are pre-compressed by the tensioning rod. Let the diameter of the tension rod be $D_{rod} = 6.0$ mm and choose the outer and inner radii of the piezoelectric rings to be $a_{out} = 2.0$ cm and $a_{in} = 1.5$ cm. The length of the stack will be $L = 15.0$ cm. Using the previous values of Young's modulus and mass

Fig. 4.9 The "ribbed piston" shown here was designed by Dr. R. W. M. Smith and Eric Mitchell, in 2013, for a different application [11]. They fabricated the piston from a high-performance plastic. The ribs increase stiffness to resist flexure without requiring an excessive amount of material, hence reducing the piston's moving mass

density for steel and PZT, the stiffnesses and masses are as follows: $K_{rod} = 3.67 \times 10^7$ N/m, $K_{PZT} = 1.83 \times 10^8$ N/m and $K_{total} = 2.20 \times 10^8$ N/m; $m_{rod} = 32.7$ gm and $m_{PZT} = 0.618$ kg, so $m_s = 0.651$ kg.

Since the two masses are not equal, we will need to properly apportion the moving mass, m_s, and K_{total} of the PZT stack and steel rod about the vibration node. We will use the equality of the frequencies on either side of the node, as we did in Sect. 3.6.2, to determine the location of the node and yield the antisymmetric normal mode frequency. If we let the node be a distance, b, from the attachment point of the head mass, m_h, then the separation between the attachment point of the tail mass, m_t, and the node must be $L - b$.

We have used the fact that the stiffness-length product is a constant before (see Sect. 2.2.2), although now we see it also as a direct consequence of our definition of Young's modulus in Eq. (4.1): $KL = FL/\Delta l = ES$. That allows us to write the stiffness of the spring between the node and the head mass as $K_h = (K_{total}L)/b$ and the stiffness of the spring between the tail mass and the node as $K_t = (K_{total}L)/(L\text{-}b)$. Since $m_t = 5m_h$, we expect $5(L\text{-}b) \cong b$. If we also scale the moving mass of the spring, then the equality of the normal mode frequencies can be solved to provide the value of b.

$$\omega_h^2 = \left(\frac{K_{total}L}{b}\right)\frac{1}{m_h + \frac{bm_s}{3L}} = \left(\frac{K_{total}L}{L-b}\right)\frac{1}{m_t + \frac{(L-b)m_s}{3L}} = \omega_t^2 \tag{4.21}$$

Solving Eq. (4.21) will provide the b/L ratio.

$$\frac{b}{L} = \frac{m_t + \frac{m}{3}}{m_t + m_h + \frac{2m_s}{3}} \tag{4.22}$$

For the values used in this example, $b/L = 0.802$. (This would also tell us where we would want to support this projector so that none of its vibrations are communicated through the support structure.) Plugging b/L back into Eq. (4.21) gives $f_h = f_t = 2.82$ kHz. If the speed of sound in water is $c_{water} = 1500$ m/s, then $2\pi a = 31.4$ cm and $\lambda_{fluid} = c/f_h = 53.2$ cm $> 2\pi a$, so the radiation mass assumption was valid. This result is close to what would have been the normal mode frequency, $f_a = \sqrt{K_{total}/\mu}/2\pi = 3.1$ kHz, using the reduced mass $\mu = (m_1 m_2)/(m_1 + m_2) = 0.583$ kg and assuming that the spring (the piezoelectric stack and tension rod) was massless.

These three examples of the direct use of solids as springs have been restricted to solid samples with a small aspect ratio, which is $l \lesssim w \cong h$, for the accelerometers, and to forces on the spring being applied longitudinally by the motor mechanism (i.e., the piezoelectric stack) and inertial effects in the *Tonpilz* case. It is rare to see long thin rods or bars being used as linear springs because the forces necessary to bend such a sample are smaller than the forces that would produce an equivalent longitudinal compression. This restriction will be quantified in the next sub-section. Also, long thin bars that are subject to compression can spontaneously buckle (i.e., collapse catastrophically) if the force exceeds a threshold value as discussed in Sect. 4.3.4.

4.3.2 Flexure Springs

Cantilevered beams are used as springs in applications that range from automotive suspensions to micromachined silicon sensors (like the beam from the atomic force microscope shown in Fig. 2.36). If you hold a meter stick with both hands at eye level,[10] with the numbers facing upward or downward, it

[10] You will also want to employ some protective eyewear.

is possible to apply torques to the two ends, as shown in Fig. 4.10, that will bend the meter stick into a smooth curve that corresponds (approximately) to the arc of a circle. If the numbered surface of the meter stick is facing toward you, then it will be very difficult to bend the meter stick vertically. Given enough torque, it might squirm (rotate into the previous orientation) or possibly fracture, although it does bend easily and reversibly if it is held flat (number side up or down). The radius of curvature, R, as expressed in Eq. (4.33), will be infinite before the meter stick is bent. As the bending increases, the radius will become shorter. In Fig. 4.10, $R \cong 2.5\ L$, where L is the length of the meter stick.

What is happening to the material when the stick is bent and what is providing the restoring force that returns the stick to its straight condition after the torques are removed? When the stick is curved, as shown in Fig. 4.10, the material near the top of the stick is placed in tension and the material near the bottom of the stick is placed in compression. Since the stresses change sign going from the top of the stick to the bottom, there must be some surface near the middle that is unstressed. That surface is called the *neutral plane*. Figure 4.11 (left) shows a small segment of the bent beam (meter stick). The dashed line corresponds to the position of that neutral plane.

Fig. 4.10 A beam (meter stick) is bent into an approximate arc of a circle by the application of opposite torques to the ends. Before bending, the radius of the arc is infinite. As more torque is applied, the bending increases and the beam's radius of curvature, R, moves in from infinity. For the bend shown here, the radius of curvature is about two and a half times the unbent length: $R \cong 2.5\ L$

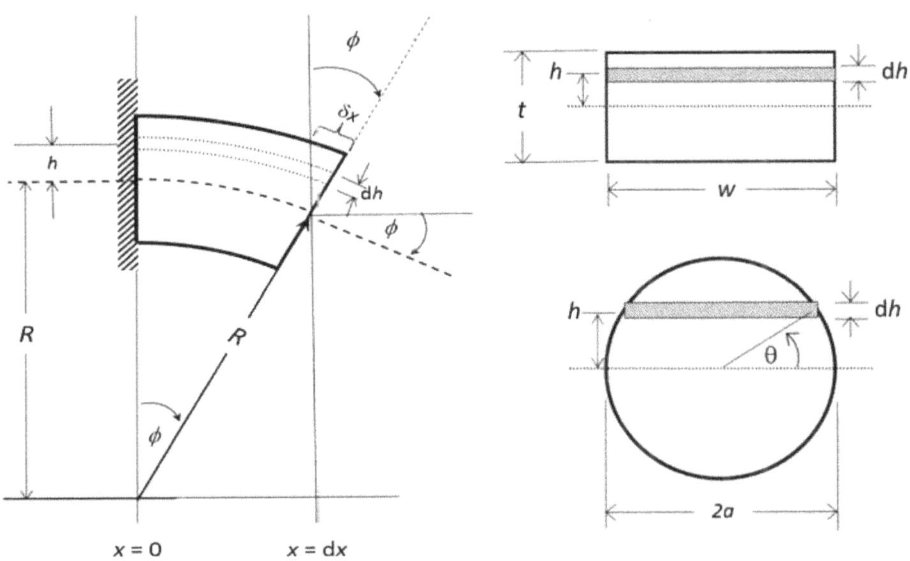

Fig. 4.11 (*Left*) Diagram of a small segment of a bent beam. The dashed line represents the neutral surface. Filaments above that surface experience tension that is proportional to the distance, h, from the neutral surface. Those filaments below the neutral surface are in compression, again in proportion to their distance below the neutral surface. (*Right*) The geometry used to calculate the radius of gyration for a beam with a rectangular or circular cross-section

If there are no additional tensile or compressive forces applied to the beam, just the applied torque, then below the neutral plane, the compressive strain is proportional to the distance, h, below the neutral plane. Above the neutral plane, the tensile strain is also proportional to the distance, h, above the neutral plane.

$$\frac{\Delta l}{l} = \frac{h}{R} \tag{4.23}$$

To work out the forces so they can be summed (integrated) to calculate the bending moment (i.e., the restoring torque), we will consider a differential segment of a rectangular beam that is w wide, t thick, and rigidly clamped at $x = 0$, as shown schematically in Fig. 4.11 (left). When $h = +(t/2)$, the uppermost surface of the element is stretched from its equilibrium length by an amount, $\Delta l = +\delta x$, and the lowermost surface at $h = -(t/2)$ is compressed by $\Delta l = -\delta x$.

By our definition of Young's modulus in Eq. (4.1), the stress (force per unit area) is also proportional to the distance from the neutral plane. Letting $dS = w\, dh$, the force, δF, produced by the filament, dh, thick and w wide, a distance, h, above the neutral plane, is related to the strain by Young's modulus, E.

$$\frac{dF}{dS} = E\frac{\delta x}{dx} \tag{4.24}$$

The strain, $\delta x/dx$, can be calculated using Garrett's First Law of Geometry (as stated in Sect. 1.1).

$$\frac{\delta x}{dx} = \frac{h \tan\phi}{dx} \cong h\frac{\phi}{dx} \quad \Rightarrow \quad \frac{dF}{dS} = E\frac{h\phi}{dx} \tag{4.25}$$

In Fig. 4.11 (left), the forces cancel but their *bending moment* about the neutral plane, $\mathfrak{M}$, is non-zero. That moment can be calculated by integration of the force times the distance from the neutral plane.

$$\mathfrak{M} = \int h\,dF = \frac{E\phi}{dx}\int h^2\,dS \equiv \frac{ES\phi\kappa^2}{dx} \quad \text{where} \quad \kappa^2 = \frac{\int h^2\,dS}{S} \tag{4.26}$$

The square of radius of gyration, κ^2, will depend upon the shape of the beam. κ^2 is calculated below for the two cases shown in Fig. 4.11 (right).

$$\begin{aligned} \kappa^2_{bar} &= \frac{w\int_{-t/2}^{t/2} h^2\,dy}{wt} = \frac{t^2}{12} \\ \kappa^2_{rod} &= \frac{4\int_0^{\pi/2}(a^2\sin^2\theta)\ (a\cos^2\theta)\,a\,d\theta}{\pi a^2} = \frac{a^2}{4} \end{aligned} \tag{4.27}$$

The stiffness of a beam in flexure depends not only upon the material but also its distribution with respect to the neutral axis. The effects of this material distribution are quantified by the square of the *radius of gyration*, κ^2. That is why steel used in construction frequently has a cross-section that is in the shape of an "I" or "H" to place more material farther from the neutral plane. This strategy will be limited because too much material separated from the neutral plane by too little "webbing" will allow the shape to become distorted, or twisted, when loaded, like trying to bend the meter stick against the tall dimension (the numbers facing you, instead of the edge facing you) in our first bending example.

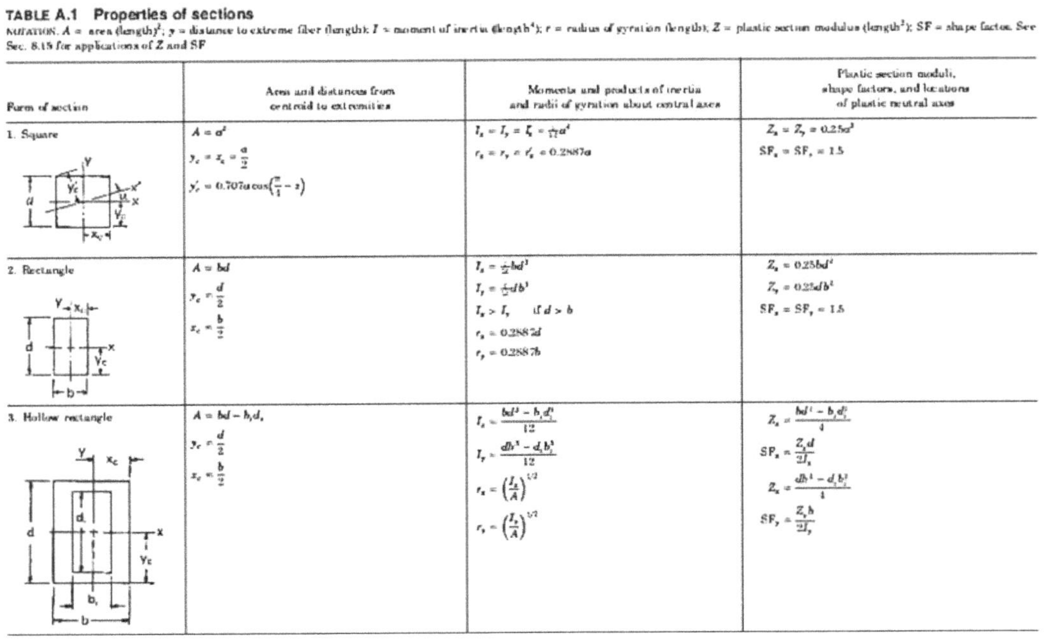

802 Formulas for Stress and Strain [APP. A

TABLE A.1 Properties of sections

NOTATION: A = area (length)^2; y = distance to extreme fiber (length); I = moment of inertia (length^4); r = radius of gyration (length); Z = plastic section modulus (length^3); SF = shape factor. See Sec. 8.15 for applications of Z and SF

Form of section	Area and distances from centroid to extremities	Moments and products of inertia and radii of gyration about central axes	Plastic section moduli, shape factors, and locations of plastic neutral axes
1. Square	$A = a^2$ $y_c = x_c = \frac{a}{2}$ $y_c' = 0.707a\cos\left(\frac{\pi}{4} - \alpha\right)$	$I_x = I_y = I_{x'} = \frac{1}{12}a^4$ $r_x = r_y = r_{x'} = 0.2887a$	$Z_x = Z_y = 0.25a^3$ $SF_x = SF_y = 1.5$
2. Rectangle	$A = bd$ $y_c = \frac{d}{2}$ $x_c = \frac{b}{2}$	$I_x = \frac{1}{12}bd^3$ $I_y = \frac{1}{12}db^3$ $I_x > I_y$ if $d > b$ $r_x = 0.2887d$ $r_y = 0.2887b$	$Z_x = 0.25bd^2$ $Z_y = 0.25db^2$ $SF_x = SF_y = 1.5$
3. Hollow rectangle	$A = bd - b_id_i$ $y_c = \frac{d}{2}$ $x_c = \frac{b}{2}$	$I_x = \frac{bd^3 - b_id_i^3}{12}$ $I_y = \frac{db^3 - d_ib_i^3}{12}$ $r_x = \left(\frac{I_x}{A}\right)^{1/2}$ $r_y = \left(\frac{I_y}{A}\right)^{1/2}$	$Z_x = \frac{bd^2 - b_id_i^2}{4}$ $SF_x = \frac{Z_xd}{2I_x}$ $Z_y = \frac{db^2 - d_ib_i^2}{4}$ $SF_y = \frac{Z_yb}{2I_y}$

Fig. 4.12 A portion of the tabulation from the "Properties of Sections" taken from the extremely useful book, *Roark's Formulas for Stress and Strain*. Only the first three of 29 cases are shown here. The radii of gyration for the shapes shown in the *left column* are provided in the *middle column*

Due to the critical importance of the moment of inertia[11] and the radius of gyration for structural engineering, there are numerous handbooks that provide the results for these integrations over nearly any imaginable cross-section. My favorite compilation for such results, as well as for the stresses on pressure vessels, bending of beams and plates, vibration of structures, etc., is *Roark's Formulas for Stress and Strain* [12]. A portion of one page from Appendix A.1, "Properties of Sections," taken from that book, is shown in Fig. 4.12.

For small deflections, application of Garrett's First Law of Geometry suggests that the angle, ϕ, is the same as the negative of the difference in the slope of the neutral plane at $x = 0$ and at $x = dx$, which can be estimated using the first two terms in a Taylor series.

$$\phi = -\left[\left(\frac{\partial y}{\partial x}\right)_{x+dx} - \left(\frac{\partial y}{\partial x}\right)_{x}\right] = -\left(\frac{\partial^2 y}{\partial x^2}\right)dx \tag{4.28}$$

In Fig. 4.11 (left), $(\partial y/\partial x)_{x=0} = 0$, but Eq. (4.28) would be correct even if that were not the case. Substitution of Eq. (4.28) for ϕ back into Eq. (4.26) provides the relation between the bending moment, $\mathfrak{M}$, and the curvature of the differential element that will be useful for the calculation of the flexural vibration of beams in Sect. 5.3.

[11] The moment of inertia $I = \int \rho y^2 dA$ is related to the square of the radius-of-gyration since the radius-of-gyration is equivalent to I for a material of unit mass density, if divided by cross-sectional area S. The square of the radius-of-gyration is also equivalent to the second moment of area.

$$\mathfrak{M} = -ES\kappa^2 \frac{\partial^2 y}{\partial x^2} \tag{4.29}$$

The work done to bend the differential element in Fig. 4.11 (left) can be calculated in terms of the angle of the bend, again by use of Garrett's First Law of Geometry, $\phi = dx/R$. The bending moment, $\mathfrak{M}$, in Eq. (4.26) can be expressed in terms of that angle.

$$\mathfrak{M} = \frac{ES\kappa^2}{dx}\phi \tag{4.30}$$

The amount of work, dW, required to bend that differential element from its equilibrium condition (straight) into an arc of angle, ϕ, will be the integral of the moment.

$$dW = \int_0^{\phi} \mathfrak{M}(\phi')\, d\phi' = \frac{1}{2}\frac{ES\kappa^2}{dx}\phi^2 \tag{4.31}$$

Substituting the expression for ϕ, derived in Eq. (4.28), into Eq. (4.31) provides an expression for the change in potential energy that is done by the work of bending the differential element that is in a convenient form for integration along bars.

$$d(PE) = -dW = \frac{ES\kappa^2}{2}\left(\frac{\partial^2 y}{\partial x^2}\right)^2 dx \tag{4.32}$$

Let's now apply Eq. (4.26) to the cantilevered beam that was drawn by Galileo and shown in Fig. 2.27. That (wooden) beam is built into a wall that constrains the slope of the beam to be zero as it leaves the wall. We can let $y(x)$ be the transverse deflection of the beam from equilibrium as a function of distance, x, from the wall. The vertical component of the force, F_v, which causes the deflection, will be the weight of the mass, M, attached at the beam's end $x = L$, with $F_v = Mg$. Equation (4.33) provides the reciprocal of the radius of curvature of the beam, R^{-1}, at any location, x.

$$\frac{1}{R} = \frac{\frac{d^2y}{dx^2}}{\left[1 + \left(\frac{dy}{dx}\right)^2\right]^{2/3}} \cong \frac{d^2y}{dx^2} \quad \text{if} \quad \left(\frac{dy}{dx}\right)^2 << 1 \tag{4.33}$$

Since only small deflections are being considered, the approximate expression for the radius of curvature can substituted into Eq. (4.26).

$$\mathfrak{M}(x) = Mg(L - x) = -\frac{ES\kappa_{beam}^2}{R} = -\frac{E(wt)t^2}{12}\frac{d^2y}{dx^2} \tag{4.34}$$

This provides an ordinary differential equation for the deflection, $y(x)$.

$$\frac{d^2y(x)}{dx^2} = -\frac{12Mg}{Ewt^3}(L - x) \cong \frac{1}{R} = \frac{1}{h}\frac{\Delta l}{l} \tag{4.35}$$

The right-hand side of Eq. (4.35) is just a polynomial; hence, its integration (twice) from the wall at $x = 0$ to the end at $x = L$ is easy to perform.

$$y(x) = -\frac{6Mg}{Ewt^3}\left(Lx^2 - \frac{x^3}{3}\right) \tag{4.36}$$

The leading minus sign shows that the deflection is downward for the addition of a positive mass. It is worth noticing that this solution satisfies the boundary conditions that the wall imposes on the beam:

$y(0) = 0$ and $(dy/dx)_{x=0} = 0$. The deflection of the beam's end is cubic in its ratio of length to thickness. The beam's effective stiffness, K_{beam}, is thus also cubic in the ratio of thickness to length.

$$y(L) = -\frac{4Mg}{Ew}\frac{L^3}{t^3} \quad \text{and} \quad K_{beam} = \left|\frac{Mg}{y(L)}\right| = \frac{Ew}{4}\left(\frac{t}{L}\right)^3 \tag{4.37}$$

4.3.3 Triangularly Tapered Cantilever Spring*

The expression for the deflection of a cantilevered beam of constant cross-sectional thickness, t, and constant width, w, was calculated, resulting in Eq. (4.37). In spring design, one of the most important constraints is the tensile strength of the spring material, particularly when a spring is subjected to cyclic strain that fully reverses every half-cycle [6]. For a rectangular beam, Eq. (4.35) demonstrates that the magnitude of the surface stress at $y = \pm(t/2)$ is greatest at $x = 0$ and goes to zero at the tip, $x = L$. That makes for a rather inefficient use of the material, since it would be preferable for all of the material to be at roughly the same state of stress, if possible, so that all parts of the beam make an equal contribution to its stiffness.

At the design limit of the spring's deflection, the stresses at all points on the surface should be at the material's stress limit (within some safety factor [6]) if the material's stiffness is being used efficiently. If you have ever used a fishing rod, you have seen this strategy in action. As shown in Fig. 4.13, the tapered fishing rod is bent into a circular arc (hence, constant radius of curvature) by the load (presumably, produced by a recalcitrant fish) applied to the tip of the rod.

It is clear from Eq. (4.35) that a beam with a linear taper (variable width), $w(x) = (w(0)/L)(L - x)$, would cancel the same linear dependence in the moment term and provide a constant radius of curvature.[12]

Fig. 4.13 Photograph of Prof. Richard Packard, UC Berkeley Physics Department, on his boat, the *Puffin*, in Alaska, showing that the tapered rod that is his fishing pole being bent into a nearly a circular arc when loaded (by a large halibut) at its tip

[12] In principle, the width could be kept constant and the thickness could be tapered so $t(x) = [t(0)/L^{1/3}](L - x)^{1/3}$. Since spring steel is available in constant thickness sheets, it is frequently more convenient to provide a linear taper of the width.

$$\frac{1}{R} \cong \frac{d^2y}{dx^2} = \frac{12Mg}{Et^3}\frac{L(L-x)}{w(0)(L-x)} = \frac{12MgL}{Ew(0)t^3} \tag{4.38}$$

Since the radius of curvature is constant, the double integration of Eq. (4.38) produces a new deflection curve, $y(x)$, and a new stiffness, $K_{triangle}$, for a beam with a linear taper.

$$y(x) = \frac{6MgLx^2}{Ew(0)t^3} \quad \text{and} \quad K_{triangle} = \frac{Mg}{z(L)} = \frac{Ew(0)}{6}\left(\frac{t}{L}\right)^3 \tag{4.39}$$

It is instructive to compare the stiffness of a tapered cantilever in Eq. (4.39) to the stiffness of a beam of constant width in Eq. (4.37). The stiffness of the constant width beam is larger by 6:4, so the tapered beam would need an initial width that is 50% larger to provide the same stiffness. Since the area of a triangle is half the base times the height, the triangular beam of the same stiffness and equal length uses 25% less material. Comparison of Eqs. (4.35) and (4.38) shows that the maximum stress is 50% lower for the triangular beam, again for the same stiffness. The triangular-shaped cantilever is clearly a more efficient use of the spring material.

Figure 4.14 shows an example of high-performance spring design that joins two triangular cantilevers. It was used to augment the stiffness of a 2 kW moving-magnet linear motor raising the natural frequency from about 20 Hz to the desired operating frequency near 60 Hz [13]. That spring was also the critical component in a test fixture (linear dynamometer) used to evaluate the performance of smaller moving-magnet linear motors [14]. Each spring in Fig. 4.14 consists of 16 leaves. Each leaf was composed of two triangular cantilever beams that are joined tip-to-tip. Because the spring in Fig. 4.14 was designed to accommodate linear displacements of ± 1.0 cm, with an overall diameter of about a half-meter, the beams were also bent into an S-shape in an orthogonal plane to relieve the longitudinal stress that would be produced by the need to lengthen each leaf as it went to its extreme

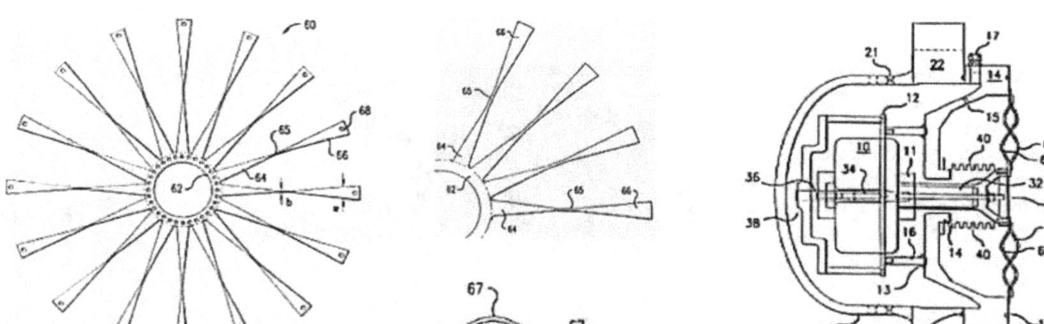

Fig. 4.14 (*Left*) High-performance flexure spring that uses 16 beams made of pairs of triangular cantilevers that are joined tip-to-tip. The central ring is attached to a piston and the outer edges are clamped to the motor housing. (*Center*) The beams are also curved to relieve longitudinal stresses (i.e., tension) that would have caused failure due to the lengthening of the beams as they moved to their extreme transverse displacements. (*Right*) Two such springs (60 and 61) are shown installed on a motor housing (18) that contains a moving-magnet electrodynamic linear motor (10). (Unlike the moving-coil electrodynamic loudspeaker in Fig. 2.12, the voice coil is wound around a laminated steel core, like the cores of electrical transformers, and is stationary. The piston is attached to an armature that supports several magnets that move due to the oscillatory magnetic forces produced by the alternating currents through the stationary coils. Such moving-magnet linear motors can be far more efficient than the moving-coil version (see Table 10.4), but at the price of reduced bandwidth. Since the coils are stationary, the electrical leads are not subject to fatigue failure which is an important failure mode for moving-coil loudspeakers. Each spring's central hub is attached to a piston (30) and a flexible metal bellows (40). That bellows [17] provide a dynamic gas seal between the resonant load (not shown, but located to the right of the motor) and the back volume that contains the motor mechanism's coils and magnets

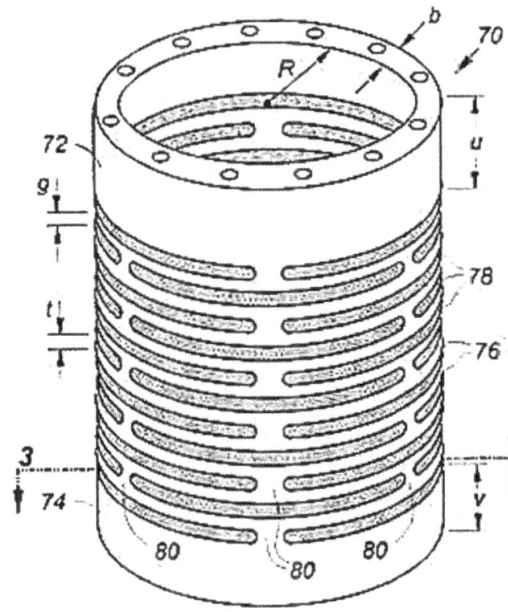

Fig. 4.15 A spring can be made from a cylinder that has been cut to provide a stack of cantilevers that are t thick, b wide, and spaced from each other by separation, g. The force is applied by each cantilever spring to the center of the spring below it through the "posts" (80). If the spaces (78 and 76) are filled with an elastomer, then the spring also becomes a flexure seal, similar to bellows [16]

transverse displacement (i.e., the change in hypotenuse of a right triangle with the displacement as its height and the unstrained spring's length as its base).

That spring was machined from a flat sheet of 17-7 PH stainless steel. That alloy is as ductile as lead in its annealed state. Due to precipitation hardening (PH) during heat treatment [15], it becomes stiff, strong, and nearly lossless. The molds used create the S-bends, shown in Fig. 4.14 (center), and hold the steel sheet during heat treatment, were designed and fabricated by Dr. R. W. M. Smith.

Another method for producing flexure springs capable of large displacement is to make cuts into a hollow cylinder that stacks cantilevers on top of each other [16]. Such an arrangement is shown in Fig. 4.15.

4.3.4 Buckling

There are two reasons that long thin bars or rods are not used as springs in compression. One reason becomes obvious if we compare the change in length, ΔL, of a rod due to the application of a compressive force, F, on one end of area, πa^2, to the transverse deflection, $z(L)$, due to the same force applied perpendicular to the length, L, of the rod as diagrammed schematically in Fig 4.16. According to our definition of Young's modulus in Eq. (4.1), $\Delta L = (FL)/(\pi a^2 E)$. We can find the displacement of a cantilevered rod subject to a force of the same magnitude, but applied at right angles to the end, by substituting $\kappa_{rod}^2 = a^2/4$ into Eq. (4.34) to produce an expression for the transverse deflection, $z(L)$, of a rod with circular cross-section and radius, a.

$$z(L) = \frac{8FL^3}{3Ea^4} \tag{4.40}$$

The ratio of the compressive displacement, ΔL, to the transverse displacement, $z(L)$, depends only upon the *slenderness ratio*, L/a.

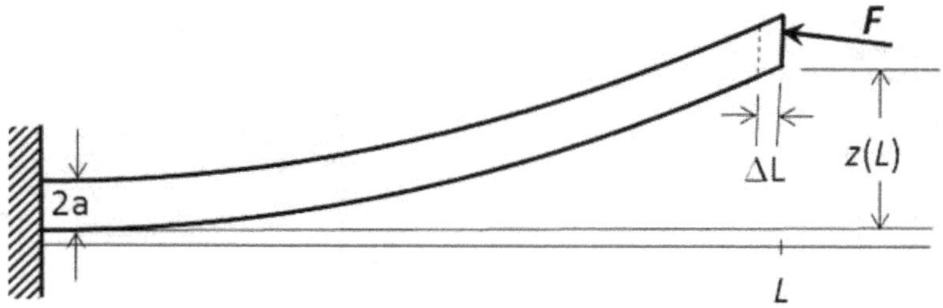

Fig. 4.16 A bar of circular cross-section with diameter, 2*a*, and length, *L*, is subject to a compressive force, ***F***, which is not exactly aligned with its undeformed direction. The force produces both a compression, **Δ*L***, and a transverse displacement, *z* (*L*)

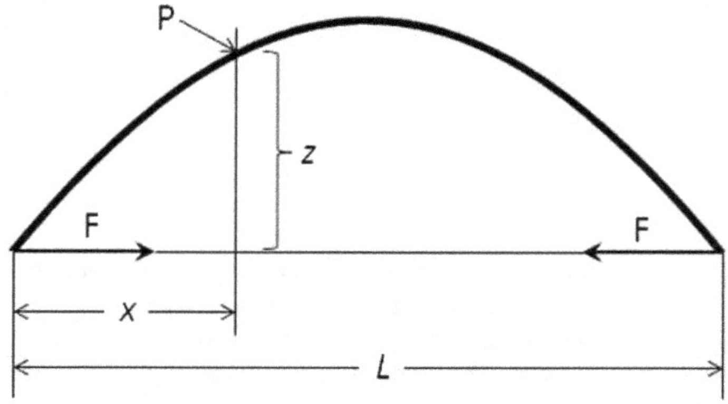

Fig. 4.17 Coordinate system for a buckled beam. The perpendicular lever arm for the force, *F*, about point, *P*, is *z*(*x*)

$$\frac{z(L)}{\Delta L} = \left(\frac{8\pi}{3}\right)\left(\frac{L}{a}\right)^2 \cong 8\left(\frac{L}{a}\right)^2 = \cot\theta \cong \frac{1}{\theta} \tag{4.41}$$

For a rod that has a diameter of 1.0 cm and is 50 cm long, $(L/a) = 100$. This corresponds to $\theta = 12.5$ μrad = 0.0007 degrees. That result implies that if the angle the force, *F*, makes with the axis of the rod is greater than θ, the transverse deflection will exceed the longitudinal compression. It would be nearly impossible to arrange the force applied to the end of a slender rod to produce only compression and no flexure.

An even bigger problem with slender rods or beams is that they will buckle if the force is greater than some threshold value, F_E, known as the *Euler force*. With a sufficiently large force, a slender structure (beam or column) will bend and then collapse, rather than compress. We can use the diagram in Fig. 4.17 to calculate that buckling force threshold. If we let the deflection of the rod from its straight (unloaded) condition be $z(x)$, then the bending moment that the applied force, *F*, creates on a piece of the rod located at point, *P*, is $\mathfrak{M}(x) = F\,z(x)$. That moment can be equated to the bending moment of the beam given in Eq. (4.26).

$$\mathfrak{M}(x) = Fz(x) = -\frac{ES\kappa^2}{R} = -ES\kappa^2\left(\frac{d^2z}{dx^2}\right) \tag{4.42}$$

This produces an ordinary second-order differential equation that is now rather familiar.

$$\left(\frac{d^2z}{dx^2}\right) + \left(\frac{F}{ES\kappa^2}\right)z = 0 \tag{4.43}$$

The solution is a sine curve.

$$z(x) = C \sin \frac{\pi x}{L} \quad \Rightarrow \quad \left(\frac{d^2z}{dx^2}\right) = -\left(\frac{\pi}{L}\right)^2 z(x) \tag{4.44}$$

Substituting this result back into Eq. (4.43) provides the critical force, F_E, that will result in the buckling of the beam if exceeded.

$$F_E = ES\kappa^2 \left(\frac{\pi}{L}\right)^2 \tag{4.45}$$

This result is independent of $z(x)$. Once the beam starts to bend, upon application of force, $F > F_E$, the reaction force, F_E, is constant, so the curvature increases until the beam collapses catastrophically.

This result for F_E pertains to the situation diagrammed in Fig. 4.17, where both ends of the beam were allowed to have a non-zero slope (called a "hinged" boundary condition). If one end of the beam is clamped, so that $(dz/dx)_{x=L} = 0$, then we see that by treating the clamped beam as being half as long as the beam that is hinged at both ends, we achieve the required result for the critical force that buckles the clamped-hinged beam, $F_{cantilever} = 4F_E$. As expected, by constraining one end of the beam to remain straight, the threshold for buckling is increased substantially.

This result is even more important for the stability of columns that support multistory buildings or rockets that need to trade off rigidity for launch weight. However, buckling is also a second important consideration for spring design that cautions against the use of slender beams as longitudinal springs.

4.3.5 Torsional Springs

As with flexure springs, torsional stiffnesses are applied in a range of sizes from the shafts that connect propellers to gas turbines in naval warships to thin wires used to make sensitive measurements, like the determination of Newton's gravitational constant, G, that measured a force of only 1.7×10^{-7} N. Using masses supported by a thin wire torsional spring, at the end of the eighteenth century, Cavendish was able to make the first determination of G [18]. Quartz fibers no thicker than human hairs have been used to provide the linear restoring torque for mirrored galvanometers (see Fig. 2.6) that have sensitivities limited only by their own temperature, as discussed in Sect. 2.4.4. It is now our goal to relate torsional stiffness to the material's shear modulus and to its geometry.

Figure 4.18 shows a rod of length, L, and circular cross-section, with radius, a, that is clamped at $x = 0$. It is being twisted at $x = L$ by a torque that produces rotation by an angle, ϕ. We can consider the rod to be composed of many concentric thin cylindrical shells of thickness, Δr, and then determine the overall stiffness by integrating from some inner radius, $a_{in} < a$, to the outer radius, $a_{out} = a$. For a solid rod, we can let $a_{in} = 0$. If we focus our attention on a small patch on the thin cylindrical shell with mean radius, r, and thickness Δr, then we see in Fig. 4.18 that the patch has been sheared by an angle θ, when the end is twisted by an angle ϕ.

$$\theta = \frac{r\phi}{L} \tag{4.46}$$

The shear stress is related to the distortion angle, θ, by the shear modulus of the material, as expressed in Eq. (4.17).

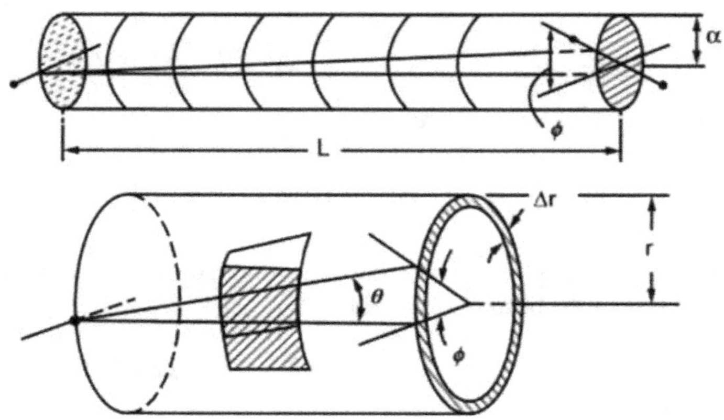

Fig. 4.18 (*Above*) Torque is applied to a rod of circular cross-section and length, L, that is rigidly fixed at $x = 0$. This produces a rotation of the end at $x = L$ by an angle, $\boldsymbol{\phi}$. (*Below*) A thin ring of material of thickness, $\boldsymbol{\Delta r}$, a distance, r, from the axis of the rod has a patch of material that has been sheared by an angle, $\boldsymbol{\theta}$. The small patch has a height, $\boldsymbol{\Delta l}$

$$\sigma_{yx} = \frac{\Delta F}{\Delta r \Delta l} = G\theta = G\frac{r\phi}{L} \tag{4.47}$$

The force, ΔF, produces a torque, ΔN, in conjunction with the "lever arm," r.

$$\Delta N = r\Delta F = rG(\Delta l \Delta r)\theta = G(\Delta l \Delta r)r^2\phi/L \tag{4.48}$$

If we integrate ΔN for the patch around the entire circumference of the cylindrical shell, the sum of the Δl's becomes $2\pi r$ and the total torque becomes $dN(r) = rG(2\pi r)\Delta r$ for the thin shell.

$$|N(r)| = \left|\vec{F} \times \vec{r}\right| = r\sigma_{xy}2\pi r\Delta r = 2\pi G r^3 \Delta r \frac{\phi}{L} \tag{4.49}$$

Integrating over Δr from an inner radius, a_{in}, to the outer radius, a, provides the torsional stiffness, K_{tube}, of the entire hollow tube of length, L.

$$N = \pi G\left(a^4 - a_{in}^4\right)\frac{\phi}{2L} \Rightarrow K_{tube} = \frac{N}{\phi} = \frac{\pi G\left(a^4 - a_{in}^4\right)}{2L} \tag{4.50}$$

For a solid rod of circular cross-section, $K_{rod} = \pi G a^4/2\,L$.

4.3.6 Coil Springs

When the word "spring" is mentioned, the most common image that word conjures is a helical coil spring. A coil spring is another efficient design that transforms extensions and compressions into shear stresses in the "wire" that is wound into a helix. Although our definition of shear stress in Eq. (4.17) "can be applied to slightly curved bars without significant error" [12], the wire of helical springs is very strongly curved, and the influence of that curvature must be included in the derivation of a relationship between the helical coil spring's stiffness and its geometry [19].

The necessary calculations have been published in a book by Wahl [20]. A few of his results that are based on the spring geometry sketched in Fig. 4.19 are reproduced here. The stiffnesses of helical coil springs made with wire of circular, square, and rectangular cross-sections are provided in Eqs. (4.51), (4.52), and (4.53) with their parameters as defined in the caption of Fig. 4.19.

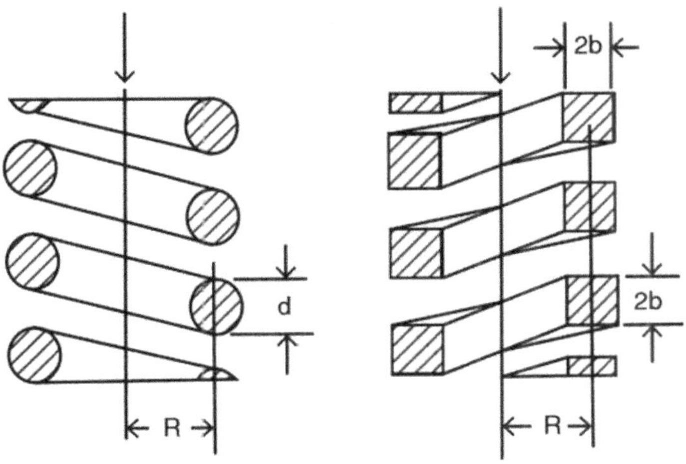

Fig. 4.19 Geometry of helical coil springs, using circular or rectangular wire, is shown at the right. Their linear stiffness is related to their geometry and to the shear modulus, G, of the spring's material. R is the mean radius of the coil. The pitch, P, not labeled in the diagram, is the distance between the centers of adjacent coils at that radius, and the pitch angle is given by $\alpha = \tan^{-1}(P/R)$. The total length of the spring will be $L = nP$, where n is the number of coils [12].

$$\mathrm{K}_{round} = G\frac{d^4}{64nR^3}\left[1 - \frac{3}{64}\left(\frac{d}{R}\right)^2 + \frac{3+\nu}{2(1+\nu)}(\tan\alpha)^2\right]^{-1} \quad (4.51)$$

$$\mathrm{K}_{square} = 0.3586\ G\frac{b^4}{nR^3} \quad \text{if} \quad \frac{R}{b} > 3 \quad (4.52)$$

$$\mathrm{K}_{rectangle} = G\frac{8b^4}{3\pi nR^3}\left\{\frac{a}{b} - 0.627\left[\tanh\left(\frac{\pi b}{2a}\right) + 0.004\right]\right\} \quad \text{where } a > b \quad (4.53)$$

Although helical coil springs are efficient in their use of material, they do couple torques, as well as the intended axial restoring forces, to their attached loads. A rather amusing demonstration of that coupling is provided by the Wilberforce pendulum [21], shown schematically in Fig. 4.20. For the round wire case, the axial (Hooke's law) stiffness is given by Eq. (4.51). For large (R/d), $\mathrm{K}_{round} \cong Gd^4/64nR^3$. The torsional stiffness can be expressed in terms of the longitudinal stiffness [22].

$$\mathrm{K}_{torsional} = \mathrm{K}_{round}R^2\left(1 + \nu\cos^2\alpha\right) \quad (4.54)$$

The twisting torque produced by the spring is related to its extension, so the longitudinal and torsional vibrations are coupled. This coupling is produced by a non-zero value of Poisson's ratio, $\nu \neq 0$. The coupled oscillator equations (see Sect. 2.7) for this case were written down by Sommerfeld [23].

$$\begin{aligned} \ddot{y} + \omega_v^2 y + (\mathrm{K}_{round}R/m)(\nu\sin\alpha\cos\alpha)\theta &= 0 \\ \ddot{\theta} + \omega_t^2\theta + (\mathrm{K}_{torsional}R/I)(\nu\sin\alpha\cos\alpha)y &= 0 \end{aligned} \quad (4.55)$$

The values of ω_t and ω_v in the caption for Fig. 4.20 are close to those provided by Sommerfeld.

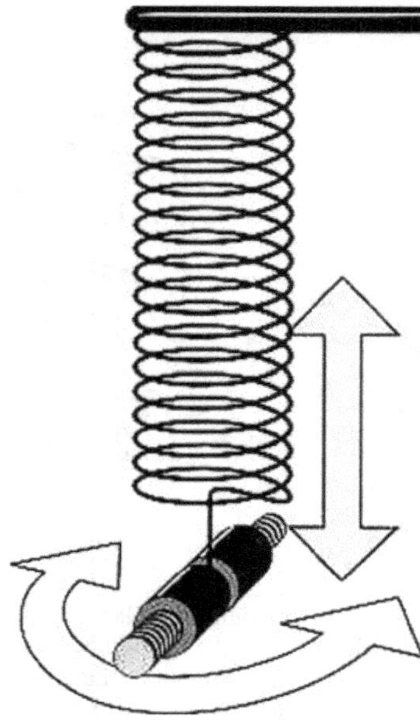

Fig. 4.20 The Wilberforce pendulum is a helical coil spring that supports a mass, m, which has a moment of inertia, I. If the extensions and twisting were uncoupled, then the normal mode frequency for vertical vibrations would be $\omega_v \cong (K_{round}/m)^{1/2}$. The torsional normal mode would have a frequency of $\omega_t \cong (K_{torsional}/I)^{1/2}$. Since vertical displacements cause the spring to twist and twisting causes vertical motion, the two modes are coupled. If $\omega_t \cong \omega_v$, and the spring is initially displaced vertically, it will start vibrating up and down but will slowly begin twisting until all of the oscillatory motion becomes angular. After the vertical displacements cease, the angular oscillations will drive the vertical motion that will increase until the angular motion decays to zero and the cycle is repeated

$$\omega_v^2 = \left(\frac{K_{round}}{m}\right)\left(1 + \nu \cos^2\alpha\right) \quad \text{and} \quad \omega_t^2 = \left(\frac{R^2 K_{round}}{I}\right)\left(1 + \nu \cos^2\alpha\right) \tag{4.56}$$

If those frequencies are set equal to each other, the moment of inertia for the mass, m, must be $I = m\kappa^2$, where $\kappa = R(1 + \nu)^{1/2}$ be the radius of gyration for the mass [24]. In principle, the measurement of I and m provides a means for determining the Poisson's ratio of the spring's material. Experimental measurements using this technique [22] have produced a reasonable value for $\nu_{steel} \cong 0.23$.

In addition to the twisting caused by the compression and expansion of a helical coil spring, the spring will also tilt. In some applications, neither of these "side effects" (no pun intended) are problematic, but if a helical coil spring is used to supplement the stiffness of a linear motor, the twisting or tilting can cause the magnets to touch the laminated steel around which the coil is wound, causing failure by rubbing in loudspeakers.

Such twisting and tilting can be mitigated by designing two concentric coil springs that are each double helices (like a DNA molecule). Such a coil spring pair is shown in Fig. 4.21 (left). With two concentric coil springs, it is possible to select their dimensions such that the twist produced by the outer spring is cancelled by the inner spring if their coils have opposite "handedness." Since each individual coil is composed of two helices that start at positions that are 180 ° apart, their symmetry also cancels the tilting.

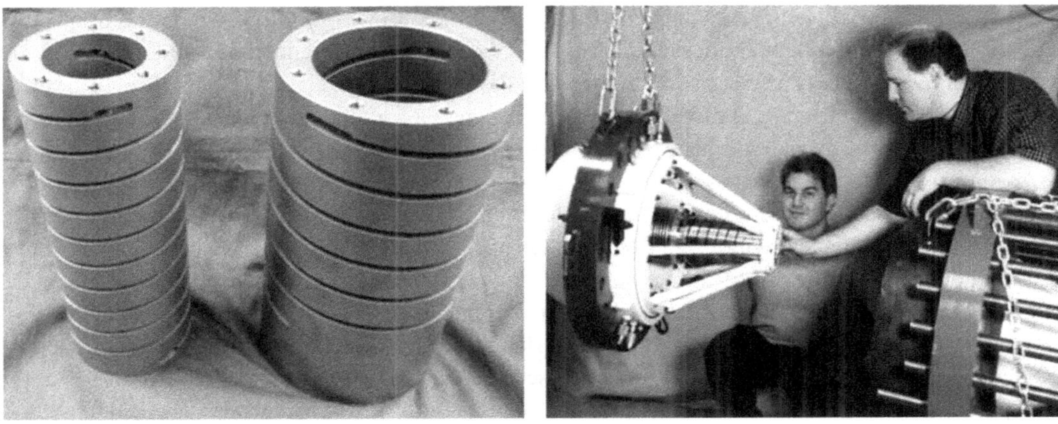

Fig. 4.21 (*Left*) Photograph of two coil springs that are each a double-start helical coil. The two starting points for the helices are visible at the tops of both springs. The double helix design keeps the springs from tilting when displaced. The helices of the larger and smaller coils have opposite "handedness;" one coil advances to the right and the other advances to the left. They have been designed so that the twisting torques produced by equal compressions are equal and opposite. These springs were machined from a solid tube of maraging steel [15]. (*Right*) The larger diameter spring is visible in this photo of an assembled sound source which incorporates a 10 kW linear motor that is contained within the hemi-elliptical cap at the rear. A flexible metal bellow that provides a dynamic gas seal around the piston is (partially) visible between the spring and the housing. One end of the double-Helmholtz resonator [25] is also visible at the right of the photo. A cross-sectional diagram of the entire resonator is shown in Fig. 8.26. John Heake, then a graduate student, and Dr. R. W. M. Smith, the spring and bellows' designer, are also in this photograph

4.4 Viscoelasticity

The expressions for the generalization of Hooke's law that involve the various moduli introduced in this chapter all share a common assumption: the strain produced by the stress (and vice versa) occurs instantaneously. We examined a similar assumption in Sect. 2.2.2, which used similitude (see Sect. 1.7) to calculate a "characteristic speed,"$c \propto L\sqrt{K/m}$, for the various parts along a helical spring of length, L, to influence each other. That perspective explained the introduction of a "quasi-static approximation," justifying the use of a static spring stiffness, K, in the dynamical equation (i.e., Newton's Second Law) for analysis of a simple, mass-spring harmonic oscillator (at sufficiently low frequencies). The quasi-static approximation also provides a basis for adding one-third of the spring's mass to the "lumped" mass attached to the spring, derived in Eq. (2.27), since the displacement of each coil was assumed to be proportional to the distance from its fixed end, as illustrated for the static case by the Gerber scale in Fig. 2.2.

We now need to revisit that quasi-static assumption to understand the behavior of springs made from rubberlike elastomeric materials.[13] For such materials, their stiffnesses are frequency dependent, as is their internal energy dissipation [26]. Thus far, we have assumed the elastic moduli were frequency-independent and lossless. In this section, a simple model will be developed that can describe the viscoelastic behavior of rubberlike materials. The model is based on a single exponential relaxation time, τ_R. The same model can be applied to many other physical systems such as the attenuation of sound in humid air (see Sect. 14.5.1) or seawater (see Sect. 14.5.2), to name just two. That *single*

[13] The terms elastomeric and rubberlike will be used interchangeably in this chapter. While all rubbers are elastomers, not all elastomers are rubbers. The distinction is codified in the American Society for Testing and Materials (ASTM) Standard D 1566, which is based on the length of time required for a deformed sample to return to its shape after removal of the deforming force, as well as the extent of that recovery [26].

relaxation time model leads to a complex stiffness, $\mathbf{K}(\omega) = K'(\omega) + jK''(\omega)$, where the frequency-dependent real part, $K'(\omega)$, quantifies the stiffness and the frequency-dependent imaginary part, $K''(\omega)$, quantifies the dissipation. It will also be shown that the two components of the complex stiffness, or equivalently the complex elastic modulus, are not independent and that their relationship is an entirely general feature of the causality inherent in any *linear response theory*.

As long as the "cause" precedes the "effect," the real and imaginary components of any *generalized susceptibility* [27], like a complex elastic modulus, the index of refraction of transparent optical materials, the dielectric susceptibility of electrical insulators [28], the speed and attenuation of sound [29], the gain and phase in electrical filter and amplifier circuits [30], the relationships between the real part (radiation resistance) and imaginary part (radiation reactance or effective hydrodynamic mass) of the radiation impedance [31], and even the absorption of sound by porous media in superfluid helium [32], etc., all obey the *Kramers-Kronig relations* (see Sect. 4.4.4) that were discovered in studies of the propagation and attenuation of X-rays during the first quarter of the twentieth century [33].[14]

4.4.1 The Maxwell (Relaxation Time) Model

We start the development of our model describing the response of viscoelastic materials by considering the behavior of a spring that is placed in series (mechanically) with a dashpot. We have already devoted a considerable amount of effort to describing the effects of a spring and dashpot that were placed in parallel, as shown in Fig. 2.6, when we examined the damped harmonic oscillator in Sect. 2.4. Figure 4.22 shows the spring and dashpot in series inside a "black box" that allows us access only to the end of the spring that is not attached to the dashpot. The displacement of the exposed end of the spring from its equilibrium position will be designated x_1.

The other end of the spring is attached to a dashpot that is inside the black box. The displacement of that junction from its equilibrium position will be designated x_2. We let the other end of the dashpot be fixed. Since we do not have physical access to x_2, it will act as a "hidden variable" that we can use for calculation of the response of x_1 to forces applied at x_1, the only location in which we have the ability to access from the outside of the black box.

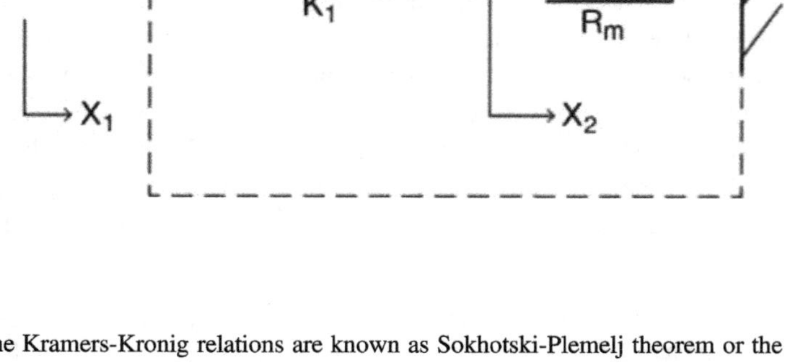

Fig. 4.22 A spring and dashpot are placed inside a "black box" (inside the *dashed lines*) that provides us with access only to one end of the spring whose displacement from equilibrium will be labeled x_1. Inside the box, attached to the other end of the spring, is a dashpot. The displacement of the junction between the "hidden" end of the spring and the dashpot is designated x_2

[14] In mathematics and signal processing the Kramers-Kronig relations are known as Sokhotski-Plemelj theorem or the Hilbert transform

Before producing a mathematical analysis of our "black box," it pays to think about the behavior of x_1 in the high- and low-frequency limits. As before, similitude (see Sect. 1.7) will be able to guide our determination of the frequency, ω_R, that separates the regimes of high- and low-frequency behavior. There is only one combination of stiffness, K_1 [N/m = kg/s^2], and mechanical resistance, R_m [N-s/m = kg/s], that has the units of frequency (or its inverse, time). Although we are unable to determine any numerical pre-factor, similitude guarantees that $\omega_R \propto (K_1/R_m)$. If x_1 is driven at frequencies well above ω_R, then $x_2 \ll x_1$, because the dashpot (producing a force proportional to the velocity) will not move easily at high frequencies. At high frequencies, x_1 will seem to obey Hooke's law with $F(x_1) \cong -K_1 x_1$.

At frequencies well below ω_R, it will be much easier to compress the dashpot than the spring. From outside the black box, x_1 will appear to be connected directly to the dashpot at sufficiently low frequencies, $\omega \ll \omega_R$. In that case, $x_1 \cong |x_2|$, so $F(\dot{x}_1) \cong -R_m \dot{x}_1$.

The behavior of x_1 for all frequencies can be calculated by writing an equation for the force applied at x_1, $F(x_1)$. Although the displacements, x_1 and x_2, may not be equal, the force through the series combination must be continuous. All of the force must end up being applied to the rigid boundary at the end of the dashpot that is not connected to the spring.

$$\mathbf{F}(x_1) = -K_1(\mathbf{x_1} - \mathbf{x_2}) = -R_m\dot{\mathbf{x}}_2 = -j\omega R_m \mathbf{x_2} \tag{4.57}$$

The right-hand version of Eq. (4.57) assumes that the applied force at x_1 is time-harmonic at a single frequency, $\mathbf{F}(x_1) = F_1 e^{j\omega t}$. Equation (4.57) can be solved for the ratio of the two displacements.

$$\frac{\mathbf{x_2}}{\mathbf{x_1}} = \frac{1}{1 - j\omega\tau_R} = \frac{1 + j\omega\tau_R}{1 + (\omega\tau_R)^2} \tag{4.58}$$

A *relaxation time*, $\tau_R = R_m/K_1$, has been introduced. It is the reciprocal of the relaxation frequency, $\omega_R = \tau_R^{-1}$, calculated previously from similitude. That displacement ratio behaves as we expected in the high- and low-frequency limits, now expressed as $\omega\tau_R \gg 1$ and $\omega\tau_R \ll 1$, respectively.

$$\lim_{\omega\tau_R \to 0}\left[\frac{\mathbf{x_2}}{\mathbf{x_1}}\right] = 1 \quad \text{and} \quad \lim_{\omega\tau_R \to \infty}\left[\frac{\mathbf{x_2}}{\mathbf{x_1}}\right] = \frac{j}{\omega\tau_R} = 0 \tag{4.59}$$

Equation (4.58) can be used to substitute x_2 into the expression for the force, given in Eq. (4.57), that is applied to the spring at $\boldsymbol{x_1}$.

$$\mathbf{F}(x_1) = -K_1(\mathbf{x_1} - \mathbf{x_2}) = -K_1\mathbf{x_1}\left(1 - \frac{1 + j\omega\tau_R}{1 + (\omega\tau_R)^2}\right) = \frac{-K_1\mathbf{x_1}\omega\tau_R}{1 + (\omega\tau_R)^2}(\omega\tau_R - j) \tag{4.60}$$

If we make the analogy to Hooke's law, $F = -Kx$, then Eq. (4.60) suggests that the equivalent spring constant of the spring-dashpot combination is a frequency-dependent complex number, $\mathbf{K_1}(\omega)$.

$$\mathbf{K_1}(\omega) = K_1'(\omega) + jK_1''(\omega) = \frac{K_1(\omega\tau_R)^2}{1 + (\omega\tau_R)^2}\left[1 - \frac{j}{(\omega\tau_R)}\right] \tag{4.61}$$

In the high-frequency limit, $\omega\tau_R \gg 1$, the black box looks just like a spring with $F(x_1) = -K_1 x_1$. In the opposite frequency limit, x_1 appears to be attached to a dashpot, once we replace τ_R by (R_m/K_1) in Eq. (4.62).

$$\lim_{\omega\tau_R \to 0}[\mathbf{F}(\mathbf{x_1})] = \lim_{\omega\tau_R \to 0}\left[-(K_1\mathbf{x_1})\frac{\omega\tau_R}{1 + (\omega\tau_R)^2}(\omega\tau_R + j)\right] = -j\omega R_m \mathbf{x_1} \tag{4.62}$$

When a force is applied to x_1, only the dashpot can dissipate power. Using Eq. (1.73), the time-averaged power dissipation, $\langle\Pi(t)\rangle_t$, is derived from the product of two complex quantities.

$$\begin{aligned}\langle\Pi(t)\rangle_t &= \frac{1}{2}\Re e[\mathbf{F}^*\dot{\mathbf{x}}_2] = \frac{1}{2}\Re e\left[\widehat{\mathbf{F}}_1 e^{-j\omega t} j\omega\widehat{\mathbf{x}}_2 e^{j\omega t}\right] = \frac{1}{2}\Re e\left[\widehat{\mathbf{F}}_1\widehat{\mathbf{x}}_1\frac{j\omega-\omega^2\tau_R}{1+(\omega\tau_R)^2}\right] \\ &= \frac{-\omega|\widehat{\mathbf{x}}_1|\left|\widehat{\mathbf{F}}_1\right|}{2}\left(\frac{\omega\tau_R}{1+(\omega\tau_R)^2}\right) = \frac{-\left|\dot{\widehat{\mathbf{x}}}_1\right|\left|\widehat{\mathbf{F}}_1\right|}{2}\left(\frac{\omega\tau_R}{1+(\omega\tau_R)^2}\right)\end{aligned} \tag{4.63}$$

The power is a negative number because it is being dissipated. By taking the derivative of the last expression with respect to $(\omega\tau_R)$, we see that the maximum dissipation occurs at $(\omega\tau_R) = 1$.[15]

Equation (4.60) for the force and Eq. (4.63) for the dissipation exhibit the expected behavior in the high- and low-frequency limits. Now that we have explicit expressions for their complete frequency dependence, their behavior can be plotted. To make such plots "universal," it is useful to scale (i.e., nondimensionalize) the stiffness and the dissipation. Since we know the high-frequency stiffness, $K_\infty = K_1$, we can plot the negative of the real part of Eq. (4.60) divided by K_∞ as a function of $\omega\tau_R$. The time-averaged power dissipation can be plotted as the energy dissipated per cycle, $\langle\Pi\rangle_t T = \langle\Pi\rangle_t/f = 2\pi\langle\Pi\rangle_t/\omega$, divided by the potential energy stored in the spring in the high-frequency limit, $(PE)_\infty = (½)\left|\widehat{\mathbf{F}}_1\right||\widehat{\mathbf{x}}_1|$, again as a function of $\omega\tau_R$.

$$\frac{|\mathbf{K}(\omega\tau_R)|}{K_\infty} = \frac{(\omega\tau_R)^2}{1+(\omega\tau_R)^2} \quad \text{and} \quad \frac{\langle\Pi\rangle_t}{\left(\frac{\left|\widehat{\mathbf{F}}_1\right||\widehat{\mathbf{x}}_1|}{2}\right)f} = \pi\frac{(\omega\tau_R)}{1+(\omega\tau_R)^2} \tag{4.64}$$

These universal curves are plotted in Fig. 4.23. The peak in the normalized dissipation has a value of $\pi/2$ and occurs at $(\omega\tau_R) = 1$.

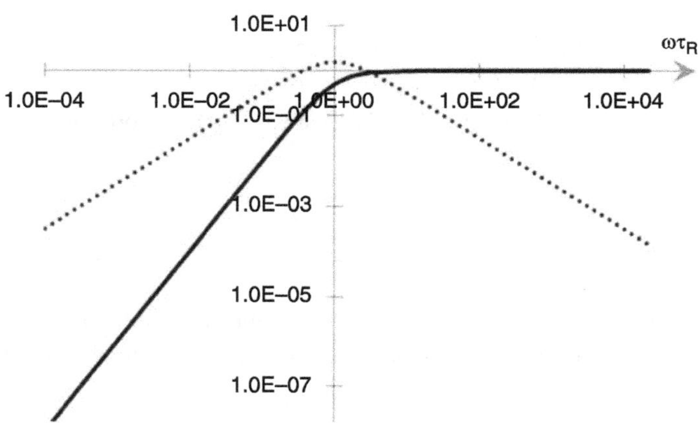

Fig. 4.23 Logarithmic plot of the scaled stiffness, $K(\omega\tau_R)/K_\infty$ (*solid line*), and the scaled dissipation per cycle (*dotted line*) for the Maxwell model from Eq. (4.64). The maximum scaled (i.e., nondimensionalized) dissipation per cycle is equal to $\pi/2$ and occurs when $\omega\tau_R = 1$. The scaled stiffness drops to zero at $\omega\tau_R = 0$

[15] For an expression like that in Eq. (4.63), it is actually easier to solve for the minimum of the inverse than to solve for the maximum:

$$\frac{d}{dx}\left(\frac{1+x^2}{x}\right) = \frac{d}{dx}\left(\frac{1}{x}+x\right) = 0$$

This series combination of a stiffness and dashpot is sometimes called the *Maxwell model* and can be used to describe materials that "creep." It is a good description of Silly Putty® (a toy made from a silicone polymer that will bounce like a rubber ball but will flow over times on the order of several minutes or hours), solutions of corn starch and water, and very viscous fluids, like warm roofing tar. All of those materials will flow slowly over longer time scales but behave elastically over times that are short compared to the relaxation time, $t \ll \tau_R$.

4.4.2 Standard Linear Model (SLM) of Viscoelasticity

The Maxwell model is not a good representation of a rubber spring; the rubber will exhibit non-zero stiffness even when $\omega \cong 0$. If that were not true, the use of rubber in springs as vibration isolators would be impossible since they must support the static load as well as isolate vibrations. The stiffness of a rubber spring increases with increasing frequency from K_o at $\omega = 0$ to some limiting high-frequency value K_∞. We can incorporate the stiffness, K_o, at zero frequency into the Maxwell model by placing a spring with stiffness, K_o, in parallel with our Maxwellian spring-dashpot combination. That combination is known as the *standard linear model* (SLM) for viscoelastic materials and is shown schematically in Fig. 4.24, again inside a "black box."

The results of Sect. 4.4.1 can be applied to the SLM to calculate the apparent stiffness and the dissipation per cycle. This time we will calculate the input mechanical impedance presented to the attachment point at x_1 by adding the Maxwellian combination in parallel to the spring with stiffness, K_o.

$$\mathbf{Z_{mech}} = \frac{F}{\mathbf{v_1}} = \frac{F}{\dot{\mathbf{x}}_1} = \frac{F}{j\omega \mathbf{x_1}} = -\frac{K_o}{j\omega} + \tau_R K_1 \left(\frac{1 + j(\omega\tau_R)}{1 + (\omega\tau_R)^2} \right) \tag{4.65}$$

As the frequency approaches zero, the first term dominates. At high frequencies, when the dashpot is immobilized, the stiffnesses of the two springs add in parallel (mechanically).

$$\lim_{\omega\tau_R \to \infty} [\mathbf{Z_{mech}}] = -\frac{K_o}{j\omega} + \frac{j\tau_R K_o}{(\omega\tau_R)} = -\frac{(K_o + K_1)}{j\omega} \tag{4.66}$$

The stiffness as a function of frequency transitions smoothly from the low-frequency limit, K_o, to the high-frequency limit, $K_\infty = K_o + K_1$.

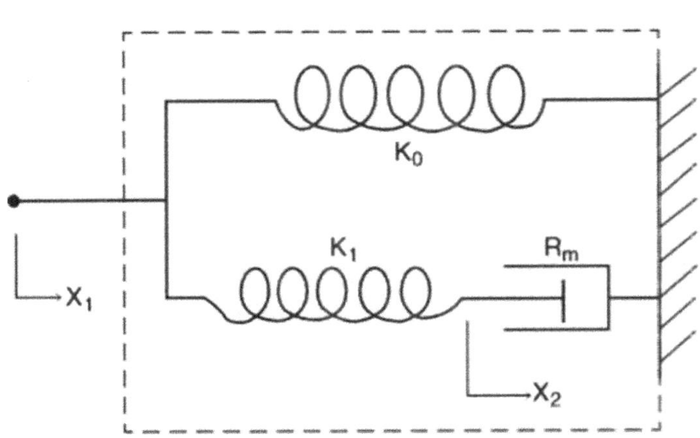

Fig. 4.24 A spring of stiffness, K_o, is placed in parallel with the series spring-dashpot of the Maxwell model in Fig. 4.22 to produce the standard linear model (SLM) of a viscoelastic material. The springs and dashpot are again shown inside the "black box," indicated by the dashed lines, to emphasize that only the terminal designated x_1 is accessible and that x_2 is a "hidden" variable

$$K(\omega\tau_R) = K_o + \frac{K_1(\omega\tau_R)^2}{1+(\omega\tau_R)^2} \tag{4.67}$$

The power dissipated in the dashpot will again be due entirely to $v_2 = j\omega x_2$. Since the force $\boldsymbol{F_1} = -j\omega \boldsymbol{x_2} R_m$, the time-averaged power dissipation can be expressed in terms of x_1 using Eq. (4.58).

$$\langle \Pi \rangle_t = \frac{1}{2}\Re e\left[-\omega^2 \frac{R_m}{2}\mathbf{x}_2^*\mathbf{x}_2\right] = -R_m \frac{\omega^2|\mathbf{x}_1^2|}{2}\left(\frac{1}{1+(\omega\tau_R)^2}\right) \tag{4.68}$$

At high frequencies, this time-averaged power dissipation approaches a constant, so the energy dissipated per cycle is proportional to ω^{-1}. The total maximum potential energy stored in both springs, E_{stored}, is just the sum of the individual stored potential energies.

$$E_{stored} = \frac{1}{2}K_o x_1^2 + \frac{1}{2}K_1(x_1 - x_2)^2 = \frac{x_1^2}{2}\left[K_o + K_1 \frac{(\omega\tau_R)^2}{1+(\omega\tau_R)^2}\right] \tag{4.69}$$

As before, we can form the dimensionless ratio of the magnitude of the time-averaged power, $\langle \Pi(t) \rangle_t$, dissipated per cycle, divided by the energy stored in the springs.

$$\begin{aligned} \frac{2\pi\langle \Pi \rangle_t}{\omega E_{stored}} &= 2\pi \frac{K_1(\omega\tau_R)}{K_o\left[1+(\omega\tau_R)^2\right] + K_1(\omega\tau_R)^2} \\ &= 2\pi \frac{(K_\infty - K_o)(\omega\tau_R)}{K_o\left[1+(\omega\tau_R)^2\right] + (K_\infty - K_o)(\omega\tau_R)^2} \end{aligned} \tag{4.70}$$

This result is rather interesting. The magnitude of the normalized dissipation depends only upon the limiting values of the stiffness and not on R_m, except through the relaxation time, $\tau_R = R_m/K_1$, that determines the frequency at which the dissipation reaches its maximum value. The relaxation time, τ_R, does not influence the magnitude of the dissipation maximum. The point, $(\omega\tau_R)_{max}$, where the dissipation reaches its peak value, and the value of the dissipation at that peak, can be determined from Eq. (4.70). They also depend only upon the limiting values of stiffness. This is not a coincidence; it is an inevitable consequence of linear response theory and causality, as will be demonstrated in Sect. 4.4.4.

$$(\omega\tau_R)_{\max} = \sqrt{\frac{K_o}{K_\infty}} \quad \text{and} \quad \left(\frac{2\pi\langle \Pi \rangle_t}{\omega E_{stored}}\right)_{\max} = 2\pi\frac{K''}{K'} = \frac{\pi K_1}{\sqrt{K_o^2 + K_o K_1}} = \frac{\pi(K_\infty - K_o)}{\sqrt{K_o K_\infty}} \tag{4.71}$$

A plot of the stiffness and normalized dissipation, as a function of $\omega\tau_R$, are provided in Fig. 4.25 for the case where $2K_o = K_1$, so $(\omega\tau_R)_{\max} = \sqrt{K_o/K_\infty} = 1/\sqrt{3} \cong 0.577$.

4.4.3 Complex Stiffnesses and Moduli*

The expressions derived in the first portion of this chapter for the moduli that relate stress and strain assumed that the material's response was instantaneous. The Maxwell model and the standard linear model for viscoelastic behavior both include a "relaxation time," τ_R. Thus far, we have focused only on the response of these spring and damper systems to time-harmonic excitation forces, but if we applied a

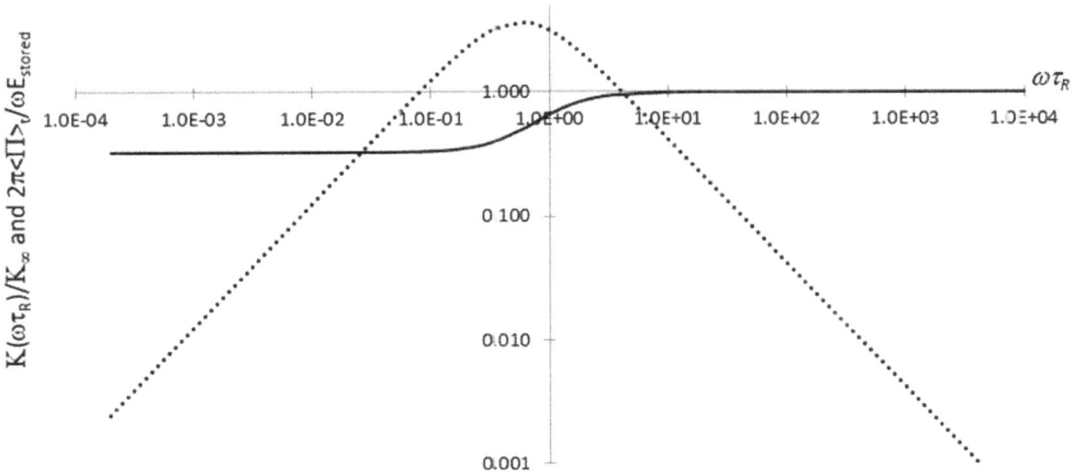

Fig. 4.25 A logarithmic plot of the stiffness ratio, $K(\omega\tau_R)/K_\infty$, shown as the *solid line*, and the normalized dissipation (power dissipated per cycle divided by the energy stored), shown as the *dotted line*, as a function of the product of angular frequency, ω, and relaxation time, τ_R, for the standard linear model with $K_1/K_o = 2$, so $(\omega\tau_R)_{max} = 1/\sqrt{3} = 0.577$

step force at time $t = 0$, then the response at times $t > 0$ would change over times on the order of τ_R. A more general linear response equation can be written that incorporates the possibilities of time-dependent behavior.

$$\left(a_o + a_1\frac{d}{dt} + a_2\frac{d^2}{dt^2} + \cdots + a_n\frac{d^n}{dt^n}\right)\boldsymbol{\sigma} = \left(b_o + b_1\frac{d}{dt} + b_2\frac{d^2}{dt^2} + \cdots + b_n\frac{d^n}{dt^n}\right)\boldsymbol{\varepsilon} \tag{4.72}$$

Introducing a real constant, χ, with the dimensions of inverse length [m^{-1}], and a generic complex modulus, Ξ [Pa], we can identify the constant terms in (4.72) with various combinations of springs and dashpots. For a single spring of stiffness K_o, $\sigma = (\chi\Xi)\varepsilon$, so a_o and b_o are non-zero, but all other $a_n = b_n = 0$ for $n \geq 1$. A single dashpot, with mechanical resistance, R_m, can also be cast into the form of Eq. (4.72).

$$\boldsymbol{\sigma} = \chi R_m\left(\frac{d}{dt}\right)\boldsymbol{\varepsilon}$$

For this dashpot, $a_o \neq 0$ and $b_1 \neq 0$, but $b_o = 0$, as do all other $a_n = 0$, with $n \geq 1$ and $b_n = 0$ for $n \geq 2$.

The response of the SLM can also be written in the form of Eq. (4.72).

$$\left[K_1 + R_m\frac{d}{dt}\right]\boldsymbol{\sigma} = \chi\left[K_oK_1 + (K_o + K_1)R_m\frac{d}{dt}\right]\boldsymbol{\varepsilon}$$

For the response of the SLM, characterized by the mechanical impedance expressed in Eq. (4.65), a_o, b_o, a_1, and b_1 are non-zero, but all other $a_n = b_n = 0$ for $n \geq 2$. The coefficients of the stress and the strain in this form of Hooke's law are now linear operators (see Sect. 1.3) instead of just constants, as they were in Eqs. (4.1), (4.8), (4.14), and (4.17).

Returning to our solution for the input mechanical impedance of the SLM configuration in Eq. (4.65), we can write the stiffness as a complex quantity, $\mathbf{K} = K' + jK''$, where $K' = \Re e[\mathbf{K}]$ and $K'' = \Im m[\mathbf{K}]$, as we did previously for the Maxwell model in Eq. (4.61).

$$\mathbf{K} = \mathrm{K}' + j\mathrm{K}'' = \mathrm{K}_o + \frac{\mathrm{K}_1(\omega\tau_R)}{1 + (\omega\tau_R)^2}[(\omega\tau_R) - j] \tag{4.73}$$

Taking the ratio of the negative of the imaginary part of this complex stiffness to the real part of the complex stiffness, it should become clear why I chose to scale the dissipation in the way shown in Eq. (4.70).

$$-\frac{\Im m[\mathbf{K}]}{\Re e[\mathbf{K}]} = \frac{\mathrm{K}_1(\omega\tau_R)}{\mathrm{K}_o\left[1 + (\omega\tau_R)^2\right] + \mathrm{K}_1(\omega\tau_R)^2} \tag{4.74}$$

Remembering that $\mathrm{K}_\infty - \mathrm{K}_o = \mathrm{K}_1$, Eq. (4.74) differs from Eq. (4.70) by a factor of 2π. This is due to the definition of the dissipation per cycle that was used in Eq. (4.70) rather than the dissipation per radian cycle time, ω^{-1}, that is the reciprocal of the radian frequency.

If we assume time-harmonic solutions to the operator form of Hooke's law in Eq. (4.72), the time-dependent stress-strain relation can be expressed in terms of the complex stress, $\boldsymbol{\sigma}$, and the complex strain, $\boldsymbol{\varepsilon}$.

$$\left(a_o + a_1(j\omega) + a_2(j\omega)^2 + \cdots + a_n(j\omega)^n\right)\boldsymbol{\sigma} = \left(b_o + b_1(j\omega) + b_2(j\omega)^2 + \cdots + b_n(j\omega)^n\right)\boldsymbol{\varepsilon} \tag{4.75}$$

The complex ratio of stress to strain can then be written as a complex elastic modulus.

$$\frac{\boldsymbol{\sigma}}{\boldsymbol{\varepsilon}} \equiv \boldsymbol{\Xi} = \frac{[b'(\omega) + b''(\omega)]}{[a'(\omega) + a''(\omega)]} = |\Xi|(1 + j\delta) \quad \text{where} \quad \delta = \frac{\Im m[\Xi]}{\Re e[\Xi]} \tag{4.76}$$

The fact that the ratio of two complex numbers is also a complex number allows us to introduce the *loss factor* or *damping factor*, δ, so that the generic complex modulus, $\boldsymbol{\Xi}$, can be represented by its magnitude, $|\boldsymbol{\Xi}|$. In this form, the strain lags the stress by a phase angle, $\theta = \tan^{-1}\delta$, or by a time delay, $\tau = \delta/\omega$.

4.4.4 Kramers-Kronig Relations

"A few examples may promote the comprehension of a law, whose extreme generality is not unlikely to convey an impression of vagueness." J. W. Strutt (Lord Rayleigh) [34]

The Kramers-Kronig (K-K) relations are very different from any equations provided thus far, since the K-K relationships are not "local"; the value of the real part of the response function is related to the imaginary part of the response function integrated over all frequencies, $0 \le \omega < \infty$, and vice versa. If we express our generic response function as $\boldsymbol{\Xi}(\omega) = \Xi'(\omega) + j\Xi''(\omega)$, then the following integrals relate their real and imaginary parts.[16]

$$\Xi'(\omega) = \frac{2}{\pi}\int_0^\infty \frac{\omega'\Xi''(\omega')}{\omega'^2 - \omega^2}\,d\omega' \tag{4.77}$$

[16] The fact that these integrals are a consequence of linear response theory and causality can be proven by the use of complex integration of functions that are "analytic in the upper half-plane" and the Cauchy residue theorem.

$$\Xi''(\omega) = \frac{-2\omega}{\pi} \int_0^\infty \frac{\Xi'(\omega')}{\omega'^2 - \omega^2} d\omega' \tag{4.78}$$

In these expressions, the integration is carried out over the dummy variable, ω', so that either the real or the imaginary parts of the response function at any single frequency, ω, will be dependent upon the behavior of the other part over all frequencies, in principle, from DC ($\omega = 0$) to daylight ($\omega = \infty$); hence, the relations are "nonlocal." Although the determination of one part at any frequency depends upon the other part's behavior over a range of frequencies, the critical realization is that the two parts are not independent.

To develop confidence in their applicability, we will employ the Kramers-Kronig relations to express the real and imaginary components of the complex stiffness, $\mathbf{K}(\omega) = K'(\omega) + jK''(\omega)$, in Eq. (4.73) that were obtained from our single relaxation time model for Hooke's law, $\mathbf{F_1}(\omega) = -\mathbf{K}(\omega)\,\mathbf{x_1}(\omega)$. Before calculating the relationship between real and imaginary components, based on the Kramers-Kronig relations, it is helpful to remember that we have an expression for the instantaneous response of the system, K_∞. The integration of the real part of the stiffness, using Eq. (4.78), to determine the imaginary part of the stiffness, is simplified if we subtract this constant from the real part of the stiffness and change the integration variable to $x = \ln(\omega'/\omega)$ [35].

$$K''(\omega) = -\frac{2}{\pi} \int_{-\infty}^{\infty} \frac{K'(x) - K_\infty}{e^x - e^{-x}} dx = -\frac{1}{\pi} \int_{-\infty}^{\infty} \frac{K'(x) - K_\infty}{\sinh x} dx \tag{4.79}$$

Integration by parts (see Sect. 1.1.2) converts Eq. (4.79) into an expression that involves the derivative of the real part of the stiffness.

$$K''(\omega) = -\frac{1}{\pi} \int_{-\infty}^{\infty} \frac{dK'(x)}{dx} \ln\ \coth\left(\frac{|x|}{2}\right) dx \tag{4.80}$$

This form is particularly convenient for our single relaxation time model for two reasons: First, the derivative with respect to frequency of the real part of the stiffness, taken from Eq. (4.67), is non-zero only over a small range of frequencies near $\omega\tau_R \cong 1$, or $x = 0$. This is also evident from inspection of the solid line in Fig. 4.25; the slope is only non-zero near $\omega\tau_R \cong 1$. Second, the natural logarithm of the hyperbolic cotangent is also sharply peaked around $x = 0$, as shown in Fig. 4.26.

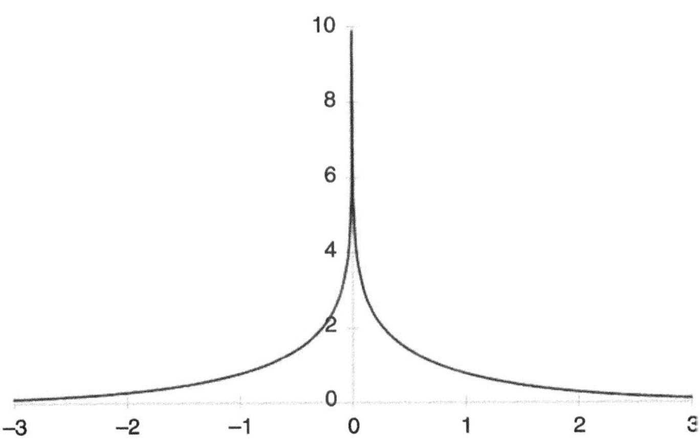

Fig. 4.26 The natural logarithm of the hyperbolic cotangent of $|x|/2$ is plotted on the vertical axis for $-3 \leq x \leq 3$. That function appears inside the integral in Eq. (4.80). Because it is singular at $x = 0$, it emphasizes the value of dk'/dx at $x = 0$, corresponding to $\omega\tau_R \cong 1$

Since the ln (coth |x|/2) has a singularity at $x = 0$, it heavily weights values of the derivative near $x = 0$; therefore any function, $G(x)$, that is multiplied by ln coth |x/2| can be taken outside the integral and evaluated at $x = 0$.

$$K''(\omega) = -\frac{1}{\pi}\int_{-\infty}^{\infty} G(x) \ln \coth\left(\frac{|x|}{2}\right) dx \tag{4.81}$$

It is then useful to expand $G(x)$ in a Taylor series about $x = 0$, recognizing that ln (coth |x|/2) is symmetric about $x = 0$, so only even-order terms in the Taylor series will produce non-zero results upon integration from $x = -\infty$ to $x = +\infty$.

$$G(x) \cong G(0) + \left(\frac{d^2G(x)}{dx^2}\right)_{x=0}\frac{x^2}{2!} + \left(\frac{d^4G(x)}{dx^4}\right)_{x=0}\frac{x^4}{4!}\cdots \tag{4.82}$$

This transforms Eq. (4.81) into an infinite series.

$$K''(\omega) = -\frac{2}{\pi}\sum_{n=0}^{\infty}\frac{1}{(2n)!}\frac{d^{2n}G(0)}{dx^{2n}}\int_{-\infty}^{\infty} x^{2n} \ln \coth\left(\frac{|x|}{2}\right) dx \tag{4.83}$$

The integration over the sharply peaked function ln (coth |x|/2) converts this integral to an algebraic result since $(0)^{2n+1} = 0$ for all n.

$$\begin{aligned} K''(\omega) &= -\frac{4}{\pi}\left(\frac{\pi^2}{8}G(0) + \frac{\pi^4}{96}\frac{d^2G(0)}{dx^2} + \cdots\right) \\ &= -\frac{\pi}{2}\left(\frac{dK'(x)}{dx}\right)_{x=0} - \frac{\pi^3}{24}\left(\frac{d^3K'(x)}{dx^3}\right)_{x=0} - \cdots \end{aligned} \tag{4.84}$$

At this point, all that remains is the variable transformation back to ω from x and to evaluate the slope of the real part of the stiffness at $x = 0$.

$$\left(\frac{dK'}{dx}\right)_{x=0} = \frac{dK'(\omega)}{d\omega}\left(\frac{d\omega}{dx}\right)_{x=0} = \omega\left(\frac{dK'(\omega)}{d\omega}\right)_{x=0} \tag{4.85}$$

Letting $\omega\tau_R = z$, the slope can be calculated from Eq. (4.73) using $d(u/v) = (v\ du - u\ dv)/v^2$ (see Sect. 1.1.2).

$$\frac{dK'(\omega)}{d\omega} = \tau_R K_1 \frac{d}{dz}\left[\frac{z^2}{1+z^2}\right] = \tau_R K_1\left(\frac{2z}{(1+z^2)^2}\right) \tag{4.86}$$

At $z = \omega\tau_R = 1$, or $x = 0$, $dK'/d\omega = \tau_R K_1/2$. Substituting back into Eq. (4.84) provides the imaginary part of the stiffness from the slope of the real part of the stiffness.

$$K''(\omega) = -\frac{\pi}{2}\left(\frac{dK'(x)}{dx}\right)_{x=0} = -\frac{\pi\omega}{2}\left(\frac{dK'(\omega)}{d\omega}\right)_{x=0} = -\frac{\pi\omega}{2}\frac{\tau_R K_1}{2} \tag{4.87}$$

This can be compared with Eq. (4.74) by taking the negative of the ratio of $K''(\omega)$ to $K'(\omega)$.

$$-\left(\frac{K''(\omega)}{K'(\omega)}\right)_{\omega\tau_R=1} = \frac{\pi}{4}\frac{K_1}{2K_o + K_1} = \frac{\pi}{8}\frac{(K_\infty - K_o)}{\left(\frac{K_\infty + K_o}{2}\right)} \cong 0.4\frac{(K_\infty - K_o)}{\left(\frac{K_\infty + K_o}{2}\right)} \tag{4.88}$$

Again, we see that the normalized attenuation at $\omega\tau_R = 1$ depends only upon the limiting stiffnesses, K_∞ and K_o, through their difference $(K_\infty - K_o)$ and their average $(K_\infty + K_o)/2$. Comparison to the exact results for the maximum in the normalized dissipation in Eq. (4.71) shows that the proper "average" is the geometric mean of K_∞ and K_o rather than half their sum, since the maximum does not occur exactly at $\omega\tau_R = 1$.

$$-\left(\frac{K''(\omega)}{K'(\omega)}\right)_{max} = \left(\frac{\langle\Pi\rangle_t}{\omega E_{stored}}\right)_{max} = \frac{1}{2}\frac{K_1}{\sqrt{K_o^2 + K_o K_1}} = \frac{1}{2}\frac{(K_\infty - K_o)}{\sqrt{K_o K_\infty}} \tag{4.89}$$

The peak in the $K''(\omega)$ occurs at the maximum in the slope of $K'(\omega)$ as shown Fig. 4.25. We can find the maximum slope from Eq. (4.86): $(dK'/d\omega)_{max} \cong 0.65$ N-s/m, and it occurs at $(\omega\tau_R)_{max} = 0.577$. Substituting all of these back into Eq. (4.84) and retaining only the first term in the Taylor series, the value of $K''(\omega)$ is equal to the exact result in Eq. (4.73) to within 2%.

$$\begin{aligned}[K''(\omega)]_{max} &= -\frac{\pi}{2}\left(\frac{dK'(x)}{dx}\right)_{max} = -\frac{\pi\omega}{2}\left(\frac{dK'(\omega)}{d\omega}\right)_{max} \\ &\cong -\frac{0.65\tau_R\pi\omega}{2}K_1 \cong 1.02(\omega\tau_R)K_1\end{aligned} \tag{4.90}$$

4.5 Rubber Springs

Rubberlike materials are used to make springs for vibration isolation. There is an extraordinary variety of commercial vibration mounts that use rubber and provide fixtures (threaded studs, mounting flanges, brackets, etc.) to isolate loads from vibrating foundations and vice versa. There are several reasons why rubber is so popular in such isolators. One of the most important reasons is that rubber can tolerate strains in excess of 100% (it can change its dimensions by a factor of two) while neither failing (fracture or tear) nor exhibiting significant inelasticity. Rubber also simultaneously provides both damping and stiffness (and does not "drip" like the viscous liquid in our idealized dashpots). As will be demonstrated in this section, damping is essential to control the behavior of the isolator if there is an excitation at the natural frequency determined by the mass of the load and the stiffness of the isolator(s).

When a machine is mounted on an isolator, the natural frequency of the isolator should be well below the operating frequency of the isolated machine. Of course, machines are started and stopped, and therefore during start-up, the frequency changes from zero to the operating frequency, so they must pass through the isolator's resonance frequency. Clearly, the isolator will have to provide sufficient damping so that the vibrations are limited as the machine passes through the isolator's natural frequency.

Another attractive feature of rubber is that chemists are very skilled at formulating rubberlike compounds in ways that can tailor its properties to produce the desired combination of damping and stiffness,[17] as well as make those rubberlike materials resistant to various environmental irritants (solvents, ozone) and mechanical degradation (tear and abrasion resistance).

In this section, we will combine our understanding of the simple harmonic oscillator and of viscoelasticity to provide a framework for understanding the vast literature addressing the applications of vibration isolators using rubberlike materials. A few calculations for a single degree-of-freedom

[17] Carbon black (soot) is a common material that is mixed with rubber to increase its stiffness and strength. The shear modulus of natural rubber (latex) can increase by an order of magnitude if it is mixed with carbon black.

vibration isolator will demonstrate the utility of the internal damping provided by rubberlike materials to the control of vibrations of an isolator that is excited by a frequency component that coincides with the isolator's natural frequency, ω_o. Extensions of this approach to multistage isolators, Lanchester dampers, dynamic absorbers, etc., are provided in handbooks on shock and vibration control [36] and in Snowdon's textbook [39].

4.5.1 Effective Modulus

In the earliest sections of this chapter, the various (lossless) moduli used to describe the response of isotropic elastic solids were defined and related to each other (see Table 4.1). Rubberlike materials have values of Poisson's ratio that are very close to one-half, $\nu_{rubber} \cong \frac{1}{2}$, corresponding to the volume-preserving behavior of liquids derived in Eq. (4.3). In the limit that ν_{rubber} approaches one-half, it can be shown that the effects of the bulk modulus, B, of rubber dominate the effects of the shear modulus, G, in the determination of the modulus of unilateral compression, D. Young's modulus, E, will be approximately three times the shear modulus for rubberlike materials.

$$\lim_{\nu\to\frac{1}{2}}\left[\frac{B}{G}\right] = \lim_{\nu\to\frac{1}{2}}\left[\frac{2(1+\nu)}{3(1-2\nu)}\right] = \infty \tag{4.91}$$

Since G can be neglected in comparison to B if $\nu_{rubber} \cong \frac{1}{2}$, the modulus of unilateral compression, D (aka, dilatational modulus), is nearly equal to B. The modulus of unilateral compression determines the longitudinal wave speed, $c_{long} = (D/\rho)^{\frac{1}{2}}$.

$$\lim_{\nu\to\frac{1}{2}}[D] = \lim_{\nu\to\frac{1}{2}}\left[B + \left(\frac{4G}{3}\right)\right] = B \tag{4.92}$$

As shown in Eq. (4.91), $B \gg G$, so in this limit, Young's modulus, E, is three times the shear modulus, G.

$$\lim_{\nu\to\frac{1}{2}}[E] = \lim_{\nu\to\frac{1}{2}}\left[\frac{9BG}{(3B+G)}\right] = 3G \tag{4.93}$$

In many vibration isolation applications, rubber springs are adhesively bonded to metal plates, so the actual constraints that led to the calculation of the moduli in Table 4.1 do not always reflect their use in practice. If a rubber sample is placed in pure shear, as shown in Fig. 4.27 (left), then the shear stress and shear strain are simply related by the shear modulus, G, as shown in Eq. (4.17). If the same combination of plates bonded to rubber is subjected to compressive stresses, as shown in Fig. 4.27 (right), then the strains are less than predicted by Young's modulus, E, but more than predicted by the modulus of unilateral compression, D, for the same stress.

Empirical relations have been developed that accommodate these constraints on the boundary between the rubber and the mounting plates commonly used for rubber vibration isolators. Such rubber springs have aspect ratios that are close to unity and produce an effective (apparent) modulus, $E_a > E$.

$$E_a = \left(1 + \beta S^2\right)E = 3\left(1 + \beta S^2\right)G \tag{4.94}$$

In Eq. (4.94), S is called a dimensionless *shape factor* that is defined as the ratio of the area of one loaded face, such as the plate in Fig. 4.27 (right), to the total area of the unloaded faces. The values of β have been calculated analytically and measured experimentally. It is found that for samples that are

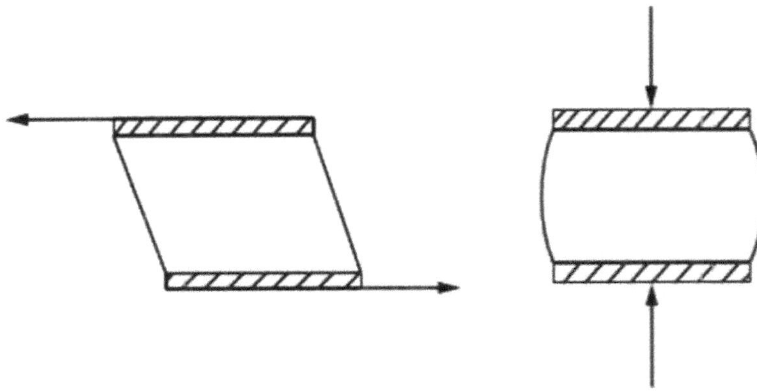

Fig. 4.27 A rubberlike sample is adhesively bonded to two metal plates. (*Left*) The sample is subjected to shear stress. (*Right*) The sample is subjected to compressive stress

circular, square, or modestly rectangular (width and length are not too dissimilar), then $\beta \cong 2$. For the cylindrical sample of radius, a, shown schematically in Fig. 4.27 (right), $S = (\pi a^2)/(2\pi a h) = a/2\,h$, so $E_a \cong 3[1 + 2(a/2\,h)^2]\,G$. If $h = 2a$, producing a cylinder of unity aspect ratio, then $E_a \cong 3.4G$.

4.5.2 Rubber-to-Glass Transition (Type I and Type II Rubbers)

The standard linear model (SLM) for viscoelasticity was predicated on the existence of a single relaxation time, $\tau_R = R_m/K_1$. At frequencies well below $\omega\tau_R = 1$, the dynamic stiffness, K', is constant, and the normalized dissipation increases linearly with frequency but is always small: $K''/K' \ll 1$. This behavior is apparent in Fig. 4.25, where the normalized dissipation is plotted as $2\pi\omega E_{stored}/\langle\Pi(t)\rangle_t = 2\pi(K''/K')$ [26]. For a rubberlike material, this behavior corresponds to a frequency-independent *dynamic shear modulus*, G, and a small dissipation factor, $\delta \ll 1$. For rubbers, this is designated Type I (low damping) behavior, typical of natural rubber (latex) and neoprene rubber.

At frequencies near $\omega\tau_R \cong 1$, the dynamic stiffness, K', is increasing with frequency (linearly in the SLM), and the normalized dissipation is nearly constant with $K''/K' \cong 1$. For a rubberlike material, this behavior corresponds to a frequency-dependent dynamic shear modulus, $G(\omega)$, and a nearly constant (frequency independent) dissipation factor, $\delta \cong 1$. For rubbers, that is Type II (high damping) behavior. Synthetic rubberlike materials like Thiokol RD, plasticized polyvinyl butyl resin (PVB), plasticized polyvinyl acetate (PVA), and filled (with carbon black) butyl rubber (like auto tire tubes) are examples of rubbers exhibiting Type II behavior.

The relaxation time, τ_R, clearly plays an important role in the behavior of rubberlike materials that are well-represented by the SLM, since it scales the excitation frequency. The relaxation time can be strongly temperature dependent. Figure 4.28, taken from Capps [26], shows Type I behavior in the "Rubbery Region" and Type II behavior in the "Glass-Rubber Transition," where Capps uses *storage modulus* for the real part of the elastic modulus and *loss factor* or *loss tangent* for the dimensionless imaginary part.

The variation in the dynamic Young's modulus and normalized damping, represented by the ratio of the real and imaginary parts of Young's modulus, is sketched by Nolle [37] as a function of both temperature and frequency for nitrile (buna-N) rubber in Fig. 4.29. Those sketches provide a graphical representation of the mutual effects of frequency and temperature on the dynamic modulus and normalized loss.

This relationship between temperature and frequency for viscoelastic materials can be viewed as reflecting the mobility of the material's molecular arrangements at the microscopic level [26]. At a

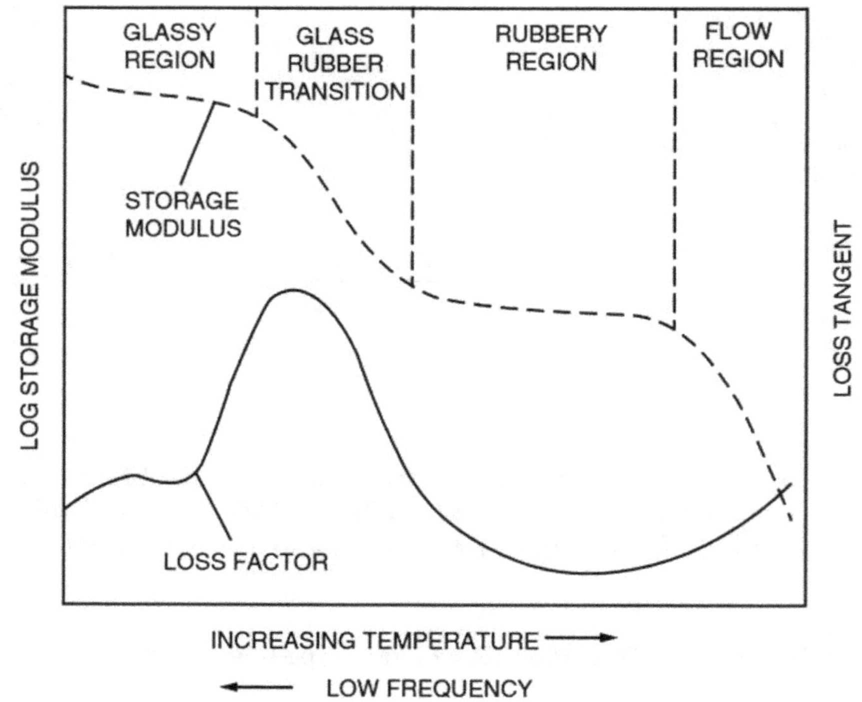

Fig. 4.28 The behavior of rubberlike materials is a strong function of both frequency and temperature. This figure is taken from Capps [26]. At the highest frequencies and/or the lowest temperatures, rubbers are "glassy" with higher dynamic moduli and low loss. At higher temperatures and/or lower frequencies, rubberlike materials have low damping and lower moduli (Type I). In the transition region, where $\omega\tau_R \cong 1$, the moduli change with frequency and the damping (loss factor) reaches its peak (Type II). At sufficiently high temperatures and/or low frequencies, rubberlike materials can creep or flow. See Fig. 1.16 for creep of an elastomeric loudspeaker surround

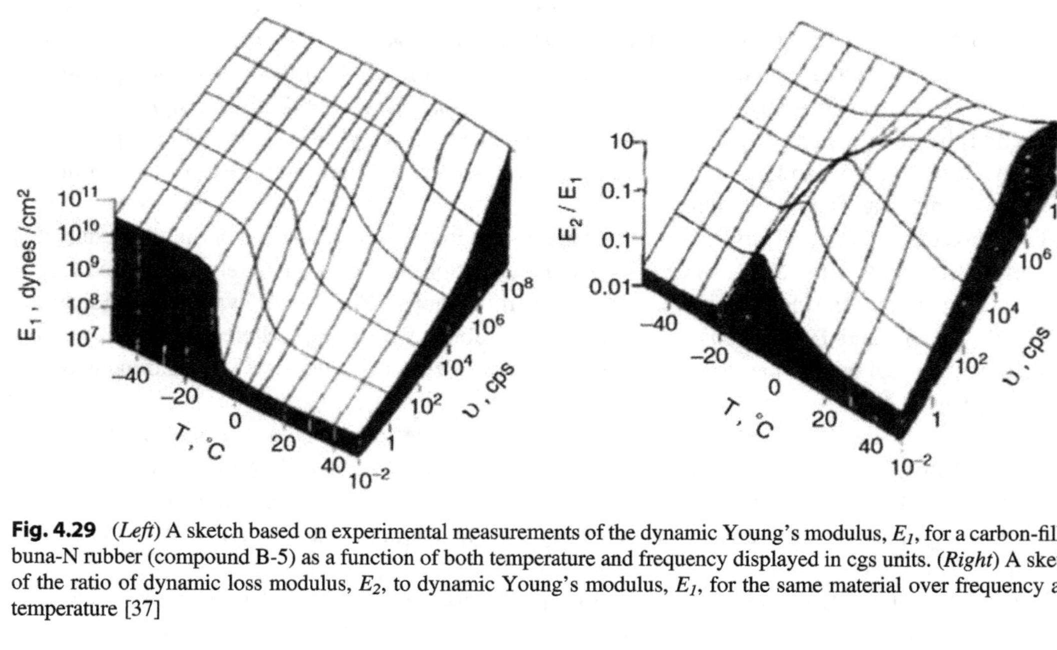

Fig. 4.29 (*Left*) A sketch based on experimental measurements of the dynamic Young's modulus, E_1, for a carbon-filled buna-N rubber (compound B-5) as a function of both temperature and frequency displayed in cgs units. (*Right*) A sketch of the ratio of dynamic loss modulus, E_2, to dynamic Young's modulus, E_1, for the same material over frequency and temperature [37]

constant excitation frequency, the internal molecular arrangement changes to a more mobile configuration as the temperature increases and the material becomes more compliant. Conversely, a reduction in temperature causes a reduction in molecular mobility,[18] so the material stiffens.

If the temperature is held constant and the excitation frequency is varied, a related but inverse phenomenon can be postulated. For low frequencies, the period of the excitation is long enough that the molecules have time to coil and uncoil, resulting in rubberlike behavior (like when the rubber is warm). At high frequencies, the molecules do not have sufficient time to relax and the material is stiff (like when the rubber is cold).

For viscoelastic materials, there is a correspondence between the frequency and temperature that can be combined using the time-temperature superposition by the application of the Williams-Landel-Ferry equation [38]. Utilization of this method for producing "master curves" that plot scaled temperature and frequency on a single axis will be postponed until Chap. 5 when the measurement of complex modulus using the resonances of thin bars will be presented.

4.5.3 Transmissibility of Rubberlike Vibration Isolators

We can now combine our understanding of simple harmonic oscillators and viscoelastic materials to calculate the transmissibility of rubberlike vibration isolators. As before, we start with a differential equation for a single degree-of-freedom mass and spring, but this time, we will allow the spring's stiffness to be complex, reflecting both its elasticity and its internal damping. As we proceed, we will have in mind a displacement-driven system (see Sect. 2.5.6) with a foundation that vibrates harmonically at a single frequency with an amplitude, $x_1(t) = x_1 \cos(\omega t)$, and we will calculate the displacement of the mass from its equilibrium position, $\mathbf{x_2}(t) = \widehat{\mathbf{x}}_2 e^{j\omega t}$, where $\mathbf{x_2}$ is considered to be a complex number to reflect the possibility that the phases of $\mathbf{x_1}$ and $\mathbf{x_2}$ will be different, as well as their amplitudes. By letting the amplitude, x_1, be a real scalar, we are setting that displacement as the phase reference for $\widehat{\mathbf{x}}_2$.

$$m\frac{d^2\mathbf{x_2}}{dt^2} = -\omega^2 m\mathbf{x_2} = \gamma\,\Xi(\omega)(x_1 - \widehat{\mathbf{x}}_2)e^{j\omega t} \tag{4.95}$$

The generic, frequency-dependent (complex) elastic modulus, $\Xi(\omega) = |\Xi(\omega)|(1 + j\delta)$, has been introduced along with a scalar geometrical factor, γ, that has the dimensions of length, to convert the modulus into a stiffness (see Sect. 4.4.3). For a shear isolator, like that shown in Fig. 4.27 (left), $\gamma = A/h$, where A is the area of the plate and h is the thickness of the rubber, and $|\Xi| = G$. For the compressional spring, shown in Fig. 4.27 (right), $\gamma = (3A/h)(1 + \beta S^2)$, again with $|\Xi| = G$.

The complex ratio of $\mathbf{x_2}/x_1$ follows directly from Eq. (4.95).

$$\frac{\mathbf{x_2}}{x_1} = \frac{\gamma\,\Xi(\omega)}{(\gamma\,\Xi(\omega) - \omega^2 m)} = \frac{1 + j\delta(\omega)}{[1 - (\omega^2 m/\gamma\,|\Xi(\omega)|) + j\delta(\omega)]} \tag{4.96}$$

The square of the normal mode frequency, ω_o, for the undriven harmonic oscillator will be given by the usual expression, although we must take care to evaluate the frequency-dependent modulus, $\Xi(\omega)$, at the normal mode frequency. At that frequency, the generic modulus will be designated $\Xi_o = |\Xi(\omega_o)|$.

[18] Said another way, the molecular mobility is "frozen out."

$$\omega_o^2 = \frac{\gamma \Xi_o}{m} \tag{4.97}$$

With that definition, the complex ratio of $\mathbf{x_2}/x_1$ can be expressed in terms of a frequency ratio, $\Omega = \omega/\omega_o$.

$$\frac{\mathbf{x_2}}{x_1} = \frac{(1 + j\delta(\omega))}{\left[1 - \Omega^2\left(\frac{|\Xi_o|}{|\Xi(\omega)|}\right)^2 + j\delta(\omega)\right]} \tag{4.98}$$

For subsequent analyses, it will be convenient to define a transmissibility, T, as the magnitude of the complex ratio, $|\mathbf{x_2}/x_1|$, and a phase angle, θ.

$$T \equiv \left|\frac{\mathbf{x_2}}{x_1}\right| = \frac{\left(1 + \delta^2(\omega)\right)^{1/2}}{\sqrt{\left[1 - \Omega^2\left(\frac{|\Xi_o|}{|\Xi(\omega)|}\right)\right]^2 + \delta^2(\omega)}}$$

$$\theta = \tan^{-1}\left(\frac{I}{R}\right) = \tan^{-1}\left\{\frac{-\delta(\omega)\Omega^2\left(\frac{|\Xi_o|}{|\Xi(\omega)|}\right)}{\left[1 - \Omega^2\left(\frac{|\Xi_o|}{|\Xi(\omega)|}\right)\right] + \delta^2(\omega)}\right\} \tag{4.99}$$

$$\text{with } R + jI = (1 + j\delta(\omega))\left[1 - \Omega^2\left(\frac{|\Xi_o|}{|\Xi(\omega)|}\right) - j\delta(\omega)\right]$$

These expressions can now be applied to Type I and Type II rubberlike materials and to the SLM for viscoelastic materials characterized by a single relaxation time, τ_R.

Type I rubberlike materials exhibit behavior similar to a viscoelastic material driven at frequencies that make $\omega\tau_R \ll 1$, resulting in a shear modulus that is frequency independent. Although viscoelastic materials in that frequency regime have normalized damping that increases linearly with frequency, the damping is always small. For Type I rubberlike materials, the dissipation factor will typically range from $0.02 \le \delta_I \le 0.20$ and is usually considered to be frequency independent. Under those assumptions, Eq. (4.99) can be written for this typical Type I transmissibility, T_I, with the phase angle, θ_I.

$$T_I = \sqrt{\frac{1 + \delta_I^2}{\left[\left(1 - \Omega^2\right)^2 + \delta_I^2\right]}} \quad \text{and} \quad \theta_I = \tan^{-1}\left[\frac{-\delta\Omega^2}{1 - \Omega^2 + \delta_I^2}\right] \tag{4.100}$$

In the limit of high frequencies, $\Omega \gg 1$, such a Type I isolator will have transmissibility that decreases as the square of the frequency.

$$\lim_{\Omega \to \infty}[T_I] = \frac{1}{\Omega^2} \tag{4.101}$$

This inverse quadratic frequency dependence is often specified as −40 dB/decade or −12 dB/octave. At resonance, the transmissibility, $T_I(\omega_o)$, depends only upon the damping.

$$T_I(\omega_o) = \frac{\sqrt{1 + \delta_I^2}}{\delta_I} \cong \frac{1 + \frac{\delta_I}{2}}{\delta_I} \cong \frac{1}{\delta_I} \quad \text{for} \quad \delta_I \ll 1 \tag{4.102}$$

Figure 4.30 shows a plot of Eq. (4.100) for $\delta_I = 0.02$ and $\delta_I = 0.20$.

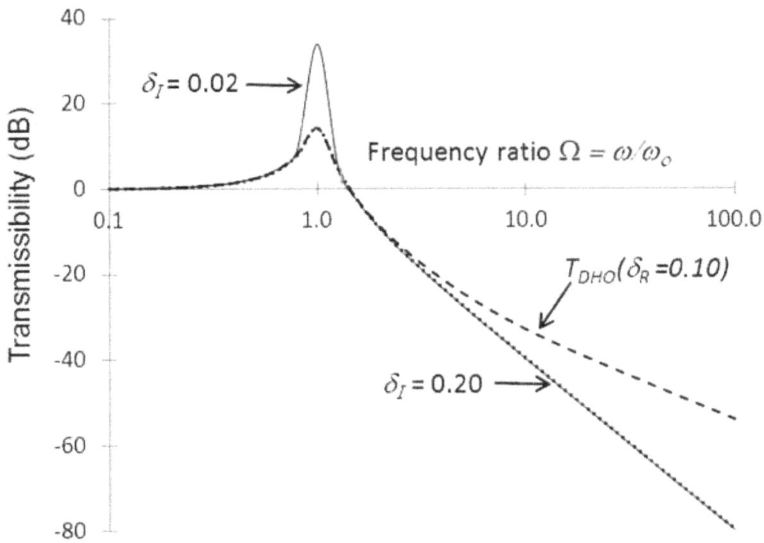

Fig. 4.30 Transmissibility for a Type I rubberlike single degree-of-freedom mass-spring system with $\delta = 0.02$ (*solid line*) and $\delta = 0.20$ (*dotted line*) plotted as a function of frequency ratio, Ω. For comparison, the *dashed line* represents a damped, displacement-driven, harmonic oscillator (DHO) with its damping adjusted to match the peak at ω_o for the $\delta = 0.20$ Type I case. At frequencies well above all this isolator's natural frequency, ω_o, both Type I isolators have transmissibilities that decrease with the square of the frequency (−40 dB/decade), but T_{DHO} only falls off at −20 dB/decade

For comparison, the transmissibility, T_{DHO}, of a mass supported by a mechanically parallel spring and dashpot combination (i.e., damped harmonic oscillator) is also shown in Fig. 4.30 as the dashed line with the mechanical resistance of the dashpot, R_m, chosen to match the Type I transmissibility for $\delta_I = 0.20$, at ω_o. To express Eq. (2.92) for the displacement-driven damped harmonic oscillator in the form of Eq. (4.100), the damping factor, δ_I, will be replaced by the damped harmonic oscillator's damping ratio, $\delta_R = \tau/\tau_{crit} = R_m\omega_o/2K$. The exponential decay time for critical damping (see Sect. 2.4.3) is $\tau_{crit} = \omega_o^{-1}$.

$$T_{DHO} = \sqrt{\frac{1 + (2\Omega\delta_R)^2}{(1 - \Omega^2)^2 + (2\Omega\delta_R)^2}}$$
$$\theta_{DHO} = \tan^{-1}\left[\frac{-(2\Omega\delta_R)\Omega^2}{1 - \Omega^2 + (2\Omega\delta_R)^2}\right] \tag{4.103}$$

The value of $\delta_R = 0.10$ was chosen to match the peak in the transmissibility at ω_o in Fig. 4.30 for $\delta_I = 0.20$.

Although the damped harmonic oscillator is frequently used as a model for typical vibration isolators, Fig. 4.30 illustrates that it is not truly representative of an isolator that uses a Type I rubberlike material as the spring. In our assumptions for the behavior of a Type I material (frequency-independent damping and stiffness), the damping factor was a constant. Inspection of Eq. (4.103) shows that the equivalent damping factor, $2\Omega\delta_R$, for the damped harmonic oscillator increases linearly with frequency. This behavior of a damped harmonic oscillator (DHO) accounts for the inverse linear decrease (−20 dB/decade) in T_{DRO} at frequencies well above ω_o, rather than the inverse quadratic frequency dependence (−40 dB/decade) for the Type I curves.[19]

Type II rubberlike materials exhibit behavior that is similar to a viscoelastic material described by the standard linear model of Sect. 4.4.2 around frequencies close to the relaxation frequency, $\omega\tau_R \cong 1$. In that frequency range, the stiffness is an approximately linear function of frequency. As is evident from Fig. 4.25, viscoelastic materials in that frequency range have normalized damping factors that are nearly frequency-independent and large, $\delta_{II} \gtrsim 0.5$, if $K_o \neq K_\infty$. Under those assumptions, Eq. (4.99) can be written for this typical Type II transmissibility, T_{II}, with the phase angle, θ_{II}.

$$T_{II} = \sqrt{\frac{1+\delta_{II}^2}{\left[(1-\Omega)^2+\delta_{II}^2\right]}} \quad \text{and} \quad \theta_{II} = \tan^{-1}\left[\frac{-\delta_{II}\Omega}{1-\Omega+\delta_{II}^2}\right] \tag{4.104}$$

Figure 4.31 shows a plot of Eq. (4.104) for $\delta_{II} = 0.50$ and $\delta_{II} = 1.0$. For comparison, the transmissibility of the damped harmonic oscillator, T_{DHO}, is also shown as the dashed line with the mechanical resistance of the dashpot, R_m, chosen to match the Type II transmissibility for $\delta_{II} = 0.50$ at ω_o.

The difference in the shape of T_{DHO} in Fig. 4.31 from that of the two T_{II} curves is related to the fact that the stiffness of the damped harmonic oscillator is constant, whereas the stiffness of the Type II isolator is assumed to increase linearly with frequency. It is that changing stiffness that broadens the peak transmissibilities around $T_{II} > 0$ dB.

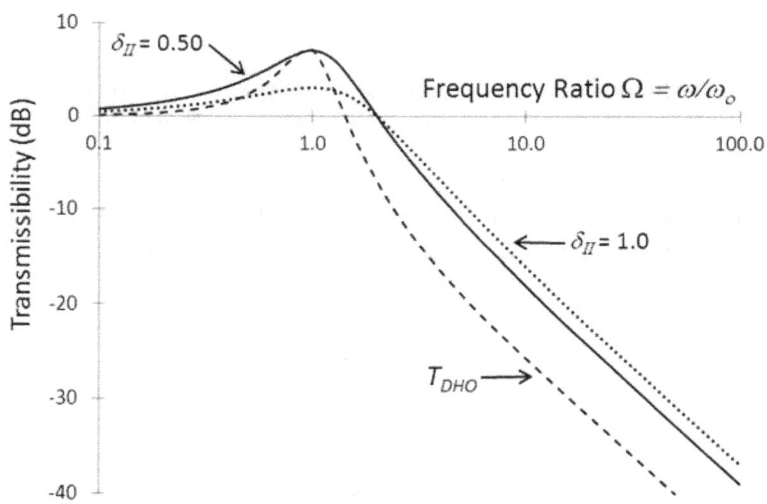

Fig. 4.31 Transmissibility for a Type II rubberlike mass-spring system with $\delta_{II} = 0.5$ (*solid line*) and $\delta_{II} = 1.0$ (*dotted line*) is plotted against the frequency ratio, Ω. For comparison, the *dashed line* represents a damped, displacement-driven, harmonic oscillator with its damping adjusted to match the peak at ω_o for the Type II case, $\delta_{II} = 0.50$. At frequencies well above the isolator's natural frequency, both Type II isolators and the damped harmonic oscillator have transmissibilities that decrease linearly with increasing frequency (i.e., −20 dB/decade)

[19] Recall from Sect. 2.5.2 that decibel is defined in terms of a power or energy ratio (see Eq. 2.69). Since transmissibility, T, is a ratio of linear quantities, as defined in Eq. (2.96), it can be expressed as a decibel only when squared: $\text{dB} = 10 \log_{10} T^2 = 20 \log_{10} T$.

The transmissibility of an isolator, based on a complex stiffness which behaves according to the SLM viscoelastic model, can also be calculated. The two important considerations are the tuning of the transition frequency given in Eq. (4.71) as $\omega_{max} = \tau_R^{-1}\sqrt{K_o/K_\infty}$, to the isolator's natural frequency (based on the low-frequency limiting stiffness, K_o), $\omega_o \equiv (K_o/m)^{1/2}$. The difference between the zero-frequency stiffness, K_o, and the infinite-frequency stiffness, K_∞, determines the maximum damping factor, as shown in Eq. (4.89). Following Snowdon [39], the transmissibility is parameterized by the tuning ratio, $\aleph \equiv \omega_{max}/\omega_o$, and the stiffness ratio, $\alpha \equiv K_\infty/K_o$, which determines the maximum value of the damping factor, δ_{max}.

$$\delta_{\max} = \frac{\alpha - 1}{2\sqrt{\alpha}} = \frac{1}{2}\frac{K_\infty - K_o}{\sqrt{K_\infty K_o}} \quad \text{or} \quad \alpha \equiv \frac{K_\infty}{K_o} = 1 + 2\delta_{\max}\sqrt{1 + \delta_{\max}^2} + 2\delta_{\max}^2 \tag{4.105}$$

Substitution into the general transmissibility equation of Eq. (4.99) yields the transmissibility for an SLM viscoelastic isolator that includes the exact frequency dependencies of both damping and stiffness for such a single relaxation time model.

$$T_{SLM} = \sqrt{\frac{\left(\Omega^2 + \aleph^2\right)^2 + \left(2\aleph\Omega\delta_{\max}\right)^2}{\left[\left(\Omega^2 + \aleph^2\right) - \Omega^2\left(\Omega^2 + \alpha\aleph^2\right)\left(\frac{1+\aleph^2}{1+\alpha\aleph^2}\right)\right]^2 + \left(2\aleph\Omega\delta_{\max}\right)^2}} \tag{4.106}$$

The behavior of the SLM viscoelastic transmissibility, T_{SLM}, is shown in two graphs presented as Fig. 4.32. First, the transmissibility is calculated for three different ratios of the limiting stiffnesses: $\alpha = K_\infty/K_o = 1.5$, 3.0, and 6.0, corresponding to $\delta_{max} = 0.20$, 0.58, and 1.02, respectively. For all three cases, the high-frequency behavior shows an inverse quadratic dependence on frequency in the transmissibility corresponding to −40 dB/decade.

Shown for comparison in Fig. 4.32 as the dash-dotted line is also the transmissibility of a damped harmonic oscillator with a damping ratio, $\delta_R = 0.29$, that matches the peak transmissibility of the viscoelastic isolator with $\delta_{max} = 0.58$. As shown in Fig. 4.32, the transmissibility of the viscoelastic

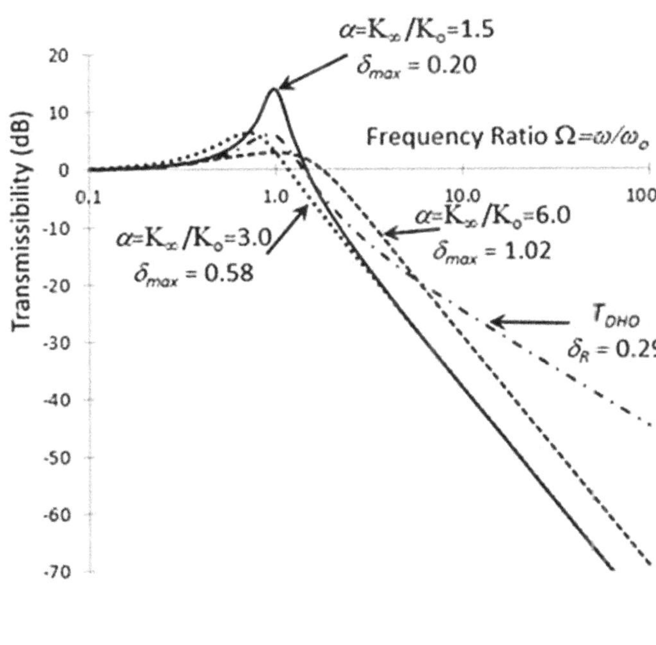

Fig. 4.32 The transmissibility, T_{SLM}, of a single degree-of-freedom isolator based on the SLM viscoelastic model is calculated for three different ratios of the limiting stiffnesses: $\alpha = K_\infty/K_o = 1.5$ (*solid*), 3.0 (*dotted*), and 6.0 (*dashed*), corresponding to $\delta_{max} = 0.20$ (*solid*), 0.58 (*dotted*), and 1.02 (*dashed*). All three cases have $\omega_o = \omega_t = \tau_R^{-1}$, so the tuning parameter, $\aleph = 1.0$. For comparison, the transmissibility, T_{DHO}, of a damped harmonic oscillator with $\delta_R = 0.29$ is shown (*dash-dotted*) which matches the peak transmissibility of the $\delta_{max} = 0.58$ case

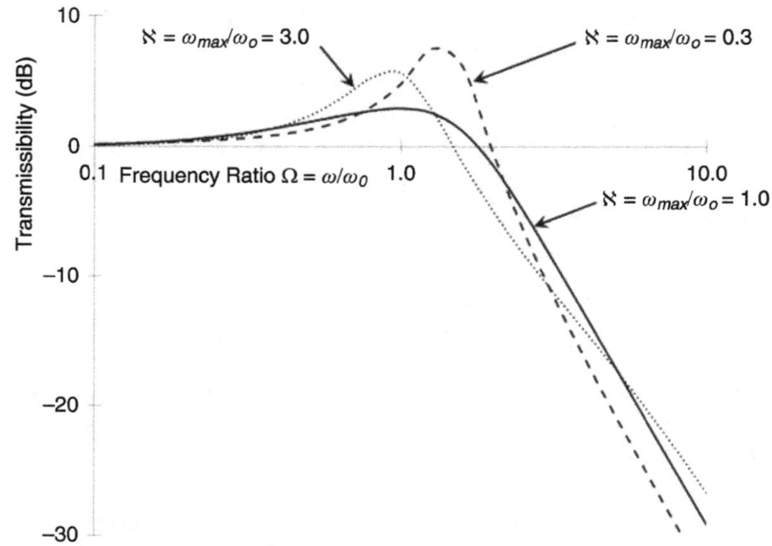

Fig. 4.33 The effects of "tuning" the transition frequency are illustrated for the SLM viscoelastic isolator. The *solid line* represents $\aleph \equiv \omega_{max}/\omega_o = 1.0$ with $\boldsymbol{\alpha} = K_\infty/K_o = 6.0$, corresponding to $\boldsymbol{\delta}_{max} = 1.02$. for all curves. The other two curves represent a "flat" detuning (peak transmissibility at a lower frequency) with $\aleph = 3.0$ (*dotted*) and a "sharp" detuning with $\aleph = 0.3$ (*dashed*)

rubber isolator falls off more rapidly with frequency than the damped simple harmonic oscillator since the rubber stiffness is constant at $\Omega \gg 1$ and the damping is decreasing with frequency, whereas the damped harmonic oscillator's damping is increasing with frequency although δ_R is constant.

The viscoelastic transition frequency, $\omega_t = {\tau_R}^{-1}$, is a strong function of temperature for most viscoelastic materials. This needs to be considered in isolator design since the temperature dependence of ω_t will shift the "tuning" parameter, $\aleph \equiv \omega_{max}/\omega_o$. In Fig. 4.33, all three transmissibility curves have $\alpha = K_\infty/K_o = 6.0$ corresponding to $\delta_{max} = 1.02$. They differ in the tuning of the harmonic oscillator natural frequency, ω_o, to the peak in the dissipation at ω_{max}. The solid line in Fig. 4.33 and the solid line in Fig. 4.32 represent $\aleph \equiv \omega_{max}/\omega_o = 1$. The other two curves in Fig. 4.33 represent a "flat" detuning (peak transmissibility at a lower frequency) with $\aleph = 3.0$ (dotted line) and a "sharp" detuning with $\aleph = 0.3$ (dashed line). The tuning, which will be a function of temperature, affects both the maximum damping and the transmissibility at high frequencies.

4.6 Anisotropic (Crystalline) Elasticity*

There are many materials that are not isotropic, both naturally occurring and synthetic. Wood is a composite material that exhibits different elastic behavior depending on whether forces are applied along the grain or across it. The same is true for many composite materials, whether they are carbon fiber composites or resin-soaked fiberglass. Natural crystals can have very complicated anisotropy (see Table 4.2). When a material has its own internal preferred "coordinate system," whether it is defined by crystalline axes or the orientation of a fibrous matrix, it is possible that stresses in one direction lead to strains in other directions. This was also true of isotropic materials as illustrated by the transverse displacements caused by longitudinal stress that were transferred by a non-zero Poisson's ratio.

Table 4.2 The number of independent stiffness matrix elements, or combinations of matrix elements and angles, required to completely specify the elastic response of crystals varies with the symmetry of the crystals

Crystal class	Symmetry group	Independent elements and angles	Minimum elements
Triclinic	C_1, C_i	21	18
Monoclinic	C_s, C_2, C_{2h}	13	12
Orthorhombic	C_{2v}, D_2, D_{2h}	9	9
Tetragonal	C_4, S_4, C_{4h}	7	6
Tetragonal	C_{4v}, D_{2d}, D_4, D_{4h}	6	6
Rhombohedral	C_3, S_6	7	6
Rhombohedral	C_3, D_3, D_{3d}	6	6
Hexagonal	C_6	5	5
Cubic		3	3

The most general form of Hooke's law that incorporates all known anisotropic effects can be expressed as a square *stiffness matrix* that relates stresses and strains.

$$\begin{pmatrix} \sigma_{xx} \\ \sigma_{yy} \\ \sigma_{zz} \\ \sigma_{xy} \\ \sigma_{yz} \\ \sigma_{zx} \end{pmatrix} = \begin{pmatrix} \sigma_{11} \\ \sigma_{22} \\ \sigma_{33} \\ \sigma_{23} \\ \sigma_{31} \\ \sigma_{12} \end{pmatrix} \equiv \begin{pmatrix} \sigma_1 \\ \sigma_2 \\ \sigma_3 \\ \sigma_4 \\ \sigma_5 \\ \sigma_6 \end{pmatrix} = \begin{pmatrix} s_{11} & s_{12} & s_{13} & s_{14} & s_{15} & s_{16} \\ s_{21} & s_{22} & s_{23} & s_{24} & s_{25} & s_{26} \\ s_{31} & s_{32} & s_{33} & s_{34} & s_{35} & s_{36} \\ s_{41} & s_{42} & s_{43} & s_{44} & s_{45} & s_{46} \\ s_{51} & s_{52} & s_{53} & s_{54} & s_{55} & s_{56} \\ s_{61} & s_{62} & s_{63} & s_{64} & s_{65} & s_{66} \end{pmatrix} \begin{pmatrix} \varepsilon_1 \\ \varepsilon_2 \\ \varepsilon_3 \\ \varepsilon_4 \\ \varepsilon_5 \\ \varepsilon_6 \end{pmatrix} \tag{4.107}$$

The stresses can be resolved into forces applied along a coordinate axis direction (*x*, *y*, or *z*), normal to the surface of the sample, to produce the longitudinal (compressive) stresses σ_{xx}, σ_{yy}, or σ_{zz}. Shearing forces can also be applied along a coordinate axis, but orthogonal to the surface normal directions producing the shear stresses, σ_{xy}, σ_{yz}, or σ_{zx}.

To simplify the matrix representation, the coordinate axes can be specified by integer indices: $i = 1$ for *x*, $i = 2$ for *y*, and $i = 3$ for *z*. Further simplification in the nomenclature can be achieved if the shear stresses are encoded as $i = 4$ for *yz*, $i = 5$ for *xz*, and $i = 6$ for *xy*. This simplified indexing scheme, known as *Voigt notation*, has been applied to the strains, ε_i, as well in Eq. (4.107).

The six components of stress, σ_i, and the six components of strain, ε_i, are related by the stiffness matrix shown in Eq. (4.107). In principle, that square matrix contains 36 independent elements. Due to the symmetry of the stresses and strains ($\sigma_{xy} = \sigma_{yx}$, etc.), the matrix is symmetric, with stiffness elements, $s_{ij} = s_{ji}$, so there are only 21 potentially independent elements.

The actual number of elements required to completely specify the elastic response of a crystal depends upon the symmetry of the crystal lattice. We already know that an isotropic solid requires only two independent elements. A cubic crystal requires three. The triclinic crystal requires all 21, although that number can be reduced by suitable choice of coordinate axes. For a triclinic crystal, the non-zero moduli are reduced to 18 by alignment of the coordinate axes with the principle directions of the crystal lattice. Table 4.2 summarizes the number of matrix elements required for complete specification of all crystal groups.

Figure 4.34 provides examples of the stiffness matrix for crystalline solids with cubic and hexagonal symmetry.

These individual matrix elements can be related to the moduli of isotropic solids where there must only be two independent matrix elements, s_{11} and s_{44}.

$$\begin{pmatrix} s_{11} & s_{12} & s_{12} & 0 & 0 & 0 \\ s_{12} & s_{11} & s_{12} & 0 & 0 & 0 \\ s_{12} & s_{12} & s_{11} & 0 & 0 & 0 \\ 0 & 0 & 0 & s_{44} & 0 & 0 \\ 0 & 0 & 0 & 0 & s_{44} & 0 \\ 0 & 0 & 0 & 0 & 0 & s_{44} \end{pmatrix} \qquad \begin{pmatrix} s_{11} & s_{12} & s_{13} & 0 & 0 & 0 \\ s_{12} & s_{11} & s_{13} & 0 & 0 & 0 \\ s_{13} & s_{13} & s_{33} & 0 & 0 & 0 \\ 0 & 0 & 0 & s_{44} & 0 & 0 \\ 0 & 0 & 0 & 0 & s_{44} & 0 \\ 0 & 0 & 0 & 0 & 0 & s_{66} \end{pmatrix}$$

Fig. 4.34 Stiffness matrices for a cubic crystal (*left*) contain three independent stiffnesses: s_{11}, s_{12}, and s_{44}. The stiffness matrix for a hexagonal crystal (*right*) has five independent stiffnesses: s_{11}, s_{33}, s_{44}, $s_{66} = (s_{11} - s_{12})/2)$, s_{12}, and s_{13}

$$\begin{aligned} G &= s_{44} \quad \text{and} \quad E = \frac{s_{44}(3s_{11} - 4s_{44})}{s_{11} - s_{44}} \\ B &= s_{11} - \frac{4}{3}s_{44} \quad \text{and} \quad \nu = \frac{s_{11} - 2s_{44}}{2(s_{11} - s_{44})} \end{aligned} \tag{4.108}$$

Polycrystalline materials behave as though they are isotropic. Unfortunately, there is no way to calculate the moduli for a polycrystalline material based on the single-crystal moduli unless the single crystals are nearly isotropic [40]. Ledbetter has compared several techniques for averaging the monocrystalline moduli under some simplifying assumptions (e.g., small grain-to-specimen sizes, very thin grain boundaries, random grain orientations, etc.) [41].

Other physical constants (e.g., thermal expansion coefficients, dielectric constants) that describe anisotropic crystalline substances can also have different values along different directions. Piezoelectric constants are necessarily anisotropic since the absence of an inversion symmetry plane is a prerequisite for the piezoelectric behavior of crystalline materials [42].

4.7 There Is More to Stiffness Than Just "K"

Many of the introductory textbooks on sound and vibration are content to define a spring constant, K, and then proceed with calculation of the consequences of this symbolic "source" of a linear restoring force. Obviously, I think that choice is shortsighted. Springs are carefully engineered products with elaborate rules for design that allow spring performance to best exploit material properties and survive the rigors of millions, if not billions, of fully reversing stress cycles, often close to the elastic limits.[20] Only a few strategies have been introduced here, but they were intended to expose the philosophical choices and physical limitations that constrain the extent to which one can select a value for K.

The use of elastomeric materials for springs, particularly for application to vibration isolators, introduces the behavior of real materials that combine both stiffness and damping. It also provided us with exposure to the single relaxation time model which is ubiquitous in the description of the responses of linear systems. Such systems range from light waves in dielectric materials to the sound speed and attenuation in systems where the relaxation time might be determined by the collision of molecules, the absorption and desorption of vapors in equilibrium with their liquids, or the chemical

[20] An automotive valve spring in an engine that operates at 2000 rpm will be stressed 240 million times during the operational lifetime of 2000 h. A typical home refrigerator/freezer may last 15 years. (In America, refrigerators are replaced more often as an interior decorating choice than due to product failure.) If a refrigerator uses a spring that vibrates at 60 Hz, then the number of fully reversing stress cycles accumulated in 15 years of service would be 28 billion cycles if the compressor operated continuously.

equilibrium of ionic species dissolved in seawater. These are only a few examples of systems that this model describes so elegantly.

We will exploit this relaxation time model again in our study of sound in fluids, but it seems much simpler for me to understand and visualize, let alone calculate the consequences of a relaxation mechanism, when the model is applied to springs and a dashpot. That analysis also led us to the rather remarkable fact that the real and imaginary components of the linear response function (the complex stiffness) are not independent; specification of the limiting values of stiffness dictates the maximum value of the dissipation, in addition to the frequency dependencies of that dissipation.

Finally, this chapter has attempted to demystify definitions of elastic moduli and their interrelationships. As will be demonstrated in the next chapter, that knowledge provides a firm foundation for the understanding of wave propagation in solids in general as well as the modes of vibration of thin bars and plates. That understanding will then be inverted to exploit the measurement of the normal mode frequencies of bars and other structures to accurately determine the elastic moduli of materials.

Talk Like an Acoustician

Strain	Bending moment
Stress	Radius of gyration
Young's modulus	Slenderness ratio
Poisson's ratio	Euler force
Bulk modulus	Single relaxation time model
Volumetric strain	Linear response theory
Modulus of unilateral compression	Kramers-Kronig relations
Dilatational modulus	Generalized susceptibility
Shear modulus	Relaxation time
Auxetic materials	Maxwell model
Accelerometer	Standard linear viscoelasticity model
Test mass	Dynamic modulus
Seismic mass	Storage modulus
Proof mass	Loss modulus
Ferroelectric ceramic	Loss factor or damping factor
Tonpilz transducers	Shape factor
Reduced mass	Stiffness matrix
Radiation mass	Voigt notation
Neutral plane	

Exercises

1. ***Tonpilz* via Rayleigh**. The normal mode frequency for the *Tonpilz* example was based on a nodal position that made both halves of the assembly oscillate at the same frequency. Use Rayleigh's method (see Sect. 3.4.1) to calculate the same normal mode frequency by imposing a linear "trial function" and minimizing the frequency with respect to the distance from the head mass to the nodal location, b (see Fig. 4.8).
2. **Effective mass correction for a vibrating cantilevered beam**. The effective moving mass of a vibrating cantilever beam can be calculated if the displacement curve, $z(x)$, is known. For a beam of thickness, t, and length, L, that is made from a material with mass density, ρ, and Young's modulus, E, calculate the ratio of the moving mass to the total mass, m_s, for the two cases below:
   (a) *Uniform beam.* A uniform beam of thickness t and width w.
   (b) *Triangular cantilever.* A triangular cantilever of uniform thickness with the same stiffness as the beam in part (a) above.

3. **Torsionally resonant toroidal standing wave oscillator**. Instead of creating high-amplitude standing waves in a straight tube with a loudspeaker at one end and a rigid boundary at the other (see Sect. 10.7.5), it is possible to generate standing waves within a gas inside a torus that contains one rigid boundary within if the toroid oscillates about its axis at a frequency that makes one half-wavelength of the sound in the gas within the toroid equal to the circumference of the toroid [43]. Using Fig. 4.35 for this exercise, let the diameter of the torus be $D = 0.50$ m and the inside diameter of the gas-filled toroidal waveguide be 10 cm and the wall thickness be 1.5 mm. The solid plate that is welded to the torus and to the torsion tube is also 1.5 mm thick. The torsion tube has a wall thickness of 2.5 mm with an ID = 8.0 cm and length of $L = 12.0$ cm.
   (a) *Frequency*. Determine the resonance frequency of this torsional oscillator if all of the components are made from steel with a density, $\rho_{steel} = 7700$ kg/m^3, and a shear modulus, $G_{steel} = 140$ GPa.
   (b) *Fatigue failure*. What is the peak angular displacement θ_1 of the torus if the fatigue strength of the steel is $G_{max} = 30$ MPa?

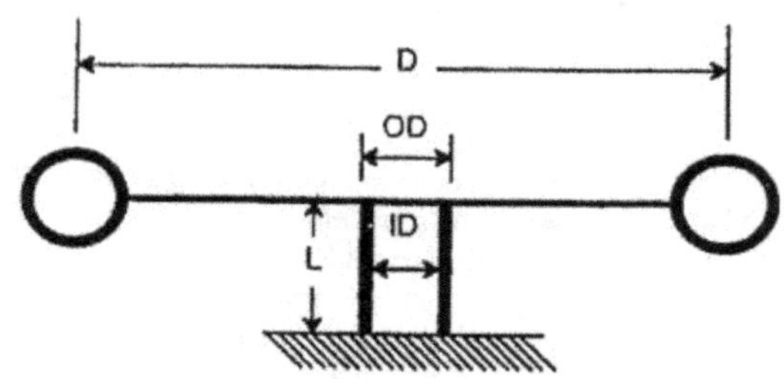

Fig. 4.35 Cross-sectional diagram of a torsional oscillator consisting of a torus supported by a hollow cylindrical tube acting as a torsional spring

4. **Torsional stiffness of a rectangular ribbon**. Most galvanometers use a thin metallic ribbon as the torsional stiffness element that provides the restoring force for the current-carrying coil and pointer. Calculate the torsional stiffness for a thin ribbon of thickness, t; width, w, with $w \gg t$; and length, L, by assuming that torsion of the ribbon changes its rectangular cross-section into a parallelogram.
   (a) *Torsional stiffness*. Show that the expression below provides the torsional stiffness, K_{ribbon}.

$$K_{ribbon} = G\frac{wh^3}{3L}$$

   (b) *Torsional resonance frequency*. A mirrored galvanometer consists of a mirror and a current-sensing coil (see Fig. 2.7). The coil is suspended in a magnetic field produced by a permanent magnet structure by a ribbon of beryllium (Be) that is 100 μm thick, 2.0 mm wide, and 15 cm long. The shear modulus of beryllium is $G_{Be} = 132$ GPa. The moment of inertia of the mirror and coil is $I_{mirror} = 3 \times 10^{-8}$ kg-m^2. What is the natural frequency of torsional vibration?
   (c) *Noise*. Using the Equipartition Theorem as discussed in Sect. 2.4.4, determine the root-means-squared angular displacement, $\theta_{rms} = \sqrt{\langle\theta^2\rangle}$, of the mirror and coil at room temperature, $T = 20$ °C = 293 K.
5. **Cantilever springs**. Consider a thin steel beam is 20.0 cm long, 2.0 cm wide, and 2.0 mm thick. Let Young's modulus of the steel be $E_{Steel} = 195$ GPa. Assume that one end of the beam is rigidly clamped at $x = 0$, so $(\partial y/\partial x)_{x=0} = 0$, and a load of 1.0 kg is applied to the end at $x = L = 20.0$ cm.
   (a) *Uniform beam stiffness*. By how much will the end of the beam be deflected from its equilibrium (straight) shape by the $M = 1.0$ kg load if $g = 9.8$ m/s^2?

(b) *Maximum stress and strain for the uniform beam.* At what value of x is the stress in the beam largest? At what value of h, the distance from the beam's neutral axis, is the stress largest at that value of x? What is the maximum strain and stress that is experienced by the loaded uniform beam?

(c) *Triangular cantilever.* What is the deflection of a similar cantilever spring, using the same amount of steel, which is also 20.0 cm long and is clamped at $x = 0$, but which has a base of $w = 4.0$ cm and the mass, $M = 1.0$ kg, is applied at the vertex of the triangle located at $x = 20.0$ cm?

(d) *Maximum stress and strain for the triangular beam.* At what value of x is the stress on the beam largest? At what value of h, the distance from the beam's neutral axis, is the stress largest at that value of x? What is the maximum strain and stress that is experienced by the loaded triangular beam?

6. **Compression of a column**. An aluminum rod has a length of 10.0 cm and a diameter of 5.0 mm. Assume Young's modulus for aluminum is $E_{Al} = 71$ GPa.
   (a) *Compression.* By how much will the beam be shortened if it is clamped at the bottom and a 5.0 kg mass is placed on the top?
   (b) *Buckling.* How much mass would be required to cause such a beam buckle?
7. **Microphone vibration isolator**. Figure 4.36 shows a piezoelectric microphone (Endevco Model 2510) on a (symmetrical) vibration mount that supports the microphone and its clamping fixture, attached to an aluminum base ($5 \times 1.5 \times 0.4$ cm) on two silicone rubber pads ($1.5 \times 1.5 \times 0.5$ cm). The rubber has a shear modulus of $G = 350$ kPa. The mass of the microphone and its mounting plate is 60 gm.
   (a) *Effective Young's modulus.* Using the dimensions of a single rubber pad in Fig. 4.36 and assuming $\beta = 2.0$, calculate the effective Young's modulus, E_a, for compression.
   (b) *Compressional stiffness.* Using the value of E_a, and the dimension of a single rubber pad in Fig. 4.36, calculate the compressional stiffness for a single pad.
   (c) *Natural frequency.* Ignoring the counterweight below the mounting bracket that attaches to the Space Shuttle's frame, calculate the microphone's natural frequency of vibration for up and down motion.

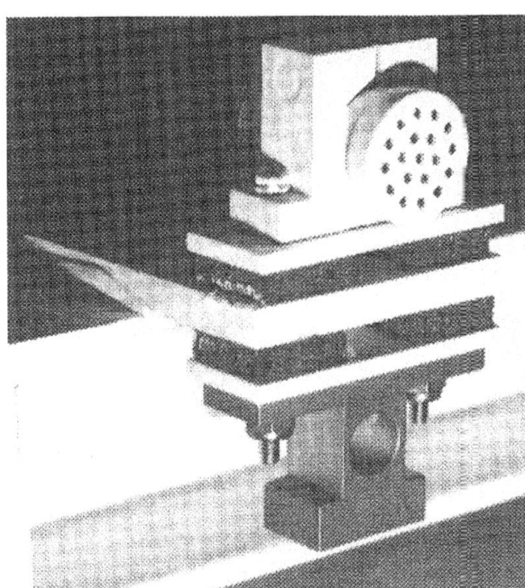

Fig. 4.36 Photograph of a vibration isolator mounted in the cargo bay of a Space Shuttle [44]

(d) *Shear vibration*. The microphone can also undergo shear vibration. Calculate the shear stiffness provided by the two rubber pads and use that result to calculate the natural frequency for side-to-side vibration.

8. **Strain transformer**. The cylindrical elastomeric mandrel shown in Fig. 4.37 is wrapped by a glass optical fiber and is capped at each end by rigid plates. The fiber is 80 μm in diameter and Young's modulus of the glass is $E = 62$ GPa. Poisson's ratio for the elastomer is $\nu = ½$, so the rubber mandrel transforms compression of the mandrel into the lengthening of the optical fiber. The diameter of the cylinder is $D = 3.98$ cm, and the height is $h = 1.92$ cm. The mandrel is wrapped with $N = 85$ turns of optical fiber.
   (a) *Glass stiffness-length product*. Based on Young's modulus of glass and the diameter of the optical fiber, what is the stiffness-length product, $K_{glass}L$, of the fiber?
   (b) *Glass stiffness*. Based on the diameter of the mandrel and the number of turns of optical fiber, what is the stiffness of the total length of fiber, K_{fiber}?
   (c) *Potential energy*. The volume of the mandrel is $V = \pi D^2 h/4$. Keeping in mind that the elastomer is volume-conserving, since $\nu = ½$, write the change in potential energy of the glass fiber for a change in the mandrel height, δh.
   (d) *Mandrel stiffness*. Assuming that all of the compressional stiffness of the mandrel is due to the stretching of the glass fiber during compression, use your expression for the potential energy in part (c) to determine the effective stiffness of the mandrel.
   (e) *Moving mass*. The mass density of the elastomer is $\rho_{2CN} = 1020$ kg/m^3. If the displacement of the mandrel is linear with distance from the fixed end (like the Gerber scale of Fig. 2.2), what is the effective moving mass of the mandrel?
   (f) *Natural frequency*. If one end of the mandrel is fixed and a proof mass, $m_{test} = 100$ gm, is placed on the other end, what is the natural frequency of vibration, f_o, if the contribution of the mandrel's moving mass is included?

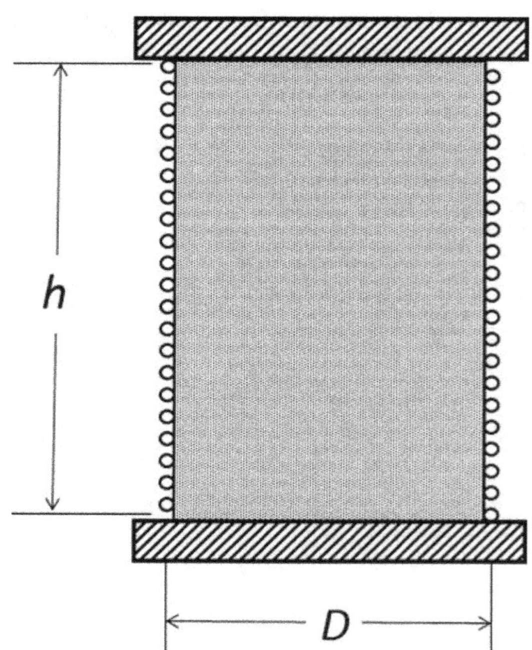

Fig. 4.37 Optical fiber strain transformer [45]

References

1. L. D. Landau and E. M. Lifshitz, Theory of Elasticity, 2nd English edition (Pergamon, 1975). See pg. 17; ISBN 0 08 006465 5
2. G.G. Marquez, *One Hundred Years of Solitude* (Harper Collins, New York, 2003); ISBN 0-19-515544-4
3. E.H. Love, *A Treatise on the Mathematical Theory of Elasticity*, 4th edn. (Dover reprint, 1944); ISBN 0-486-60172-9
4. R.S. Lakes, Foam structures with a negative Poisson's ratio. Science **235**(4792), 1038–1040 (1987)
5. S. R. Lampman (ed.), *ASM Handbook, Vol. 19 – Fatigue and Fracture* (ASM International, Metals Park, 1996); ISBN 0-87170-385-8
6. J.E. Shigley, C.R. Mischke, *Mechanical Engineering Design*, 5th edn. (McGraw-Hill, Boston, 1989); ISBN 0-07-056899-5
7. O.B. Wilson, *Introduction to Theory and Design of Sonar Transducers* (Peninsula, Los Altos, 1988)
8. S. L. Ehrlich and A. F. Medeiros, Wide band transducer, U. S. Patent No. 3,860,901 (January 14, 1975)
9. D. Boucher, Process to increase the power of the low frequency electro acoustic transducers and corresponding transducers, US Patent No. 5,070486 (December 3. 1991)
10. L.D. Landau, E.M. Lifshitz, *Mechanics* (Pergamon, Oxford, 1960). See §24
11. M.E. Poese, R.M. Keolian, R.W.M. Smith, E. Mitchell, S.L. Garrett, Performance estimates for a helium-based thermoacoustic Stirling chiller. Int. J. Low Carbon Technol. **8**, 79–84 (2013)
12. W.C. Young, R.G. Budynas, *Roark's Formulas for Stress and Strain*, 7th edn. (McGraw-Hill, New York, 2002); ISBN 0-07-072542-X
13. S. L. Garrett, R. M. Keolian and R. W. M. Smith, High-efficiency moving magnet loudspeaker," U. S. Patent No. 6,307,287 (October 23, 2001)
14. J. Liu, S. Garrett, Characterization of a small moving-magnet electrodynamic linear motor for use in a thermoacoustic refrigerator. J. Acoust. Soc. Am. **118**(4), 2289–2295 (2005)
15. H.E. Boyer, T.L. Gall, *Metals Handbook: Desk Edition* (American Society for Metals, Metals Park, 1985); ISBN 0-81170-188-X
16. S. L. Garrett, Cylindrical spring with integral gas seal, US Patent No. 6,755,027 (June 29, 2004)
17. EJMA, *Standards of the Expansion Joint Manufacturers Association*, 10th edn. (EJMA, Tarrytown, 2017)
18. H. Cavendish, Experiments to determine the Density of the Earth. Philos. Trans. R. Soc. Lond. Part II **88**, 469–526 (1798)
19. C.J. Ancker Jr., J.N. Goodier, Pitch and curvature correction for Helical Springs. ASME J. Appl. Mech. **25**(4), 471–483 (1958)
20. M. Wahl, *Mechanical Springs*, 2nd edn. (McGraw-Hill, New York, 1963)
21. L.R. Wilberforce, On the vibrations of a loaded spiral spring. Philos. Mag. **38**, 386–392 (1896)
22. R.E. Berg, T.S. Marshall, Wilberforce oscillations and normal modes. Am. J. Phys. **56**(1), 32–38 (1991)
23. A. Sommerfeld, *Mechanics of Deformable Bodies: Lectures on Theoretical Physics*, vol II (Academic, New York, 1950), pp. 308–314
24. N. Feather, *Vibration and Waves* (Penguin, London, 1964), pp. 59–67
25. J. Wilhelm, K.A. Gillis, J.B. Mehl, M.R. Moldover, An improved Greenspan acoustic viscometer. Int. J. Thermophys. **21**, 983–999 (2000)
26. R. N. Capps, Elastomeric Materials for Acoustical Applications (Naval Research Laboratory, Underwater Sound Reference Detachment, 15 September 1989)
27. L.D. Landau, E.M. Lifshitz, *Statistical Physics*, 2nd edn. (Addison-Wesley, Reading, 1969), pp. 384–391, See §125; ISBN 0-201-04167-7
28. L.D. Landau, E.M. Lifshitz, *Electrodynamics of Continuous Media* (Pergamon, London, 1960); ISBN 0-08-09105-9. §62
29. B. Bhatia, *Ultrasonic Absorption: An Introduction to the Theory of Sound Absorption and Dispersion in Gases, Liquids, and Solids* (Oxford, 1967). Reprinted (Dover, 1985); ISBN 0-486-64917-2
30. H. Bode, Relations between attenuation and phase in feedback amplifier design. Bell Syst. Tech. J. **19**, 421 (1940)
31. V. Mangulis, Kramers-Kronig or dispersion relations in acoustics. J. Acoust. Soc. Am. **36**, 211–212 (1964).; Relation between the radiation impedance, pressure in the far field, and baffle impedance, J. Acoust. Soc. Am. **36**, 212–213 (1964)
32. S. Garrett, S. Adams, S. Putterman, I. Rudnick, Resonant nonlinear mode conversion in He II. Phys. Rev. Lett. **41**(6), 413–416 (1978)
33. R. Kronig, H.A. Kramers, Absorption and dispersion in X-rays spectra. Z. Phys. **48**, 174 (1928)
34. J. W. Strutt (Lord Rayleigh), *The Theory of Sound*, vol I, 2nd edn. (Macmillan, London, 1894), p. 153; Reprinted (Dover, 1945); ISBN 486-60292-3
35. M. O'Donnell, E.T. Jaynes, J.G. Miller, Kramers-Kronig relationships between ultrasonic attenuation and phase velocity. J. Acoust. Soc. Am. **69**(3), 696–701 (1981)

36. C.M. Harris, A.G. Piersol, *Harris' Shock and Vibration Handbook*, 5th edn. (McGraw-Hill, New York, 2002); ISBN 0-07-137081-1
37. W. Nolle, Dynamical mechanical properties of rubberlike materials. J. Polym. Sci. **5**(1), 1–54 (1950)
38. J.D. Ferry, *Viscoelastic Properties of Polymers*, 3rd edn. (Wiley, New York, 1980.; ISBN: 978-0-471-04894-7
39. J.C. Snowdon, *Vibration and Shock in Damped Mechanical Systems* (Wiley, New York, 1968)
40. L. D. Landau and E. M. Lifshitz, Theory of Elasticity, 2nd English edition (Pergamon, 1975). §10, see pp. 41–42- ISBN 0 08 006465 5
41. H. Ledbetter, Chapter 11: Monocrystal-polycrystal elastic-constant models, in *Handbook of Elastic Properties of Solids, Liquids, and Gases: Vol. III – Elastic Properties of Solids: Biological and Organic Materials, Earth and Marine Sciences*, ed. by M. Levy, H. Bass, R. Stern, (Academic, 2000). ISBN 0-12-445763-0
42. W.G. Cady, *Piezoelectricity* (McGraw-Hill, New York/London, 1946)
43. S. L. Garrett, Torsionally resonant toroidal thermoacoustic refrigerator," U.S. Patent No. 5,953,921 (September 21, 1999)
44. C. D. Stehle, Vibration isolation of a microphone, U.S. Defense Technical Report No. AD A 161 018 (October 1985)
45. D.L. Gardner, T. Hofler, S.R. Baker, R.K. Yarber, S.L. Garrett, A fiber-optic interferometric seismometer. J. Lightwave Technol. **5**(7), 953–960 (1987)

5

Wave Propagation in Bars

The perspectives and techniques that have been developed in the previous chapters will now be applied to calculation of wave propagation in solids. Their application to longitudinal and shear waves will be both familiar and simple. What you will find to be even more satisfying is the success of those same techniques for finding solutions for waves in a system that does not obey the wave equation and whose solutions are not functions of $x \pm ct$.

This chapter is focused mainly on waves in bars that are "thin," meaning that the square root of their cross-sectional area is much less than their length, $\sqrt{S} \ll L$. The frequencies of the normal modes of such thin bars will be used to determine the bars' elastic constants to high precision. The results for thin bars will also lead to the understanding of waves in samples with dimensions much greater than the wavelength of the sound. Examples of short wavelength propagation in "bulk" samples range from ultrasonic waves in crystals and for biomedical diagnosis and therapy to seismic waves that penetrate our Earth and the Sun.

Before initiating these next investigations, it may be worthwhile to review the sequence of steps we have taken thus far to calculate the behavior of harmonic oscillators and waves on strings:

- Determine the relationship that governs forces and displacements in our system of interest (e.g., Hooke's law, the equation of state, or the "constitutive equation").
- Examine the behavior of a differential element and determine the net forces (or net torques) acting on that element.
- Assume a sufficiently small displacement from equilibrium (i.e., "linear" behavior) and apply a Taylor series expansion to evaluate the net forces (or net torques) acting on the differential element.
- Use Newton's Second Law of Motion to calculate the acceleration of the differential element.
- Ignore any contributions to the equation relating accelerations to restoring forces that involve second-order quantities, such as products of two or more first-order deviations from equilibrium, like displacements or velocities, since such quadratic products are small compared to the linear contributions.
- Obtain a wave propagation speed.
- Apply the "machinery" to extract solutions under specific circumstances (e.g., harmonic analysis, superposition, boundary conditions, initial conditions, Fourier's theorem, mechanical impedance at steady state).

5.1 Longitudinal Waves in Thin Bars

There are three independent types of waves that can be excited in thin bars of solid materials at frequencies that are sufficiently low that the wavelengths of these waves are much greater than the cross-sectional dimensions (i.e., $\lambda \gg S^{1/2}$): (*i*) longitudinal waves of compression and expansion, (*ii*) torsional waves that produce twisting, and (*iii*) flexural waves that cause the bar to bend. Having examined the elastic moduli for isotropic solids, it should be clear that for longitudinal waves in thin bars, it is Young's modulus that quantifies the relationship between the longitudinal stresses and strains for a solid whose sides are unconstrained. Thus, the edges are free to "bulge" when compressed and to "neck down" when expanded. Assuming a bar of rectangular cross-section, as shown in Fig. 5.1, then $(\Delta w/w) = (\Delta h/h) = -\nu(\Delta\xi/dx)$.

The longitudinal strain, ε_{xx}, will be the ratio of the change in the length of the differential element from its original undisturbed length, dx, to its expanded length, $\xi(x + dx) - \xi(x)$, divided by its original length, dx. As before, a Taylor series expansion about $\xi(x)$ will be used to simplify the expression for the longitudinal strain.

$$\varepsilon_{xx} = \frac{\xi(x+dx) - \xi(x)}{dx} \cong \frac{\left(\frac{\partial \xi}{\partial x}\right) dx}{dx} = \left(\frac{\partial \xi}{\partial x}\right) \tag{5.1}$$

Using Eq. (4.1), longitudinal stress, σ_{xx}, will be related to that longitudinal strain, ε_{xx}, by Young's modulus, E, for the bar's material, allowing $S_x(x)$ to be the cross-sectional area of the bar as a function of position along the bar.

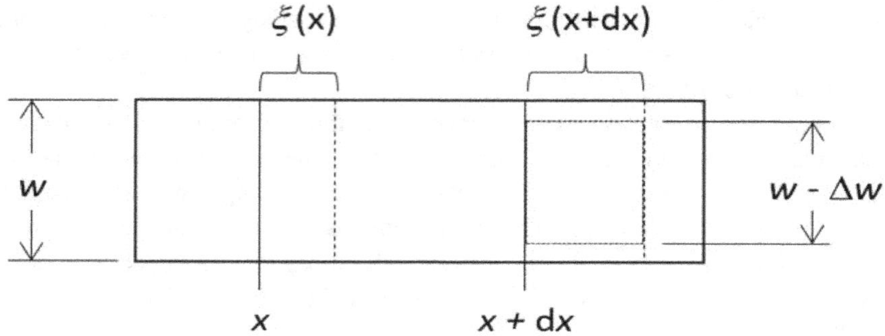

Fig. 5.1 Coordinate system for the displacement from equilibrium, $\xi(x)$, caused by a longitudinal wave passing through a thin bar. At the moment shown, the differential element of the bar has expanded, so its width (and height, not shown) will decrease in accordance with Poisson's ratio

$$\sigma_{xx} = \frac{F_x(x)}{S_x(x)} = -E(x)\frac{\partial \xi(x)}{\partial x} \Rightarrow F_x = -ES_x\frac{\partial \xi}{\partial x} \tag{5.2}$$

In the right-hand version, we have assumed that neither E nor S_x is a function of position, although the dependence of S_x on strain, ε_{xx}, if $\nu \neq 0$, will be given in Eq. (5.4).

To determine the net longitudinal force, dF_x, acting on the differential element of length, dx, we expand Eq. (5.2) in a Taylor series about $F_x(x)$ and assume that Young's modulus is a constant.

$$dF_x = F_x(x) - F_x(x+dx) \cong \frac{\partial F_x(x)}{\partial x}dx = E\frac{\partial}{\partial x}\left(S_x\frac{\partial \xi}{\partial x}\right)dx \tag{5.3}$$

If the cross-sectional area of the bar is also a constant, it too can be taken outside of the derivative. Having introduced Poisson's ratio in Eq. (4.2), we know that the cross-sectional area will not remain constant. Again, assume a bar of rectangular cross-section, so $S = wh$, and then use logarithmic differentiation to calculate the relative change in cross-section.

$$\frac{\delta S_x}{S_x} = \frac{\delta w}{w} + \frac{\delta h}{h} = -2\nu\varepsilon_{xx} = -2\nu\frac{d\xi}{dx} \tag{5.4}$$

Substitution of Eq. (5.4) into Eq. (5.3) provides an expression for the net longitudinal force, dF_x, acting on the differential element.

$$dF_x = E\frac{\partial}{\partial x}\left[S_x\left(1 - 2\nu\frac{\partial \xi}{\partial x}\right)\frac{\partial \xi}{\partial x}\right]dx \cong ES_x\frac{\partial^2 \xi}{\partial x^2}dx \tag{5.5}$$

Expanding the term in square brackets, we see that the correction to the cross-sectional area, due to Poisson's ratio, is second order in the strain, thus can be neglected in a first-order (linear) analysis. Since we are making a linear approximation, the strains are assumed to be small, $\partial\xi/\partial x = \varepsilon_{xx} \ll 1$, so we can ignore the Poisson term because it is proportional to ${\varepsilon_{xx}}^2 \ll \varepsilon_{xx}$.

The net longitudinal force, dF_x, is equal to the acceleration of the mass of the differential element, $\rho S_x\, dx$, where ρ is the mass density [kg/m^3] of the bar's material at x.

$$\rho S_x dx \frac{\partial^2 \xi}{\partial t^2} = ES_x \frac{\partial^2 \xi}{\partial x^2} dx \quad \Rightarrow \quad \frac{\partial^2 \xi}{\partial t^2} - \frac{E}{\rho} \frac{\partial^2 \xi}{\partial x^2} = 0 \tag{5.6}$$

We know that the solution to this wave equation will be the same superposition of a left- and right-going traveling wave as appeared in Eq. (3.18).

$$\xi(x, t) = Ae^{j(\omega t - kx)} + Be^{j(\omega t + kx)} \tag{5.7}$$

Equation (5.6) is a wave equation that reveals a propagation speed (phase speed) for longitudinal waves in thin bars, c_B.

$$c_B = \frac{\omega}{k} = f\lambda = \sqrt{\frac{E}{\rho}} \tag{5.8}$$

This speed of longitudinal wave propagation depends upon the ratio of the material's modulus to its mass density. It should also be noted that neither the cross-sectional area of the bar, S_x, nor the shape of that cross-section influences c_B, as long as the wavelength, λ, is much greater than the width, height, or diameter of the bar: $\lambda = c_B/f \ll \sqrt{S_x}$. Denser and softer metals, like lead, will have slower propagation speeds, while stiffer, less dense metals, like beryllium, will have faster speeds. For pure lead, $c_B \cong$ 1200 m/s and for beryllium, $c_B \cong$ 12,870 m/s.

Using our prior experience with standing waves on strings of finite length (see Sect. 3.3.1), the process of imposing ideal boundary conditions should be familiar. Experimentally, the easiest boundary condition to apply to a thin bar of length, L, to be excited in its longitudinal mode, is that the displacements of both ends are unconstrained or "free." If there are no forces on the end at $x = 0$ or the end at $x = L$, then Eq. (5.2) can be used to express these "free" boundary conditions.

$$F_x = -ES_x \left(\frac{\partial \xi}{\partial x}\right)_{x=0} = -ES_x \left(\frac{\partial \xi}{\partial x}\right)_{x=L} = 0 \tag{5.9}$$

The derivative of Eq. (5.7) with respect to x is simplified by our choice of exponential functions of space and time.

$$F_x(x, t) = -ES_x \left(\frac{\partial \xi}{\partial x}\right) = -jkES_x \left[Be^{j(\omega t + kx)} - Ae^{j(\omega t - kx)}\right] \tag{5.10}$$

The force at $x = 0$ can be evaluated by setting Eq. (5.10) to zero. This boundary condition can be satisfied by the superposition of traveling waves in Eq. (5.7) if $A = B$.

$$\begin{aligned} \xi(x, t) &= \Re e\left[B\left(e^{j(\omega t + kx)} + e^{j(\omega t - kx)}\right)\right] \\ &= \Re e\left[Be^{j\omega t}\left(e^{jkx} + e^{-jkx}\right)\right] = \Re e\left[\widehat{\mathbf{C}} e^{j\omega t} \cos kx\right] \end{aligned} \tag{5.11}$$

In Eq. (5.11), we have used the fact that $2\cos x = (e^{jx} + e^{-jx})$ and have absorbed $2B$ into the new complex amplitude constant (phasor), $\widehat{\mathbf{C}}$, that will not be determined until the initial conditions have been specified.

As was done before for strings, setting $F_x(L, t)$ to zero quantizes the values of the wavenumber that satisfy both boundary conditions simultaneously.

$$F_x(L,t) = -ES_x\left(\frac{\partial\xi}{\partial x}\right)_{x=L} = ES_x k_n C_n \sin(k_n L) e^{j\omega_n t} = 0 \tag{5.12}$$

This results in the same set of normal mode frequencies as the case for the fixed-fixed string in Eq. (3.20). These correspond to integer multiples of half-wavelengths between the ends of the bar.

$$\begin{aligned} \xi_n(x,t) &= \Re e\left[\widehat{\mathbf{C}}_\mathbf{n} \cos(k_n x) e^{j\omega_n t}\right] \\ \text{with } k_n &= \frac{n\pi}{L} \Rightarrow f_n = n\frac{c_B}{2L}; \quad n = 1,2,3,\ldots \end{aligned} \tag{5.13}$$

Once again, there is a harmonic series of normal mode frequencies. Of course, in this case, there are displacement and velocity anti-nodes at both ends of the bar instead of velocity nodes at the ends, as was the case for the fixed-fixed string. For our standing wave solutions in Eq. (5.13), the displacement is proportional to cos $(k_n x)$, but for the standing wave on a fixed-fixed string, $y_n(x) \propto \sin(k_n x)$, as shown in Eq. (3.21).

The fundamental free-free longitudinal mode of a very large aluminum bar, 3 m in length, with a mass of 2300 kg, was used in an attempt to detect gravitational waves.[1] [1] Using Young's modulus and the mass density of aluminum at room temperature, the speed of longitudinal waves $c_L = 5510$ m/s, making $f_1 \cong 860$ Hz. In operation, the bar was cooled below 4.2 K to permit the use of a superconducting quantum interferometer (SQUID) that measured the motion of a simple harmonic oscillator, tuned to f_1. The SQUID was attached to one end of the bar through a simple harmonic oscillator attached to one "free" end to amplify its motion by the Q of the harmonic oscillator. That raised the resonance frequency to 904 Hz, due to the increase in Young's modulus with decreasing temperature. The quality factor of the cryogenically cooled, freely suspended bar was enormous: $Q \cong 10^9$ [2].

Although harder to achieve in practice, an ideal fixed-fixed bar, with $\xi(0) = \xi(L) = 0$, would also have integer multiple half-wavelengths between the ends of the bar, resulting in a harmonic series of normal mode frequencies. The displacement functions, $\xi_n(x, t)$, have the same functional form as the fixed-fixed string given in Eq. (3.12).

Like the solutions for the fixed-free string in Eqs. (3.23) and (3.24), a clamped-free bar that executes longitudinal vibrations will exhibit the same series of only odd-harmonic standing wave modes, corresponding to odd integer multiples of a quarter wavelengths between the fixed end at $x = 0$ and the free end at $x = L$, where $(\partial\xi/\partial x)_L = 0$.

$$\begin{aligned} \frac{\partial\xi(L,t)}{\partial x} &= k_n\widehat{\mathbf{C}}_n \cos(k_n L) e^{j\omega_n t} = 0 \\ \cos k_n L &= 0 \Rightarrow k_n L = \left(\frac{2n-1}{2}\right)\pi \quad \Rightarrow \quad \omega_n = \left(\frac{2n-1}{2}\right)\frac{\pi c_B}{L} \\ \Rightarrow \lambda_n &= \frac{4L}{2n-1} \quad \text{or} \quad L = (2n-1)\frac{\lambda_n}{4} \quad \text{for} \quad n = 1,2,3,\ldots \end{aligned} \tag{5.14}$$

[1] Such gravitational wave detectors are called "Weber bars" in honor of the first attempt to use a longitudinally resonant bar to detect gravitational waves that was made by J. Weber. Weber claimed to detect gravitational waves in an article entitled "Gravitational-Wave-Detector Events," Phys. Rev. Lett. **20**, 1307–1308 (1968), although his claim could not be substantiated.

5.1.1 Longitudinal Waves in Bulk Solids

As the ratio of the wavelength of the sound to the transverse dimensions of the bar becomes smaller, the compressive stresses associated with the longitudinal wave cannot be (partially) relieved by the bulge of the bar induced through Poisson's ratio. When the transverse dimensions of the sample substantially exceed the wavelength of the longitudinal wave, the modulus that provides the restoring force in Eq. (5.2) is no longer Young's modulus, E, but becomes the modulus of unilateral compression (also known as the dilatational modulus), $D > E$ (see Fig. 4.4). The propagation speed for a longitudinal wave in "bulk material," c_L, is higher than the bar speed.

$$c_L = \frac{\omega}{k} = \sqrt{\frac{D}{\rho}} > c_B \tag{5.15}$$

Expressing D in terms of E and ν (see Table 4.1) allows the ratio of the bulk longitudinal wave speed, c_L, to the thin bar speed, c_B, to be expressed in terms of Poisson's ratio.

$$\frac{c_L}{c_B} = \frac{\sqrt{D/\rho}}{\sqrt{E/\rho}} = \sqrt{\frac{(1-\nu)}{(1+\nu)(1-2\nu)}} \tag{5.16}$$

Poisson's ratio for most common metals is about $\nu \cong 1/3$, so the sound speed ratio in metals is $c_L/c_B \cong (3/2)^{1/2} \cong 1.2$, about a 20% higher speed for longitudinal waves in bulk materials, where transverse stresses cannot be relieved, in part, by "bulge."

5.1.2 The Quartz Crystal Microbalance

The quartz crystal microbalance (QCM) is used to "weigh" very thin films, such as those deposited on microelectronic circuits by vacuum evaporation of metals (frequently gold). This is accomplished by monitoring the change in the resonance frequency of the vibrating piezoelectric sample. As will be shown in this section, the deposition of thin films that have thicknesses that are on the order of a single atomic layer can be detected by monitoring the resonance frequency shift.

In exact analogy with the mass-loaded string (see Sect. 3.6), a bar or plate with a mass "attached" to the free face will require that the bar or plate provide the force necessary to accelerate the mass load adsorbed or plated onto the free face that is moving with longitudinal displacement, $\xi(L)e^{j\omega t}$.

$$F_x(L,t) = M\left(\frac{\partial^2\xi}{\partial t^2}\right)_{x=L} = -DS_x\left(\frac{\partial\xi}{\partial x}\right)_{x=L} \tag{5.17}$$

In Eq. (5.17), the modulus of unilateral compression, D, is used to relate stress and strain because we will assume that the quartz plate (usually a disk) has a diameter, $2a$, that may be one hundred times the plate's thickness, $t = L$. Since the wavelength for the fundamental vibrational mode of a free-free plate is $\lambda = 2\,t$, and $t \ll a$, it is clear that the use of Young's modulus in this case would be incorrect. We will treat the "rear" of the disk as being suspended in a vacuum, hence free at $x = 0$.

$$\xi(x,t) = \Re e\left[\hat{\mathbf{C}}\cos(kx)e^{j\omega t}\right] \tag{5.18}$$

The face at $x = L$ will be mass-loaded by the deposition of a thin film with mass, M_{film}, so that Eq. (5.18) can be substituted into Eq. (5.17) to determine the quantization condition on k_n.

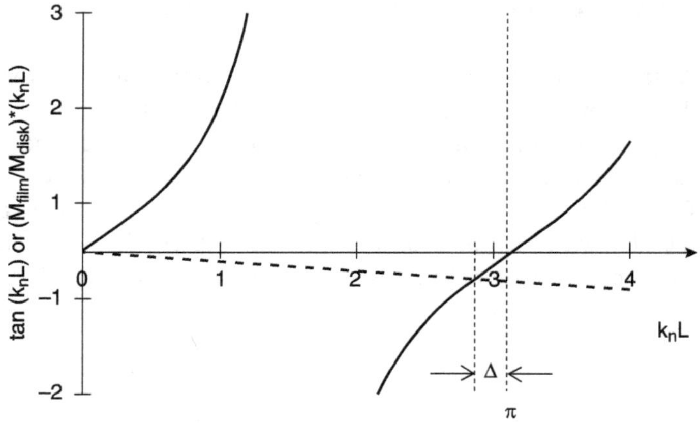

Fig. 5.2 Graph of the transcendental Eq. (5.20) over a small region near the origin with $M_{film}/M_{disk} = 0.10$ is shown as the *dashed line*. The tangent function, shown as the *solid line*, has a slope = 1.0 near $(k_n L) = 0$ or $(k_n L) = \pi$, so the first intersection of the *dashed line* and the tangent occurs at a value of $(k_n L)$ that is slightly less than π by a small amount, Δ

$$-\omega^2 M_{film} C_n \cos(k_n L) e^{j\omega_n t} = D S_x k_n C_n \sin(k_n L) e^{j\omega_n t} \tag{5.19}$$

Using the fact that $c_L{}^2 = \omega^2/k^2 = D/\rho$, Eq. (5.19) can be written in a form similar to the quantization condition for the wavenumbers, k_n, of a fixed, mass-loaded string that was given in Eq. (3.68).

$$\tan(k_n L) = \frac{-\omega^2 M_{film}}{D S_x k_n} = -\frac{k_n^2 M_{film} D}{k_n D S_x \rho} \frac{L}{L} = -\frac{M_{film}}{M_{disk}}(k_n L) \tag{5.20}$$

This transcendental equation is plotted in Fig. 5.2 with $M_{film}/M_{disk} \ll 1$.

For a very small mass loading, $M_{film}/M_{disk} \ll 1$, the first intersection of tan $(k_n L)$ and the straight line, $(-M_{film}/M_{disk})(k_n L)$, occurs at a value of $k_n L$ that is slightly less than π by an amount Δ, as shown in Fig. 5.2. We expect this decrease in the wavenumber (hence, resonant frequency), since the mass-loaded frequency should be lower than the normal mode frequency for its free-free value, $(k_1 L) = \pi$. Since the slope of tan x at the zero-crossings is unity, $\Delta \cong \pi(M_{film}/M_{disk}) \Rightarrow \delta k_1 \cong -(\pi/L)(M_{film}/M_{disk})$. Using this approximation (which, of course, becomes progressively better as the thickness of the film decreases), the relative shift in the wavenumber of the fundamental mode is $\delta k_1/k_1$, where $k_1 \cong (\pi/L)$. The relative frequency shift, $\delta f_1/f_1$, can be related to the film thickness, if the mass densities of the disk, ρ_{disk}, and the deposited film material, ρ_{film}, are known.

$$\frac{\delta f_1}{f_1} = \frac{\delta k_1}{k_1} = -\frac{\pi}{L}\frac{M_{film}}{M_{disk}}\left(\frac{L}{\pi}\right) = -\left(\frac{\rho_{film}}{\rho_{disk}}\right)\frac{t_{film}}{L} \tag{5.21}$$

This result is rather easy to interpret. This frequency shift is equivalent to the frequency change that would result if the disk was made longer by the thickness, δL, of the adsorbed film, if the adsorbed film had the same density as the density of the disk: $\delta L = (\rho_{film}/\rho_{disk})\, t_{film}$, so $-\delta f_1/f_1 = \delta L/L$.

A simple example will illustrate the sensitivity of this method. We start by calculating the thickness of a monolayer of gold atoms. The atomic weight of gold is $M_{Au} = 197$ a.m.u. = 197 gm/mole, and its mass density is $\rho_{Au} = 19.3$ gm/cm^3. The cube root of the number of gold atoms $(N_{Au})^{1/3}$ in a volume V of a golden cube that is 1 cm along each edge will yield the spacing between layers of gold atoms, where Avogadro's number is $N_A \equiv 6.02214076 \times 10^{23}$ particles/mole.

$$\sqrt[3]{N_{Au}} = \left(\frac{\rho_{Au} V}{M_{Au}} N_A\right)^{1/3} = 3.89\ x\ 10^7/\mathrm{cm} \tag{5.22}$$

With 3.89×10^7 layers per centimeter, the average thickness of a single gold monolayer is $t_{Au} = 2.57 \times 10^{-10}$ m = 2.57 Å.[2]

Assume that a disk of piezoelectric quartz is used that is $L = t = 0.50$ mm thick and is vibrating in its fundamental free-free mode in a vacuum. The frequency of that mode is $f_1 = c_L/2\ L$, where $c_L = 5750$ m/s for quartz, so $f_1 = 5.75$ MHz. Equation (5.21) can be used to calculate the decrease in the frequency of the fundamental mode, $-\delta f_1$, due to adsorption of a single monolayer of gold.

$$\delta f_1 = f_1\left(\frac{\rho_{Au}}{\rho_{quartz}}\right)\left(\frac{t_{Au}}{L_{quartz}}\right) = -5.75MHz\left(\frac{19.3}{2.65}\right)\left(\frac{2.57x10^{-10}}{5x10^{-4}}\right) = -21.5Hz \tag{5.23}$$

A frequency shift of 21.5 Hz is easy to detect. For this example, $-\delta f_1/f_1 = 3.74\ x\ 10^{-6}$, or 3.74 parts per million (ppm) for addition of a single atomic layer of gold.

One common example is the control of metal deposition on integrated circuit "chips." Gold is often "evaporated" in a vacuum system to deposit conduction paths between circuit elements and pin pads used to connect the integrated circuit to a printed circuit board. Monitoring the change in frequency of a quartz microbalance is an accurate way to determine when the desired thickness has been achieved.

5.1.3 Bodine's "Sonic Hammer"

At this point in the analysis, we would normally address the response of the driven system. Since the analogies between longitudinal vibrations in a thin bar and the transverse vibrations of a string are seen to be exact, we will focus on an interesting application instead. At the opposite extreme in size from the use of longitudinal resonance to weigh thin films with better than microgram precision, we will now focus on a method of driving pilings weighing several metric tons by exciting a standing wave resonance in "thin" bars. Shown below in Fig. 5.3 (left) is a construction site where a piling is to be driven into the ground by excitation of the longitudinal standing wave resonance of the pile.

An ordinary pile driver raises and releases a mass to drive (i.e., hammer) the pile into the ground, effectively launching a longitudinal pulse that travels down the pile until it reaches the end that is in contact with the earth and where the pulse reflection transmits a force that promotes penetration of the pile. Such "drop hammers" used for pilings of the size shown in Fig. 5.3 (right) will require masses in the range of several metric tons, typically 1500–4000 kg, dropped from heights of 2 to 3 m above the piling. Each impact releases between 30 kJ and 120 kJ per blow. Assuming about one blow every 2 s, using an average of 60 kJ/blow, deposits an average power of 30 kW ≅ 40 hp to the pile.[3]

A "sonic hammer," exciting the pile at its longitudinal resonance frequency, stores vibrational (or strain) energy in the pile and only has to replace the energy delivered to the earth to maintain steady-state operation at constant vibratory amplitude. The maximum stored potential energy $(PE)_{max}$ can be calculated in the usual way for a bar of cross-sectional area, S_x; length, L; and volume, $V = S_xL$, using Eq. (1.22) or Eq. (2.14). The maximum vibrational strain amplitude is ε_{xx} and $\xi = \varepsilon_{xx}dx$ is defined in Eq. (5.1).

[2] This is a slight underestimate because the crystalline structure of gold is face-center cubic. The accepted atomic diameter for gold is 2.88 Å.

[3] One horsepower [hp] is defined as 745.7 W.

Fig. 5.3 (*Left*) Job site during preparation to drive a pile using resonant excitation provided by two engines from a Sherman tank. The engines provided counterrotation of eccentric masses (see Fig. 5.4) that drove the pile at its longitudinal standing wave resonance frequency. (*Right*) A 140 kW (200 HP) resonant driver, mounted on a 60 cm diameter steel pile, that uses a high-speed hydraulic valve to drive the pile at resonance. (Photographs courtesy of Matthew Janes)

$$(PE)_{\max} = -\int_0^L F_x d\xi = S_x \int_0^L E\varepsilon_{xx}^2 dx = S_x E \int_0^L \varepsilon_{xx}^2 \cos^2\left(\frac{n\pi x}{L}\right) dx = \frac{E\varepsilon_{xx}^2}{2} V \tag{5.24}$$

This analysis will be applied to a steel pipe with wall thickness of 22 mm and outside diameter of 30.0 cm, that is, 17.8 m long, with a mass of 2668 kg [3]. With resonant excitation, the pipe penetrated over 9 m in 14 min. A drop hammer could sink the first few meters of pipe with 60 blows/m, but at greater depths, 150–200 blows/m were required. For 9 m penetration, over 1000 blows were required taking over 40 min.

More importantly, for construction in dense urban locales, the ground vibrations due to the use of a drop hammer exceed the limit that can cause damage to existing structures nearby. In a similar experiment, ground vibrations from a drop hammer measured 10 m from the pile showed ground accelerations of 1.75 m/s^2, while the resonant drive produced accelerations of 0.25–0.30 m/s^2 at the same measurement distance [3].

At the job site, construction workers can measure the amplitude of vibration near a velocity antinode by taking a chalk stick and drawing a line along the pile's circumference. Since the surface of the pile is vibrating up and down, this draws a sinusoid and provides a visual representation of the peak-to-peak vibrational displacement, $2\xi_o$. For this type of installation, $2\xi_o \cong 6$ cm, and the resonance frequency is roughly 100 Hz. The maximum fully reversing strain, $\varepsilon_{xx} \cong 0.03/17.8 = 1.7 \times 10^{-3}$, is well below the elastic limit of steel. The total volume of steel in the pipe is $V = 0.37$ m^3. Using $E_{steel} = 195$ GPa, the total stored potential energy given by Eq. (5.24) is $(PE)_{\max} \cong 100$ kJ, fairly close to the energy in a single hammer drop, except this energy is available continuously.

The rate at which this energy is delivered to the ground depends upon the quality factor of the resonance, usually dominated by the losses at the entry cavity. In this demonstration, the average power delivered to the pipe $\langle\Pi\rangle_t \cong 50$ kW.[4] Using Eq. (2.70), the average quality factor for the pipe's longitudinal vibration can be calculated.

$$Q = \frac{2\pi f E_{stored}}{\langle\Pi\rangle_t} \cong 1300 \tag{5.25}$$

In addition to the reduction in surrounding vibrations for the pile driving application, this resonant vibrational strategy has other applications. In particular, since the resonant method does far less damage to the surrounding material, it is a good way to take geological core samples. It is particularly well-suited to ice core sampling, since very little damage is done to the ice and it takes less time to cut the core (which is important in Arctic weather conditions). This technique has also been used for tunneling in the construction of the Bay Area Rapid Transit (BART) subway system around San Francisco, Oakland, and Berkeley in California. A similar approach has also been developed for rock crushing [5], although that device used resonant excitation of the chute to shake the crushing balls against the material, not longitudinal waves in bars.

There have been problems with these longitudinally resonant pile drivers. The early drivers were prone to failure due to the enormous cyclic loading of the bearings and gears shown in Fig. 5.4. In some cases, crane operators were responsible for the failure because they could not believe that the piles would sink so quickly and did not lower the drive mechanism quickly enough for the ground to absorb the vibrational energy. The reduced cutting load caused the Q of the pile to become so large that the increased vibrational amplitude damaged the drivers. Hydraulic actuators have largely removed that difficulty. Since the resonance frequency changes continuously as the pile or pipe enters the ground, a control system can be employed to keep the system "tuned" to the longitudinal bar resonance as its length changes and as it encounters differing soils.

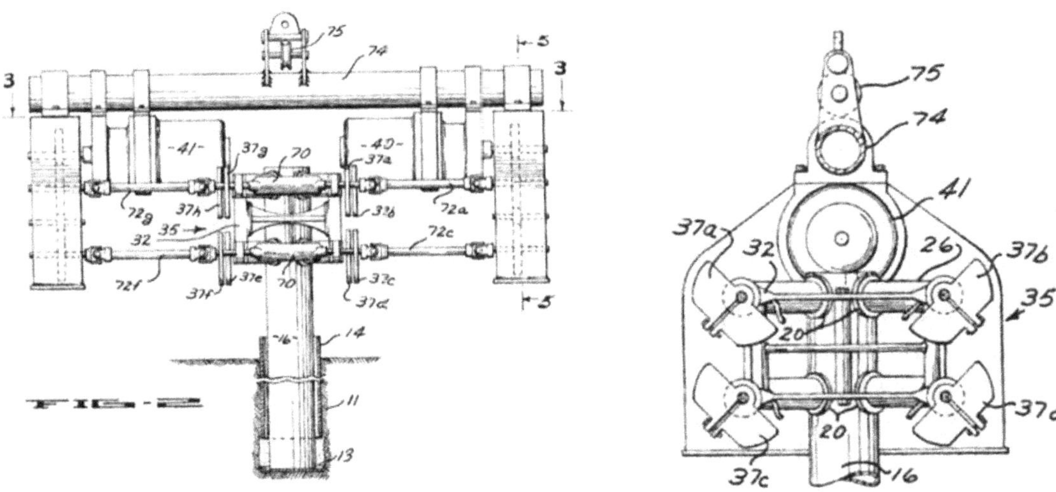

Fig. 5.4 (*Left*) Mechanical driving system attached to the top of the piling for resonant excitation of longitudinal standing waves in a construction. Motors (40 and 41) drive shafts (72), through gearboxes that are connected to the eccentrically mounted masses (37) to provide the longitudinal excitation to the top of the piling (16). (*Right*) View showing the rotating eccentric masses (37) [4]

---

[4] Using hydraulic fluid at 17 MPa, with an average flow rate of $U = 4.5$ l/s $= 4.5 \times 10^{-3}$ m^3/s, the hydraulic power $\langle\Pi\rangle_t = (\Delta p)\,U \cong 76$ kW.

5.2 Torsional Waves in Thin Bars

The same procedure that has been used for the longitudinal waves can be applied used to calculate the speed of torsional wave propagation in thin bars or tubes of circular cross-section and to calculate the speed of shear waves in bulk solids. Since we have already calculated the torque required to twist a circular rod or tube of length, L, by an angle, ϕ, in Eq. (4.50), that result can be applied to a differential length, dx, of a solid rod of radius, a. The difference in the twist at x and at $x + dx$ is $d\phi$, so ϕ/L becomes $\partial\phi/\partial x$.

$$N(x) = G\frac{\pi a^4}{2}\frac{\partial\phi}{\partial x} \tag{5.26}$$

The difference in the torque between the ends of the differential element is then approximated by a Taylor series.

$$dN(x) = N(x+dx) - N(x) = \left(\frac{\partial N}{\partial x}\right)dx = G\frac{\pi a^4}{2}\left(\frac{\partial^2\phi}{\partial x^2}\right)dx \tag{5.27}$$

For a circular rod with constant mass density, ρ, the mass of the differential element is just $dm = \pi\rho a^2 \mathrm{d}x$. The *polar moment of inertia*, I_{disk}, of a circular disk is $ma^2/2$, so for the differential element, $I_{disk} = (\pi/2)\rho a^4 dx$.

Newton's Second Law guarantees that the net torque is equal to the moment of inertia times the angular acceleration.

$$dN(x) = G\frac{\pi a^4}{2}\left(\frac{\partial^2\phi}{\partial x^2}\right)dx = dI_{disk}\left(\frac{\partial^2\phi}{\partial t^2}\right)dx = \frac{\pi}{2}\rho a^4\left(\frac{\partial^2\phi}{\partial t^2}\right)dx \tag{5.28}$$

This result can be rearranged to yield a torsional wave equation for a rod of circular cross-section that introduces the shear wave speed, c_S.

$$\frac{\partial^2\phi}{\partial x^2} - \frac{\rho}{G}\frac{\partial^2\phi}{\partial t^2} = 0 \Rightarrow c_S = \sqrt{\frac{G}{\rho}} \tag{5.29}$$

For a circular bar of radius, a, the propagation speed, c_S, for torsional waves does not depend upon the radius of the bar. In fact, since the volume is unchanged by shear deformation, it does not even require that the radius, a, be much less than λ, as was the case for longitudinal waves. Since there is no volume change, there is no "bulge" regardless of the ratio of the sample size to the wavelength. Bulk shear waves also propagate at c_S.

The torsional wave speed in a thin bar does depend on the shape of the bar's cross-section. We can generalize the torque equation that was written in Eq. (5.26) to shapes that are not circularly symmetric by introducing a *torsional rigidity*, J, which will depend upon the bar's cross-sectional shape [6].

$$N(x) = J\frac{\partial\phi}{\partial x} \tag{5.30}$$

Values for J/G are provided in the second column of Table 5.1, so for our circular bar example, $J_{disk} = \pi G a^4/2$, as expressed in Eq. (5.26). Similarly, the polar moment of inertia, I, can be evaluated for the corresponding cross-sectional shape as provided in the third column of Table 5.1. Both J and I can be substituted into the wave equation to produce a torsional wave speed, c_{SS}, that is dependent upon the cross-sectional shape of a thin bar with transverse dimensions that are much less than the length [7].

Table 5.1 Both the torsional rigidity, J, and the moment of inertia, I, of a rod depend upon the shape of its cross-section

Cross-section	$NL/G \equiv J/G$	I	Torsional speed
Circular Radius $= a$	$\frac{\pi a^4}{2}$	$\frac{\pi a^4}{2}$	c_S
Ring Outer $= a_{out}$ Inner $= a_{in}$	$\frac{\pi\left(a_{out}^4-a_{in}^4\right)}{2}$	$\frac{\pi\left(a_{out}^4-a_{in}^4\right)}{2}$	c_S
Elliptical Semi-major $= a$ Semi-minor $= b$	$\frac{\pi a^3 b^3}{a^2+b^2}$	$\frac{\pi}{4}ab\left(a^2+b^2\right)$	$\frac{2ab}{a^2+b^2}c_S$
Square Edge $= w$	$9w^4/64$	$w^4/6$	$\sqrt{\frac{27}{32}}c_S \cong 0.92c_S$
Rectangular Thickness $= t$ Width $= w$	$\frac{wt^3}{16}\left[\frac{16}{3} - 3.36\frac{t}{w}\left(1 - \frac{t^4}{12w^4}\right)\right]$	$\frac{tw\left(t^2+w^2\right)}{12}$	$\frac{2t}{w}\ c_S$ if $t << w$
Equilateral triangle Edge $= a$	$\frac{a^4\sqrt{3}}{80}$	$\frac{\sqrt{3}a^4}{80}$	$\sqrt[4]{\frac{3}{5}}c_S \cong 0.88c_S$

$$\frac{\partial^2\phi}{\partial x^2} - \frac{J}{\rho I}\frac{\partial^2\phi}{\partial t^2} = 0 \quad \Rightarrow \quad c_{SS} = \sqrt{\frac{J}{\rho I}} \tag{5.31}$$

Table 5.1 summarizes the speed of torsional waves on bars of various cross-sectional shapes by taking the ratio of the torsional stiffness-length product that relates static torque, N, to the shear modulus, G, and dividing by the polar moment of inertia, I. These factors for a rod and a hollow tube were calculated in Sect. 4.3.5 and given in Eq. (4.45). The other stiffness-length products were taken from Table 20 in Roark [8].

The torsional stiffness-length product (or ratio of torsional rigidity to shear modulus) is provided by the ratio of the torque, N, divided by the shear modulus, G. The ratio of torsional stiffness-length product to the moment of inertia, I, provides the constant multiplying the shear wave speed, c_S, for each bar cross-sectional shape.

5.3 Flexural Waves in Thin Bars

The last of the three independent vibrational modes of thin bars mentioned in the introduction is the flexural or bending modes. We will again follow the same procedure to determine their behavior, but that path will take us to a description that is fundamentally different from the results we have derived for transverse waves on strings or the longitudinal and torsional waves in thin bars. As was done for the case of transverse waves, we seek an expression for the vertical force, $F_y(x)$, on a differential element that can be related to its inertia via Newton's Second Law. We start from the expression for the bending moment, $\mathfrak{M}$, in the limit of small displacements, which is related to the curvature in Eq. (4.29).

$$\mathfrak{M} = -ES\kappa^2\frac{\partial^2 y}{\partial x^2} \tag{5.32}$$

The square of the radius of gyration, κ^2, is defined in Eq. (4.29).

Using the fact that the moment is the torque produced by an applied force times its perpendicular distance from the reference point, we can calculate the vertical force that creates the moment in Eq. (5.32). It can be calculated by placing the reference location at $x = 0$. In that way, the vertical force applied at $x = 0$ has no moment. The torque due to the shear force at $x + dx$ will be the product of $F_y(x + dx)$ and its "lever arm," dx.

$$F_v(x+dx)dx = \mathfrak{M}(x+dx) - \mathfrak{M}(x) \tag{5.33}$$

By our choice of $x = 0$ as the reference location, $\mathfrak{M}(x) = 0$. A Taylor series can then be exploited to express the moment of the vertical force, $F_y(x + dx)dx$, that will counteract the moment.

$$F_v(x+dx)dx = \frac{\partial \mathfrak{M}}{\partial x}dx = -ES_x\kappa^2\frac{\partial^3 y}{\partial x^3}dx \tag{5.34}$$

This expression for the vertical component of the force plays the same role as Eq. (3.2) that relates the slope of the string to the vertical force on a string element, or Eq. (5.2) that relates the stress in the bar to the longitudinal force on the bar element. Of course, Eqs. (3.2) and (5.2) related these forces to a first derivative with respect to position of the displacement from equilibrium, while Eq. (5.34) relates the vertical force to the third derivative of the displacement.

We determine the net vertical force acting on our differential element in the usual way.

$$dF_v = F_v(x+dx) - F_v(x) = \frac{\partial F_v}{\partial x}dx = -ES_x\kappa^2\frac{\partial^4 y}{\partial x^4}dx \tag{5.35}$$

This result is also known as the Euler-Bernoulli beam equation after Leonhard Euler and Daniel Bernoulli who first derived this result in the 1750s [9]. The combination of Young's modulus for the beam multiplied by the area and the radius of gyration is known as the *flexural rigidity*, $ES_x\kappa^2$, since it is a measure of the moment of the force necessary to deform the bar.

The mass of the differential element is $\rho S_x dx$, so Newton's Second Law produces the differential equation describing the bar's dynamic response to flexure.

$$\rho S dx\frac{\partial^2 y}{\partial t^2} = -ES_x\kappa^2\frac{\partial^4 y}{\partial x^4}dx \Rightarrow \frac{\partial^2 y}{\partial t^2} + \kappa^2 c_B^2\frac{\partial^4 y}{\partial x^4} = 0 \tag{5.36}$$

The right-hand expression makes use of the propagation speed of longitudinal waves in thin bars, $c_B^{\,2} = E/\rho$, after cancellation of the common factors dx and S_x.[5]

5.3.1 Dispersion

Equation (5.36) is a linear, fourth-order, homogeneous, partial differential equation that is clearly **not** the wave equation. Its solutions will not be just any functions of the argument $x \pm ct$, where c is a constant (independent of frequency or wavelength) representing the propagation speed for all waves. Fortunately, Eq. (5.36) contains only even-order partial derivatives, so substitution of a complex exponential will provide a relation between frequency and wavenumber that is a real number.[6] This relationship can be demonstrated by substitution of our usual rightward-going traveling wave solution, $y(x,t) = \Re e\left[\widehat{\mathbf{C}} e^{\,j(\omega t - kx)}\right]$, into Eq. (5.36) to solve the time-independent Helmholtz equation.

[5] This equation is only approximately correct since the moment is also opposed by the rotary inertia of the bar. The approximation is very good when $\lambda \gg a$, which is our current focus. The complete equation for the dynamics, which includes rotary inertia and shear, is derived in several references, for example, D. Ross, *The Mechanics of Underwater Noise* (Pergamon, 1976); ISBN 0-08-021182-8.

[6] As will be shown in Chap. 9 on dissipative hydrodynamics, this is not the case for equations that contain both odd- and even-order derivatives. The Navier-Stokes equation, describing viscous dissipation; the Fourier heat diffusion equation, describing thermal conduction losses; and the Schrodinger equation of quantum mechanics all contain both odd- and even-order derivatives that require complex numbers to relate frequency and wavenumber.

$$-\omega^2 y + \kappa^2 c_B^2 k^4 y = 0 \;\Rightarrow\; k = \pm\sqrt{\frac{\omega}{\kappa c_B}} \tag{5.37}$$

The most useful equation in acoustics is the relationship between frequency and wavelength: $f\lambda = \omega/k = c$. Up to this point, c has been treated as a constant in all systems, irrespective of frequency or wavelength. This is not the case for solutions to Eq. (5.36). We can solve Eq. (5.37) for the ratio of frequency to wavenumber.

$$\frac{\omega^2}{k^2} = \kappa^2 c_B^2 k^2 \quad \Rightarrow \quad \frac{\omega}{k} = \pm(\kappa c_B)k = \pm 2\pi \frac{\kappa c_B}{\lambda} \tag{5.38}$$

The propagation speed for harmonic flexural waves of frequency, ω, is inversely proportional to the wavelength of the disturbance.

In all our previous solutions to the dynamical response for systems described by the wave equation, ω/k is equal to a constant, c, that is identified as the speed of propagation for all wavelike disturbances, whether they are pulses or harmonic waves, of either the traveling or standing variety, independent of their frequency or wavelength. Equation (5.38) demonstrates that the ratio of frequency to wavenumber is no longer a constant that is independent of frequency for flexural waves on bars.

$$c_{ph} \equiv \frac{\omega}{k} = \pm\sqrt{\kappa\omega c_B} \tag{5.39}$$

Because there is no longer a single speed of propagation for all disturbances, we have defined the *phase speed*, $c_{ph} \equiv \omega/k$. This is the speed of propagation of wave fronts of any disturbance that is a *pure tone*, having a single frequency, ω, and a single wavelength, λ, thus containing only one Fourier component. A wavelike disturbance that contains more than one Fourier component will not retain its shape as it propagates under the influence of Eq. (5.36), which is still a linear equation. Since the different frequency components of such a disturbance will travel with different speeds, Eq. (5.39) implies that the higher-frequency components will get ahead of the lower-frequency components, so the waveform produced by their superposition will change over time and distance.

The dependence of propagation speed on frequency or wavelength is known as *dispersion*. Dispersion of light waves was demonstrated experimentally by Newton [10] who used a glass prism to separate white light into its different constituent wavelengths. The angle of refraction of light passing through a transparent medium depends upon the ratio of the propagation speeds, in Newton's case, between air and glass. For glass, that propagation speed depends upon the wavelength of the light. In effect, the prism provided a Fourier analysis (64 years before the birth of Fourier) of the different frequency components (different colors) that superimpose to create white light.[7] The genius of Isaac Newton was displayed by his use of a second prism to recombine the different colors to produce white light, thus demonstrating that the colors were not an experimental artifact.

Capillary waves on the surface of water have the same dispersion relation as flexural waves on thin bars [11], $c_{ph} = (2\pi\sigma/\rho\lambda)^{1/2}$, where σ is the surface tension of water.[8] Like flexural waves on thin bars,

---

[7] The laws of electromagnetism, known as Maxwell's equations, which govern the propagation of light, are also linear differential equations.

[8] The full dispersion relation for surface waves on fluids produces the dispersion relation (below) that also includes the force of gravity g, which dominates the restoring force at long wavelengths, as well as surface tension (capillarity). When the wavelengths are long compared to the depth, h, so $kh \ll 1$, $\tanh(kh)$ is proportional to kh and the speed is again dispersionless. On water, capillarity (surface tension, σ) dominates gravity for wavelengths less than about one-half centimeter.

$$c_{ph}^2 = \frac{\omega^2}{k^2} = \left(\frac{g}{k} + \frac{\sigma}{\rho}k\right)\tanh(kh)$$

Fig. 5.5 This photograph of a water drop's splash clearly illustrates the dispersion of waves that have a phase velocity, c_{ph}, that is inversely related to the wavelength. The components of the disturbance with the shortest wavelengths have moved ahead of the longer-wavelength components. The dispersion for surface waves on water with wavelengths less than about 5 mm is dominated by surface tension.[9]

shorter wavelengths (i.e., higher frequencies) propagate faster than longer wavelengths. This is visible in Fig. 5.5, where a drop of water launches cylindrically diverging surface waves with the short wavelengths moving away from the source faster than the longer-wavelength components produced by the splash.[9]

5.3.2 Flexural Wave Functions

Capitalizing on our success with the substitution of a complex traveling waveform, $y(x,t) = \Re e\left[\widehat{C}e^{\,j(\omega t - kx)}\right]$, into Eq. (5.36) to produce the dispersion relations of Eqs. (5.37), (5.38), and (5.39), we will now seek a complete solution to Eq. (5.36) by the method known as *separation of variables*. By setting $y\,(x,\,t) = Y\,(x)\,T(t)$, the spatial dependence of the solution depends only upon $Y\,(x)$, and the temporal behavior depends only upon $T\,(t)$. Since Eq. (5.36) is a linear equation, we can still construct more complicated waveforms by the superposition of harmonic waves, so we will let $T\,(t) = e^{j\omega\, t}$. Substitution of $T\,(t)$ into Eq. (5.36) converts it from a partial differential equation to an ordinary (forth-order) differential equation for $Y\,(x)$.

$$\frac{d^4Y}{dx^4} = \frac{\omega^2}{\kappa^2 c_B^2}Y = \frac{\omega^4}{c_{ph}^4}Y \tag{5.40}$$

Letting $Y\,(x) = Y\,e^{kx}$, a simple quartic expression is generated for the wavenumber k: $k^4 = (\omega/c_{ph})^4$. By taking the fourth root (see Sect. 1.5.2), we generate four solutions for $k = \pm\omega/c_{ph}$ and $\pm\, j\omega/c_{ph}$. Those four solutions are superimposed in Eq. (5.41).

$$\mathbf{y}(x,t) = \left[C_1 e^{(\omega x/c_{ph})} + C_2 e^{-(\omega x/c_{ph})} + C_3 e^{\,j(\omega x/c_{ph})} + C_4 e^{-j(\omega x/c_{ph})}\right]e^{j\omega t} \tag{5.41}$$

[9] For ripple-tank demonstrations, the depth of the water is about 5 mm. At that depth, the restoring forces of gravity and surface tension balance to produce a frequency-independent surface wave velocity, $c \cong 23$ cm/s. [See M. J. Lighthill, *Waves in Fluids* (Cambridge, 1978); ISBN 0 521 21689 3. Sec. 1.8 (Ripple-tank simulations)]

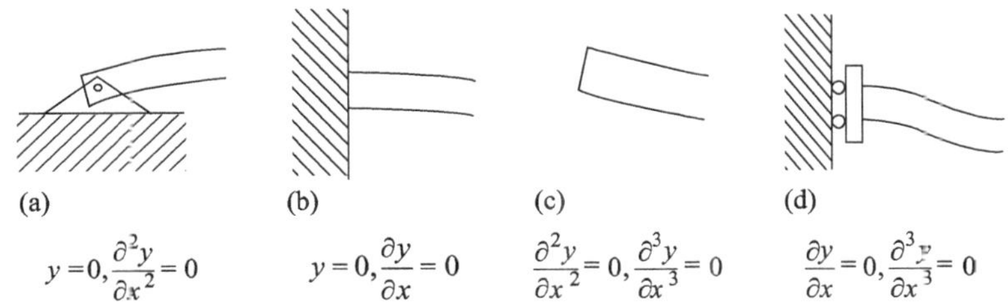

Fig. 5.6 Schematic representation of four possible boundary conditions for flexure of a bar: (**a**) hinged, (**b**), clamped, (**c**) free, and (**d**) guided or sliding. From Han et al. [12]

This result confirms our first solution, a right-going traveling wave, that was used to produce Eq. (5.37), corresponding to $C_1 = C_2 = C_3 = 0$ and $C_4 \neq 0$. Linearity allows us to combine these exponential solutions to produce a solution that can be expressed as four independent trigonometric functions with real constants.

$$A = (C_1 + C_2)/2, \;\; B = (C_1 - C_2)/2, \;\; C = (C_3 + C_4)/2, \text{and } D = (C_3 - C_4)/2j.$$
$$y(x,t) = \left[A\cosh\left(\frac{\omega x}{c_{ph}}\right) + B\sinh\left(\frac{\omega x}{c_{ph}}\right) + C\cos\left(\frac{\omega x}{c_{ph}}\right) + D\sin\left(\frac{\omega x}{c_{ph}}\right)\right]\cos(\omega t + \phi) \tag{5.42}$$

Since these are solutions to a fourth-order differential equation, four boundary conditions must be specified to determine the four amplitude coefficients. The four most common "perfect" boundary conditions are clamped, free, hinged, and guided. They are summarized in Fig. 5.6.

If the bar is clamped, then the end cannot move in the vertical direction, so $y = 0$. The clamped end of the bar must also approach that boundary with zero slope, so $(\partial y/\partial x) = 0$. If the bar is free, then the boundary can exert no moments (torques), so from Eq. (5.32), $(\partial^2 y/\partial x^2) = 0$. The free boundary cannot provide any vertical forces, so from Eq. (5.34), $(\partial^3 y/\partial x^3) = 0$.

A hinged boundary (or knife-edged clamp) will also make $y = 0$. Due to the hinge, the boundary can exert no torques, so $(\partial^2 y/\partial x^2) = 0$. Of course, the hinge applies vertical forces to keep $y = 0$, so $(\partial^3 y/\partial x^3) \neq 0$. A guided or sliding boundary allows the beam to move in the vertical direction (i.e., $y \neq 0$) but constrains both the slope and the vertical force to be zero.

5.3.3 Flexural Standing Wave Frequencies

In Sect. 4.3.2, we calculated the deflection curve for a horizontal cantilevered beam that was clamped at $x = 0$ ($y = 0$ and $\partial y/\partial x = 0$) and was loaded by a vertical force (e.g., a weight) at $x = L$ (see Galileo's illustration in Fig. 2.27). We will start this section by calculating the normal modes of oscillation for a cantilevered beam that is clamped at $x = 0$ and is free at $x = L$. Using the trigonometric form of the general solution in Eq. (5.42) and applying the boundary condition at $y(0) = 0$, sinh (x) and sin (x) are automatically zero, but cos (x) and cosh (x) are both equal to 1, so $A = -C$. For the slope to vanish at $x = 0$, $B = -D$, since $d(\sinh(x))/dx = \cosh(x)$ and $d(\sin(x))/dx = \cos(x)$. Suppressing the time dependence temporarily (no pun intended), the wave function for the cantilever beam can be written with only two arbitrary constants when it is clamped at $x = 0$.

$$y(x) = A\left[\cosh\left(\frac{\omega x}{c_{ph}}\right) - \cos\left(\frac{\omega x}{c_{ph}}\right)\right] + B\left[\sinh\left(\frac{\omega x}{c_{ph}}\right) - \sin\left(\frac{\omega x}{c_{ph}}\right)\right] \tag{5.43}$$

Evaluation of the second and third derivatives of Eq. (5.43) at $x = L$ will provide the quantization of the allowable modal (normal mode) frequencies, ω_n, and wavenumbers, k_n.

$$A\left[\cosh\left(\frac{\omega_n L}{c_{ph}}\right) + \cos\left(\frac{\omega_n L}{c_{ph}}\right)\right] + B\left[\sinh\left(\frac{\omega_n L}{c_{ph}}\right) + \sin\left(\frac{\omega_n L}{c_{ph}}\right)\right] = 0$$
$$A\left[\sinh\left(\frac{\omega_n L}{c_{ph}}\right) - \sin\left(\frac{\omega_n L}{c_{ph}}\right)\right] + B\left[\cosh\left(\frac{\omega_n L}{c_{ph}}\right) + \cos\left(\frac{\omega_n L}{c_{ph}}\right)\right] = 0 \quad (5.44)$$

Both of the equations in Eq. (5.44) can be used to solve for B in terms of A.

$$B = A\frac{\sin\left(\frac{\omega_n L}{c_{ph}}\right) - \sinh\left(\frac{\omega_n L}{c_{ph}}\right)}{\cos\left(\frac{\omega_n L}{c_{ph}}\right) + \cosh\left(\frac{\omega_n L}{c_{ph}}\right)} = -A\frac{\cos\left(\frac{\omega_n L}{c_{ph}}\right) + \cosh\left(\frac{\omega_n L}{c_{ph}}\right)}{\sin\left(\frac{\omega_n L}{c_{ph}}\right) + \sinh\left(\frac{\omega_n L}{c_{ph}}\right)} \quad (5.45)$$

After cancellation of the common factor of A, cross-multiplication of the left and right fractions in Eq. (5.45) produces a single equation that quantizes the normal mode frequencies, ω_n.

$$\sinh^2\left(\frac{\omega_n L}{c_{ph}}\right) - \sin^2\left(\frac{\omega_n L}{c_{ph}}\right) = \left[\cosh\left(\frac{\omega_n L}{c_{ph}}\right) + \cos\left(\frac{\omega_n L}{c_{ph}}\right)\right]^2 \quad (5.46)$$

Application of the trigonometric identities $\cos^2(\theta) + \sin^2(\theta) = 1$ and $\cosh^2(\theta) - \sinh^2(\theta) = 1$ produces a workable transcendental equation that can be solved for the normal mode frequencies.

$$\cosh\left(\frac{\omega_n L}{c_{ph}}\right)\cos\left(\frac{\omega_n L}{c_{ph}}\right) = -1 \quad (5.47)$$

The solution of Eq. (5.47) is inconvenient because the hyperbolic cosine will diverge exponentially. The use of the following half-angle formulas can afford further simplification.

$$\tan^2\left(\frac{\theta}{2}\right) = \frac{1 - \cos\theta}{1 + \cos\theta} \quad \text{and} \quad \tanh^2\left(\frac{\theta}{2}\right) = \frac{\cosh\theta - 1}{\cosh\theta + 1} \quad (5.48)$$

Since the limits of the hyperbolic tangent for large arguments are ± 1, the following forms are both easier to graph and avoid the exponential divergence of the hyperbolic cosine.

$$\coth^2\left(\frac{\omega_n L}{2c_{ph}}\right) = \tan^2\left(\frac{\omega_n L}{2c_{ph}}\right) \quad \text{or} \quad \cot\left(\frac{\omega_n L}{2c_{ph}}\right) = \pm\tanh\left(\frac{\omega_n L}{2c_{ph}}\right) \quad (5.49)$$

Figure 5.7 is a plot of the two functions on the right-hand side of Eq. (5.49) vs. their argument $x = \omega L/2c_{ph}$. For arguments of $x > 4$, the value of $\pm\tanh(x) \cong \pm 1$. $\text{Cot}^{-1}(x) = \pm 1$ for $x = (2n - 1)(\pi/4)$, so for $n \geq 3$, $\omega L/2c_{ph} = (2n - 1)(\pi/4)$. An equation solver can be used to calculate the lowest two solutions to Eq. (5.49) that are $x_1 \cong 1.1937(\pi/4)$ and $x_2 \cong 2.9884(\pi/4)$.

Combining the solutions to Eq. (5.49) with the square of the phase speed in Eq. (5.39), the flexural normal mode frequencies, f_n, of the clamped-free bar can be written as a function of the geometry of the bar (L and κ), Young's modulus, and the mass density of the beam's material, $c_B = (E/\rho)^{1/2}$.

$$f_n = \frac{\omega_n}{2\pi} = \frac{\pi}{8}\frac{c_B\kappa}{L^2}\left(1.1937^2,\ 2.9884^2,\ 5.0005^2,\ 7^2,\ 9^2,\ \ldots\right) \quad (5.50)$$

It is worthwhile to point out the significant differences in this expression from those for the normal mode frequencies of the vibrating string and other systems that obey wave equations of the form of

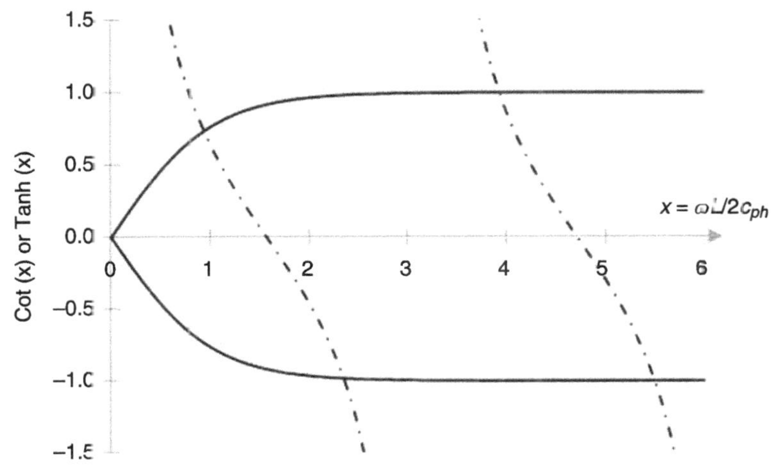

Fig. 5.7 The two terms on the right-hand side of Eq. (5.49) are plotted as a function of $x = \omega L/2c_{ph}$. The hyperbolic tangents are shown as *solid lines* and the *dash-dot line* is the cotangent. Their intersections define the normal mode frequencies, ω_n, for a clamped-free (cantilevered) beam

Eq. (5.6) or Eq. (5.29). Even with "ideal" boundary conditions, the normal mode frequencies of the flexing bar are not harmonically related. Also, for flexural modes, frequency is inversely proportional to the length of the bar squared. The frequencies also depend explicitly on the cross-sectional shape of the bar through the radius of gyration, κ, as defined in Eq. (4.26). Measurement of the modal frequencies and damping has been used to determine Young's modulus of small samples [13]. However, at least two independent modes have to be measured to determine both independent moduli for an isotropic solid [14].

5.3.4 Flexural Standing Wave Mode Shapes

The displacement amplitudes of the bar undergoing flexural vibrations in one of its normal modes can be determined by substituting the quantized wavenumber, $k_n = \omega_n/c_{ph}$, corresponding to the frequencies, f_n, of Eq. (5.50), into the mode shapes of Eq. (5.43).

$$y(x) = A_n[\cosh(k_n x) - \cos(k_n x)] + B_n[\sinh(k_n x) - \sin(k_n x)] \tag{5.51}$$

Following Eq. (5.45), B_n can be expressed in terms of A_n.

$$-B_n = A_n \frac{\cosh(k_n L) + \cos(k_n L)}{\sinh(k_n L) + \sin(k_n L)} = A_n \frac{\sinh(k_n L) - \sin(k_n L)}{\cosh(k_n L) + \cos(k_n L)} \tag{5.52}$$

Since the amplitude is arbitrary until the initial conditions are specified, we can choose to normalize the amplitudes by requiring that the integral over the length of the bar of the squared amplitude be $(L/2)$, just as it would be if the amplitude functions were just the sine and cosine functions used for standing wave solutions to the wave equation.

$$\int_0^L y^2(x)dx = \frac{L}{2} \tag{5.53}$$

Substitution of $k_n L$ into Eq. (5.52) with all $A_n = 1/\sqrt{2}$ makes $B_1 = -0.518$, $B_2 = -0.721$, and all subsequent $B_n = -1/\sqrt{2}$ for $n \geq 3$.

The mode shapes for $1 \leq n \leq 4$ are shown in Fig. 5.8 and are compared to the solutions for the longitudinal modes of a fixed-free bar having the same wavelength. The ratio of the modal frequencies,

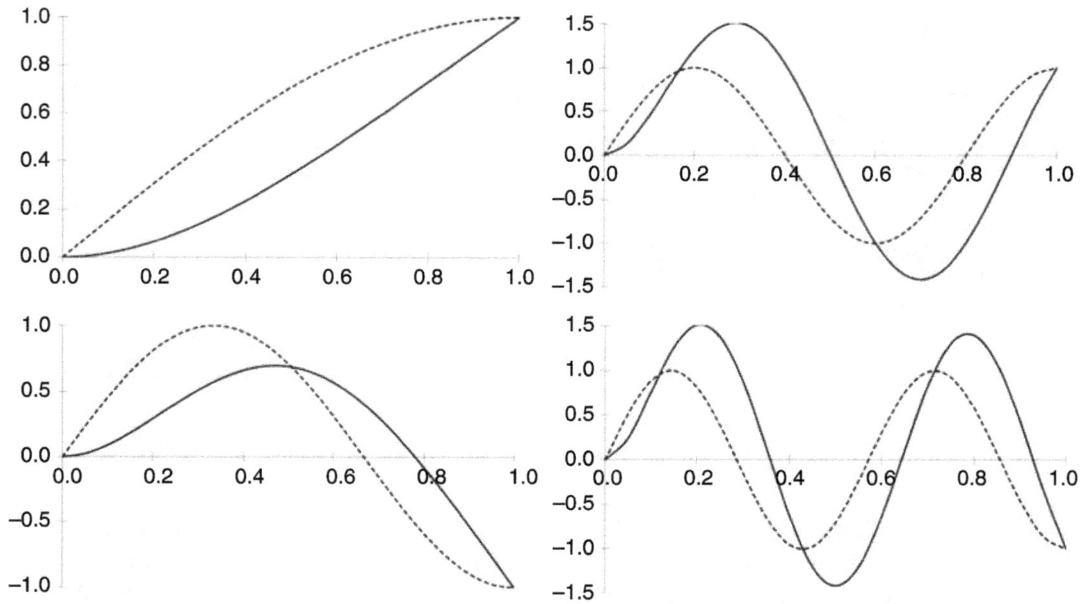

Fig. 5.8 The mode shapes for flexural vibration of clamped-free bars (*solid line*) are compared to the longitudinal mode shapes of a fixed-free bar having the same wavelength (*dashed line*) for the four lowest modes, normalized to match the displacements of both modes at $x = L$. Frequency ratios and the location of the zero-crossings are provided in Table 5.2

Table 5.2 Clamped-free (cantilevered) beam modal frequency ratios, f_n/f_1; normalized wavelengths, λ_n/L; and normalized location of standing wave displacement nodes as measured from the clamped end

Mode	f_n/f_1	λ_n/L	First node/L	Second node/L	Third node/L
1	1.0000	3.3508			
2	6.2669	1.3385	0.7806		
3	17.5475	0.7999	0.5017	0.8998	
4	34.3863	0.5714	0.3584	0.6428	0.9286

f_n, to the fundamental mode frequency, f_1, is provided in Table 5.2, along with the normalized wavelengths, λ_n/L, and the normalized location of the displacement nodes as measured from the clamped end.

A bar with clamped boundaries on both ends can be solved in the same way, starting with Eq. (5.43), which satisfies the boundary conditions at $x = 0$. Using the same procedure to fit the clamped conditions at $x = L$, the wavenumbers are again quantized by a transcendental equation that is plotted in Fig. 5.9.

$$\cosh(k_n L)\cos(k_n L) = 1 \quad \text{or} \quad \tan\left(\frac{k_n L}{2}\right) = \pm\tanh\left(\frac{k_n L}{2}\right) \tag{5.54}$$

The resulting values of $k_n L$ for the normal solutions of Eq. (5.54) can be combined with the phase speed in Eq. (5.39) to yield the flexural normal mode frequencies, f_n, of the clamped-clamped bar as a function of the geometry of the bar (L and κ) and the properties of the bar's material, $c_B = (E/\rho)^{1/2}$.

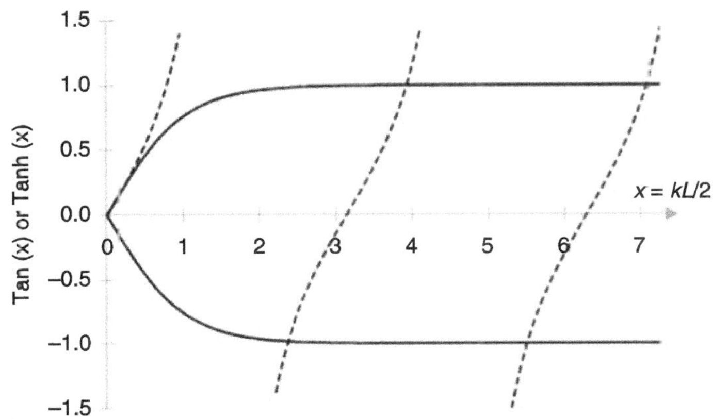

Fig. 5.9 The two terms on the right-hand side of Eq. (5.54) are plotted as a function of $x = kL/2 = \omega_n L/2c_{ph}$. The hyperbolic tangents are shown as solid lines and the dashed lines are the tangent. Their intersections provide the normal mode frequencies, ω_n, for a clamped-clamped or free-free bar

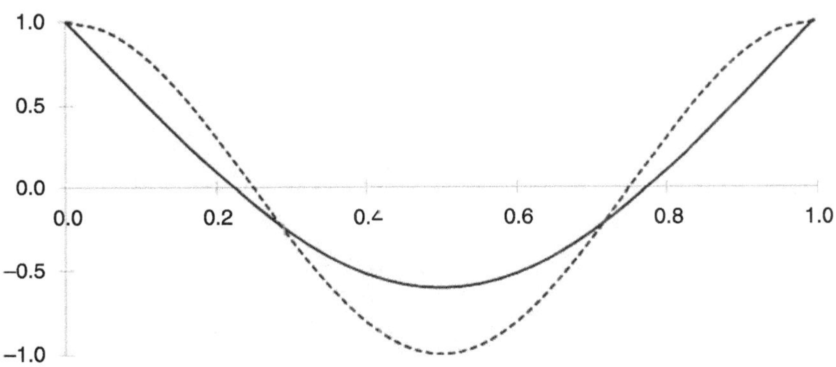

Fig. 5.10 The flexural mode shape for a free-free bar (*solid line*) is compared to the longitudinal mode shapes of a free-free bar (*dashed line*) for the fundamental vibrational mode. Normalized to match the displacements of the flexural mode at $x = 0$ and $x = L$

$$f_n = \frac{\omega_n}{2\pi} = \frac{\pi}{8}\frac{c_B \kappa}{L^2}\left(3.01124^2,\ 4.99951^2,\ 7.00002^2,\ 9^2,\ 11^2,\ \ldots\right) \tag{5.55}$$

The normal mode shapes are obtained, as before, by determining the B_n's, again assuming $A_n = 1/\sqrt{2}$.

$$B_n = A_n \frac{\sin(k_n L) + \sinh(k_n L)}{\cos(k_n L) - \cosh(k_n L)} = -A_n \frac{\cos(k_n L) - \cosh(k_n L)}{\sin(k_n L) - \sinh(k_n L)} \tag{5.56}$$

Substitution of $k_n\ L$ into Eq. (5.52) with all $A_n = 1/\sqrt{2}$ makes $B_1 = -0.6947$, and all subsequent $B_n = -1/\sqrt{2}$ for $n \geq 2$.

The flexural mode shapes for a free-free bar can be obtained by double differentiation with respect to x of the clamped-clamped mode shapes, resulting in the rather surprising fact that the normal mode frequencies for both free-free and clamped-clamped bars in flexure are identical.

The fundamental mode shape for the $n = 1$ free-free bar is shown in Fig. 5.10. Again, it is compared to the solutions for the fundamental longitudinal free-free mode of a thin bar having the same length. The ratio of the modal frequencies, f_n, to the fundamental mode frequency, f_1, is provided in Table 5.3.

Table 5.3 Clamped-clamped or free-free bar modal frequency ratios, f_n/f_1, and normalized wavelengths, λ_n/L. The remaining columns to the right of λ_n/L are the normalized location of standing wave displacement nodes for the free-free case

Mode	f_n/f_1	λ_n/L	First node/L	Second node/L	Third/L	Fourth/L	Fifth/L
1	1.0000	1.3284	0.2244	0.7734			
2	2.75654	0.8001	0.132	0.500	0.868		
3	5.40392	0.5714	0.094	0.356	0.644	0.906	
4	8.93295	0.4444	0.073	0.277	0.500	0.723	0.927

Fig. 5.11 (*Left*) Traditional tuning forks are shown above with its fundamental mode of vibration shown below. (*Right*) This "tonometer" contains 670 tuning forks ranging from 16 Hz to 4096 Hz (eight octaves) and was built by the Prussian physicist K. Rudolph Koenig (1832–1901). It was displayed at the London Exhibition, in 1852, where it was awarded the gold medal for scientific instruments and was transported to the Philadelphia Exposition where it appeared in 1876. (Photo courtesy of the Smithsonian National Museum of American History)

The free-free flexural modes of bars are the percussive tone generators in the glockenspiel and many children's toy instruments (e.g., Pixiphone).[10] By supporting the resonant bars at the nodal locations for the fundamental mode (see Table 5.3), the higher overtones will be present when the bar is struck, providing those instruments with a characteristically dramatic attack, which decays quickly, leaving the fundamental mode to linger.

A very common configuration for a free-free bar vibrating in its flexural mode is the tuning fork, shown in Fig. 5.11 (left) along with the "tonometer" that was commonly used to check the pitch of musical instruments. You may use an electronic instrument to tune your guitar or musical instrument, but the frequency standard inside the electronic device is likely to be a tiny quartz tuning fork. In fact, if

[10] The glockenspiel has bars of uniform cross-section, so the ratio of their overtones to the fundamental is given in Table 5.3. The underside of the bars in a xylophone is thinned near their center to make their overtones the ratio of integers: $f_2/f_1 = 3$ and $f_3/f_1 = 6$.

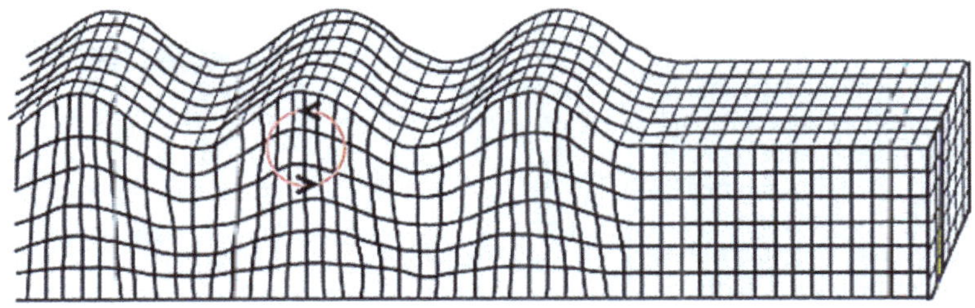

Fig. 5.12 Distortion of the free surface of a solid due to the propagation of a Rayleigh wave. Arrows indicate the motion of the material due to the wave

you own a "smartphone," it probably has a tuning fork with tines that are about 2 mm long that is used as a gyroscope (angular rate sensor). The same is true for games, automotive anti-skid and navigation sensors, remote-controlled aircraft (drones), digital cameras, etc.

When the tines of a tuning fork are oscillating, as shown in Fig. 5.11, and the entire fork is rotated with some angular velocity, $\vec{\Omega}$, then the Coriolis force [15] can be expressed as $\vec{F} = 2M\left(\vec{v} \times \vec{\Omega}\right)$, where M is the effective moving mass of the vibrating tines and $\vec{v}$ is the velocity of the tines. This force will cause the tines to develop a component of their motion in a plane that is orthogonal (due to the cross-product) to the original plane of vibration. This out-of-plane motion is then sensed to determine the rate of rotation [16].

5.3.5 Rayleigh Waves*

In principle, the phase speed in Eq. (5.39) could exceed the speed of light for large enough values of frequency, ω. At such frequencies, the assumption that the transverse dimensions of the bar are much smaller than the wavelength ($S^{1/2} \ll \lambda$) would be violated. As the frequency increases, these flexural waves do not bend the entire bar but become localized deformations at the surface. Such waves are known as Rayleigh waves [17], and their associated deformations are diagrammed in Fig. 5.12. A derivation of their propagation speed would take us too far from our development of waves on bars and is available elsewhere [7]. The speed of Rayleigh waves is about 10% slower than the shear wave speed, c_S, and is nondispersive.

The Rayleigh wave solutions are not contained in Eq. (5.36) because that equation does not include the contributions of rotary inertia and shear that become important at higher frequencies [18].

5.4 Resonant Determination of Elastic Moduli

> "*One of the most precise ways of measuring the elastic constants of a substance is by measuring the density of the material and the speeds of two kinds of waves.*" R. P. Feynman. [14]

Having derived the normal mode frequencies for longitudinal, torsional, and flexural vibrations of a thin bar, this section will describe a simple and inexpensive apparatus that can selectively excite and detect these modes. In addition to the pedagogical value of such an apparatus, it is also an excellent way to determine the elastic moduli of materials that are not ferromagnetic [19]. Sample data will be presented and analyzed for a composite of high tensile strength epoxy [20] with 20% glass fibers (by weight) to determine its Young's and shear moduli. In Sect. 5.4.5, the technique will be used to illustrate the viscoelastic behavior of an elastomer.

The (intentionally crude) apparatus, shown in Fig. 5.13, suspends the ends of the bar sample in between the pole pieces of two strong horseshoe magnets. Both ends of the bar are free and have coils

Fig. 5.13 Photograph of an apparatus for selective excitation and detection of longitudinal, torsional, and flexural resonance of a thin bar. The bar is supported by crossed rubber bands attached to ring stands that suspend the bar with its ends in the gap between "horseshoe" magnets. The sample in the photo is about 25 cm long and 1.2 cm in diameter

of wire glued to their free ends that are used as the electrodynamic transducers for all three modes. In this apparatus, the gap between the magnets' pole pieces is 2.0 cm and the magnitude of the magnetic induction, $\left|\vec{B}\right|$, at the center of the gap between the pole pieces is about 2.0 kilogauss = 0.20 tesla.

The transducers are just two coils (about ten turns each) of ordinary insulated solid copper magnet wire that were wound around the handle of a screwdriver.[11] #28 gauge copper wire (376 micron diameter) has a current-carrying capacity of about 300 mA, which is more than adequate for the driven coil.[12] It is convenient to leave about 50 cm of wire at both ends of each coil that can be used to attach the coil to the drive amplifier or the measurement system (e.g., low-noise pre-amplifier,[13] oscilloscope, spectrum analyzer, or typically, all three).

Before attaching the coils, the bar should be weighed and its physical dimensions determined as accurately as possible. Those measurements will be used to determine the mass density of the sample. After weighing, the coils can be attached to the ends of the sample using any appropriate adhesive. Care must be taken to be sure that the plane defined by both coils is aligned, since the magnets' fields are usually both in the same direction. For most samples, a simple glue, like that used to assembly model airplanes (e.g., Duco™ cement), or even clear nail polish,[14] is entirely adequate since the forces the drive coil will be modest, typically less than 20 mN. After the coils have been glued to the sample, the assembly should be weighed again so that an effective length correction can be added to the physical length of the sample to correct for the mass of the coils and their adhesive added to the ends of the bar (see Sect. 5.4.4).

[11] A screwdriver handle makes an ideal mandrel since the handle has grooves to improve grip that provide spaces to weave the last turn in and out of the grooves to hold the coil together when it is slipped off the screwdriver's handle.

[12] Almost no current flows through the detection coil, so it could be much smaller gauge wire, but it is often convenient to make both coils from the same wire.

[13] If a low-noise voltage pre-amplifier is available (e.g., PAR 113, Ithaco 1201, or SRS 560), they usually also provide some adjustable low-pass filtering capabilities that can remove low-frequency seismic vibrations if the apparatus is on a table that is not rigid.

[14] Clear nail polish has lower viscosity, making attachment of the coils easier since capillarity will draw the polish into the coil and concentrate where the coil contacts the bar.

5.4.1 Mode-Selective Electrodynamic Excitation and Detection

The torsional mode can be excited preferentially by orienting both coils horizontally with the pole pieces also horizontally, as shown in Fig. 5.14 (top right). This technique was first used by Barone and Giacomini [21] to study the modes of bars having variable cross-section and then by Leonard [22] who measured the attenuation of torsional modes to disprove the existence of the "Fitzgerald effect." [23, 24][15] The short portion of the coil, which crosses the end of the bar along a diameter, should stick out slightly beyond the pole pieces. When the alternating electrical current, $I(t) = i_1 \cos(\omega t)$, flows in one direction through the drive coil, a Lorenz force, $\left|\vec{F}(t)\right| = (B\ell)I(t)$, will be downward on one side of the coil and upward on the other, where ℓ is the total length of wire in the coil subjected to the magnetic field on one side of the bar. When the current changes direction, so will the forces. This twisting moment will selectively excite the torsional modes of the bar.

The other coil, on the opposite end of the bar, is in the same orientation with respect to its magnet and will generate a voltage (*emf*) that is proportional to the time rate of change of the magnetic flux, Φ,

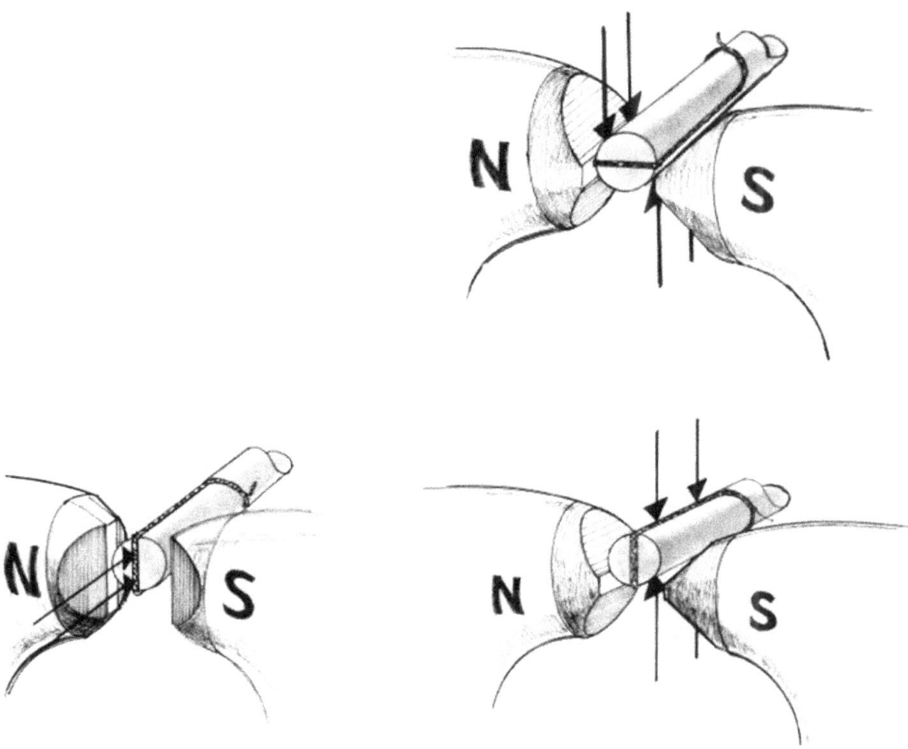

Fig. 5.14 Arrangements of the electrodynamic transducer coils in the magnet gap for excitation and detection of the torsional (*top right*), flexural (*bottom right*), and longitudinal (*bottom left*) modes of a thin bar. The arrows show the strength and direction of the forces produced by the current flowing through the coils that are functioning as the drive transducers

[15] Fitzgerald had thought he discovered a new attenuation mechanism, but Leonard showed that Fitzgerald's "effect" was absent, and Fitzgerald had just measured an artifact of the attachment of piezoelectric transducers to his sample.

through the receiver coil. The flux, Φ, is the product of the projection of the surface area of the coil, $A_\perp = A\cos\phi$, times the magnitude of the magnetic induction, $\left|\vec{B}\right|$, times the number of turns or wire, N. Here A is the area within the coil, and ϕ is the angle between a vector normal to the plane of the coil and the magnetic field lines.

$$emf = -\frac{d\Phi}{dt} = -N\left|\vec{B}\right|\frac{dA_\perp}{dt} = -N\left|\vec{B}\right|A\frac{d(\cos\phi)}{dt} \tag{5.57}$$

With the coil oriented as shown in Fig. 5.14 (top right), $\phi \cong 90°$, so $A_\perp \cong 0$, and $\Phi \cong 0$. When a torsional wave is excited, the end of the bar will experience a time-harmonic twisting, so $\phi(t) = \Re e[\phi_1 e^{j\omega t}]$, generating a voltage (*emf*) given by Faraday's law in Eq. (5.57). With $\phi \cong 90°$, the sensitivity of the transducer to the twisting will be its greatest. The *emf* will be generated at the same frequency, ω, as the twisting that is caused by the torsional resonance and will also be proportional to the amplitude, ϕ_1.

Flexural resonances can be excited and detected by simply rotating the bar (with its attached coils) by 90° from the torsional mode orientation and raising (or lowering) the bar by one bar radius. This places one portion of each coil in the strongest part of the magnetic field and lifts the other part of the coil to a region of weaker magnetic field, as shown in Fig. 5.14 (bottom right). Now the vertical forces on each long section of the coil will be in opposite directions. Because of the stronger magnetic field is applied to the lower portion of the coil (as shown), there will be a net vertical force that will create a harmonic bending of the bar and excite the flexural modes of vibration.

In this orientation, the coil on the opposite end of the bar will be "dipping" in and out of the stronger regions of magnetic field. In this case $\phi \cong 0°$, so $A_\perp \cong A$, but the magnitude of the magnetic induction, $\left|\vec{B}\right|$, will be changing with time, $B(t) = B_o + \Re e[B_1 e^{j\omega t}]$, as the receiver coil moves toward and away from the strongest regions of magnetic field. Now Eq. (5.57) will produce an $emf = NA(dB/dt)$ that will be at the same frequency as the flexural vibrations and proportional to their amplitudes.

Finally, the longitudinal modes can be selectively excited and detected by reorienting the pole pieces to make the gap vertical, as shown in Fig. 5.14 (bottom left). This concentrates the magnetic field in the vertical direction. The end of the coil that is along the diameter at the end of the bar is placed in the strongest portion of that magnetic field. In this orientation, the driven end of the coil will be putting oscillatory longitudinal forces on the end of the bar thus exciting the longitudinal modes. Since the length of that portion of the coil is considerably shorter than the portions of the coil along the perimeter of the sample, the longitudinal modes will not be excited as strongly as the torsional and flexural modes for the same electrical current amplitude, i_1, flowing through the driving coil. Detection of the longitudinal mode is the same as the flexural mode. The magnetic flux, $\Phi(t)$, is modulated by the coil's motion, being pushed into and out of the strongest portions of the magnetic field by the motion of the end, $\xi(t) = \left[\hat{\xi}e^{j\omega t}\right]$.

5.4.2 Bar Sample Size and Preparation

Sample data acquired by this method are displayed in Fig. 5.16 for the epoxy-glass composite described at the beginning of Sect. 5.4. The bar was fabricated by mixing the epoxy and then adding glass fibers to the mix. The mixture was then poured into a hollow plastic tube that had been coated on the inside with a mold release agent to facilitate removal of the sample after the epoxy had cured. The sample was pushed out of the tube and the ends were cut straight using a band saw.

Before the transducer coils were attached, the bar was weighed, $M = 55.723 \pm 0.001$ g, and the bar's dimensions were measured. The length $L = 33.60 \pm 0.02$ cm, but the cross-section was found to be slightly elliptical with major and minor diameters ranging from $d_{max} = 13.17 \pm 0.02$ mm to $d_{min} = 12.77 \pm 0.20$ mm, yielding an average diameter of $d = 12.97 \pm 0.20$ mm. These physical measurements produced an average mass density, $\rho = 1256 \pm 4$ kg/m^3. It is not unusual to have this degree of ellipticity in a sample which was cast in a plastic tube. The cross-section of glass tubes is more circular, but also more dangerous to fabricate, since the sample can seldom be removed without breaking the glass.

The transducer coils were attached as described previously using Duco™ cement. With a bar that has an elliptical cross-section, it is preferable to orient the coils along either the major or minor diameter so that the polarization of the flexural mode is known. For this sample, the major and minor diameters differed by about 3% corresponding to a 6% difference in the free-free flexural mode frequencies caused by the difference in the κ^2 value that appears in Eq. (5.55).

After attachment of the coils, the bar was weighed again and found to be 1.5 ± 0.1 g heavier. It is difficult to be sure exactly how much the motion of the wires that connect the coil to the instrumentation contribute to the measurement, but in this case, the accuracy is not degraded significantly since the uncertainty in the effective length of the bar (see Sect. 5.4.4) is only about ½%.

5.4.3 Measured Resonance Spectra

Figure 5.16 displays three resonance spectra that were acquired using a dynamic signal analyzer to measure the bar's response only in a narrow band around the frequency at which it was being driven. The setup is diagrammed schematically for the torsional mode in Fig. 5.14 (top right). The analyzer also had an internal curve-fitting program, so the measured spectral response is shown as the solid line and the curve-fit to that response is shown as the dashed line (Fig. 5.16).

The five lowest-frequency torsional modes are shown in Fig. 5.16 (upper right). They appear to be harmonically related, with uniform spacing between the resonances. That appearance is confirmed by the resonance frequencies listed in Table 5.4. It is also worthwhile to note that the *dynamic range* (the ratio of the largest measured signal to the smallest) of that measurement is about 70 dB, corresponding to an amplitude ratio in excess of 3000:1. At the minima in the spectrum (i.e., at frequencies between the resonances), the curve-fit dashed line alternates between being larger and smaller than the measurements. That is because the coils also act as an electrical transformer coupling the oscillating magnetic flux in the drive coil (weakly) into the receiver coil. This could be eliminated by orienting the planes of the two coils to be mutually perpendicular and rotating one set of magnets by 90° or by placing a magnetic shield (iron is good, μ-metal is better) around the middle of the bar without

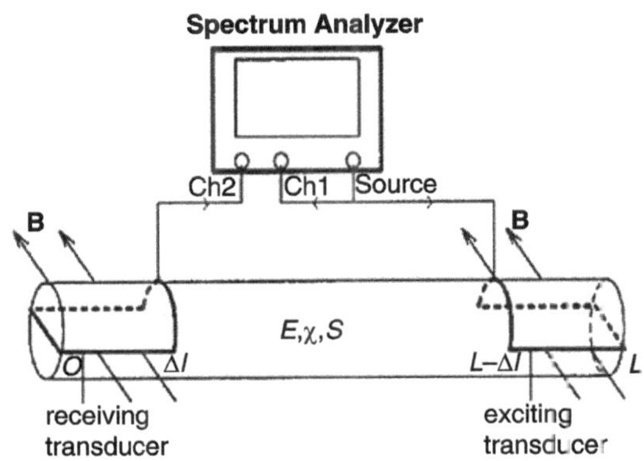

Fig. 5.15 Schematic representation of the connection of the dual-channel spectrum analyzer that measured the transfer function between the drive signals, monitored in Channel 1, and the bar's resonance responses monitored in Channel 2. As shown, the apparatus was set up to measure the torsional modes of the bar [25]

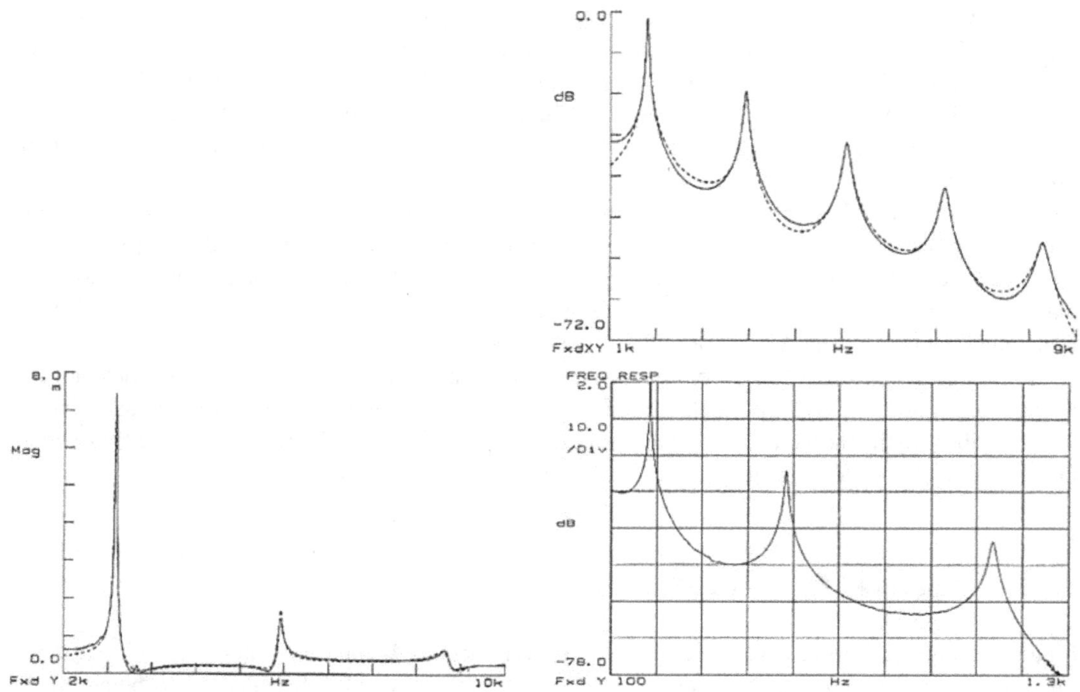

Fig. 5.16 Resonance spectra for the lowest-frequency torsional (*upper right*), flexural (*lower right*), and longitudinal (*lower left*) resonances. The solid lines are the measured spectra, and the dashed lines were pole-zero fits to those measured spectra

Table 5.4 Summary of the measured normal mode frequencies of the thin bar sample of an isotropic glass-epoxy composite. The error is the standard deviation of the average of the normalized frequencies, f_n/f_1, for each mode

Mode number [*n*]	Frequency [Hz]	Normalized frequency [f_n/n]	Average uncertainty [Hz (%)]
Torsional			
1	1644	1644	
2	3341	1670	
3	5080	1693	
4	6750	1688	
5	8446	1689	
Average		1677	±20 (±1.2%)
Flexural			
$(3.01124)^2$	203.1	22.40	
$(4.99951)^2$	560.4	22.42	
$(7.00002)^2$	1104	22.53	
Average		22.45	±0.07 (±0.2%)
Longitudinal			
1	2941	2941	
2	5936	2968	
3	8896	2965	
Average		2958	±15 (±0.4%)

touching the bar. Those methods of eliminating the *electromagnetic cross talk* between the drive and receiver coils were not implemented because the signal-to-noise ratio was more than adequate to accurately determine the peak frequencies that were not measurably influenced by the electromagnetic cross talk.

Table 5.4 lists the measured resonance frequencies of the five lowest-frequency torsional modes. Since the modes are expected to be related harmonically, the table also forms the ratio between the frequencies, f_n, of the nth mode, to the fundamental frequency, f_1. For harmonic modes of a free-free bar, that ratio should be a constant. The ratio of the standard deviation of the five modal frequencies to their average is ±1.2%. Since this is in the range of our relative uncertainty of the mass density (±0.3%), and the ellipticity (±3%), there was no motivation to improve or to understand the source of that small anharmonicity.

The three lowest-frequency flexural modes of the bar are shown in the spectrum in Fig. 5.16 (lower right). It is obvious from inspection of that spectrum that those modes are not harmonically related. The frequency difference between the first and second modes is smaller than the difference between the second and third, as would be expected from Eq. (5.55). It is also possible to see a small amount of noise in the spectrum that appears in the "spectral tail" above 1.2 kHz. The dynamic range of that spectrum is 80 dB, corresponding to an amplitude ratio of 10,000:1. The frequencies of those modes are listed in Table 5.4, as is their normalized frequency, f_n/f_1, which has a relative uncertainty of ±0.2%.

Since the shear modulus can be related to the torsional mode frequencies and Young's modulus related to the flexural mode frequencies, we have enough information to determine the complete elastic response of our sample that is assumed to be isotropic.[16] Since the longitudinal modes were also measured, they can be used to estimate the accuracy of the method for the determination of Young's modulus.

The spectrum displaying those longitudinal modes is shown in Fig. 5.16 (lower left). As mentioned, the amplitudes of the longitudinal modes are smaller than of the other two modes because the transduction is less efficient. The small bump to the right of the largest longitudinal mode is a weak excitation of the second torsional mode at 3341 Hz. Since the side portions of the coils are so much longer than the short section of coil at the end, it is possible to excite a small amount of torsional vibration even though the magnets and coils are arranged to couple preferentially to the longitudinal mode. The third resonance at nearly 9 kHz hardly looks like a resonance at all, again due to phase cancellation between the superposition of the acoustical resonance signal and the electromagnetic cross talk signal. Despite those limitations, the relative uncertainty of the normalized longitudinal mode resonance frequencies in Table 5.4 is only ±0.4%, which is consistent with the precision of the measurements of the frequencies of the other two modes.

The epoxy-glass composite used in this example was chosen because it does not have a particularly large quality factor, Q, making measurement of its modal frequencies a bit more challenging. Wooden bars will produce similar results. If this same apparatus is used to measure the resonances of metal bars (e.g., aluminum or brass) or ceramics (e.g., alumina) or glass, the Q will be significantly larger, as will be the signal-to-noise ratio. Consequently, the results for the moduli will be even more precise.

[16] Although this sample is a composite, the glass fibers in the epoxy matrix are short (about 800 microns long) and randomly oriented. The sample behaves isotropically upon length scales that are on the order of the diameter and the resonance wavelengths.

5.4.4 Effective Length Correction for Transducer Mass

If any addition to a bar be made at the end, the period of vibration is prolonged. J. W. Strutt (Lord Rayleigh). [26]

Before calculating the moduli from the experimental data, we can eliminate one source of systematic error. Because the coils and their adhesives add mass to the end of the bar, we can calculate an effective length correction to account for this, just as was done for the quartz crystal microbalance in Sect. 5.1.2, except this time we will use Rayleigh's method instead of an approximation to the transcendental equation diagrammed in Fig. 5.2. Such effective length corrections were calculated by Rayleigh in *Theory of Sound* for the free-free longitudinal case in Vol. I, §155. The free-free flexural case is covered in Vol. I, §186. Rayleigh did not do the calculation for mass loading of the free-free torsional modes [26].

The mass of the transducers is concentrated very close to the free ends of the bar, and the length of the coils is much shorter than the length of the bar, so we can assume that all of the mass of the coil is concentrated at the end, at least for the lower-frequency modes, where the wavelength of the resonance exceeds the length of the bar or is a substantial fraction of the bar's length. We will ignore the stiffness added by the coils since they are attached close to the stress-free ends.[17] The coils only affect the bar's mass, so we can apply Rayleigh's method (see Sect. 3.2.2) and only concern ourselves with the changes in the kinetic energy, $\delta\,(KE)$, that cause changes in frequency, $\delta\omega = 2\pi\delta f$, due to the mass loading of the coils. Since $\omega^2 \propto (PE/KE)$, the torsional and longitudinal modal frequencies are inversely proportional to the length of the bar, and it is possible to express the length correction, δL, in terms of the change in kinetic energy, $\delta(KE)$.

$$\frac{\delta f}{f} = \frac{\delta\omega}{\omega} = -\frac{\delta L}{L} = -\frac{1}{2}\left(\frac{\delta KE}{KE}\right). \tag{5.58}$$

The effective length, L_{eff}, of the bar that compensates for the mass loading of the coils is related to the original length, L: $L_{eff} = L + \delta L$. If the mass of the bar prior to the attachment of the coils is M and the mass added by the coils is $2\,m$, then after the coils are attached, $M_{tot} = M + 2\,m \equiv M + \Delta M$, where $(\Delta M)g$ is the change in the weight of the bar after the transducer coils are attached.

The calculation for the effective length of the bar, in the lowest-frequency longitudinal mode, is simplified if the symmetry of that mode is exploited. Since the center of the bar is a node, the entire left-half of the bar can be replaced by a rigid (fixed) end at $x = 0$. The fundamental longitudinal resonance can be treated as the fundamental resonance of a fixed-free bar of length, $L' = L/2$. The displacements, $\xi(x,\,t)$, of the fixed-free half-bar can be written in terms of the maximum displacement amplitude, ξ_1, of the free end, with $\omega/k = c_B$ and k_1 given by Eq. (5.14).

$$\xi(x,t) = \xi_1 \sin\left(\pi\frac{x}{2L}\right)\sin(\omega t) \tag{5.59}$$

The kinetic energy of the half-bar undergoing longitudinal oscillations, $(KE_1^{L}/2)$, without the added mass, can be determined by integration along the length of the half-bar, $L' = L/2$, where the half-mass of the half-bar is $M' = M/2$.

---

[17] The stiffness contributions of the transducer coils have been calculated by Guo and Brown [25].

$$\frac{KE_1^L}{2} = \frac{\rho S\omega^2 \xi_1^2}{2} \int_0^{L'} \sin^2\left(\pi \frac{x}{2L}\right) dx = \frac{1}{4} M' \omega^2 \xi_1^2 \tag{5.60}$$

The one coil of mass, m, at the end of the bar, will move with the velocity of the end of the bar, $\omega\xi_o \sin \omega t$, and will have a maximum kinetic energy, $KE_{coil} = (m/2)\omega^2 x_1^2$. The relative shift in the kinetic energy of the half-bar will be $\delta(KE)/(KE) = KE_{coil}/(KE_o^L/2) = 2m/M'$.

$$\left(\frac{\delta L}{L}\right)_L = -\frac{\delta f}{f} = \frac{1}{2} \frac{KE_{coil}}{(KE_o^L/2)} = \frac{m}{M'} = \frac{2m}{M} = \frac{\Delta M}{M} \Rightarrow L_{eff}^L = L\left(1 + \frac{\Delta M}{M}\right) \tag{5.61}$$

The same style of argument can be applied to the torsional case, except that the location of the added mass with respect to the axis of the bar is important because it is the moment of inertia, I, of the coil that will load the bar's torsional oscillations. For this case, the kinetic energy is due to rotation, so that $KE_{coil} = (½)I_{coil}(\omega\theta_1)^2$, where θ_1 is the maximum angular displacement at the end of the bar in the fundamental mode. The moment of inertia of a disk of mass, δm, and diameter, d, is $I = (1/8)(\delta m)d^2$, so the kinetic energy of torsional oscillations, $(KE_1^T/2)$, of the half-bar is again given by integration over the half-length of the bar, L'.

$$\frac{KE_1^T}{2} = \frac{\rho S\omega^2 \theta_1^2}{2} \frac{d^2}{8} \int \sin^2\left(\pi \frac{x}{2L}\right) dx = \frac{M' \omega^2 \theta_1^2 d^2}{32} \tag{5.62}$$

For an ellipse, $d^2 = d_{max}\, d_{min}$.

Since the single coil mass, m, is primarily located at the bar's radius, its moment of inertia is $I_{coil} = (¼)md^2$,[18] and the change in the kinetic energy of rotation caused by the addition of one coil to the end of the half-bar is $KE^T_{coil} = (1/8)m(d\omega\vartheta_1)^2$.

$$\left(\frac{\delta L}{L}\right)_T = -\frac{\delta f}{f} = \frac{1}{2} \frac{KE_{coil}}{(KE_o^T/2)} = \frac{2m}{M'} = \frac{4m}{M} = \frac{2\Delta M}{M} \Rightarrow L_{eff}^T = L\left(1 + \frac{2\Delta M}{M}\right) \tag{5.63}$$

The mass of the coils have twice the effect on the frequency of the torsional modes, hence producing twice the effective length correction, L_{eff}^T, as that required for the longitudinal modes, L_{eff}^L, in Eq. (5.61).

The exact solution for the fundamental flexural mode of a free-free bar is again given by the superposition of four trigonometric functions with $k_1 L = 4.73$, $A_1 = 0.5$, and $B_1 = -0.982A_1$.

$$y(x,t) = A_1[\cosh(k_1 x) + \cos(k_1 x)] + B_1[\sinh(k_1 x) + \sin(k_1 x)] \sin(\omega_1 t) \tag{5.64}$$

The kinetic energy of the bar without coils, vibrating in its flexural mode, KE_1^F, is related to the integral of its transverse velocity, $(\partial y/\partial t)$, over its length.

$$KE_1^F = \frac{\rho S}{2} \int_0^L \left(\frac{\partial y}{\partial t}\right)^2 dx \tag{5.65}$$

Rather than solving that integral for the kinetic energy of the unloaded bar in its fundamental mode using Eq. (5.64), we can create a second-order polynomial function, $y_{poly}(x, t)$, to approximate the

[18] Since the coil is actually on the surface of the bar, its diameter is slightly larger than d. We will neglect this difference by arguing that the part of the coil that crosses the bar's end has a lower moment of inertia. It would be a small correction to an already small correction (so works the rationalizations in the mind of an experimentalist).

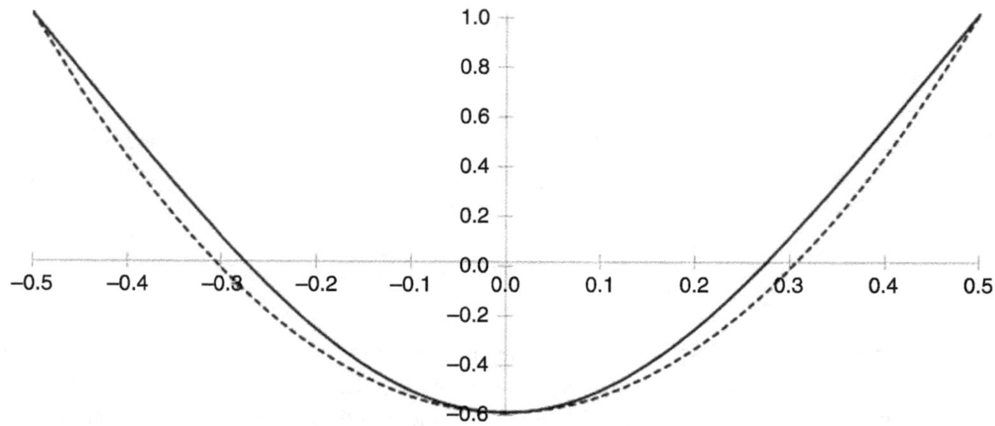

Fig. 5.17 Comparison of the exact solution for a free-free bar vibrating in its fundamental flexural mode (*solid line*) to the second-order polynomial approximation (*dashed line*) that is matched to the maximum $y(0) = -0.608$ at $x = 0$ and the at the free ends where $y(\pm L/2) = 1$

actual displacements that will be easier to integrate. To exploit the symmetry of the fundamental mode shape, we can place the origin of the coordinate system at the center of the bar.

$$y_{poly}(x,t) = Ax^2 + C \tag{5.66}$$

At the center of the bar, located at $x = 0$, the slope, $(\partial y_{poly}/\partial x)_0 = 0$. At the ends of the bar, $x = \pm L/2$, $(\partial^2 y_{poly}/\partial x^2)_{\pm L/2} = A/2 \neq 0$, so the second-order polynomial approximation does not meet the requirement that there be no torques applied to the free end. Since $(\partial^3 y_{poly}/\partial x^3)_{\pm L/2} = 0$, the second-order polynomial approximation does satisfy the boundary condition that requires that there also be no vertical forces.

The displacement of the polynomial approximation can be fit to Eq. (5.64) at $x = 0$ by making $C = -0.608$ and fit to the two ends, where $y\,(\pm L/2) = 1$, by making $A = 4(1 - C) = 6.432$. Figure 5.17 shows both the exact solution of Eq. (5.64) and the second-order polynomial approximation of Eq. (5.66). The kinetic energy of the unloaded bar can be integrated over $-½ \leq x \leq ½$, remembering that the vertical motion of the free ends has unit amplitude.

$$KE_o^F \cong \frac{\rho LS\omega^2}{2}\int_{-L/2}^{L/2} y_{poly}^2(x)\,dx = \left[\frac{A^2x^5}{5} + \frac{2ACx^3}{3} + C^2x\right]_{-L/2}^{L/2} = \frac{0.235M\omega^2}{2} \tag{5.67}$$

Rayleigh's method will always overestimate the energy, and the $|y_{poly}(x)| > |y(x)|$ everywhere, as can be seen in Fig. 5.17, so I will let $(0.235)^{-1} = 4.26 \cong 4$. The kinetic energy of each coil of mass, m, at both ends, also with vertical motion of unit amplitude, is $KE_{coils} = (½)(2m)\omega^2 y_{poly}^2$. From Eq. (5.55), we see that the frequency is inversely related to the square of the length of the bar.

$$\frac{\delta L}{L} = -\frac{1}{2}\frac{\delta f}{f} = -\frac{1}{2}\left(-\frac{1}{2}\frac{KE_{coils}}{KE_o^F}\right) = \frac{1}{4}\left(\frac{4\Delta M}{M}\right) = \frac{\Delta M}{M} \Rightarrow L_{eff}^F = L\left(1 + \frac{\Delta M}{M}\right) \tag{5.68}$$

This is in agreement with Rayleigh's result for the flexural mode: "If the load be at the end, its effect is the same as lengthening of the bar in the ratio *M:M+dM*." [26]

With the effective lengths removing the systematic error that would be introduced by the mass loading of the transducer coils on the ends of the bar, the moduli can be calculated. The mass ratio is

$\Delta M/M$ = (1.5 grams/55.723 grams) = 2.7 x 10^{-2}. For the longitudinal and flexural modes, $L_{eff}^{\,L} = L_{eff}^{\,F} \equiv L_{eff} = 34.50$ cm. Young's moduli calculated from both methods are determined by Eqs. (5.13) and (5.55).

$$E_L = 4\rho L_{eff}^2\left(\frac{f_n^L}{n}\right)^2 \quad \text{and} \quad E_F = \left(\frac{32}{\pi}\right)^2\frac{\rho L_{eff}^4}{d^2}\left(\frac{f_n^F}{n^2}\right)^2 \tag{5.69}$$

Before evaluating E_L and E_F, it will be useful to propagate the relative errors (see Sect. 1.8.4) to obtain an estimate of the accuracy of the results using log differentiation (see Sect. 1.1.3). It is clear that errors in ρ, L_{eff}, and f_n/n are all statistically independent, so the errors are combined in a Pythagorean sum.

$$\frac{\delta E_L}{E_L} = \left[\left(\frac{\delta\rho}{\rho}\right)^2 + \left(2\frac{\delta L_{eff}}{L_{eff}}\right)^2 + \left(2\frac{\delta(f_n^L/n)}{(f_n^L/n)}\right)^2\right]^{½} \tag{5.70}$$

Defining the random error as being one standard deviation, $\delta\rho/\rho = \pm 0.32\%$, $\delta(L_{eff})/|L_{eff}| = \pm\, 0.06\%$, and $\delta(f_n/n)/|f_n/n| = \pm\, 0.5\%$. This results in a relative overall uncertainty of $\delta E_L\,/E_L = \pm 1.1\%$. Equation (5.69) can then be evaluated: $E_L = 5.23 \pm 0.06$ GPa.

The same procedures can be applied to the calculation of E_F and its relative uncertainty.

$$\frac{\delta E_F}{E_F} = \left[\left(\frac{\delta\rho}{\rho}\right)^2 + \left(4\frac{\delta L_{eff}}{L_{eff}}\right)^2 + \left(-2\frac{\delta d}{d}\right)^2 + \left(2\frac{\delta(f_n^F/n^2)}{(f_n^F/n^2)}\right)^2\right]^{½} \tag{5.71}$$

Fortunately, the relative uncertainty in effective length is the smallest of the four pieces because it is also the most heavily weighted in Eq. (5.71). The difficulty is the assignment of $\delta d/d$ to account for the ellipticity of the sample's cross-section. I will estimate it to be half the difference in the major and minor diameters over their average, $\delta d/d = \pm\,(d_{max} - d_{min})/(d_{max} + d_{min}) = \pm\, 1.54\,\%$. This makes $\delta E_F/E_F = \pm 3.2\%$, so from Eq. (5.69), $E_F = 5.53 \pm 0.18$ GPa.

Although the lower limit of $E_F \geq 5.35$ GPa almost overlaps the upper limit of $E_L \leq 5.29$ GPa, their discrepancy is slightly outside our estimate of experimental error. This is probably due to the fact that we represented the elliptical cross-section as a circle with an average diameter $d = (d_{max} + d_{min})/2$. The entire discrepancy could possibly be removed if the polarization of the flexural vibrations were rotated by about 30° toward the d_{min} orientation. Under the circumstances, if I needed a value of Young's modulus and could not repeat the experiment, I'd use the error-weighted average of the two values and append the larger of the two uncertainties: $E = 5.31 \pm 0.18$ GPa.

To evaluate the shear modulus from the measured torsional mode frequencies, we use Eq. (5.63) to write $L_{eff}^{\,T} = 35.4$ cm.

$$G = 4\rho\left(L_{eff}^T\right)^2\left(\frac{f_n^T}{n}\right)^2 \tag{5.72}$$

Since this is the same relationship as for the longitudinal standing wave modes, the error propagation is again given by Eq. (5.70), now with $\delta\,(f_n/n)/(f_n/n) = \pm 0.6\%$ resulting in $\delta\, G/G = \pm 2.4\%$, so from Eq. (5.72), $G = 1.77 \pm 0.04$ GPa. Since $E \cong 3G$, Poisson's ratio is very close to one-half (see Eq. 4.93).

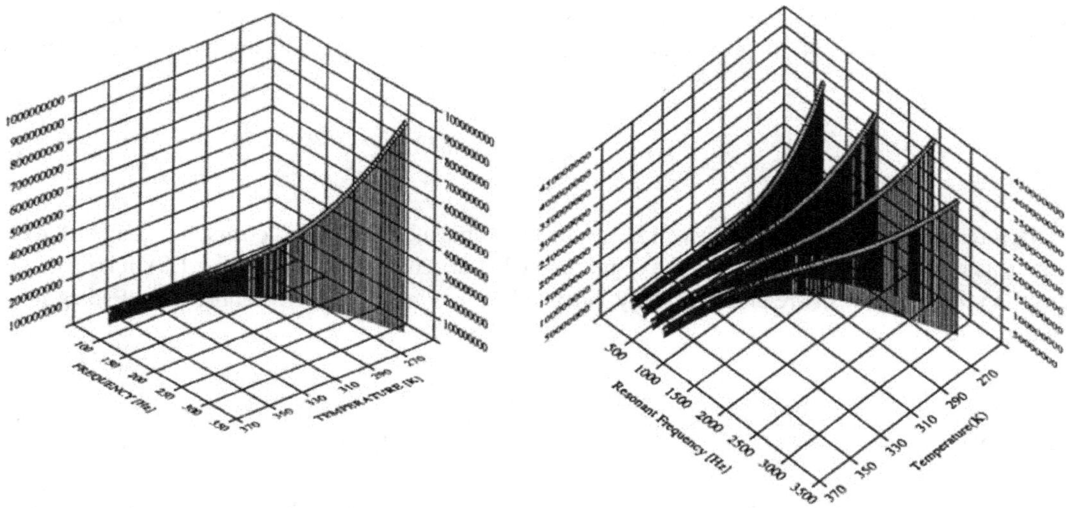

Fig. 5.18 The results of the automated measurement of the lowest-frequency flexural mode (*left*) and the four lowest-frequency shear modes (*right*) of an elastomeric bar made of PR 1592 [27], an encapsulant used for hydrophones, over a temperature range from 260 K (−13 °C) to 360 K (87 °C). This temperature range includes the material's glass transition temperature. Each point in the temperature-frequency plane represents one measured resonant mode with the height above the point being the elastic modulus corresponding to that temperature and frequency [28]. The viscoelastic behavior is obvious due to the low modulus values at low frequency and high temperature and the high modulus at high frequencies and low temperatures

5.4.5 Modes of a Viscoelastic Bar

The previous example demonstrated the simplicity and accuracy of extracting elastic moduli from resonance frequencies of thin bars. Often, it is necessary to determine how those elastic constants vary with some environmental parameter like temperature. Data for a bar sample cast from a viscoelastic elastomer is presented in Fig. 5.18.

The measurements of the resonance frequencies were automated by use of a phase-locked loop frequency tracking system (see Sect. 2.5.3) shown in the block diagram of Fig. 5.19. The sample, with its support structure and the magnets (like Fig. 5.13, but with fine wires providing sample support, replacing the rubber bands) was placed in an environmental control chamber to vary the sample's temperature. The bar is driven by a voltage-controlled oscillator (VCO) that produces a sinusoidal voltage that is amplified and applied to the drive coil. The signal from the receive coil is then passed through a pre-amplifier that has a high-pass filter to remove low-frequency noise and vibration. Of course, all of the electronics instruments are outside the environmental chamber.

The frequency of the VCO is controlled to keep the bar vibrating at one of its free-free normal mode frequencies by controlling the relative phase between the drive voltage and the received signal voltage. The force on the driven coil, $F(t) = (B\ell)I(t)$, is in phase with the current or voltage, since the coil's electrical impedance is primarily resistive. The output of an electrodynamic transducer is governed by Faraday's law, as expressed in Eq. (5.57), so the receive coil's velocity (linear or angular) and the force should be in-phase at resonance; hence, those two voltages should be in-phase.[19]

[19] Unlike the example in Sect. 2.5.3, which controlled a single degree-of-freedom simple harmonic oscillator, the bar has standing wave modes, so the phase relation between force and velocity will alternate by 180 ° between adjacent modes.

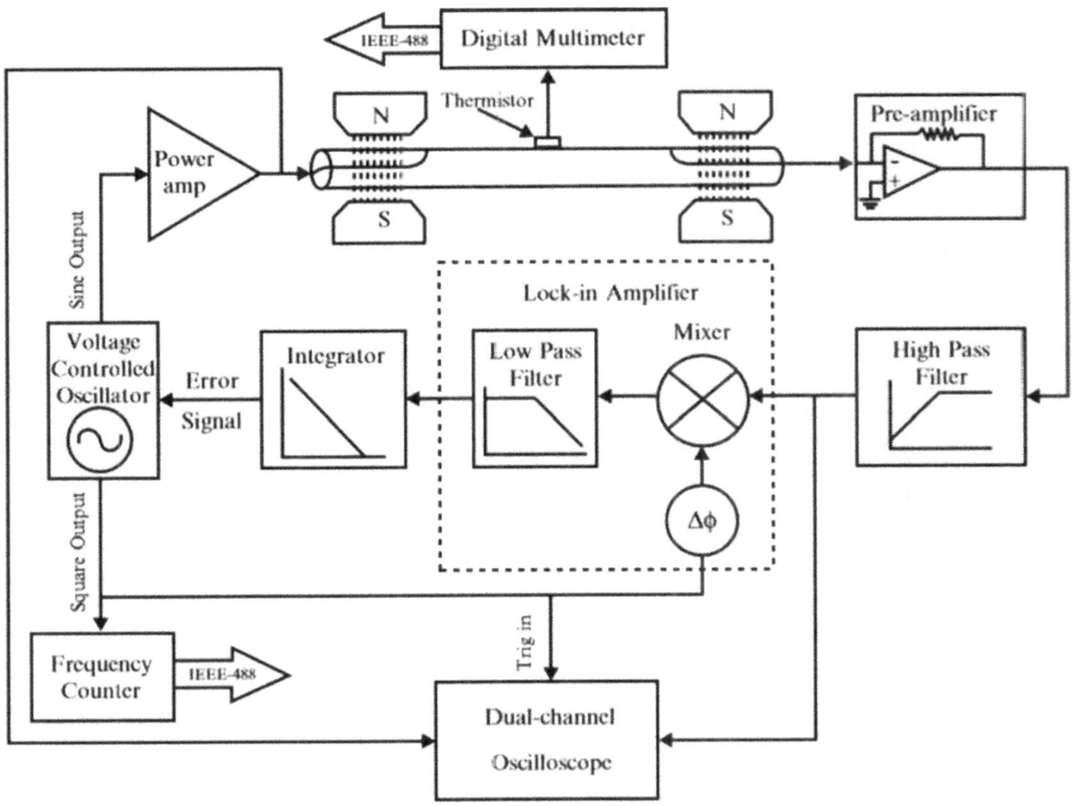

Fig. 5.19 Block diagram of the instrumentation and electronics [19] used to automatically track the resonance frequencies and amplitudes for a vibrating bar to produce results like those shown in Fig. 5.18

A two-phase lock-in amplifier [29] was used in this application. The algorithm could be based on a simple pair of analog multiplier chips, such as the analog Devices AD633, or it could be accomplished entirely in software. One phase will generate the error signal that is integrated and fed back to the input of the VCO. Integration of the error signal allows the integrator to "store" the required voltage to tune the VCO to the resonance frequency, since the error signal voltage will be zero when the frequency-tracking system is locked onto the resonance frequency.

The other (quadrature) output signal from the lock-in will be proportional to the amplitude, since the relative phases of the two lock-in outputs will differ by 90°. With the feedback network forcing the output of the frequency control channel to zero, the entire signal will appear at the *quadrature* output from the other lock-in channel. Because of the low-pass filters on the outputs of the mixers (i.e., multipliers), the effective equivalent noise bandwidth $(\Delta f)_{EQNB}$ (see Sect. 2.5.2), even for a single-stage low-pass (RC) filter, with a roll-off of −6 dB/octave, will have $(\Delta f)_{EQNB} = (\pi/2)(\Delta f)_{-3dB}$ (see Fig. 2.8). As more stages are added, the coefficient decreases from $\pi/2$ to 1. For a two-stage filter (i.e., −12 dB/octave), the $(\Delta f)_{EQNB} = 1.22(\Delta f)_{-3dB}$. For measurements that are slow, due to long thermal equilibration times, the time constant, τ_{-3dB}, of the low-pass filter can be set to 10 s, so $(\Delta f)_{-3dB}$ is $(2\pi\tau_{-3dB})^{-1} = 16$ mHz making $(\Delta f)_{EQNB} = 25$ mHz, even for a single-stage low-pass filter.

Inspection of the data in Fig. 5.18 shows the "cleanliness" that can be achieved with this resonance frequency tracking strategy. There are some challenges in setting up the system so that the gains of the amplifiers and their associated phase shifts (an inevitable consequence of the Kramers-Kronig relations), as well as the phase shifts caused by the low-pass filters and integrator, do not lead to positive feedback and oscillatory behavior in the control system. For a laboratory system, it is easy to adjust the amplifier gains and the filter time constants since these are just knobs or rotary switches on the front panel of the lock-in amplifier. For a production system, it is useful to do a careful gain-phase analysis for all the components [30].

If we hold the current that is applied to the driven coil constant, so that its amplitude is independent of frequency, then the driving force (or torque) will remain constant. In that case, the magnitude of the response measured by the receiver transducer will be proportional to the quality factor, Q, at resonance and inversely proportional to the internal dissipation of the material. Such amplitude data was collected and recorded simultaneously with the modal frequencies of Fig. 5.18.

The data shown in Fig. 5.18 combines both the modulus and the damping from the resonance frequency and amplitude measurements made on the PR 1592 [27] elastomeric bar sample. The frequency data for a torsional mode was converted to the shear modulus, as was done for the epoxy-glass composite sample using Eq. (5.72). Similarly, the amplitude data was converted to loss tangent. Since the data was acquired over a range of temperatures that included the glass transition temperature, T_{glass}, the effective range of frequencies for this measurement is over ten orders of magnitude, although the actual frequencies only varied by about a factor of four, as seen in Fig. 5.18.

The horizontal axis in Fig. 5.20 is labeled *reduced frequency*, not frequency. As seen in our analysis of viscoelastic materials in Sect. 4.4, the influence of frequency on the response of a viscoelastic material only shows up in the equations as a product with the relaxation time, τ_R. The behavior depends upon $\omega\tau_R$ rather than on ω directly. The reason that the range of reduced frequencies is so enormous in Fig. 5.20 is that viscoelastic relaxation time, τ_R, is a very strong function of temperature.

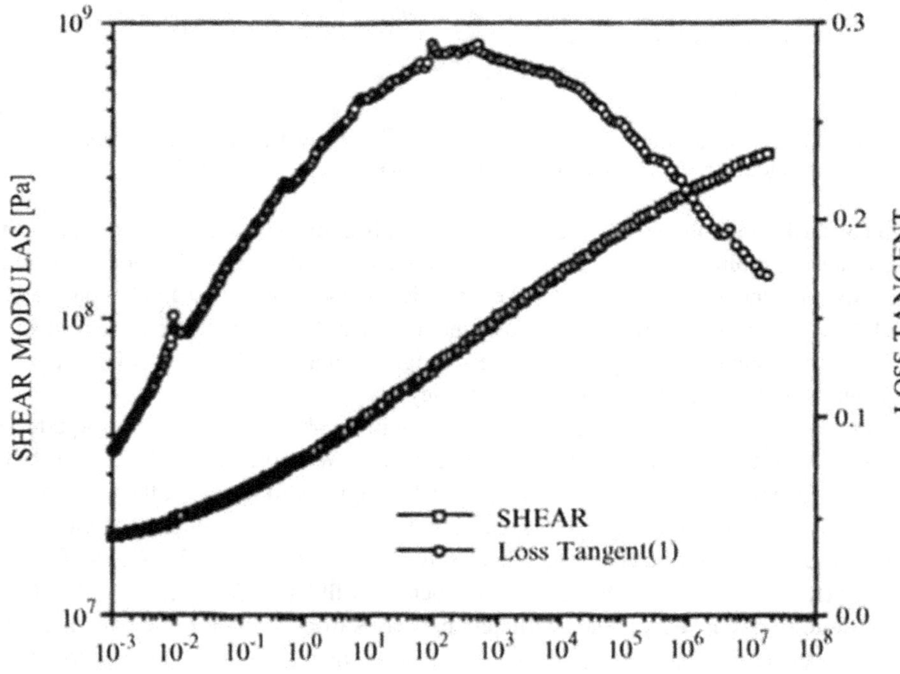

Fig. 5.20 Master curve for the magnitude of the complex shear modulus and the associated loss tangent plotted vs. the reduced frequency. This plot combines both frequency and temperature variations using the Williams-Landel-Ferry transformation of Eq. (5.73) to produce over ten decades span in reduced frequency. This plot clearly illustrates the single relaxation time behavior introduced in Sect. 4.4. The reference temperature for this graph is 10 °C. At that temperature, the actual frequencies and reduced frequencies are identical

As shown in Fig. 4.28, most of the changes in both modulus and loss occur in the regions that are close to the glass transition temperature, being the temperature where $\omega\tau_R \cong 1$. This dependence on both frequency and temperature can be converted to a single parameter that is proportional to $\omega\tau_R$ rather than to either ω or $\tau_R(T)$, which is what was done to produce the reduced frequency that is plotted on the horizontal axis of Fig. 5.20. Such a curve is called the *master curve*, since it can provide the complex modulus at any frequency, if the temperature is known, or the effect a change in temperature will make on the complex modulus at a fixed frequency.

It is obvious from inspection of the plots of moduli in Fig. 5.18 that the measured moduli are much smaller at low frequencies and high temperatures, while the moduli are greatest at high frequencies and low temperatures. There are several algorithms that can be used to convert a frequency measurement made at a temperature, T, to reduced frequency, so it can be plotted with other frequency measurements made at other temperatures. The Williams-Landel-Ferry (WLF) equation [31] was used to produce the graph in Fig. 5.20 by creating a frequency shift factor, α_T, so that data acquired at a frequency, f, and at a temperature, T, are plotted on the reduced frequency axis at a reduced frequency, $f' = \alpha_T f$.

$$\ln \alpha_T = \frac{-C_1\left(T - T_{glass}\right)}{C_2 + \left(T - T_{glass}\right)} \tag{5.73}$$

The constants C_1, C_2, and T_{glass} are different for different polymers, but, as the form of Eq. (5.73) suggests, the dependence of $\tau_R\ (T)$ is nearly an exponential function of the difference between the measured temperature, T, and T_{glass}.

5.4.6 Resonant Ultrasound Spectroscopy*

"So, apparently due to a mathematical fortuity that may have occurred during a lapse in Murphy's vigilance, the displacement vectors, u_i, which are solutions to the elastic wave equation with free boundary conditions on S, are just those points in function space at which L is stationary." W. M. Visscher [32]

The measurement of the elastic moduli of a long thin bar, as described in the previous sections, was simplified because the longitudinal, torsional, and flexural modes of such a bar were distinct. Furthermore, the modes could be excited and detected selectively by the orientation of the electrodynamic transducers with respect to the magnetic field. If a sample is not available in the shape of a long thin bar, it is still possible to determine the elastic moduli from measured resonance frequencies, but the extraction of the moduli from the resonance spectrum is much more difficult and computationally intensive.

In the early 1990s, a technique was developed that allows the normal mode frequencies of an object of arbitrary shape, and possibly anisotropic elasticity, to be accurately approximated if the object is freely suspended (hence, the quote that started this section regarding the "lapse in Murphy's vigilance"). More importantly, measurement of the resonance frequencies of such normal modes of vibration can be inverted to extract the elastic moduli of the sample. This technique has become known as *resonant ultrasound spectroscopy* (RUS) [33].

Figure 5.21 shows an apparatus that can excite and detect the resonances of a very small solid rectangular parallelepiped sample (in this case). The normal modes of oscillation for such a rectangular sample are represented in Fig. 5.22. As we have done before, the "trick" is to use polynomial functions that are products of powers of the three Cartesian coordinates, where $\lambda = \ell$, m, and n is a set of three nonnegative integers.

$$\Phi_{\ell,m,n} = x^{\ell} y^m z^n \equiv \Phi_\lambda \tag{5.74}$$

These functions are then used to express the displacements, u_i, along the three Cartesian directions.

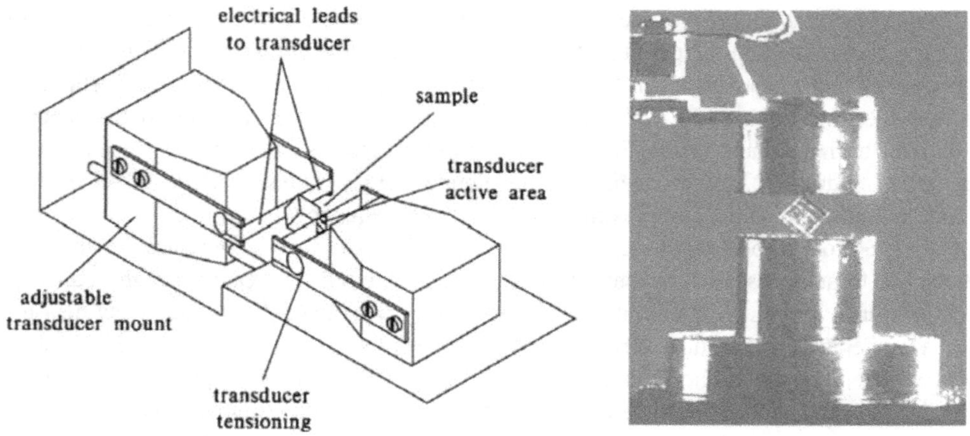

Fig. 5.21 (*Left*) Schematic diagram of an apparatus that is capable of holding a very small and fragile solid sample between two piezoelectric polymer film (PVDF) transducers, 0.50 mm wide, that only touch the sample at the corners and can excite and detect the sample's normal modes of vibration [33, 34]. (*Right*) Photograph of a room temperature RUS apparatus that excites the sample using ordinary piezoelectric transducers

$$u_i = \sum_{\lambda \in \Omega} a_{i\lambda} \Phi_\lambda \tag{5.75}$$

The number of such functions must be sufficient for the number of modes to be determined, so that the energy can be minimized through the proper choice of the $a_{i\lambda}$ using standard matrix methods. This polynomial approach will work regardless of the shape of the sample, as long as the motions of its boundaries are not constrained. The paper by Visscher et al. provides a rather entertaining description of the variety of such shapes: "including spheres, hemispheres, spheroids, ellipsoids, cylinders, eggs, shells, bells, sandwiches[20], parallelepipeds, cones, pyramids, prisms, tetrahedra, octahedra, and potatoes." [32]

The execution of these matrix operations was originally rather time-consuming or required a supercomputer. The examples shown in the paper by Visscher et al., in 1991, required a Cray-1 computer. Only 20 years later, the necessary calculations can be executed on a laptop computer. The results can be inverted to provide values of the elastic constants and their uncertainties, based on the measured resonance frequencies.

RUS has been extraordinarily successful in measuring the elastic properties of very small samples. Spoor and Maynard were able to measure the elastic moduli of a sample of quasicrystalline AlCuLi that had a mass of only 70 micrograms [34]. This is necessary because it is frequently difficult to grow large crystalline samples of exotic materials like quasicrystals or high-temperature superconductors. The simplicity and compactness of the apparatus also make RUS well-suited to measurements of the temperature dependences of elastic moduli and/or measurement on samples that are potentially dangerous.

My favorite examples are the studies on the solid phases of plutonium done by Albert Migliori and his colleagues at Los Alamos National Laboratory.[21] Plutonium exhibits four different solid phases over a relatively small range of temperatures. Shown in Fig. 5.23 are measurements of the bulk and

[20] Visscher was incorrect about "sandwiches." Polynomials cannot fit a sharp interface, so the results are too inaccurate for RUS. For a sharp interface, finite elements are required.

[21] Large samples of plutonium self-heat by radioactive decay (or worse!).

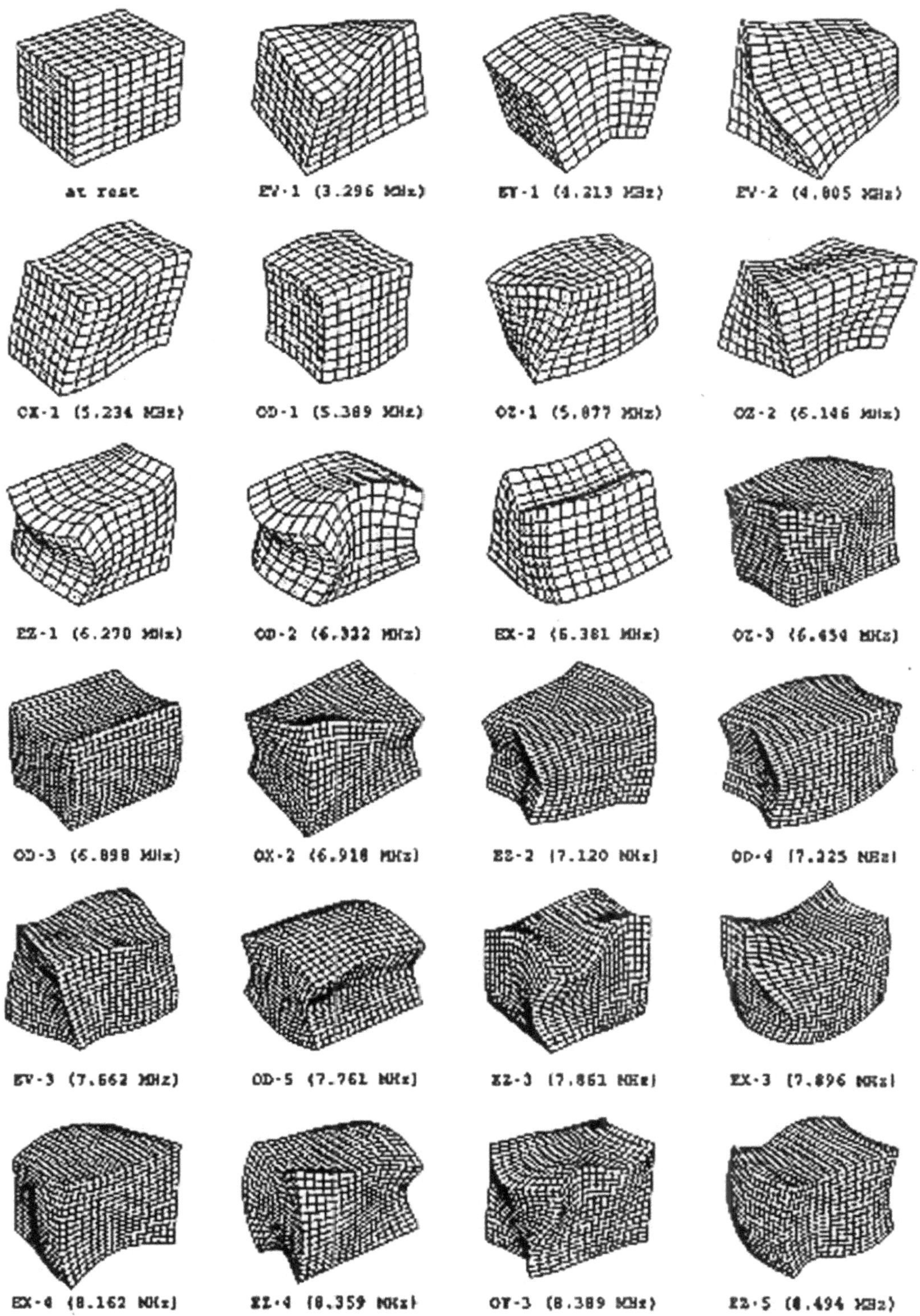

Fig. 5.22 Normal mode vibrational patterns for a rectangular parallelepiped showing that longitudinal, flexural, and torsional motions are combined in several of the modes [36]. By fitting the polynomial mode shapes, it is possible to isolate the individual elastic moduli, even for materials with anisotropic elastic properties, as discussed in Sect. 4.6, to high accuracy

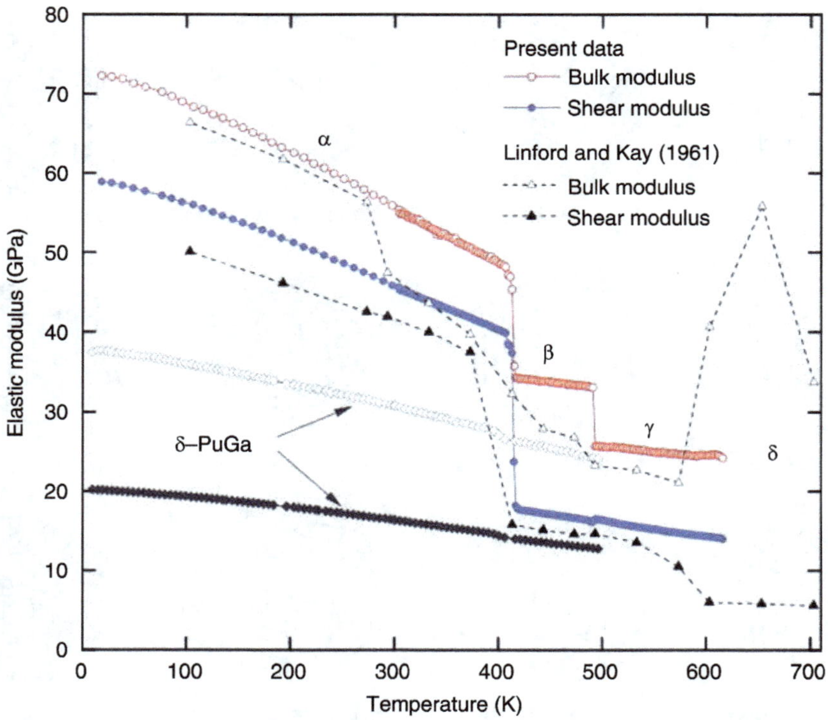

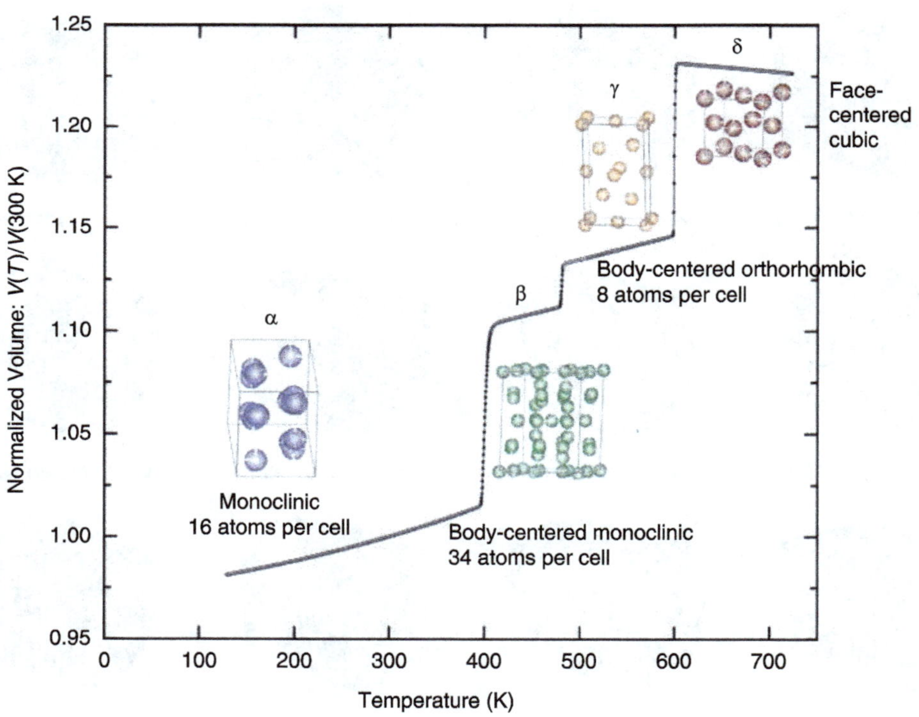

Fig. 5.23 (*Above*) Shear and bulk moduli of plutonium clearly showing the transitions between the various solid phases. (*Below*) Plutonium's density transitions for four solid phases including diagrams of the atomic arrangements for those phases [35]

shear moduli of pure Pu from 18 K to 616 K using RUS [35]. As with other resonant determinations of elastic moduli, acoustic measurements provide extraordinary data density and unmatched precision.

5.5 Vibrations of a Stiff String*

All of the calculations made in Chap. 3, which developed our understanding of the dynamics of wave motion on strings, assumed that the string was "limp," meaning that only tension provided the restoring force—the string itself had no flexural rigidity ($E = 0$, so $ES\kappa^2 = 0$). Anyone who has replaced the strings on a guitar and poked the end of a finger with the stiff string knows differently; guitar strings have sufficient rigidity to penetrate the skin (painfully). Now that we have analyzed the flexure of bars, where the only restoring force is the rigidity of the bar, we are in a position to calculate the effects that rigidity adds to the behavior of strings that are also under tension.

We have already done all of the necessary work. The vertical acceleration of a differential element of a stiff string will just be the sum of the vertical forces as calculated previously in Eq. (3.3) for the tension and in Eq. (5.36) for the flexural rigidity.

$$\rho S \frac{\partial^2 y}{\partial t^2} = \mathrm{T}\frac{\partial^2 y}{\partial x^2} - ES\kappa^2 \frac{\partial^4 y}{\partial x^4} \quad \Rightarrow \quad \frac{\partial^2 y}{\partial t^2} - c_{st}^2 \frac{\partial^2 y}{\partial x^2} + c_B^2 \kappa^2 \frac{\partial^4 y}{\partial x^4} = 0 \tag{5.76}$$

Recall that the linear mass density of a string is $\rho_L = \rho S$. The speed of transverse waves on a limp string will be designated $c_{st} = (\mathrm{T}/\rho_L)^{1/2}$, to distinguish it from the speed of longitudinal waves in thin bars, $c_B = (E/\rho)^{1/2}$.

As in Sect. 5.3.1, the assumption of a rightward-going traveling wave solution of the form $y(x,t) = \Re e\left[\widehat{\mathbf{C}} e^{j}(\omega t - kx)\right]$ will lead to an equation that yields the dispersion relation that combines tension and stiffness.

$$c_B^2 \kappa^2 k^4 - c_{st}^2 k^2 - \omega^2 = 0 \tag{5.77}$$

The solution of this characteristic equation is just that for a quadratic equation in k^2.

$$k^2 = \frac{c_{st}^2 \pm \sqrt{c_{st}^4 - 4c_B^2\kappa^2\omega^2}}{2c_B^2\kappa^2} = \frac{c_{st}^2}{2c_B^2\kappa^2}\left[1 \pm \sqrt{1 + \frac{4\omega^2 c_B^2 \kappa^2}{c_{st}^4}}\right] \tag{5.78}$$

We know that the phase speed, c_{ph}, for a flexural wave depends upon the sharpness of the bend. As derived in Eq. (5.38), $c_{ph} = (2\pi c_B \kappa)/\lambda$ and c_{st} is independent of wavelength. We expect that in the limit of very low frequencies, hence, very long wavelengths, the tension should be the dominant restoring force. A binomial expansion can be used for $(4\omega^2 {c_B}^2 \kappa^2 / {c_{st}}^4) \ll 1$ to illustrate this limit.

$$\lim_{\omega^2 \to 0}\left[k^2\right] = \frac{c_{st}^2}{2c_B^2\kappa^2}\left[1 \pm \left(1 + \frac{2\omega^2 c_B^2 \kappa^2}{c_{st}^4}\right)\right] = \frac{\omega^2}{c_{st}^2} \tag{5.79}$$

As expected, for very long wavelengths, the phase speed, $c_{ph} = \omega/k$, is constant and equal to the speed of transverse waves on a limp string, c_{st}.

In the opposite limit of very high frequencies and short wavelengths, we expect the phase speed, c_{ph}, that is characteristic of flexure waves on rigid bars that are not under tension.

$$\lim_{\omega^2\to\infty}\left[k^2\right]=\frac{\omega}{c_B\kappa}\quad\Rightarrow\quad c_{ph}\equiv\frac{\omega}{k}=c_B\kappa k=\sqrt{c_B\kappa\omega}\tag{5.80}$$

This result is identical to Eqs. (5.38) and (5.39).

The solutions to Eq. (5.76) can be found by again using separation of variables, as we did in Sect. 5.3.2, with $y(x, t) = Y(x)T(t) = Ye^{kx}Te^{j\omega t}$. This substitution converts Eq. (5.76) from a partial differential equation to an ordinary differential equation.

$$\frac{d^4Y}{dx^4}-\frac{c_{st}^2}{c_B^2\kappa^2}\frac{d^2Y}{dx^2}-\frac{\omega^2}{c_B^2\kappa^2}=0\quad\Rightarrow\quad k^4-\left(\frac{c_{st}^2}{c_B^2\kappa^2}\right)k^2-\frac{\omega^2}{c_B^2\kappa^2}=0\tag{5.81}$$

The definition of two new constants, $2\beta^2$ and γ^4, simplifies the application of the quadratic formula to Eq. (5.81).

$$k^4-2\beta^2k^2-\gamma^4=0,\quad\text{where}\quad 2\beta^2\equiv\frac{c_{st}^2}{c_B^2\kappa^2}\quad\text{and}\quad\gamma^4\equiv\frac{\omega^2}{c_B^2\kappa^2}\tag{5.82}$$

As before, taking the fourth root of k^4 will generate four values of wavenumber, although in this case, we need to use the quadratic formula to find k^2.

$$\begin{gathered}k_\pm^2=\frac{2\beta^2\pm\sqrt{4\beta^4+4\gamma^4}}{2}=\beta^2\pm\beta^2\sqrt{1+\frac{\gamma^4}{\beta^4}}\\ \text{so}\\ k_+^2=\beta^2\left[1+\sqrt{1+\frac{\gamma^4}{\beta^4}}\right]\quad\text{and}\quad k_-^2=\beta^2\left[1-\sqrt{1+\frac{\gamma^4}{\beta^4}}\right]\end{gathered}\tag{5.83}$$

Since $(\gamma/\beta)^4=4c_{ph}^4/c_{st}^4>0$, the square roots of $k_+{}^2$ will be real numbers, and the square roots of $k_-{}^2$ will be purely imaginary numbers. Substitution of these results back into $Y(x)$ produces four exponentials that are similar to those in Eq. (5.41), but with the hyperbolic trigonometric functions having $k_+ x$ in their argument and the circular trigonometric functions using $k_- x$.

$$\begin{aligned}Y(x)&=C_1e^{k_+x}+C_2e^{-k_+x}+C_3e^{jk_-x}+C_4e^{-jk_-x}\quad\text{or}\\ Y(x)&=A\cosh(k_+x)+B\sinh(k_+x)+C\cos(k_-x)+D\sin(k_-x)\end{aligned}\tag{5.84}$$

At this point, we could proceed by fitting boundary conditions, generating transcendental equations that quantize the allowable values of k_n for normal mode vibrations, and calculating the resonance frequencies, ω_n, of those modes. Having done this once before for the flexural modes of bars (and having found it to be tedious!), we will not go down that route here again. Those readers who are nostalgic for such algebraic gymnastics are referred to Morse [37], although his final result is incorrect.[22]

Since our goal is to determine the change in the modal frequencies and resulting anharmonicity of the overtones for a fixed-fixed string due to its flexural rigidity,[23] we can assume that tension is the

[22] Morse's solution for the frequency in his result for his ν_n, equal to our f_n, includes a mode-independent constant term. He is not able to produce the n^2 dependence without adding another term to his Taylor series expansion, and his result is obviously incompatible with Young's observations [38].

[23] Piano technicians compensate for this anharmonicity. Anharmonicity is present in different amounts in all of the ranges of the instrument but is especially prevalent in the bass and high treble registers. The result is that octaves are tuned slightly wider than the harmonic 2:1 ratio. The exact amount that octaves are "stretched" by a piano tuner, by tuning the octave to a match half the frequency of the second overtone instead of the first, varies from piano to piano and even from register to register within a single piano—depending on the exact anharmonicity of the strings involved. With small pianos, the anharmonicity is so significant that they are stretched by matching the triple-octave.

dominant restoring force (since tension is what is adjusted to tune the stiff string's pitch). Additionally, the transverse displacement function, $y(x, t)$, of the string is not significantly affected by the string's flexural rigidity.

By now, the previous statements should suggest to the serious reader the application of Rayleigh's method, using the mode shapes, $y_n(x, t)$, for a fixed-fixed string in Eq. (3.21), repeated here, as the trial functions.

$$y_n(x,t) = \Re e\left[C_n e^{j\omega_n t} \sin\left(n\frac{\pi x}{L}\right)\right]; \quad n = 1, 2, 3, \ldots \tag{5.85}$$

We have already demonstrated that Rayleigh's method provides the exact solution for the modal frequencies, ω_n, of limp fixed-fixed string in Eq. (3.32) by taking the square root of the ratio of the stability coefficient to the inertia coefficient. Since the addition of flexural rigidity does not change the linear mass density, ρ_L, of the string, the addition of flexural rigidity to the tension as a restoring force means that the relative change in frequency, $\delta f_n/f_n$, will be determined by the relative change in the stability coefficient (the potential energy) alone.

$$\frac{\delta f_n}{f_n} = \frac{1}{2}\frac{\delta(PE)_n}{(PE^T)_n} = \frac{1}{2}\frac{(PE^F)_n}{(PE^T)_n} \tag{5.86}$$

The expression for the potential energy due to flexure of a differential element of length, dx, given by $d(PE^F)$, that is bent by an angle, ϕ, is provided in Eq. (4.32). This expression can be integrated over the length, L, of the string using the fixed-fixed mode shapes of Eq. (5.85).

$$\begin{aligned}(PE^F)_n &= \frac{ES\kappa^2}{2}\int_0^L \left(\frac{\partial^2 y_n}{\partial x^2}\right)^2 dx = \frac{ES\kappa^2}{2}\left(\frac{n\pi}{L}\right)^4 C_n^2 \int_0^L \sin^2\left(\frac{n\pi x}{L}\right) dx \\ \Rightarrow \quad (PE^F)_n &= n^4\pi^4 C_n^2 \frac{ES\kappa^2}{4L^3}\end{aligned} \tag{5.87}$$

The maximum potential energy, $(PE^T)_n$, in mode n for the limp string was provided in Eq. (3.31) and reproduced here.

$$(PE)_n = \frac{\pi^2 \mathrm{T}L}{\lambda_n^2} C_n^2 = \frac{n^2\pi^2\mathrm{T}}{4L} C_n^2 \tag{5.88}$$

The relative frequency shift, $\delta f_n/f_n$, for any mode number, n, is given by substitution of Eqs. (5.87) and (5.88) into Eq. (5.86).

$$\frac{\delta f_n}{f_n} = \frac{1}{2}\frac{(PE^F)_n}{(PE^T)_n} = \frac{(\pi n)^2}{2}\frac{ES}{\mathrm{T}}\left(\frac{\kappa}{L}\right)^2 = \frac{(\pi n)^2}{2L^2}\frac{c_B^2\kappa^2}{c_{st}^2} \tag{5.89}$$

This solution reproduces the result of Young who claimed that the measurements were "entirely compatible with the relationship given" [38]. The effects of the string's stiffness grow with the square of the mode number, n. The dimensionless ratios, ES/T and $(\kappa/L)^2$, scale the importance of the flexural rigidity and the tension as restoring forces.

We can now do an example to estimate the anharmonicity for a piano string based on Eq. (5.89). The *speaking length*,[24] L, of the piano string, corresponding to "middle-C," with a frequency of

[24] The "speaking length" of a piano string is the distance between the bridge, located on the sound board near the hitching pin, and the *capo d'astro*, near the tuning pin. It is the speaking length that determines the distance between the fixed-fixed boundaries.

$f = 261.6$ Hz, is 63 cm. Strings are typically made from high-carbon steel (e.g., ASTM A228) with a Young's modulus of $E = 210$ GPa and mass density of $\rho = 7850$ kg/m^3. The string is tensioned to about T = 780 N. Assuming the tension dominates the flexural rigidity, the speed of transverse vibrations is $c_{st} = f\lambda = 2\,fL = 332$ m/s. The linear mass density of the string can be related to the tension and c_{st}: $\rho_L = \mathrm{T}/c_{st}^{\,2} = 7.07 \times 10^{-3}$ kg/m. Using the mass density of steel, ρ_{steel}, the diameter of the wire is also determined, $d = (4\rho_L/\pi\rho_{steel})^{½} = 1.07$ mm; hence, $\kappa^2 = a^2/4 = d^2/16 = 7.17\,x\,10^{-8}m^2$.

Young's dimensionless anharmonicity coefficient [38], b, can be expressed in terms of d using Eq. (5.89).

$$b = \frac{\pi^3}{128}\frac{E}{\mathrm{T}}\left(\frac{d^2}{L}\right)^2 \tag{5.90}$$

Substituting the nominal values calculated for the middle-C string of a piano, $b = 2.15 \times 10^{-4}$, with $\delta f_n/f_1 = n^2 b$. The first overtone, $n = 2$, at the octave above middle-C is therefore sharp by 861 ppm, or about 1.5 cents.[25] In Table 3.2, we saw that the $n = 5$ and $n = 7$ harmonics did not correspond exactly to a just-temperament interval, with $n = 5$ falling between $E^{\#}$ and F^b and $n = 7$ falling between $A^{\#}$ and B^b for a scale based on middle-C. Those harmonics will be up-shifted by 5.375×10^{-3} (9.3 cents) and 1.011% (17.5 cents), respectively.

This example set an effective lower limit on the anharmonicity of stiff piano strings, since the anharmonicity is greatest in the upper and lower registers of the piano, farthest from middle-C.

5.6 Harmonic Analysis

In this chapter, we used our understanding of elasticity from the previous chapter to describe the propagation of waves in solids and used our understanding of the modal frequencies of those standing waves in solids to evaluate the elastic moduli introduced in the previous chapter—a fair exchange. More importantly, we saw in this chapter the power of harmonic analysis (i.e., the assumption of time harmonic solutions) as augmented by a more formalized version of the approach known as separation of variables. That perspective not only was useful for solutions to the wave equations but also to the solution of higher-order partial differential equations. Imposition of boundary conditions again resulted in the quantization of normal mode frequencies in the flexural cases, though requiring four boundary conditions, rather than two required for systems that obeyed the wave equation.

We are now fully equipped to take our techniques for analyzing zero-dimensional harmonic oscillators and one-dimensional systems (strings and bars) and extend them to two-dimensional membranes and plates.

Talk Like an Acoustician

Flexural rigidity	Dynamic range
Polar moment of inertia	Electromagnetic cross talk
Torsional rigidity	Quadrature output
Phase speed	Reduced frequency
Dispersion	Master curve
Pure tone	Resonant ultrasound spectroscopy
Separation of variables	Speaking length (for piano strings)

[25] Recall from Sect. 3.3.3 that one cent is one one-hundredth of an equal-temperament semitone (in the logarithmic sense), or a frequency ratio of $2^{1/1200} = 1.000578$.

Exercises

1. **Longitudinal vibration of a fixed-mass loaded bar**. The longitudinal vibrational normal modes of a thin bar with fixed-fixed and free-free boundary conditions were calculated in Sect. 5.1. In this exercise, you will calculate the longitudinal modes of a bar of length, L, and cross-sectional area, S_x, that is fixed at $x = 0$ and has a mass, M, attached at $x = L$.
   (a) *Normal mode frequencies*. If the added mass, M, has twice the mass of the bar, $m_B = (M/2)$, calculate the first three normal mode frequencies in terms of c_B and L.
   (b) *Hooke's law limit*. Approximate the frequency of the gravest mode (i.e., the f_o mode corresponding nearly to a mass-spring oscillator) by assuming that the mass, M, is restored by a stiffness, $K = ES_x/L$. Also report these results in terms of c_B and L.
   (c) *Effective mass approximation*. Use a Taylor series expansion [similar to the technique that produced Eq. (3.71)] to determine f_o by adding the effective mass of the bar to the mass, M, at $x = L$.
2. **Elastic moduli.** A sample of syntactic foam[26] is cast into a solid rod of circular cross-section with a length of 14.96 cm and a diameter of 1.55 cm. Its mass is 17.1 gm. The fundamental torsional and longitudinal resonances of the free-free vibrations of the rod were measured to be 4192 Hz and 6952 Hz, respectively. Determine the shear modulus, Young's modulus, and Poisson's ratio of this syntactic foam sample.
3. **M'bira**. Also known as an African thumb piano, the m'bira, shown in Fig. 5.24, is an indigenous musical instrument of the Shona. The most common tuning is called Nyamaropa. It is similar to the western Mixolydian mode that is produced by playing a sequence of eight notes (no sharps or flats) starting and ending with "G" in an equally tempered scale (see Sect. 3.3.3).
   Assuming that the width and thickness of the reeds are constant, and the vibrational frequencies are that of a clamped-free bar vibrating in its fundamental fixed-free mode, produce a table for the lengths of each reed, starting with $G_4 = 392$ Hz and ending with $G_5 = 784$ Hz, if the reed for G_5 is 5.0 cm long.

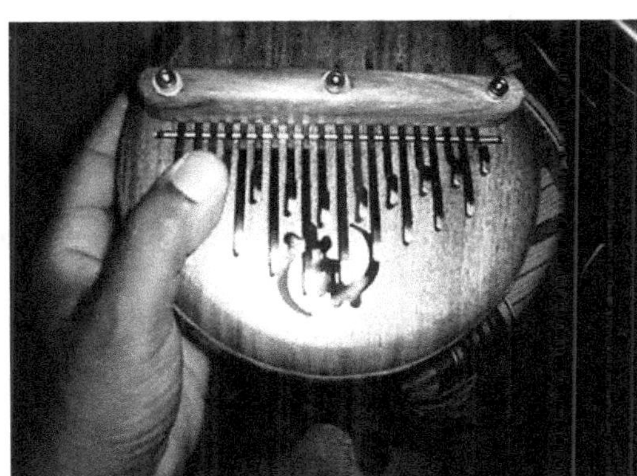

Fig. 5.24 African thumb piano

[26] Syntactic foam was developed in the 1960s to provide buoyancy for instruments deployed in the deep ocean. They are composite materials fabricated by filling a castable epoxy with hollow glass microspheres. Those microspheres (sometimes also called microballoons) are very rigid, so the foam is not crushed when subjected to large hydrostatic pressures.

Table 5.5 If you don't have easy access to the necessary items, you can use the frequencies listed at the right taken for a ruler with thickness, $t = 0.79$ mm, and width, $w = 24.0$ mm. Based on the length, $L = 32.4$ cm, and the mass, $M = 47.2$ gm, the mass density of the ruler's material is $\rho = 7680$ kg/m^3

Length [m]	Frequency [Hz]
0.29	7.09
0.28	7.51
0.25	9.53
0.22	12.31
OJO	14.96
0.18	18.13
0.16	22.97
0.14	29.98
0.12	40.67

4. **Modulus of a 12″ ruler**. Try to determine the elastic modulus of a steel machinist's ruler by clamping it to your desk and measuring the frequency of vibration when you "twang" the ruler. (Machinists call that ruler a "scale.") Since the length of a ruler is already accurately marked, it will be easy to determine the vibrating length, although ensuring that the end attached to your desk is "clamped" may be more difficult. A C-clamp will probably provide an adequate restraint. The thickness of your ruler can be determined with a micrometer or Vernier calipers. Measuring the frequency of vibration may be more challenging, although many "smartphones" have an FFT frequency analyzer "app" available. Measure the vibration frequencies, f_n, as a function of the free length, L_n, of the ruler and use Eq. (5.50) to measure Young's modulus of your ruler. This technique has been used to determine Young's modulus of metallic glasses [13].
   (a) *Bar speed.* Based on your data or the data in Table 5.5, think about how the data should be plotted and/or averaged to determine the bar speed, $c_B = [E/\rho]^{1/2}$, and its uncertainty if $|\delta t/t| = 0.6\%$, corresponding to $\pm 1\sigma$. Report your result and its relative uncertainty, providing any graphical output used to determine your result.
   (b) *Young's modulus.* Use the ruler's mass density to determine the ruler's Young's modulus and its relative uncertainty. If you use the data in Table 5.5, you can let the relative uncertainty corresponding to $\pm 1\sigma$ in the reported mass density to be $|\delta\rho/\rho| = 1.0\%$.
5. **Complex modulus measurement**. In Chap. 2, Problem 21 provided the amplitude and frequency at 22 points for the third torsional mode of a cylindrical rod, with a diameter of about ½" = 12.7 mm, made from Eastman Tritan® Copolyester TX1001. The effective length of the rod for torsional vibration that included the mass loading due to the coils is $L_{eff} = 31.94$ cm. The density of the rod is $\rho = 1.18$ gm/cm^3.
   (a) *Normal mode frequency.*[27] The measured frequencies and amplitudes are provided in Table 2.2. Determine the resonance frequency, f_3, by choosing the largest measured amplitude and the two other amplitudes adjacent to the largest value. Fit those three frequency-amplitude pairs to a second-order polynomial, and use the resulting coefficients to make a quadratic interpolation to determine the best value for f_3. Plug f_3 back into that polynomial to determine the maximum amplitude of that third mode, $A_3(1)_{max}$.

[27] If you have solved Prob. 21 in Chap. 2 by doing a least-squares fit of the data to the Rayleigh line shape of Eq. (2.63), you may use those results for f_3, $A_3(1)$, and Q_3 in parts (a) and (c) of this problem. Of course, if you just love to analyze high-quality resonance data (some of us do!), then do it both ways and compare your results. After that, make your friends and colleagues call you a "spectroscopist."

(b) *Shear modulus*. Based on L_{eff}, ρ, and f_3/n, determine the "real" part of the bar's shear modulus at f_3.

(c) *Quality factor*. Determine the two "down 3 dB" frequencies, f_+ and f_-, by finding those frequencies which correspond to an amplitude of $A_3(1)_{\max}/\sqrt{2}$. Since none of the measured amplitudes exactly equal that -3 dB amplitude, choose the two measurements that straddle $A_3(1)_{\max}/\sqrt{2}$ and fit a straight line between them to find f_- by linear interpolation. Do the same to find f_+. Use those two -3 dB frequencies to calculate the quality factor, Q_3, for that $n = 3$ torsional mode using Eq. (B.3).

(d) *Loss modulus*. Use Eq. (B.9) and your value for the elastic modulus you calculated in part (b) and Q_3 to determine the "imaginary" part of the complex shear modulus.

References

1. E. Mauceli et al., The Allegro gravitational wave detector: Data acquisition and analysis. Phys. Rev. D **54**(2), 1264–1275 (1996)
2. M. McHugh et al., Calibration of the ALLEGRO resonant detector. Classical Quantum Gravity **22**(18), S965–S973 (2005)
3. M. Janes, Sonic Pile Driving: The History and the Resurrection of Vibration Free Pile Driving. http://www.resonancetechnology.ca/res%20driver%20history%20090112.pdf. A very impressive video comparing the resonant and drop hammer techniques is also available from that site: http://www.resonancetechnology.ca/gallery%20video%201.php
4. G. Bodine, Jr., Sonic method and apparatus for driving casings through earthen formations. U.S. Patent No. 3,375,884 (April 2, 1968)
5. G. Bodine, Jr., Sonic method and apparatus for grinding rock material and the like to powder. U.S. Patent No. 3,429,512 (February 25, 1969)
6. S.P. Timoshenko, J.N. Goodier, *Theory of Elasticity*, 3rd edn. (McGraw-Hill, New York, 1970). See Chapter 11; ISBN 978-1-4020-7745-6
7. L.D. Landau, E.M. Lifshitz, *Elasticity*, 2nd edn. (Pergamon, New York, 1970).; See §18 and §25. ISBN 0 08 006465 5
8. W.C. Young, R.G. Budynas, *Roark's Formulas for Stress and Strain*, 7th edn. (McGraw-Hill, New York, 2002); ISBN 0-07-072542-X
9. S.P. Timoshenko, *History of Strength of Materials* (McGraw-Hill, New York, 1953); ISBN 0-486-61187-6
10. Newton, Opticks or a Treatise on the Reflexions, Refractions, Inflections and Colours of Light, Also Two Treatises of the Species and Magnitude of Curvilinear Figures, 1st ed. (Smith & Walford, 1704). 4th ed., with forward by A. Einstein and preface by I. B. Cohen (Dover, 1979); ISBN 978-0-486-60205-9
11. D. Heckerman, S. Garrett, G. Williams. P. Weidman, Surface tension restoring forces on gravity waves in a narrow channel. Phys. Fluids **22**, 2270–2276 (1979)
12. S.M. Han, H. Benaroya, T. Wei, Dynamics of transversely vibrating beams using four engineering theories. J. Sound Vib. **225**(5), 935–988 (1999)
13. M.B. Barmatz, H.J. Leamy, H.S. Chen, A method for the determination of Young's modulus and internal friction in metallic glasses. Rev. Sci. Instrum. **79**(6), 885–886 (1971)
14. R.P. Feynman, R.B. Leighton, M. Sands, *The Feynman Lectures on Physics*, vol II (Addison-Wesley, Reading, 1963), pp. 38–38
15. L.D. Landau, E.M. Lifshitz, *Mechanics* (Pergamon, Oxford, 1960). §39
16. N. Yazdi, F. Ayazi, K. Najafi, Micromachined inertial sensors. Proc. IEEE **86**(8), 1640–1659 (1998)
17. W. Strutt (Lord Rayleigh), On waves propagated along the plane surface of an elastic solid. Proc. London Math. Soc. **S1–17**, 4–11 (1885)
18. F. Graff, *Wave Motion in Elastic Solids* (Clarendon, Oxford, 1975). Reprinted by Dover, 1991; ISBN 0-486-66745-5
19. S.L. Garrett, Resonant acoustic determination of elastic moduli. J. Acoust. Soc. Am. **88**(1), 210–221 (1990)
20. E-Cast F-28 with Hardener 215, United Resins Corp., 1305 E. St. Gertrude Place, Unit B., Unit B, Santa Ana, CA 92705
21. A. Barone, A. Giacomini, Experiments on some electrodynamic ultrasonic vibrators. Acoustica **4**, 182–184 (1954)
22. R.W. Leonard, Attenuation of torsional waves in Teflon. J. Acoust. Soc. Am. **40**(1), 160–162 (1966)

23. E.R. Fitzgerald, Simple method for observing audio frequency resonances and sound beams in crystals. J. Acoust. Soc. Am. **36**(11), 2086–2089 (1964)
24. R.W. Leonard, Comments on the existence of the Fitzgerald effect. J. Acoust. Soc. Am. **38**(4), 673–674 (1965)
25. Q.S. Guo, D.A. Brown, Determination of the dynamic elastic moduli and internal friction using thin resonant bars. J. Acoust. Soc. Am. **108**(1), 167–174 (2000)
26. J. W. Strutt (Lord Rayleigh), *Theory of Sound*, 2nd edn. (Macmillan, London, 1894).; Reprinted by Dover, 1948, Vol. I, Ch. VII and Ch. VIII
27. Product Research & Chemicals, Corp., 5454 San Fernando Road, Glendale, CA 91203
28. H. Tan, Resonant acoustic determination of complex elastic moduli (Unclassified), Master's Thesis in Engineering Acoustics, U. S. Naval Postgraduate School, Monterey, CA (March, 1991); DTIC Report No. AD A245 058
29. P. Horowitz, W. Hill, *The Art of Electronics*, 2nd edn. (Cambridge University Press, Cambridge, 1989). §15.15; ISBN 0-521-37095-7
30. R.E. Best, *Phase-locked Loops: Design, Simulation, and Applications*, 6th edn. (McGraw-Hill, New York, 2007); ISBN 978-0-07-149375-8
31. J.D. Ferry, *Viscoelastic Properties of Polymers*, 3rd edn. (Wiley, New York, 1980); ISBN: 978-0-471-04894-7
32. W.M. Visscher, A. Migliori, T.M. Bell, R.A. Reinert, On the normal modes of free vibration of inhomogeneous and anisotropic elastic objects. J. Acoust. Soc. Am. **90**(4 Pt. 1), 2154–2162 (1991)
33. V.M. Keppens, J.D. Maynard, A. Migliori, Listening to materials: From auto safety to reducing the nuclear arsenal. Acoust. Today **6**(2), 6–13 (2010)
34. P.S. Spoor, J.D. Maynard, A.R. Kortan, Elastic isotropy and anisotropy in quasicrystalline and cubic AlCuLi. Phys. Rev. Lett. **75**(19), 3462–3465 (1995)
35. Y. Suzuki, V.R. Fanelli, J.B. Betts, F.J. Freibert, C.H. Mielke, J.N. Mitchell, M. Ramos, T.A. Saleh, A. Migliori, Temperature dependence of elastic moduli of polycrystalline β plutonium. Phys. Rev. B **84**, 064105 (2011)
36. P. S. Spoor, Elastic Properties of Novel Materials using PVDF Film and Resonant Ultrasound Spectroscopy, Ph. D. thesis in Acoustics, Penn State, December, 1996
37. P.M. Morse, *Vibration and Sound* (Acoustical Society of America, New York, 1976), pp. 166–170, §16. ISBN 0-88318-287-4
38. R.W. Young, Inharmonicity of plane wire piano strings. J. Acoust. Soc. Am. **24**(3), 267–273 (1952)

6

Vibrations of Membranes and Plates

This is the final chapter in Part I of this textbook that addresses vibration. In this chapter, we will apply the techniques developed thus far to two-dimensional vibrating surfaces. In Part II, we begin our investigations into waves in fluids. This final vibration chapter is then appropriately transitional, since most waves that are produced in fluids result from the vibrations of two-dimensional surfaces. All acoustical stringed musical instruments[1] include some mechanism to transfer the vibration of the string to a two-dimensional surface, whether that is the sounding board of a piano, autoharp, or dulcimer; the

body of a guitar, mandolin, or some member of the violin family; or the drumhead of a banjo or *erhu*. Strings are very inefficient in their ability to couple their vibrations to the surrounding fluid medium; two-dimensional surfaces are quite a good deal more efficient. Of course, drums of all kinds make sound due to the vibration of a membrane. Obviously, loudspeakers create sound in the air by the vibrations of a two-dimensional piston – there are no electrodynamic loudspeakers (see Sect. 2.5.5) that have only voice coils with no cone and/or dome attached.

In this chapter, we will first focus on vibrations of membranes and then consider vibrations of plates. The distinction between a membrane and a plate is analogous to the difference between a limp string and a rigid bar in one dimension. The restoring forces on membranes are due to the tension in the membrane, while the restoring forces for plates are due to the flexural rigidity of the plate.[2]

The transition to two dimensions also introduces some other features that did not show up in our analysis of one-dimensional vibrating systems. Instead of applying boundary conditions at one or two

---

[1] Electrical stringed instruments have a "pickup" (transducer) that converts the string's vibrations into an electrical signal. In performance, those electrical signals are amplified and applied to a loudspeaker which is a vibrating two-dimensional surface. For recorded music, those signals may be recorded electronically and the vibrations postponed until the recording is played back, again with the use of a two-dimensional vibrating surface.

points, they will have to be applied along a line or a curve. In this way, incorporation of the boundary condition is linked inexorably to the choice of coordinate systems used to describe the resultant modal shape functions. The vibration of a circular membrane could be described in a Cartesian coordinate system, since radial and azimuthal variations in the vertical displacements could be represented by an infinite superposition of plane waves, but it is much easier if we choose a polar coordinate system to describe vibrations of a circular membrane or plate.

When we applied a force to a string, the string was deformed into a triangular shape. When a membrane is excited by a point force, its displacement is infinite since a point force will represent infinite pressure.[3] Our analysis will avoid this divergence by concentrating on oscillating fluid pressures rather than forces to drive the motion of membranes and the diaphragms of microphones.

It was also true for strings, or torsional and longitudinal waves in bars, that the number of modes within a given frequency interval, Δf, was fairly constant or exactly constant for ideal fixed-fixed or free-free boundary conditions: $f_{n+1} - f_n = f_1$. The number of modes within a frequency interval, Δf, was simply $\Delta f/f_1$ if $\Delta f \gg f_1$. Under all circumstances, $f_{n+1} \neq f_n$ for strings or bars. For two-dimensional vibrators, two indices are required to uniquely specify the frequency of a normal mode, $f_{m,n}$, with the number of modes in a given frequency interval increasing in proportion to the center frequency of that interval, even though the interval represents a fixed frequency span. It is also possible that modes with different mode numbers might correspond to the same frequency of vibration, a situation that is designated as "modal degeneracy."

6.1 Rectangular Membranes

Despite the differences between one- and two-dimensional vibrating systems just mentioned, there is nothing new that we will need to employ that we have not already used in our investigations of strings to develop an equation for the transverse vibrations of a membrane or to find two-dimensional solutions to that resulting equation. In many ways, a membrane can be thought of as the layer produced

[2] The flexural rigidity of a plate is defined differently from the flexural rigidity of a bar because it is much more difficult to bend a plate along one axis if it has already been bent along another perpendicular axis. (Try this with a sheet of paper – easy to make the first bend but more difficult to make the second bend along an orthogonal axis.)

[3] Since the area of a point is zero, any force divided by zero area produces an infinite pressure. Similarly, we can think of the membrane's surface tension. When a membrane is poked by a blunt end, the force magnitude is the surface tension times the circumference of the blunt end. The point force has zero circumference, so there is no balance force and infinite displacement will result.

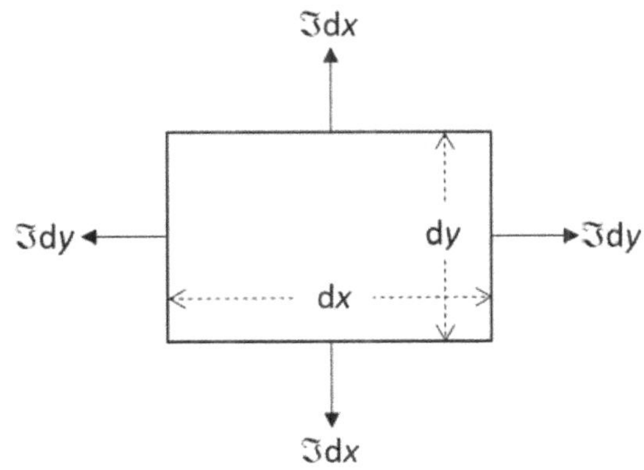

Fig. 6.1 Forces on a differential area element $dA = dx\,dy$ of a membrane as expressed in a Cartesian coordinate system

by placing an infinite number of strings side by side, although the perspective we are about to cultivate will be far more useful.

As with the string, we will begin by calculating the net vertical force, dF_z, on a differential element of a membrane with differential area, $dA = dx\,dy$, under assumptions that are analogous to those used for strings. We will assume the membrane is sufficiently thin that the membrane's material will provide no flexural rigidity—a limp membrane. We will also assume that the membrane is uniform and that it has a surface mass density (mass per unit area), ρ_S, that is independent of position on the membrane's surface and that $\rho_S = \rho t$, where ρ is the mass density of the membrane's material and t is the membrane's constant thickness. We will also assume linear behavior and neglect the changes in the membrane's tension due to its transverse displacements in the same way we ignored changes in the string's length that could affect its tension, as described in Eq. (3.28). As before, we will assume sufficiently small displacements that the last statement is approximately true.

Figure 6.1 shows a differential element of a membrane, located in the *x*-*y* plane, which is acted upon by tensile forces produced by a tension per unit length, $\Im$, applied at all the edges of the membrane. The tension on a differential element will depend upon the length of its edge, in this case either dx or dy. In Fig. 6.1, the restoring forces due to the two boundaries of width, dx, located between y and $y + dy$, are just $\pm\Im\ dx$ pulling in opposite directions. Similarly, the restoring forces due to the two boundaries of width dy, located between x and $x + dx$, are just $\pm\Im\ dy$ pulling in opposite directions.

If that differential element of the membrane is displaced in the z direction, then $F_{z,x}$ is the net vertical force that will cause the membrane element to be restored to its equilibrium position (and overshoot, due to its mass).

$$F_{z,x} = \Im\, dy\left[\left(\frac{\partial z}{\partial x}\right)_{x+dx} - \left(\frac{\partial z}{\partial x}\right)_{x}\right] = \Im\,\frac{\partial^2 z}{\partial x^2}\,dx\,dy \tag{6.1}$$

Of course, the force provided by the other pair of tensions, $F_{z,y}$, will have the same form, and the acceleration of the differential element in the z direction of membrane, due to both $F_{z,x}$ and $F_{z,y}$, will be determined by Newton's Second Law where $\rho_S = \rho t$ is the surface mass density. In that case, t is the membrane's thickness, although otherwise it will indicate time, as in the equations that follow.

$$\rho_S dx\,dy \frac{\partial^2 z(x,y,t)}{\partial t^2} = \Im \left(\frac{\partial^2 z(x,y,t)}{\partial x^2} + \frac{\partial^2 z(x,y,t)}{\partial y^2} \right) dx\,dy$$
$$\Rightarrow \quad \frac{\partial^2 z(x,y,t)}{\partial t^2} - c^2 \left(\frac{\partial^2 z(x,y,t)}{\partial x^2} + \frac{\partial^2 z(x,y,t)}{\partial y^2} \right) = 0 \tag{6.2}$$

The right-hand version of Eq. (6.2) introduced a transverse wave propagation speed, $c = \sqrt{\Im/\rho_S}$, and has the form of a wave equation, but in two dimensions.

Because we will be describing the displacement of the membrane in both Cartesian and polar coordinate systems, it is convenient to express the wave equation for membranes in terms of the *Laplacian operator*, ∇^2.

$$\nabla^2 z(x,y,t) = \frac{1}{c^2} \frac{\partial^2 z(x,y,t)}{\partial t^2} \tag{6.3}$$

In this way, the wave equation for membranes can be expressed in either coordinate system. By comparison to Eq. (6.2), it is easy to see how the Laplacian can be expressed in Cartesian coordinates. Also shown in Eq. (6.4) is the Laplacian expressed in polar coordinates, where r is the distance of a point from the origin of the coordinate system and θ is the angle that point makes with respect to the positive horizontal axis.

$$\nabla^2 = \frac{\partial^2}{\partial x^2} + \frac{\partial^2}{\partial y^2}; \quad \nabla^2 = \frac{\partial^2}{\partial r^2} + \frac{1}{r}\frac{\partial}{\partial r} + \frac{1}{r^2}\frac{\partial^2}{\partial \theta^2} \tag{6.4}$$

The polar form will be derived in Sect. 6.2.

6.1.1 Modes of a Rectangular Membrane

We will start by seeking normal mode solutions with frequencies, $\omega_{m,n}$, for a rectangular membrane that is L_x long and L_y wide and is clamped along all four edges: $z(0, y, t) = z(L_x, y, t) = z(x, 0, t) = z(x, L_y, t) = 0$. Since we seek normal mode solutions that require all parts of the membrane vibrate with the same frequency, ω, we will assume that$\mathbf{z}(x, y, t) = z(x, y)e^{j\omega t}$, thus converting Eq. (6.2) from a wave equation into a time-independent *Helmholtz equation*, with $k^2 = \omega^2/c^2$.

$$\frac{\partial^2 z(x,y)}{\partial x^2} + \frac{\partial^2 z(x,y)}{\partial y^2} + k^2 z(x,y) = 0 \tag{6.5}$$

We will seek solutions to Eq. (6.5) by separation of variables, substituting $z(x, y) = X(x)Y(y)$.

$$Y\frac{d^2X}{dx^2} + X\frac{d^2Y}{dy^2} + k^2XY = 0 \quad \Rightarrow \quad \frac{1}{X}\frac{d^2X}{dx^2} + \frac{1}{Y}\frac{d^2Y}{dy^2} = -k^2 \tag{6.6}$$

From the right-hand version of Eq. (6.6), it is clear that the left-hand term, $(1/X)$ (d^2X/dx^2), depends only upon x and is independent of y. The second term, $(1/Y)$ (d^2Y/dy^2), depends only upon y and is independent of x. Since k^2 is a constant, it is only possible to have a solution to Eq. (6.6) for all x and y if those two terms are separately equal to constants, which we will set equal to $-k_x^2$ and $-k_y^2$.

$$\frac{d^2X}{dx^2} + k_x^2 X = 0 \quad \text{and} \quad \frac{d^2Y}{dy^2} + k_y^2 Y = 0 \tag{6.7}$$

Substitution of Eq. (6.7) into Eq. (6.6) produces the *separation condition*.

$$k_x^2 + k_y^2 = k^2 = \frac{\omega^2}{c^2} \tag{6.8}$$

At this juncture, we are rather familiar with the solutions to harmonic oscillator equations like those in Eq. (6.7). By choosing sine functions for the spatial dependence of x and y, we will automatically have a solution that satisfies $\mathbf{z}(0, y, t) = \mathbf{z}(x, 0, t) = 0$.

$$z(x, y, t) = \Re e\left[\widehat{\mathbf{C}} \sin(k_x x + \phi_x) \sin(k_y y + \phi_y) e^{j\omega t}\right] \tag{6.9}$$

As before, imposition of the other two boundary conditions, $z(L_x, y, t) = z(x, L_y, t) = 0$, quantizes the allowed values of k_x and k_y.

$$k_x = \frac{m\pi}{L_x} \quad \text{and} \quad k_y = \frac{n\pi}{L_y} \quad \text{with} \quad m, n = 1, 2, 3, \ldots \tag{6.10}$$

It is worthwhile to note that neither $m = 0$ nor $n = 0$ provides an acceptable solution since $\sin(0) = 0$, so the vertical displacements would be zero everywhere if either integer index were zero.

Combining the separation restriction on the sum of k_x^2 and k_y^2 in Eq. (6.8) with their quantization conditions in Eq. (6.10) produces an expression for the normal mode frequencies, $f_{m,n}$, in terms of the dimensions of the membrane, L_x and L_y, and the speed of transverse waves on the membrane, $c = \sqrt{\Im/\rho_S}$.

$$f_{m,n} = \frac{\omega_{m,n}}{2\pi} = \frac{c}{2}\sqrt{\left(\frac{m}{L_x}\right)^2 + \left(\frac{n}{L_y}\right)^2} \tag{6.11}$$

This expression for the frequencies of the normal modes of a fixed-fixed rectangular membrane in Eq. (6.11) represents a Pythagorean sum of the modes of two fixed-fixed strings of lengths, L_x and L_y.

The mode shapes can now be calculated by substituting the normal modal frequencies of Eq. (6.11) back into our separated solution, $z(x, y) = X(x)Y(y)$. The most important aspect of Eq. (6.9) is recognition that the mode shape is the product of the two sine functions; if the value of either function is zero for some value of the function's argument, the transverse displacement is zero for that value.

This is evident in Fig. 6.2 for all mode shapes except the (1, 1) mode which only has displacement nodes along the boundaries.[4] For the (2, 1) mode, there is an obvious nodal line that bisects the membrane with the displacements on either side of the nodal line being 180° out-of-phase in time. The (1, 2) mode is just like the (2, 1) mode but with the nodal line in the orthogonal direction. The (2, 2) mode has two perpendicular nodal lines with adjacent quadrants moving 180° out-of-phase; both $X(x)$ and $Y(y)$ go to zero along both bisectors of the membrane. The frequency of the (2, 2) modes is identical to the (1, 1) mode of a membrane that is half as long and half as wide.

[4] This is analogous to the fundamental mode of a fixed-fixed string whose mode shape has no zero crossings (see Fig. 2.28).

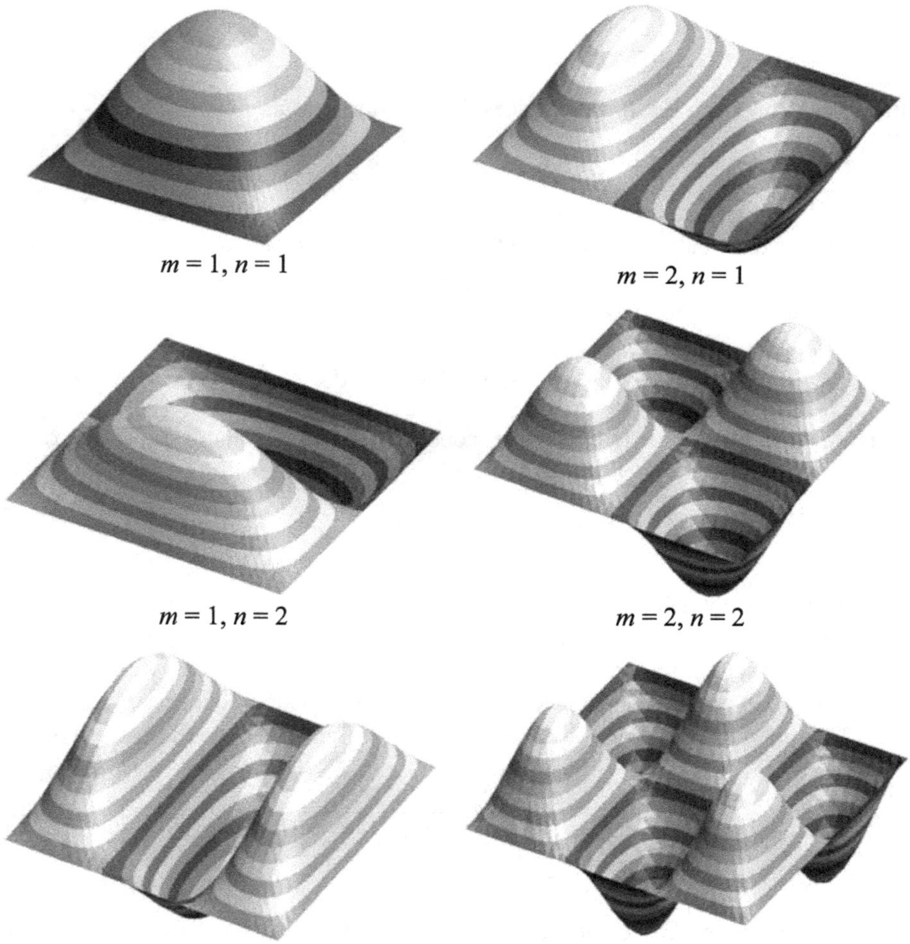

Fig. 6.2 Greatly exaggerated mode shapes for a rectangular membrane with the grey bands indicating areas having the same transverse displacement. The left column, from top to bottom, are the (1, 1), (1, 2), and (3, 1) modes. The right column, from top to bottom, are the (1, 2), (2, 2), and (3, 2) modes

6.1.2 Modal Degeneracy[5]

The normalized, normal modal frequencies in Eq. (6.11) have been provided in Table 6.1 for two rectangular membranes. One example makes $L_x = L_y\sqrt{2}$ and the other has $L_x = (3/2)\ L_y$. The normalized frequencies in Table 6.1 are the normal mode frequencies divided by the fundamental normal mode frequency, $f_{1,1}$. The normal mode frequencies for both cases are tabulated in order of increasing frequency from $f_{1,1}$ to $5f_{1,1}$. In one case, the total number of modes is 35 and the other is 34.

The first feature that is worthy of notice is that there is no apparently systematic order by which the values of m an n progress as the frequency increases. Of course, there is order imposed by Eq. (6.11), but since $L_x \neq L_y$, and in both of these examples $L_x > L_y$, a unit increase of m causes a smaller frequency

[5] Not to be confused with moral degeneracy; they are entirely unrelated.

Table 6.1 Normal mode frequencies less than or equal to $5f_{1,1}$ for two rectangular membranes are arranged in order of increasing frequency

$L_x = \sqrt{2}L_y$						$L_x = 1.5L_y$					
m	n	$f_{m,n}/f_{1,1}$	m	n	$f_{m,n}/f_{1,1}$	m	n	$f_{m,n}/f_{1,1}$	m	n	$f_{m,n}/f_{1,1}$
1	1	1.000	5	3	3.786	1	1	1.000	3	4	3.721
2	1	1.414	6	2	3.830	2	1	1.387	6	2	3.721
1	2	1.732	4	4	4.000	1	2	1.754	5	3	3.731
3	1	1.915	1	5	4.123	3	1	1.861	7	1	3.971
2	2	2.000	7	1	4.123	2	2	2.000	4	4	4.000
3	2	2.380	2	5	4.243	3	2	2.353	6	3	4.160
4	1	2.449	6	3	4.243	4	1	2.370	1	5	4.197
1	3	2.517	7	2	4.359	1	3	2.557	7	2	4.224
2	3	2.708	5	4	4.359	2	3	2.732	2	5	4.306
4	2	2.828	3	5	4.435	4	2	2.774	5	4	4.332
5	1	3.000	8	1	4.690	5	1	2.896	3	5	4.481
3	3	3.000	4	5	4.690	3	3	3.000	8	1	4.515
1	4	3.317	7	3	4.726	5	2	3.234	7	3	4.616
5	2	3.317	6	4	4.761	4	3	3.340	6	4	4.707
4	3	3.367	8	2	4.899	1	4	3.374	4	5	4.715
2	4	3.464	1	6	4.933	6	1	3.431	8	2	4.739
6	1	3.559	5	5	5.000	2	4	3.508	5	5	5.000
3	4	3.697									

The frequencies are reported as the ratio of the modal frequency, $f_{m,n}$, to the frequency of the fundamental mode: $f_{m,n}/f_{1,1}$. (*Left*) $L_x = L_y\sqrt{2}$ and (*Right*) $L_x = 1.5\ L_y$. On the left, there are six modal degeneracies: $f_{5,1} = f_{3,3} = 3.000f_{1,1}$. $f_{1,4} = f_{5,2} = 3.317f_{1,1}$, $f_{1,5} = f_{7,1} = 4.123f_{1,1}$, $f_{2,5} = f_{6,3} = 4.243f_{1,1}$, $f_{7,2} = f_{5,4} = 4.359f_{1,1}$, and $f_{8,1} = f_{4,5} = 4.67f_{1,1}$ On the right, there is only a single degeneracy: $f_{3,4} = f_{6,2} = 3.721f_{1,1}$

increase than a unit increase of n. If the membrane had the aspect ratio of a thin ribbon, with $L_x = 10L_y$, then the (17, 1) mode would have a lower frequency than the (1, 2) mode.

Another important feature of the modes in Table 6.1 is the fact that there are six pairs of modes for the $L_x = L_y\sqrt{2}$ case that have distinctly different mode shapes yet share identical normal mode frequencies. Such modes are called *degenerate modes*. If the membrane were square, so $L_x = L_y$, and then all of the $m \neq n$ modes will be double-degenerate since $f_{m,n} = f_{n,m}$. All modes of the square membrane having $m = n$ are non-degenerate since there is only one mode shape corresponding to such frequencies.

Degenerate modes have special properties. Owing to the fact that they have the same frequency, they can be superimposed, and the ratio of their amplitudes and the differences in their phases can be arbitrary.[6] In that case, the membrane can oscillate with each point undergoing simple harmonic motion at frequency, $f_{m,n} = f_{n,m}$, but with an infinite variety of mode shapes. If we consider the (1, 2) and (2, 1) modes of a square membrane, and superimpose them with identical phases, but with amplitudes controlled by a sine and cosine function, then it is possible to orient the nodal line in any direction, θ, with respect to the x axis.

$$\mathbf{z}(x, y, t) = (z_{m,n} \cos\theta + z_{n,m} \sin\theta)e^{j\omega_{m,n}t} \tag{6.12}$$

If $\theta = +45°$, then the nodal line coincides with one diagonal of the square, and if $\theta = -45°$, then the nodal line coincides with the other diagonal.

[6] Of course, since this is a linear system, any two (or more) modes can be superimposed, but since they have different frequencies, then the sum of their displacements will not be periodic unless the frequencies of the modes are commensurate (i.e., their frequencies have integer ratios).

It is also possible to have the superposition coefficients vary with time.

$$\mathbf{z}(x,y,t) = [z_{m,n}\cos(\omega_{m,n}t) \pm z_{n,m}\sin(\omega_{m,n}t)]e^{\,j\omega_{m,n}t} \tag{6.13}$$

For the superposition scheme in Eq. (6.13), the nodal line would rotate once per cycle, creating a traveling wave rotating in the clockwise direction for the minus sign and counterclockwise for the plus sign. Traveling waves on a membrane of finite extent are only possible through the superposition of degenerate modes. In our later treatment of fluid-filled toroidal resonators (see Sect. 13.5) and manipulation of acoustically levitated objects (see Sect. 15.6), traveling waves created by the superposition of degenerate modes will prove to be very useful.

The frequencies of degenerate modes will be split (i.e., $f_{m,n} \neq f_{n,m}$) if the symmetry of the membrane becomes broken. If the uniformity of the membrane is disturbed either by a tension that is not uniform or by the addition of a mass, making the surface mass density nonuniform, the degenerate modes will be split into two distinct modes with different frequencies. As an example of such *mode splitting*, consider a uniform square membrane with $L_x = L_y$, shown in Fig. 6.3, that has a small piece of putty with mass, M, stuck to its surface at $x = L_x/4$ and $y = L_y/2$.

The added mass is located on a nodal line for the (1, 2) mode. If we assume that the added mass is small (i.e., $M \ll \rho_S L_x L_y$), it will not affect the frequency, $f_{1,2}$, in the position shown in Fig. 6.3, because it is located on a node and will remain stationary when that mode is excited. On the other hand, the added mass is located at a point of maximum transverse displacement for the (2, 1) mode. Since the mass is assumed to be small compared to the mass of the membrane, the added mass will lower the frequency, $f_{2,1}$, by an amount that would be easy to calculate using Rayleigh's method. Equation (6.9) can be used as the trial function to calculate the unperturbed kinetic energy, $(KE)_{2,1}$. The kinetic energy of the mass is just $(M/2)v^2$, where $v = \dot{z}(L_x/4, L_y/2)$ is the transverse velocity of the unloaded membrane at the point $x = L_x/4$ and $y = L_y/2$.

$$\frac{\delta f_{2,1}}{f_{2,1}} \cong -\frac{Mv^2}{4(KE)_{2,1}} \tag{6.14}$$

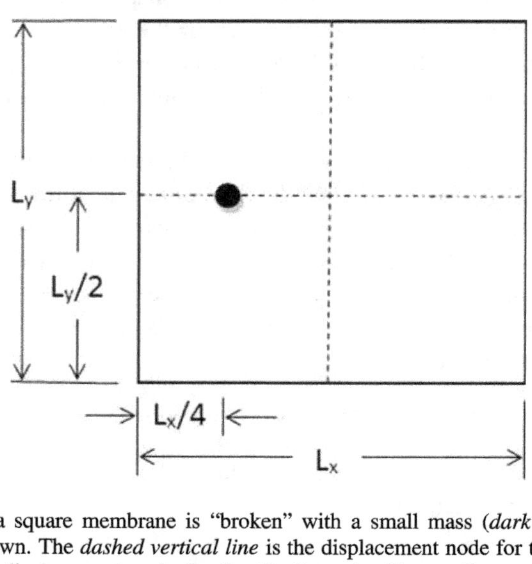

Fig. 6.3 The symmetry of a square membrane is "broken" with a small mass (*dark circle*) that is bonded to the membrane at the location shown. The *dashed vertical line* is the displacement node for the (1, 2) mode, and the *dash-dotted horizontal line* is the displacement node for the (2, 1) mode. The small mass, M, is bonded to the square membrane's surface. Without the added mass, the frequencies of those two modes are degenerate for a square membrane, $f_{1,2} = f_{2,1}$. The mass will split the degeneracy by lowering $f_{1,2}$ and leaving $f_{2,1}$ unchanged

6.1.3 Density of Modes

Another interesting feature of the modal frequencies in Table 6.1 is that the number of modes in any frequency range increases with increasing frequency. For example, in both the $L_x = L_y\sqrt{2}$ and the $L_x = (3/2)\,L_y$ examples, there are four modes with a relative frequency $f_{m,n}/f_{1,1} < 2$. For $2 \le f_{m,n}/f_{1,1} < 3$, there are 6 or 7 modes; for $3 \le f_{m,n}/f_{1,1} < 4$, there are 10 modes; and for $4 \le f_{m,n}/f_{1,1} < 5$, there are 12 or 14 modes. For a fixed-fixed string, there would be only one mode in each frequency interval, independent of the frequency.

This increase in the *modal density* is a characteristic of a two-dimensional system, not a feature that is unique only to rectangular membranes, as will be demonstrated in Sect. 6.2.2. Figure 6.4 provides a histogram that shows the total number of modes with relative frequencies, $f_{m,n}/f_{1,1}$, that are less than a given normalized frequency ratio indicated on the horizontal axis. The number of modes for both examples in Table 6.1 grows at a rate that is proportional to the square of the normalized frequency.

There is a simple geometric interpretation that explains the approximately quadratic growth in the number of modes with increasing frequency. Using the quantized values of the two wavenumbers, k_x and k_y, in Eq. (6.10), it is possible to represent each mode as a point on a plane surface whose axes are k_x and k_y, as shown in Fig 6.5.[7] The separation requirement in Eq. (6.8) dictates that $k^2 = (\omega/c)^2 = k_x^2 + k_y^2$.

At any frequency, k^2 is determined by the frequency and the speed, c, of transverse waves on the membrane. The separation requirement is equivalent to specifying the radius $k = \sqrt{k_x^2 + k_y^2}$ of a circle

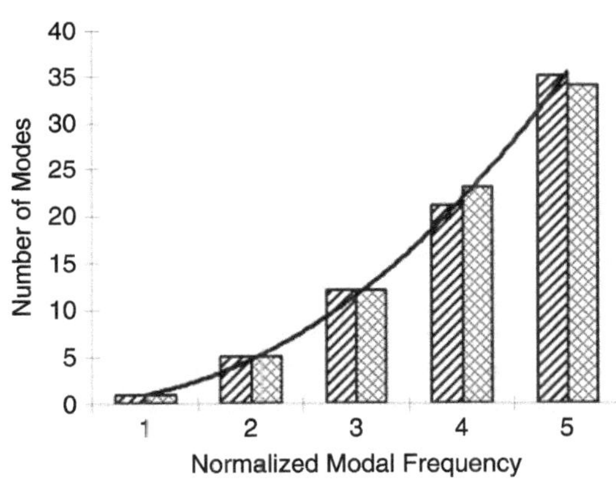

Fig. 6.4 A histogram showing the number of modes with normalized frequencies, $f_{m,n}/f_{1,1}$, less than or equal to the number on the horizontal axis. The *bars with diagonal lines* are taken from the modes of the $L_x = L_y\sqrt{2}$ membrane in Table 6.1 (*left*). The *cross-hatched bars* are for modes of the $L_x = (3/2)\,L_y$ membrane within the same table (*right*). The *solid line* is proportional to the square of the normalized modal frequency

[7] This plotting strategy creates a "*k*-space," also known as *reciprocal space*, although we are more familiar with a Cartesian coordinate system where the axes represent distance. In *k*-space, the axes have units of inverse length [m^{-1} ≡ retem], and areas have units of [m^{-2}]. If this bothers you, just multiply all the *k* values by $(c/2\pi)$ so that the axes have units of frequency [Hz]. Of course, frequency [Hz] is the "reciprocal space" of time [s]. The *k*-space representation is commonly used in solid-state physics and crystallography to represent atomic lattices (the "reciprocal lattice"). The diffraction patterns created by the scattering of X-rays through a crystalline solid will project the *k*-space pattern on a screen. This can easily be observed by shining a laser through a fine-mesh woven screen. The resulting diffraction pattern, when projected on a screen normal to the laser beam, will generate a cross of bright dots. The farther the spacing between such dots, the finer the screen mesh, just as the circles representing the modes in Fig. 6.5 are more closely spaced in the k_x direction than the circles in the k_y direction.

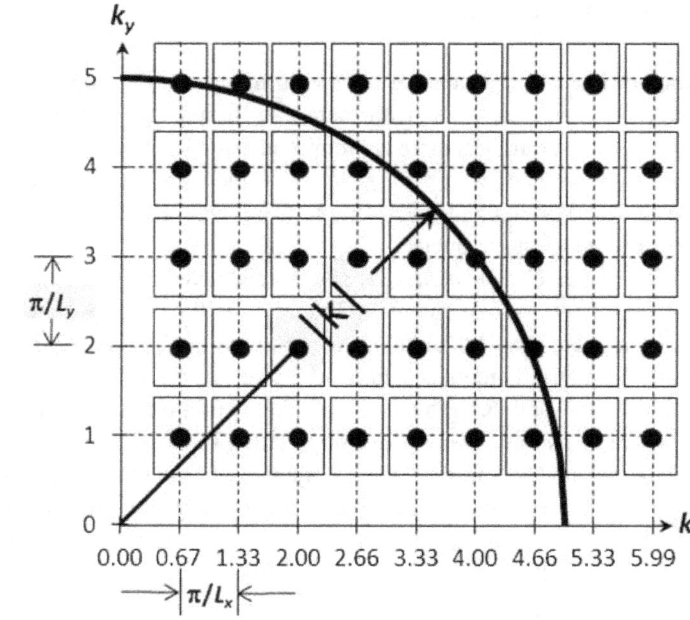

Fig. 6.5 Each black circle on this graph represents one mode of a rectangular membrane of length L_x and width $L_y < L_x$. The *solid arc* encloses one quadrant of a circle with radius $|\vec{k}|$. Each modal point is surrounded by a *rectangle* with unit reciprocal area, $\forall_{unit} = \pi^2/L_xL_y$. Those rectangles are shown slightly undersized to let them be distinguished from each other. Because there are no $(m, 0)$ or $(0, n)$ modes and no $(0, 0)$ mode, the rectangles fill all of the space except for the two strips adjacent to the k_x and k_y axes

centered on the origin in the $k_x - k_y$ plane, shown in Fig. 6.5. Since k is the distance from the origin of that Cartesian coordinate system to any pair of points (k_x, k_y), any (k_x, k_y) pair inside that circle will be a mode with frequency less than or equal to ω, and any (k_x, k_y) pair outside the circle will have a higher frequency than ω.

We can associate a k-space rectangle of width, $k_x = \pi/L_x$, and height, $k_y = \pi/L_y$, with each point representing any single mode and designate the area of one such rectangle as the reciprocal area of a unit cell, $\forall_{unit} = \pi^2/L_xL_y$. Again, in k-space, this "reciprocal area," $\forall$, has units of $[\text{m}^{-2}]$. If we attach one unit cell to each mode by placing the center of the cell at the mode point, then we can fill all k-space with such cells except for the two strips of width, $\pi/2L_x$ and $\pi/2L_y$, that are adjacent to the k_x and k_y axes.

The number of modes with frequencies less than or equal to ω can be estimated by dividing the k-space area of one quadrant of a circle with radius, k, $\forall_k = (\pi/4)\, k^2$, by the k-space area occupied by a single mode that we have designated $\forall_{unit}$. Because there are no $(m, 0)$ or $(0, n)$ modes and no $(0, 0)$ mode, we should also exclude the areas of the two strips, $\forall_{strip}$, adjacent to the k_x and k_y axes from the area, $\forall_k$. To first approximation (neglecting the double counting of the area where the k_x and k_y axial strips overlap at the origin), the number of modes, $N_{rect}(\omega)$, with frequencies less than ω, can be calculated by dividing those two areas.

$$N_{rect}(\omega_{m,n}) \cong \frac{\forall_k - \forall_{strip}}{\forall_{unit}} = \frac{\left(\frac{\pi}{4}\right)k_{m,n}^2 - \left(\frac{\pi}{2L_x} + \frac{\pi}{2L_y}\right)k_{m,n}}{\left(\frac{\pi}{L_x}\right)\left(\frac{\pi}{L_y}\right)}$$
$$= \frac{L_xL_y}{4\pi}k_{m,n}^2 - \frac{L_x + L_y}{2\pi}k_{m,n} = \frac{L_xL_y}{4\pi c^2}\omega_{m,n}^2 - \frac{L_x + L_y}{2\pi c}\omega_{m,n} \tag{6.15}$$

$$N_{rect}(f_{m,n}) \cong \frac{\pi L_xL_y}{c^2}f_{m,n}^2 - \frac{L_x + L_y}{c}f_{m,n} = \frac{\pi L_xL_y}{\lambda_{m,n}^2} - \frac{L_x + L_y}{\lambda_{m,n}}$$

At sufficiently high frequencies, so that $k_x \gg \pi/L_x$ and $k_y \gg \pi/L_y$, the first term in the approximation of Eq. (6.15) becomes more accurate and demonstrates that the number of modes, $N_{rect}(\omega)$, below a given frequency, ω, should increase almost like ω^2, as was shown in Fig. 6.4 for both cases examined in Table 6.1.

We can compare the results of Eq. (6.15) to the exact mode counts in Table 6.1. There are 34 modes with normalized frequencies, $f_{m,n}/f_{1,1} \leq 5$, for $L_x = 3L_y/2$ on the right-hand side of the table. For that case, Eq. (6.11) can be used to calculate the frequency of the (1, 1) mode, $f_{1,1}$, in terms of L_y and the speed of transverse waves, c.

$$f_{1,1} = \frac{c}{2}\sqrt{\left(\frac{4}{9}\right)\frac{1}{L_y^2} + \frac{1}{L_y^2}} = \frac{c}{L_y}\sqrt{\frac{13}{36}} \quad \Rightarrow \quad \frac{L_y^2}{c^2} = \frac{13}{36}\frac{1}{f_{1,1}^2} \tag{6.16}$$

These results can be substituted into Eq. (6.15) to calculate the number of modes with normalized frequencies, $f = f_{m,n}/f_{1,1} \leq 5$.

$$\begin{aligned} N_{rect}(f_{5,5}) &\cong \frac{3\pi L_y^2}{2c^2} f_{5,5}^2 - \frac{5L_y}{2c} f_{5,5} \quad \Rightarrow \\ N_{rect}(f = 5) &\cong \frac{3\pi}{2}\frac{13}{36} f^2 - \frac{5}{2}\sqrt{\frac{13}{36}} f = 35 \end{aligned} \tag{6.17}$$

That result is less than 3% larger than the exact result.

The approximation of Eq. (6.15) can be tested in a more rigorous way by a linearized least-squares fit (see Sect. 1.9.3) that plots the number of modes with frequencies less than or equal to $f_{m,n}$, divided by the frequency, $N(f_{m,n})/f_{m,n}$, against frequency.

$$\frac{N_{rect}(f_{m,n})}{f_{m,n}} \cong \frac{\pi L_x L_y}{c^2} f_{m,n} - \frac{L_x + L_y}{c} = m f_{m,n} + b \tag{6.18}$$

The data for $L_x = L_y\sqrt{2}$, taken from the left side of Table 6.1, is plotted against the normalized frequency, $f = f_{m,n}/f_{1,1}$, to produce the best-fit straight line in Fig. 6.6. The exact number of modes with $f \leq 5$ is 35. Equation (6.15) gives $N_{rect}(5) = 34.3$, which is only a 2% difference.

The frequency difference between modes of adjacent frequencies is also decreasing with increasing frequency. The approximate number of modes, ΔN, in a frequency interval, Δf, can be determined by differentiation of the expression in Eq. (6.15) for the number of modes, $N(f_{m,n})$, with frequencies less than $f_{m,n}$.

$$\begin{aligned} \frac{dN(f_{m,n})}{df_{m,n}} &= \frac{2\pi L_x L_y}{c^2} f_{m,n} - \frac{L_x + L_y}{c} \quad \Rightarrow \\ \Delta N(f_{m,n}) &= \left(\frac{2\pi L_x L_y}{c^2} f_{m,n} - \frac{L_x + L_y}{c}\right)\Delta f \end{aligned} \tag{6.19}$$

The value of the derivative, $dN/df_{m,n}$, is called the *modal density*. It increases linearly with frequency in any bounded two-dimensional system. The density of modes for a one-dimensional string is a constant that is independent of frequency.

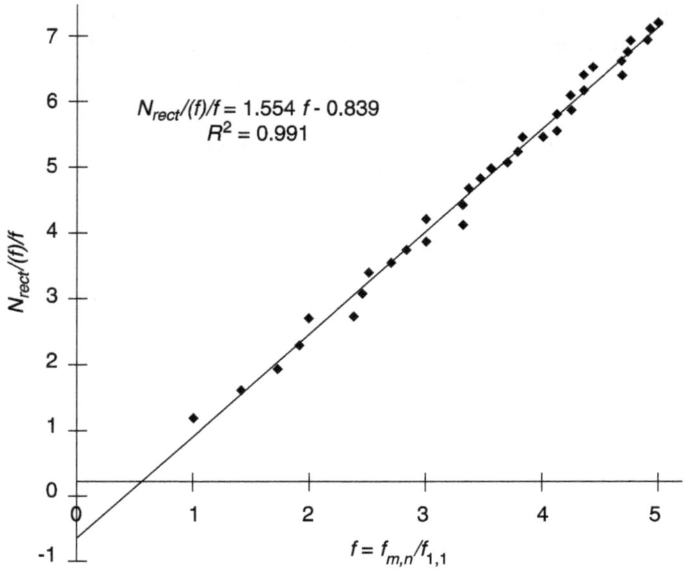

Fig. 6.6 Plot of the number of modes with normalized frequencies, $f = f_{m,n}/f_{1,1}$, less than or equal to f divided by f vs. f. The best-fit straight line should be compared to the theoretical result in Eq. (6.18). Substitution of $f = 5$ into the best-fit straight line gives N_{rect} (5) = 34.7, while the same substitution into Eq. (6.15) gives the result in Eq. (6.17): N_{rect} (5) = 34.3. The fit coefficients (i.e., slope and intercept) differ slightly from those in Eq. (6.18) because Table 6.1 data do not double-count the strip overlap

6.2 Circular Membranes

Rectangular membranes are rare due to the stresses at the corners that tend to rip the thin membrane material. Circular membranes are very common. Two ubiquitous examples are musical instruments (i.e., drums, banjos, and the *erhu*) and the microphones that are used to record and/or amplify their sounds. Although it is possible, in theory, to describe the transverse vibrations of a circular membrane using a Cartesian coordinate system, the specification of the fixed boundary condition for a circular membrane of radius, a, is difficult to execute when written as $z(x^2+y^2 = a^2, t) = 0$ and then imposed on the wave equation in Cartesian coordinates. It is much easier to use a polar coordinate system, letting the boundary condition be expressed simply as $z\,(a, \theta, t) = 0$. The price of this simplification is that we must now express the Laplacian operator in polar coordinates, as written in Eq. (6.4).

We started the derivation of the wave equation in rectangular coordinates by examination of a differential area element of the membrane, $dA = dx\, dy$, shown in Fig. 6.1, and expressed in Cartesian coordinates. We can now examine the forces acting on a differential area element, $dA = dr\,(r\, d\theta)$, of the membrane in polar coordinates using Fig. 6.7.

The net vertical force, $F_{z,\theta}$, due to tensions perpendicular to the radial direction, can be expanded in a Taylor series just as was done for the rectangular membrane element in Eq. (6.1).

$$F_{z,\theta} = \Im\, dr\left[\left(\frac{1}{r}\frac{\partial z}{\partial \theta}\right)_{\theta+d\theta} - \left(\frac{1}{r}\frac{\partial z}{\partial \theta}\right)_{\theta}\right] = \frac{\Im}{r^2}\frac{\partial^2 z}{\partial \theta^2} r\, dr\, d\theta \qquad (6.20)$$

The net for vertical force, $F_{z,r}$, along the radial direction receives similar treatment, again based on Fig. 6.7.

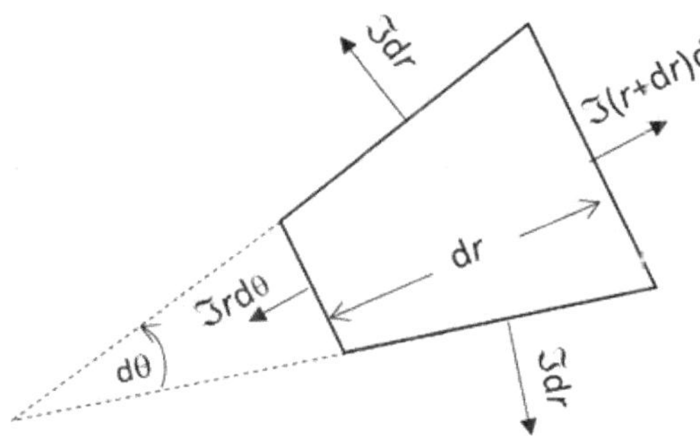

Fig. 6.7 A differential area of membrane expressed in polar coordinates

$$F_{z,r} = \Im d\theta\left[\left(r\frac{\partial z}{\partial r}\right)_{r+dr} - \left(r\frac{\partial z}{\partial r}\right)_{r}\right] = \frac{\Im}{r}\frac{\partial}{\partial r}\left(r\frac{\partial z}{\partial r}\right) r\,dr\,d\theta \tag{6.21}$$

The sum of the vertical forces can then be equated to the product of the vertical acceleration and the mass of the differential element, $\rho_S\, dA = \rho_S\,(r\,d\theta)\,dr$.

$$\frac{\partial^2 z}{\partial r^2} + \frac{1}{r}\frac{\partial z}{\partial r} + \frac{1}{r^2}\frac{\partial^2 z}{\partial \theta^2} = \nabla^2 z = \frac{1}{c^2}\frac{\partial^2 z}{\partial t^2} \tag{6.22}$$

Again, $c^2 = \Im/\rho_S$ and the Laplacian, ∇^2, are just that provided previously without justification in Eq. (6.4).

6.2.1 Series Solution to the Circular Wave Equation

We can separate Eq. (6.22) to convert it from a second-order partial differential equation into two ordinary differential equations while imposing harmonic time dependence: $\mathbf{z}(r, \theta, t) = R(r)\Theta(\theta)e^{j\omega t}$.

$$\Theta\frac{d^2 R}{dr^2} + \frac{\Theta}{r}\frac{dR}{dr} + \frac{R}{r^2}\frac{d^2\Theta}{d\theta^2} + k^2 R\Theta = 0 \tag{6.23}$$

Multiplication by $r^2/R\Theta$ produces two independent ordinary differential equations coupled by k^2.

$$\frac{r^2}{R}\left(\frac{d^2 R}{dr^2} + \frac{1}{r}\frac{dR}{dr}\right) + k^2 r^2 = -\frac{1}{\Theta}\frac{d^2\Theta}{d\theta^2} \tag{6.24}$$

Since the left-hand terms in Eq. (6.24) depend only upon r and the right-hand term depends only on θ, they must each separately be equal to a constant since the variations of r and θ are independent of each other.

The angular dependence is easy to calculate, since it is just the solution to a simple harmonic oscillator equation.

$$\frac{d^2\Theta}{d\vartheta^2} + m^2\Theta = 0 \tag{6.25}$$

The two solutions to this differential equation will be written with curly brackets to remind us that those two solutions guarantee that every solution with $m \neq 0$ will be double-degenerate.

$$\Theta_m(\theta) = \begin{Bmatrix} \cos(m\theta) \\ \sin(m\theta) \end{Bmatrix} \tag{6.26}$$

Of course, any superposition of those two solutions is also a solution, since Eq. (6.24) is a linear differential equation.

For the solutions, $\Theta_m(\theta)$, to have physical significance, they must yield the same result if the azimuthal coordinate, θ, is increased or decreased by integer multiples of 2π, since that would bring us back to the same physical location on the membrane.

$$\Theta_m(\theta) = \Theta_m(\theta \pm 2n\pi) \quad \text{with} \quad n = 0, 1, 2, 3, \ldots \tag{6.27}$$

That restriction is known as a *periodic boundary condition*, and it is only satisfied if m is an integer or is zero.

The equation for the radial dependence of the transverse displacement can be generated by substitution of Eq. (6.25) into Eq. (6.24).

$$\frac{d^2R}{dr^2} + \frac{1}{r}\frac{dR}{dr} + \left(k^2 - \frac{m^2}{r^2}\right)R = 0 \tag{6.28}$$

It is important to recognize that this equation for the radial variation in the transverse vibrations of the membrane is unique for every choice of m made to satisfy the azimuthal variation given by Eq. (6.26) for $\Theta_m(\theta)$. Equation (6.28) actually represents an infinite number of second-order differential equations, each having two solutions for every integer value of m.

A solution to Eq. (6.28) can be generated by assuming a polynomial form, $R_m(r) = a_o + a_1 r + a_2 r^2 + a_3 r^3 \ldots$, and then equating all coefficients of like powers of r that must individually sum to zero.[8] This will be demonstrated for only one such solution with $m = 1$.

$$\begin{array}{rcccccccc}
R_1 = & & & a_o & +a_1 r & +a_2 r^2 & +a_3 r^3 \\
-\dfrac{R_1}{r^2} = & -\dfrac{a_o}{r^2} & -\dfrac{a_1}{r} & -a_2 & -a_3 r & -a_4 r^2 & +a_5 r^3 \\
+\dfrac{1}{r}\dfrac{dR_1}{dr} = & & \dfrac{a_1}{r} & +2a_2 & +3a_3 r & +4a_3 r^2 & +5a_4 r^3 \\
+\dfrac{d^2R_1}{dr^2} = & & & 2a_2 & +6a_3 r & +12a_4 r^2 & +20a_5 r^3
\end{array} \tag{6.29}$$

Equating the sums of like powers to zero will allow all of the polynomial's coefficients to be expressed in terms of a_1, which can only be determined once the initial conditions have been specified.

$$\begin{array}{l}
a_0 = 0, \quad a_2 = -\dfrac{a_o}{3} = 0, \quad a_3 = -\dfrac{a_1}{8}, \quad a_4 = 0 \\
a_5 = -\dfrac{a_3}{24} = \dfrac{a_1}{8(24)}, \quad a_6 = 0, \quad a_7 = \dfrac{a_1}{16(24)(24)}, \quad \text{etc.}
\end{array} \tag{6.30}$$

Clearly, $R_1(r)$ is an odd function of r since all $a_{2n} = 0$ for $n = 0, 1, 2, 3, \ldots$. Substitution back into the polynomial provides a power series solution for the radial part of the solution to Eq. (6.28) when $m = 1$.

[8] Mathematicians know this as the Frobenius method named after German mathematician Ferdinand Georg Frobenius (1849–1917).

$$R_1(r) = \frac{a_1}{2}\left(r - \frac{2r^3}{4^2} + \frac{3r^5}{4^2 6^2} - \frac{4r^7}{4^2 6^2 8^2} + \dots\right) \tag{6.31}$$

Although this result may appear unfamiliar, it is only because we do not usually think of trigonometric functions as power series solutions to the differential equations that generated them. In fact, as shown in Eqs. (1.5 and 1.6), both sine and cosine functions can be represented by a power series. We routinely refer to those trigonometric power series solutions by their functional names (sine and cosine), and we regularly employ algebraic relationships that allow us to determine their derivatives, integrals, magnitude, and values of their arguments for maxima and minima of those functions, multiple angle formulæ, etc.

The same is true for the power series solutions to Eq. (6.28). Those equations (one for each integer value of m) are known as Bessel's equations, and their solutions are known as *Bessel functions* that we will abbreviate as J_m ($k_{m,n}$ r).[9] The subscript "m" on J_m refers to the corresponding variation in the azimuthal function, Θ_m (θ); hence each radial solution is tied to the azimuthal variation through the product, $J_m(k_{m,\,n}r)\Theta_m(\theta)$, to produce the complete solution to Eq. (6.22).

$$\mathbf{z_{m,n}}(r,\theta,t) = \widehat{\mathbf{C}}_{m,n} J_m(k_{m,n} r) \begin{Bmatrix} \cos(m\theta) \\ \sin(m\theta) \end{Bmatrix} e^{j\omega_{m,n} t} \tag{6.32}$$

Each solution is specified by two indices: m indicates the azimuthal variations, and n designates the successive zero-crossings of the Bessel function. For example, the (0, 1) mode will have no azimuthal variation, and the normal mode frequency, $f_{0,1}$, will correspond to the first zero-crossing of the J_o Bessel function. A (1,2) mode will have a single nodal diameter, although its orientation will be arbitrary, due to the superposition of the cos (θ) and sin (θ), until something breaks the azimuthal symmetry, such as a perturbation in the membrane's mass or the position of a driving force applied over a non-zero area. It also requires that the frequency of that mode, $f_{1,2}$, will be determined by the second zero-crossing of the J_1 Bessel function, excluding the zero at the origin.

Figure 6.8 provides a graph of the first three Bessel functions of integer order: $m = 0$, $m = 1$, and $m = 2$. More complete graphs for additional values of m and other useful functions are available in Junke and Emde [1]. The Bessel functions in Fig. 6.8 look similar to cosine and sine functions, although there are obvious differences. The first difference is that the Bessel functions are only plotted for positive values of their arguments $x \geq 0$; a radial location that is along the negative x axis is represented in polar coordinates as a positive r with $\theta = 180°$.

The $m = 0$ Bessel function, J_o (x), looks a bit like cos (x). It approaches the origin with a value of one and a slope of zero. On the other hand, the envelope of the amplitudes of the oscillations in J_o (x) has progressively smaller values. There is also a similarity between J_1 (x) and sin (x). The amplitudes of both are zero at the origin, and their initial slopes near the origin are both linear, although $[d\sin(x)/dx]_{x=0} = 1$ and $[dJ_1(x)/dx]_{x=0} = ½$. Again, like J_o (x), the amplitude envelope of $J_1(x)$ also is decreasing with increasing argument. The similarity of J_2 (x) and sin (x) is like that of J_1 (x) except the

[9] German astronomer F. W. Bessel (1784–1846) first achieved fame by computing the orbit of Halley's comet. In addition to many other accomplishments in connection with his studies of planetary motion, he is credited with deriving the differential equation bearing his name and carrying out the first systematic study of the general properties of its solutions (now called *Bessel functions*) in his famous 1824 memoir.

Nonetheless, Bessel functions were first discovered in 1732 by D. Bernoulli (1700–1782), who provided a series solution for the oscillatory displacements of a heavy hanging chain (see Sect. 3.4.3). Euler (1707–1783) later developed a series similar to that of Bernoulli, which was also a Bessel function, and Bessel's equation appeared in a 1764 article by Euler dealing with the vibrations of the circular drumhead. Fourier (1768–1836) also used Bessel functions in his classical treatise on heat in 1822, but it was Bessel who first recognized and documented their special properties.

Fig. 6.8 Bessel functions $J_o(x)$, $J_1(x)$, and $J_2(x)$ are plotted for $0 \leq x \leq 15$. $J_o(x)$ resembles cos (x), except that the amplitude decreases with x. $J_1(x)$ resembles sin (x) with an initially linear slope for $x < 1$ but again with decreasing amplitude. $J_2(x)$ also resembles sin (x) with decreasing amplitude but with an initially quadratic slope for $x < 1$

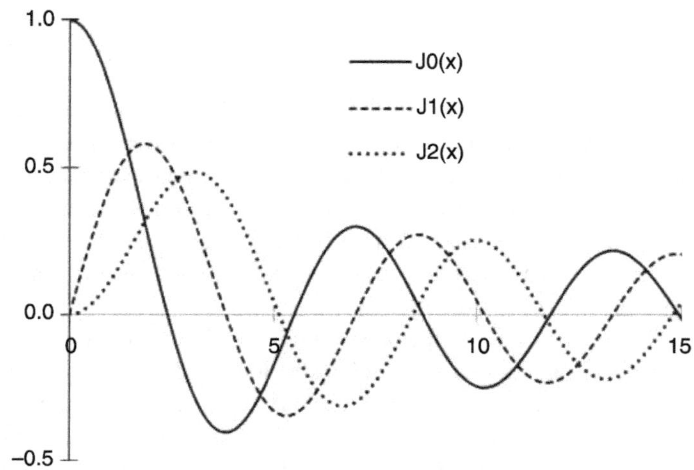

Fig. 6.9 Neumann functions $N_o(x)$, $N_1(x)$, and $N_2(x)$, are plotted for $0 \leq x \leq 15$. All Neumann functions diverge to $-\infty$ in the limit that x goes to zero

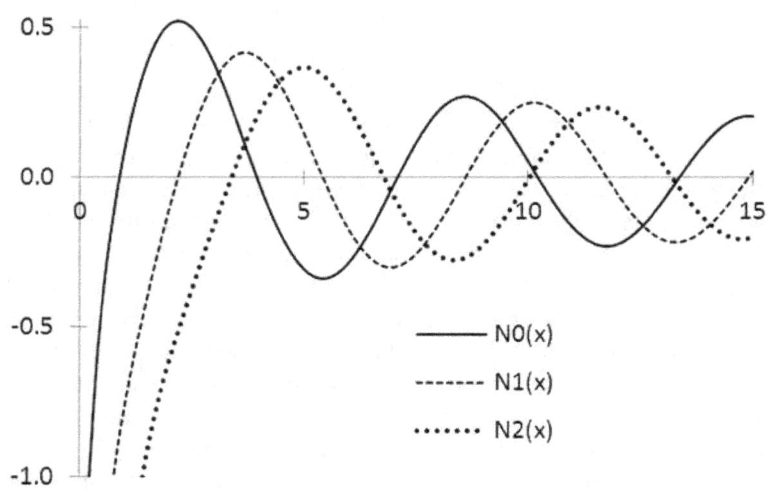

initial slope, $[dJ_2(x)/dx]_{x=0} \propto x^2$. In fact, the initial slopes of all subsequent $J_m(x)$ are proportional to x raised to the m: $[dJ_m(x)/dx]_{x=0} \propto x^m$. Many useful properties of Bessel functions are provided in Appendix C and in standard mathematical reference books [2].

Bessel's equations are all second-order differential equations, a different equation for each integer value of m in Eq. (6.28). As such, there must be two independent solutions for each value of m. The second infinite set of solutions, for various values of m, is known as the *Neumann functions*, $N_m(x)$. The first three Neumann functions, for $m = 0$, $m = 1$, and $m = 2$, are plotted in Fig. 6.9. All Neumann functions go to negative infinity as x approaches zero: $N_m(0) = -\infty$, although they do so with different functional forms (e.g., logarithmically or with an inverse power function of their argument). We therefore reject the Neumann solutions for our description of the transverse vibrations of circular membranes since the displacement at the membrane's center is never infinite.

As presented in Sect. 6.2.3, the Neumann functions, $N_m(x)$, are required to match the second boundary condition for an annular membrane with outer radius, a_{out}, and inner radius, a_{in}. For an annulus, $a_{in} > 0$, so the divergent portion of the Neumann functions do not occur in the region where the membrane exists.

6.2.2 Modal Frequencies and Density for a Circular Membrane

The values of $k_{m,n}$ are quantized by the imposition of the radial boundary condition $z_{m,n}(a, \theta, t)$ that requires $R_m(a) = 0 = J_m(k_{m,n}a)$. This boundary condition will be satisfied by locating the arguments, $x_n = k_{m,n}a$, of $J_m(x_n)$ that are the zero-crossing for $J_m(x_n)$. We do this without much thought for sine and cosine functions since their zero-crossings at x_n are periodic. This is not true for Bessel functions, as their zero-crossings do not have equal spacing, although the spacing becomes more equal as x increases. The values at which a Bessel function is zero is commonly designated $j_{m,n}$, so $J_m(j_{m,n}) = 0$. Appendix C includes a table of $j_{m,n}$ zero-crossings. A more complete compilation is provided in Table 9.5 of Abramowitz and Stegun [2].

The normal mode frequencies, $\omega_{m,n}$, are related to the speed of transverse waves, c, and the radius, a, of the membrane through the wavenumber, $k_{m,n}$.

$$k_{m,n}a = \frac{\omega_{m,n}a}{c} = \frac{2\pi f_{m,n}a}{c} = j_{m,n} \quad \Rightarrow \quad f_{m,n} = \frac{j_{m,n}c}{2\pi a} \tag{6.33}$$

The mode shapes corresponding to these normal mode frequencies for a few low-order modes are sketched in Fig. 6.10 along with the corresponding values of $j_{m,n} = k_{m,n}a$ and the ratio of the modal frequency, $f_{m,n}$, to the frequency of the lowest pure radial mode, $f_{0,1} \cong (2.40483/2\pi)(c/a) = 0.38274(c/a)$.

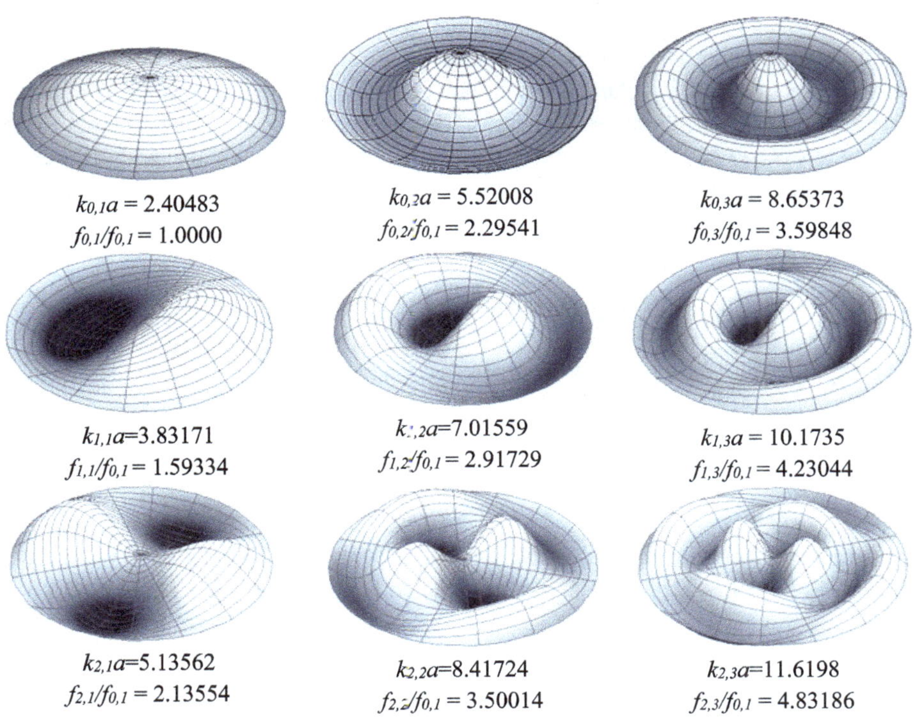

Fig. 6.10 Greatly exaggerated mode shapes for a circular membrane with radius, a. The top row contains the first "pure radial" J_0 modes: (0, 1), (0, 2), and (0, 3). The middle row shows the first three J_1 modes, each with a single nodal diameter: (1, 1), (1, 2), and (3, 1). The bottom row has the first three J_2 modes with two perpendicular nodal diameters: (2, 1), (2, 2), and (2, 3). Beneath each mode shape is their corresponding value of $j_{m,n} = k_{m,n}a$ and the ratio of the mode frequency, $f_{m,n}$, to the fundamental radial mode, $f_{0,1}$. [Mode illustrations courtesy of Daniel Russell]

Table 6.2 Normal mode frequencies less than or equal to 5.131 $f_{0,1}$ for a circular membrane are arranged in order of increasing frequency

#	m,n	$j_{m,n}$	$f_{m,n}/f_{0,1}$	#	m,n	$j_{m,n}$	$f_{m,n}/f_{0,1}$
1	0,1	2.40482	1.00000	18	3,2	9.76102	4.05893
2	1,1	3.83171	1.59334	19	6,1	9.93611	4.13174
3	1,1	3.83171	1.59334	20	6,1	9.93611	4.13174
4	2,1	5.13562	2.13555	21	1,3	10.1735	4.23044
5	2,1	5.13562	2.13555	22	1,3	10.1735	4.23044
6	0,2	5.52007	2.29541	23	4,2	11.0647	4.60104
7	3,1	6.38016	2.65307	24	4,2	11.0647	4.60104
8	3,1	6.38016	2.65307	25	7,1	11.0864	4.61005
9	1,2	7.01559	2.91730	26	7,1	11.0864	4.61005
10	1,2	7.01559	2.91730	27	2,3	11.6198	4.83188
11	4,1	7.58834	3.15546	28	2,3	11.6198	4.83188
12	4,1	7.58834	3.15546	29	0,4	11.7915	4.90328
13	2,2	8.41724	3.50015	30	8,1	12.2251	5.08357
14	0,3	8.65372	3.59848	31	8,1	12.2251	5.08357
15	5,1	8.77148	3.64745	32	5,2	12.3386	5.13077
16	5,1	8.77148	3.64745	33	5,2	12.3386	5.13077
17	3,2	9.76102	4.05893				

The frequencies are reported as the ratio of the modal frequency, $f_{m,n}$, to the frequency of the fundamental "pure radial" mode: $f_{m,n}/f_{0,1}$. Because the orientation of the nodal diameters for the $m \neq 0$ modes is arbitrary, all (m, n) modes with $m \neq 0$ are double-degenerate; thus they appear twice in this table

The lowest-frequency modes are listed in order of increasing frequency in Table 6.2 for $f_{m,n}/f_{0,1} < 5.131$. As seen with the rectangular membranes of Table 6.1, the number of modes within a fixed frequency interval, Δf, increases with increasing frequency. Rather than attempt to derive the number of modes, $N_{circ}\ (f_{m,n})$, with frequencies less than $f_{m,n}$ for a circular membrane, as we did in Sect. 6.1.3, we can attempt to mimic the geometrical factors in Eq. (6.15) by substituting for membrane's area, $L_xL_y = \pi a^2$, and half the membrane's perimeter, $L_x + L_y = \pi a$.

$$N_{circ}\left(f_{m,n}\right) \cong \left(\frac{\pi a f_{m,n}}{c}\right)^2 - \frac{\pi a f_{m,n}}{c} \tag{6.34}$$

This can be expresses in terms of the normalized frequency ratio, $f = f_{m,n}/f_{0,1}$, since Eq. (6.33) gives $(a/c) = (j_{0,1}/2\pi)\ (f_{m,n}/f_{0,1})$.

$$N_{circ}(f) \cong \frac{j_{0,1}^2}{4}\left(\frac{f_{m,n}}{f_{0,1}}\right)^2 - \frac{j_{0,1}}{2}\frac{f_{m,n}}{f_{0,1}} \cong 1.446\left(\frac{f_{m,n}}{f_{0,1}}\right)^2 - 1.202\frac{f_{m,n}}{f_{0,1}} \tag{6.35}$$

$N_{circ}\ (f = 5.131) = 32$. This is in reasonable agreement with the 33 modes in Table 6.2 below that frequency ratio. A plot of $N_{circ}\ (f)/f$ vs. f, similar to that in Fig. 6.6, is provided in Fig. 6.11 for the data in Table 6.2.

6.2.3 Mode Similarities Illustrating Adiabatic Invariance

We arrived at the results for the normal mode frequencies of rectangular and circular membranes by slightly different routes; in one case, we used solutions that were combinations of sine functions, and for the other case, we used combinations of trigonometric functions with integer-order Bessel

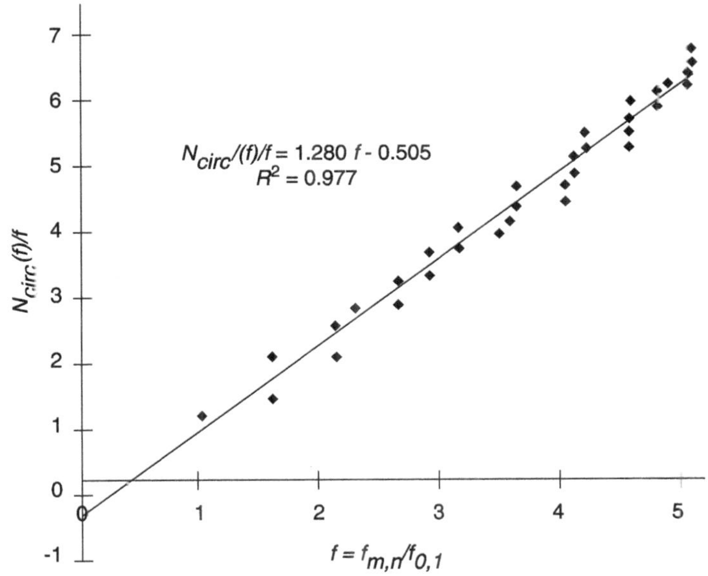

Fig. 6.11 Plot of the number of modes with normalized frequencies, $f = f_{m,n}/f_{0,1}$, less than or equal to f divided by f vs. f. The best-fit straight line should be compared to the theoretical result in Eq. (6.35). Substitution of $f = 5.131$ into the best-fit straight line gives N_{circ} (5.131) = 31.1, while the same substitution into Eq. (6.35) gives N_{circ} (5.131) = 32. The fit coefficients (slope and intercept) differ slightly from those in Eq. (6.35) because the Table 6.2 data do not double-count the strip overlap in Fig. 6.5

functions. The caption for Fig. 6.8 comments on the similarity of the integer-order Bessel functions and sine and cosine functions. Comparisons between Figs. 6.2 and 6.10 show similarities in the distribution of nodal lines that are characteristic of individual modes.

In Sect. 2.3.4, the concept of adiabatic invariance was used to demonstrate that frequency shifts, δf, were related to changes in the energy, δE, due to work being done on or work done by a vibrating system. The ratio of the frequency of a mode and the energy in that mode was a conserved quantity, E/f = constant, if the mode shape did not "hop" to another mode shape during the change [3]. We can use adiabatic invariance to examine our results for square and circular membranes by recognizing that the work done by making a deformation of the boundary could be nearly zero if the area remains unchanged.

The work we do on the mode by pushing the boundary inward and the work done by the mode when we let the vibration push the boundary outward should be nearly equal if the areas of the two membranes are equal. If we conserve that area of the square, with $L_x = L_y$, then the radius of the circle will be $a = L_x/\sqrt{\pi}$. The net work performed by pushing and pulling during that transformation, if the deformation is sufficiently slow that no additional modes are excited, will be nearly zero.

That assumption is applied to four modes with similar mode shapes in Table 6.3. The agreement between the two modal frequencies of similar mode shape is quantified by taking the difference in the modal frequencies and dividing by their average, $\Delta f/\bar{f}$. For the three modes with similar nodal line structures and symmetries, the agreement is always better than $\pm 3\%$. The agreement between the two modes that have an equal number of in- and out-of-phase segments is only about $\pm 7\%$ since the $f_{4,3}$ mode of the square does not share the same symmetry as the nearly equivalent circular $f_{2,3}$ mode.

This comparison should be instructive in three ways: (*i*) Similar mode shapes lead to similar normal mode frequencies, even if the derivations of those frequencies are quite distinct. (*ii*) The agreement between the frequencies should give us confidence that both calculations were correct. (*iii*) Most importantly, application of adiabatic invariance can provide a tool to analyze modes of membranes (and as we will see in later chapters, also standing sound waves in three-dimensional enclosures and waveguides) that do not have boundaries that allow exact solution by the technique known as separation of variables.

Table 6.3 The square membranes and the circular membranes have equal areas, so the radius of the circle is $a = L_x/\sqrt{\pi}$

Square	Adiabatic transformation	Circle	$\Delta f/\langle f\rangle$
$f_{1,1} = \frac{c}{\sqrt{2}L_x}$ $f_{1,1} = 0.707\frac{c}{L_x}$	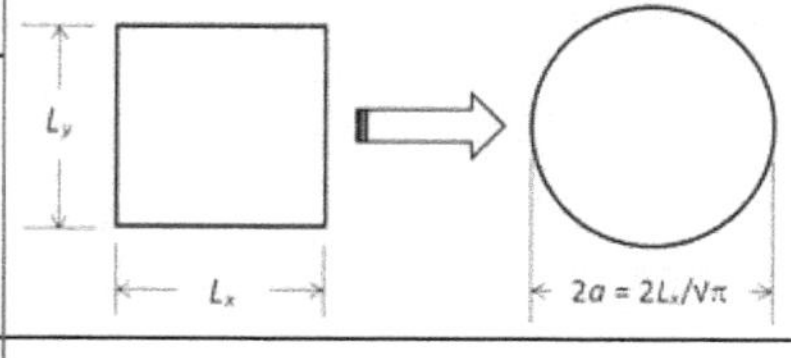	$f_{0,1} = \frac{j_{0,1}c}{2\sqrt{\pi}L_x}$ $f_{0,1} = 0.678\frac{c}{L_x}$	−4.2%
$f_{1,2} = \frac{c\sqrt{5}}{2L_x}$ $f_{1,2} = 1.118\frac{c}{L_x}$	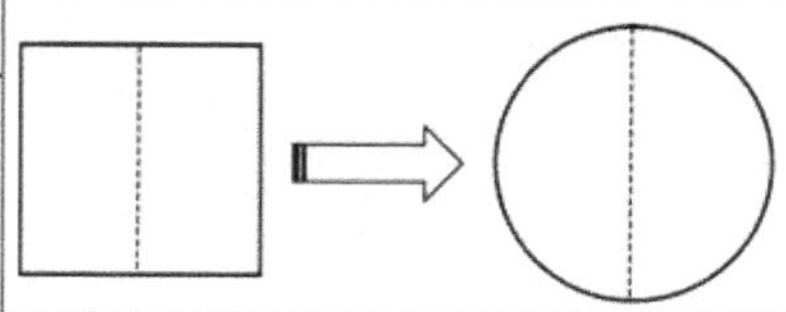	$f_{1,1} = \frac{j_{1,1}c}{2\sqrt{\pi}L_x}$ $f_{1,1} = 1.081\frac{c}{L_x}$	−3.4%
$f_{2,2} = \frac{c\sqrt{2}}{L_x}$ $f_{2,2} = 1.414\frac{c}{L_x}$	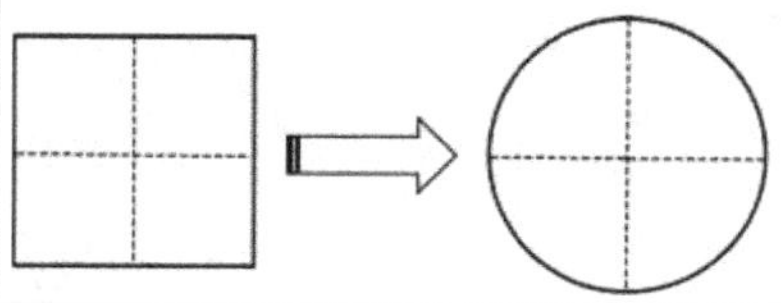	$f_{2,1} = \frac{j_{2,1}c}{2\sqrt{\pi}L_x}$ $f_{2,1} = 1.449\frac{c}{L_x}$	+2.4%
$f_{4,3} = \frac{5c}{2L_x}$ $f_{4,3} = 2.500\frac{c}{L_x}$	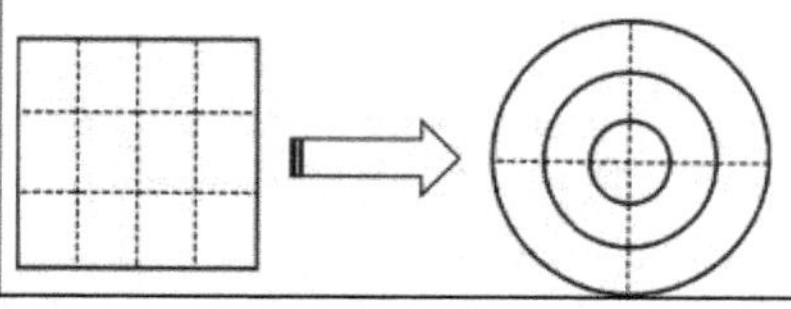	$f_{2,3} = \frac{j_{2,3}c}{2\sqrt{\pi}L_x}$ $f_{2,3} = 2.870\frac{c}{L_x}$	+13.8%

Adiabatic invariance suggests that the frequencies of the modes should be equal if the mode shapes are not too distorted by the transformation and the membrane's area is unchanged. Although the $f_{4,3}$ for the square and $f_{2,3}$ for the circle both have 12 regions vibrating in- and out-of-phase, the two membrane do not share the same symmetry

In the twenty-first century, modes of objects that do not lend themselves to analysis by separation of variables are usually treated by numerical techniques such as finite element or boundary element simulations. Although those numerical techniques are very useful and powerful, they can also produce results that are either incorrect or hard to interpret. It is always a good idea to approximate a non-separable geometry by one for which analytical results can be obtained, as we did with rectangular and circular membranes, so that mode shapes and normal mode frequencies can be compared to the numerical results. As we have demonstrated, that approach can produce results with accuracies down to the few percent level while, more importantly, providing a classification system for the modes of the non-separable geometry.

6.2.4 Normal Modes of Wedges and Annular Membranes*

The fact that a fixed boundary and a nodal line, or a nodal circle, has exactly the same influence on the solutions to the modal structure of a two-dimensional membrane means that it is possible to calculate exact, or near exact, normal mode frequencies for shapes that are not separable in any available coordinate system. This will be illustrated by an approximate calculation of the modal frequencies of a membrane that has a boundary fixed by an equilateral triangle.

To approximate an equilateral triangle, we can consider one-sixth of hexagonal membrane and then adiabatically transform the hexagon into a circular membrane vibrating in an $m = 3$ mode, as shown in Fig. 6.12. To preserve the areas of the two figures, $a^2 = (3\sqrt{3}/2\pi)\, b^2$ or $a = b\sqrt[4]{27/4\pi^2} = 0.909b$.

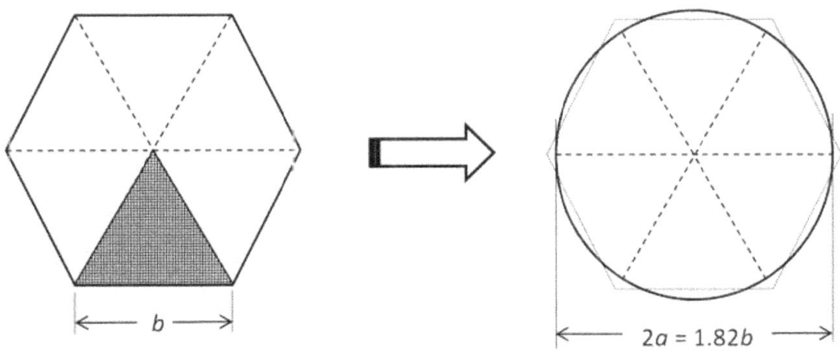

Fig. 6.12 The normal mode frequencies of the equilateral triangular membrane, shown at the left as one-sixth of a regular hexagon, can be approximated by the $m = 3$ normal mode frequencies of the circular membrane shown at the right. The two figures have equal areas. Higher-frequency modes of the triangular membrane can be approximated from the $m = 3i$ modes, where $i = 1, 2, 3, \ldots$ All $m = 3i$ modes have nodal diameters that coincide with two of the fixed boundaries of the equilateral triangle

Table 6.4 Approximate normal mode frequencies of the equilateral triangular membrane, in order of increasing frequency, approximated by imposing adiabatic invariance and treating the triangle as a 60 ° wedge of a circular membrane of area equal to the hexagon in Fig. 6.12. To maintain the triangular boundary condition, only $m = 3i$ modes can be used where $i = 1, 2, 3, \ldots$

m	n	$j_{m,n}$	$f_{m,n}/f_{3,1}$
3	1	6.380	1.000
3	2	9.761	1.530
6	1	9.936	1.557
3	3	13.015	2.040
9	1	13.354	2.093
6	2	13.589	2.130
3	4	16.223	2.543
12	1	16.698	2.617
6	3	17.004	2.665
9	2	17.241	2.702

Based on Eq. (6.33), the lowest-frequency normal mode would correspond to the $j_{3,1}$ zero-crossing of the J_3 Bessel function.

$$f_{3,1} \cong \frac{j_{3,1}c}{2\pi a} = \frac{6.38}{1.82\pi}\left(\frac{c}{b}\right) = 1.12\left(\frac{c}{b}\right) \tag{6.36}$$

As mentioned in Sect. 6.2.1, the solution for the normal modes of an annular membrane requires superposition of the Bessel and the Neumann functions to match the fixed boundary conditions at the outer and inner radii: $z(a_{out}, \theta, t) = z(a_{in}, \theta, t) = 0$. Again, sometimes an approximate solution can be found if the ratio of the radii is nearly equal to the ratio of two zero-crossings of the same Bessel function. If a_{out}/a_{in} has the same ratio of two zero-crossings of some Bessel function, then the nodal circles for those modes can be considered to coincide with the inner and outer radii of the annular membrane.

If we consider an annular membrane with $a_{out} = 12.0$ cm and $a_{in} = 10.0$ cm, corresponding to an annulus that is 2.0 cm wide, then $a_{out}/a_{in} = 1.20$. The average of the three adjacent zero-crossings for the three modal combinations in Eq. (6.37) has a ratio of 1.202 ± 0.012.

$$\frac{j_{0,6}}{j_{0,5}} = 1.210; \quad \frac{j_{1,6}}{j_{1,5}} = 1.191; \quad \frac{j_{2,5}}{j_{2,4}} = 1.214 \tag{6.37}$$

To a rather good approximation, there would be three modes that satisfy the inner and outer boundary conditions.

$$\begin{aligned} f_{2,5} &= \frac{j_{2,5}}{2\pi}\left(\frac{c}{a_{out}}\right) = \frac{17.96}{2\pi}\left(\frac{c}{a_{out}}\right) = 2.858\left(\frac{c}{a_{out}}\right) \\ f_{0,6} &= \frac{j_{0,6}}{2\pi}\left(\frac{c}{a_{out}}\right) = \frac{18.071}{2\pi}\left(\frac{c}{a_{out}}\right) = 2.876\left(\frac{c}{a_{out}}\right) \\ f_{1,6} &= \frac{j_{1,6}}{2\pi}\left(\frac{c}{a_{out}}\right) = \frac{19.616}{2\pi}\left(\frac{c}{a_{out}}\right) = 3.122\left(\frac{c}{a_{out}}\right) \end{aligned} \tag{6.38}$$

The mode corresponding to $j_{0,6}$ for a circular membrane would have no nodal radii, the mode corresponding to $j_{1,6}$ would have two nodal radii separated by 180°, and the $j_{2,5}$ mode would have four nodal radii separated by 90°. Although this is neither a complete nor an exact solution, it is sufficient to provide a good indication of what normal mode frequencies should be observed for this particular annular membrane.

6.2.5 Effective Piston Area for a Vibrating Membrane

In Sect. 12.8, we will see that the energy radiated by a vibrating rigid circular piston of radius, a, which is in contact with a fluid, depends only upon the product of the piston's normal velocity and its area, if the circumference of the piston is less than the wavelength of the sound in the fluid: $2\pi a/\lambda_{fluid} = k_{fluid}\, a \lesssim 1$. The restriction that $k_{fluid}\, a \lesssim 1$ is called the *compactness criterion*. Under such circumstances, it will be convenient to define a *volume velocity*, with units [m³/s]: $U(t) = \Re e\left[\widehat{\mathbf{U}} e^{j\omega t}\right] = \Re e\left[A_{piston}\dot{\mathbf{x}}(t)\right] = \Re e\left[A_{piston}\omega\widehat{\mathbf{x}} e^{j\omega t}\right]$. In fact, it is only the volume velocity that matters in such cases; the shape of the piston or the fact that different parts of the piston are moving with different amplitudes is irrelevant for a compact sound source,[10] as long as all parts of the radiating surface are moving at a single frequency, ω.

Since the amplitude of the normal mode vibrations of a membrane depends upon position on its surface, it is convenient to define an effective piston area, A_{eff}. For a rectangular diaphragm with fixed boundary conditions along the rim, it is easy to integrate over the surface, since there is no membrane motion at the limits of integration.

$$A_{eff} = \iint_S \sin\left(\frac{m\pi x}{L_x}\right)\sin\left(\frac{n\pi y}{L_y}\right)\, dx\, dy = \frac{L_x L_y}{2} \quad \text{if} \quad m = n = 1 \tag{6.39}$$

The effective piston area is only non-zero when both m and n are odd integers. If either m or n are even, one half of the diaphragm's area will be moving out-of-phase with the other half and will therefore produce no net volume velocity.

---

[10] There will be sound, of far smaller amplitude, radiated even if the net volume velocity is zero because the positively and negatively phased portions of the diaphragm's motion are offset from each other. This results in higher-order radiation process (e.g., dipole, quadrupole, etc.) that is far less efficient radiators and is discussed in Sect. 12.4.

This effective area can be used to calculate an average (effective) displacement amplitude, $\langle z \rangle_{eff}$, corresponding to the motion of a rigid surface with the same surface area as the diaphragm. This effective displacement would produce the same volume velocity as the actual surface that has a unit maximum displacement amplitude, $z_{max} = 1$.

$$\frac{\langle z \rangle_{eff}}{z_{max}} = \frac{A_{eff}}{A} = \frac{\frac{1}{2} L_x L_y}{L_x L_y} = \frac{1}{2} \quad \text{if} \quad m = n = 1 \tag{6.40}$$

We can repeat these calculations for a circular membrane. Since the effective area will again be zero for any mode where $m \neq 0$, due to the exact cancellation of in-phase and out-of-phase motion across nodal diameters, the integral need only be performed for the purely radial (azimuthally constant) $J_0\,(k_{0,n}\, r)$ modes.

$$A_{eff} = \iint_S J_o(k_{0,n} r)\, dS = \int_0^a J_o(k_{0,n} r)\, 2\pi r\, dr \tag{6.41}$$

That integral can be evaluated by use of one of the relations for Bessel functions of the first kind provided in Appendix C.

$$\int x J_0(x)\, dx = x J_1(x) \tag{6.42}$$

Letting $x = k_{0,n}\, r$, $r = x/k_{0,n}$, and $dr = dx/k_{0,n}$. Eq. (6.42) can be used to solve Eq. (6.41).

$$A_{eff} = \frac{1}{k_{0,n}^2} \int_0^{ak_{0,n}} J_0(x)\, 2\pi x\, dx = 2\pi \frac{a k_{0,n}}{k_{0,n}^2} J_1(k_{0,n}\, a) = 2\pi a^2 \frac{J_1(k_{0,n}\, a)}{k_{0,n}\, a} \tag{6.43}$$

For the fundamental radial mode, $k_{0,1}\, a \cong 2.4048$. The value of $J_1\,(2.4048) = 0.51915$ could have been evaluated from the series expansion in Eq. (C.5), interpolated from tables like those in Abramowitz and Stegun [2], or provided from most mathematical or spreadsheet software packages, which is how I produced the result.

$$\frac{\langle z \rangle_{eff}}{z_{max}} = \frac{A_{eff}}{\pi a^2} = 2 \frac{J_1(k_{0,1}\, a)}{k_{0,1}\, a} \cong \frac{2 \cdot 0.51915}{2.4048} = 0.432 \tag{6.44}$$

Table 6.5 provides the effective piston areas for the first four radial modes. Any mode with $m > 0$ will have no net piston area.[10] As expected, the effective areas become successively smaller for the higher-order radial modes because of the cancellation produced by adjacent regions bordered by the nodal circles that move out-of-phase with each other.

Table 6.5 Effective piston area for the four lowest-frequency purely radial normal modes of a circular membrane based on Eq. (6.43)

Mode	$k_{0,n}a$	$J_1(k_{0,n}a)$	$A_{eff}/\pi a^2$
0,1	2.40482	0.519149	0.43176
0,2	5.52008	−0.34026	−0.12323
0,3	8.65373	0.271446	0.06274
0,4	11.79153	−0.23246	−0.03943

Negative values indicate that the net volume velocity is opposite to the velocity of the membrane's center

6.2.6 Normal Mode Frequencies of Tympani

If a diaphragm is stretched over a gas-filled, sealed volume, V_o, as it is for a kettledrum (tympani) or a condenser microphone (see Sect. 6.3), then the stiffness of the gas behind the diaphragm provides an additional restoring force. This is similar to the effect that inclusion of the flexural rigidity of a string, calculated in Sect. 5.5, had on the normal mode frequencies of a string that was previously assumed to be limp.

If we assume that the compressions and expansions of the gas trapped behind the diaphragm occur adiabatically,[11] so that $pV^{\gamma} =$ constant, with *polytropic coefficient* for an ideal gas, $\gamma = c_P/c_V$, then the excess pressure, δp, behind the diaphragm will be related to the change in volume, $\delta V = A_{eff}\, z_{max}$, caused by the diaphragm's displacement. Logarithmic differentiation of the Adiabatic Gas Law (see Sect. 1.1.3) relates the changes in volume to the changes in pressure.

$$pV^{\gamma} = constant \Rightarrow \ln p + \gamma \ln V = \ln(constant) \Rightarrow \frac{\delta p}{p_m} = -\gamma \frac{\delta V}{V_o} \tag{6.45}$$

When the diaphragm is in its equilibrium position, the mean pressure, p_m, will be the same on both sides of the diaphragm.

If the speed of transverse waves on the diaphragm, $c = \sqrt{\Im/\rho_S}$, is much less than the speed of sound in the gas, c_{gas}, then all of the gas trapped in the volume behind the diaphragm will respond to the net displacement of the diaphragm (or the volume velocity) produced by the diaphragm's displacement, as calculated in Sect. 6.2.5.

As we saw for our analysis of the stiff string in Sect. 5.5, it was possible to incorporate two restoring forces into a single equation of motion. In this case, we need to calculate the vertical force on a differential surface element of the membrane, $dA = r\,dr\,d\theta$, that will just be the product of the excess pressure and the area: $dF_v = \delta p\, dA$. Adding this contribution, due to *gas stiffness*, to the vertical restoring force provided by the tension in Eqs. (6.20) and (6.21), a new version of Newton's Second Law can be written that incorporates the restoring force provided by excess pressure. The equation is simplified by the fact that only the $m = 0$ modes need to be evaluated; derivatives with respect to angle can therefore be ignored.

$$\frac{\Im}{r}\frac{\partial}{\partial r}\left(r\frac{\partial z}{\partial r}\right) r\,dr\,d\theta + (\delta p) r\,dr\,d\theta = \rho_S (r\,dr\,d\theta)\frac{\partial^2 z}{\partial t^2} \tag{6.46}$$

After cancelling common factor and letting $\mathbf{z}(r,t) = R(r)\mathbf{T}(t) = R(r)e^{j\omega t}$, we are left with an *inhomogeneous ordinary differential equation.*

$$\frac{1}{r}\frac{d}{dr}\left(r\frac{dR(r)}{dr}\right) + k^2 R(r) = -\frac{(\delta p)}{\Im} \tag{6.47}$$

Since the speed of sound in the gas is assumed to be much greater than the speed of transverse wave on the diaphragm, the excess gas pressure, δp, produced by the compression of the gas trapped behind the membrane in a sealed volume, V_o, provides the additional restoring force that depends only upon the equivalent "piston" displacement of the diaphragm. Following the same approach used in Sect. 6.2.5 and expressing the pressure using the Adiabatic Gas Law in Eqs. (6.45), (6.47) can be rewritten entirely in terms of the radial displacement function, $z(r) = R(r)$.

[11] The assumption that the compressions and expansions of the gas trapped behind the diaphragm obey the Adiabatic Gas Law will be placed on a firmer theoretical basis in Chap. 9 when the subject of thermal conduction and adiabatic temperature changes (see Sect. 1.1.3) will be related to acoustical pressure oscillations.

$$\frac{1}{r}\frac{d}{dr}\left(r\frac{dR(r)}{dr}\right) + k^2R(r) = \frac{\gamma p_m}{\Im V_o}\int_0^a R(r)\,2\pi r\,dr \tag{6.48}$$

If Eqs. (6.47) and (6.48) had been an ordinary homogeneous differential equations, we would recognize that $z(r) = R(r) = J_0\,(k_{0,n}\,r)$ is one solution, along with $R(r) = N_0\,(k_{0,n}\,r)$. The Neumann functions are rejected here for the same reasons they were rejected in Sect. 6.2.1. Because these equations are inhomogeneous, their solutions will require that we add a *particular solution* to the homogeneous solution. Since we still require that $z(a, t) = 0$, the following solution will automatically satisfy that boundary condition for any value of $k_{0,n}^{gas}$, while the amplitude constants, C_0, can later be determined by the initial conditions (see Sect. 6.2.7).

$$z(r) = C_{0,n}\left[J_0\left(k_{0,n}^{gas}r\right) - J_0\left(k_{0,n}^{gas}a\right)\right] \tag{6.49}$$

The superscript added to the wavenumber, $k_{0,n}^{gas}$, will distinguish it from the normal mode wavenumbers of Eq. (6.33) that correspond to solutions that incorporate only tension as the restoring force.

The integral that results from substitution of the solution in Eq. (6.49) into the right-hand side of Eq. (6.48) can be simplified by use of Eq. (6.42).

$$\begin{aligned}\frac{2\pi\gamma p_m C_{0,n}}{\Im V_o}\int_0^a z(r)r\,dr &= \frac{2\pi\gamma p_m C_{0,n}}{\Im V_o}\left[\frac{rJ_1\left(k_{0,n}^{gas}r\right)}{k_{0,n}^{gas}} - \frac{r^2}{2}J_o\left(k_{0,n}^{gas}a\right)\right]_0^a \\ &= \frac{\pi a^2\gamma p_m C_{0,n}}{\Im V_o}\left[\frac{2J_1\left(k_{0,n}^{gas}a\right)}{k_{0,n}^{gas}a} - J_0\left(k_{0,n}^{gas}a\right)\right]\end{aligned} \tag{6.50}$$

The recurrence relation between Bessel functions of successive orders in Eq. (C.18) can be used to express the difference of the two Bessel functions in Eq. (6.50) in terms of a single Bessel function.

$$J_{m+1}(x) = \frac{2m}{x}J_m(x) - J_{m-1}(x) \quad \Rightarrow \quad \frac{2J_1\left(k_{0,n}^{gas}a\right)}{k_{0,n}^{gas}a} - J_0\left(k_{0,n}^{gas}a\right) = J_2\left(k_{0,n}^{gas}a\right) \tag{6.51}$$

Substitution of Eq. (6.49) into the left-hand side of Eq. (6.48) is simple since $J_0(k_{0,n}r)$ is the solution to the homogeneous equation and $-J_0\,(k_{0,n}\,a)$ is just a constant.

$$\begin{aligned}-k^2J_0(k_{0,n}a) &= \frac{\pi a^2\gamma p_m}{\Im V_o}J_2\left(k_{0,n}^{gas}a\right) \;\Rightarrow\; J_0\left(k_{0,n}^{gas}a\right) = -\aleph\frac{J_2\left(k_{0,n}^{gas}a\right)}{\left(k_{0,n}^{gas}a\right)^2} \\ \aleph &= \frac{\pi a^4\gamma p_m}{\Im V_o}\end{aligned} \tag{6.52}$$

The constant, $\aleph$, provides a dimensionless measure of the relative importance of the gas stiffness behind the membrane and the membrane's tension per unit length that incorporates the radius of the membrane, a, and the volume of trapped gas behind the membrane, V_o. If the membrane is large and limp or if V_o is small, then $\aleph \gg 1$, and the gas's stiffness will dominate the membrane's restoring force. If the tension per unit length, $\Im$, is large and/or the back volume, V_o, is large, then $\aleph \ll 1$, and the tension dominates the restoring force. For that case, the solutions to Eq. (6.52) are close to those calculated in Eq. (6.33).

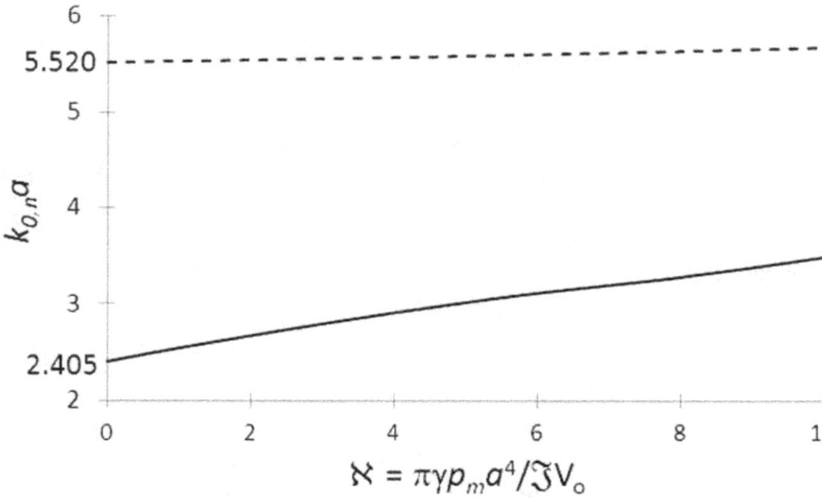

Fig. 6.13 Increase in the wavenumber $k_{0,1}^{gas}a$ (*solid line*) and $k_{0,2}^{gas}a$ (*dashed line*) due to additional stiffness provided by gas at mean pressure, p_m, with polytropic coefficient, $\gamma = c_P/c_V$, trapped in a volume, V_o, behind a diaphragm of radius, a, and tension per unit length, $\Im$. For $\aleph = 0$, the values of $k_{0,n}^{gas}a$ are the same as those in Eq. (6.33) and Table 6.2

$$\lim_{\aleph \to 0} \left[k_{0,n}^{gas}a\right] = k_{0,n}^{gas}a + \frac{2\aleph}{\left(k_{0,n}^{gas}a\right)^3} \tag{6.53}$$

As we demonstrated in Sect. 6.2.5 and in Table 6.5, the effective piston area becomes smaller as n becomes larger. For that reason, the gas stiffness has a much larger effect on the $n = 1$ normal mode vibrational frequency than on the $n > 1$ modes. Figure 6.13 plots $k_{0,n}^{gas}a$ for $n = 1$ and $n = 2$ vs. increasing values of $\aleph$, from 0 to 10. In general, Eq. (6.52) provides a transcendental equation that must be solved for $x = k_{0,n}^{gas}a$ to determine the normal mode frequencies of the membrane when both tension and gas stiffness provide the restoring forces.

$$J_0(x) = -\aleph \frac{J_2(x)}{x^2} \tag{6.54}$$

6.2.7 Pressure-Driven Circular Membranes

As mentioned in the introduction of this chapter, a point force applied to a two-dimensional membrane results in an infinite displacement at the point where the force is applied since the pressure at that point is infinite.[3] A much more interesting and useful way to displace a membrane is to apply a pressure that is different on either of the membrane's surfaces. If we again restrict our analysis to time-harmonic excess acoustic pressures, $\delta p(t) = \Re e[\widehat{\mathbf{p}}e^{j\omega t}]$, applied to one surface of the membrane by a fluid, and if we again assume that $c_{fluid} \gg c = \sqrt{\Im/\rho_S}$, so the compactness criterion is satisfied, $2\pi a/\lambda_{fluid} = k_{fluid}\, a \lesssim 1$, then the excess pressure on the membrane can be considered to be uniform over its entire surface.

With a spatially uniform driving pressure, only the $m = 0$ modes of the membrane will be excited, so Eq. (6.47) will describe the membrane's displacement. The particular solution to this inhomogeneous ordinary differential equation will be similar to Eq. (6.49) but with a constant term added to the homogeneous solution that is related to the amplitude, $|\widehat{\mathbf{p}}|$, of the time-harmonic driving pressure differential.

$$z(r) = C_{0,n} J_0(kr) - \frac{|\widehat{\mathbf{p}}|}{k^2 \mathfrak{T}} \tag{6.55}$$

Note that the wavenumber in the previous equation is not subscripted because we want to evaluate the driven membrane's displacement at all frequencies, not just the discrete normal mode frequencies. The driving pressure provides the initial condition that will determine the value of the amplitude coefficient, $C_{0,\,n}$, that must also be chosen to satisfy the boundary condition, $z(a) = 0$.

$$C_{0,n} = \frac{|\widehat{\mathbf{p}}|}{k^2 \mathfrak{T}} \frac{1}{J_0(ka)} \tag{6.56}$$

Substitution of Eq. (6.56) into Eq. (6.55) provides an expression for the driven membrane's displacement as a function of the radial distance, r, from its center.

$$z(r,t) = \Re e\left\{ \frac{\widehat{\mathbf{p}} e^{j\omega t}}{k^2 \mathfrak{T}} \left[\frac{J_0(kr) - J_0(ka)}{J_0(ka)} \right] \right\} \tag{6.57}$$

Since we have not included damping in Eq. (6.47), the amplitude of the membrane's motion will be infinite for driving frequencies corresponding to $ka = j_{0,n}$, since$J_0(j_{0,\,n}) = 0$.

At low frequencies, where $kr \ll 1$, expansion of Eq. (6.57), using only the first two terms of the power series expansion for $J_0(x)$ in Eq. (C.4), shows that the transverse displacement of the membrane, subject to a uniform time-harmonic excess pressure, has a parabolic dependence upon the distance from the membrane's center at $r = 0$.

$$\lim_{kr \to 0} \left\{ z(r) = \frac{|\widehat{\mathbf{p}}|}{k^2 \mathfrak{T}} \left[\frac{J_0(kr) - J_0(ka)}{J_0(ka)} \right] \right\} = \frac{|\widehat{\mathbf{p}}|}{4\mathfrak{T}} \left(a^2 - r^2 \right) = \frac{|\widehat{\mathbf{p}}| a^2}{4\mathfrak{T}} \left(1 - \frac{r^2}{a^2} \right) \tag{6.58}$$

The fact that the membrane's displacement is independent of frequency, at sufficiently low frequencies, suggests that a small circular membrane could provide a means of measuring acoustic pressure.

The membrane's effective piston displacement, $\langle z \rangle_{eff}$, can be calculated by integration of $\mathbf{z}(r, t)$, given by Eq. (6.57), over the membrane's surface area, as was done in Sect. 6.2.5 for a circular membrane vibrating in one of its purely radial normal modes.

$$\langle z(t) \rangle_{eff} = \frac{\Re e\left[\widehat{\mathbf{p}} e^{j\omega t}\right]}{k^2 \mathfrak{T} J_0(ka)} \frac{1}{\pi a^2} \int_0^a [J_0(kr) - J_0(ka)]\, 2\pi r\, dr \tag{6.59}$$

The integral is identical to the one solved in Eq. (6.50) and must produce the same result.

$$\langle z(t) \rangle_{eff} = \frac{\Re e\left[\widehat{\mathbf{p}} e^{j\omega t}\right] a^2}{\mathfrak{T} (ka)^2} \frac{J_2(ka)}{J_0(ka)} \tag{6.60}$$

Using the series expansions for $J_0(x)$ in Eq. (C.4) and for $J_2(x)$ in Eq. (C.6), the effective transverse displacement can be written for $ka < 1$.

$$J_0(x) \cong 1 - \frac{x^2}{4} \quad \text{and} \quad J_2 \cong \frac{x^2}{8}\left(1 - \frac{x^2}{12}\right) \quad \text{for} \quad x < 1 \tag{6.61}$$

Application of the binomial expansion in Eq. (1.9) produces an expression for that ratio of the two Bessel functions in Eq. (6.60).

$$\frac{J_2(ka)}{J_0(ka)} \cong \frac{\left(\frac{(ka)^2}{8}\right)\left(1 - \frac{(ka)^2}{12}\right)}{1 - \frac{(ka)^2}{4}} \cong \left(\frac{(ka)^2}{8}\right)\left(1 - \frac{(ka)^2}{12}\right)\left(1 + \frac{(ka)^2}{4}\right) \tag{6.62}$$

Keeping only terms up to second order in *ka* produces a useful expression for the effective diaphragm displacement caused by application of a time-harmonic uniform excess pressure.

$$\langle z(t)\rangle_{eff} = \Re e\left[\hat{\mathbf{p}}e^{j\omega t}\right]\frac{a^2}{8\mathfrak{T}}\left[1 + \frac{(ka)^2}{6}\right] \quad \text{for} \quad ka < 1 \tag{6.63}$$

This result extends the approximation made to produce the parabolic displacement of Eq. (6.58) to higher frequencies and shows that at low frequencies $\langle z\rangle_{eff} = (½)z(0)$. The dependence upon $(ka)^2$ indicates that average displacement increases as the first purely radial resonance frequency, at $ka = 2.405$, is approached, as expected. Since the membrane is still being driven within its stiffness-controlled frequency regime (see Sect. 2.5.1), resonance behavior can still be ignored even if the damping is small.

6.3 Response of a Condenser Microphone

The invention of the *condenser microphone* by Wente, in 1917 [4], marks the beginning of the electroacoustic era that took us from tuning forks, Helmholtz resonators [5], sensitive flames [6], and the Rayleigh disk (see Sect. 15.4.3) to the sophisticated electronic data acquisition and analysis tools that serve us so well today. Wente's microphone created an acoustic pressure sensing system that facilitated quantitative characterization of acoustical disturbances in air using electronic instrumentation.

Wente realized that the nearly frequency-independent displacement of a stretched diaphragm in response to pressure, expressed in Eqs. (6.58) and (6.63), could be sensed if the diaphragm formed one plate of a parallel-plate electrical capacitor with the other plate being provided by a fixed electrode (the backplate in Fig. 6.14).

Figure 6.14 provides a cut-away view of a modern condenser microphone and typical parameter values for such a one-inch condenser microphone. Based on the values in the table on the right-hand side of Fig. 6.14, the speed of transverse waves on the membrane, $c = \sqrt{\mathfrak{T}/\rho_S} = 219$ m/s $< c_{air} = 343$ m/s. The frequency of the first radial normal mode of the membrane is $f_{0,1} = 2.4048c/2\pi a = 9.43$ kHz, if the additional stiffness of the air in the back chamber is neglected.[12] Based on Eq. (6.52) and the values in Fig. 6.14, $\aleph = 1.933$, raising $k_{0,1}\, a$ from 2.4048 to $k_{0,1}^{gas} a =$

[12] The measured resonance frequency in vacuum was 9.44 kHz [7].

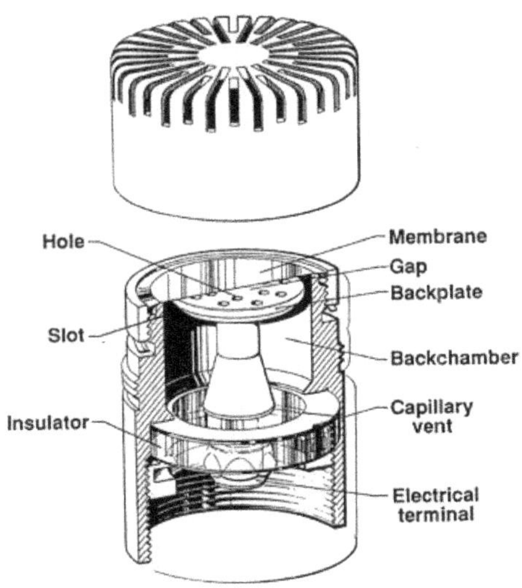

Component	Sym bol	Value	Units
Membrane:			
radius	a	8.89	mm
thickness	t	5	μm
surface mass density	ρ_S	0.0445	kg/m²
tension per unit length	$\mathfrak{I}$	2,141	N/m
Backplate electrode:			
radius	b	6.617	mm
Back chamber:			
volume	V_o	674	mm³
Air:			
gap	h_o	26.55	μm
temperature	T	19	°C
pressure	p_m	101,325	Pa
density	ρ_m	1.205	kg/m³
polytropic coefficient	γ	1.403	
sound speed	c_{air}	343	m/s

Fig. 6.14 (*Left*) Cut-away view of a modern condenser microphone with its screw-on protective grid displaced upward to make the membrane visible. [Courtesy of Brüel & Kjær Instruments.] (*Right*) Parameter values for a typical "1-inch" condenser microphone based on a B&K Type 4126, S/N: 256903 [7, 8].

2.6651, thus increasing the gas-stiffened normal mode frequency by about 10.8% to $f_{0,1}^{gas} = 10.45$ kHz.[13] Use of the approximation in Eq. (6.53) produces a slightly larger value of $k_{0,1}^{gas} a = 2.683$, although the difference would be indistinguishable in Fig. 6.13.

The electrical capacitance, C, of such a parallel-plate capacitor depends upon equilibrium gap, h_o, between the membrane of radius, a, and the backplate of radius, b. Since $a \gtrsim b$ and $h_o \ll b$ and the gap is filled with air, the capacitance of such a condenser microphone capsule can be expressed in terms of the *permittivity of free space*, $\varepsilon_o = 8.854$ pF/m.[14]

$$C(t) = \varepsilon_o \frac{\pi b^2}{\langle g(t) \rangle} \tag{6.64}$$

Based on the values of the physical specifications in Fig. 6.14 (right), the equilibrium (unpolarized) capacitance of the microphone, $C_o = \varepsilon_o \pi b^2 / h_o = 45.9$ pF.

Since the displacement of the microphone's membrane is a function of the radial distance from the center, as expressed in Eq. (6.57) or approximated at low frequencies in Eq. (6.58), it is necessary to integrate that displacement over the backplate surface area, πb^2, to calculate the time-dependent effective gap spacing, $\langle g(t) \rangle = h_o + \langle h(t) \rangle$, driven by the acoustical pressure variations. The effective gap spacing, $<g(t)>$, will be the sum of the equilibrium gap spacing, h_o, and the time-harmonic, area-

[13] That result assumes $\gamma = 7/5$, but in the small space between the diaphragm and the backplate, the compressions and expansions of the gas are nearly isothermal, corresponding to $\gamma = 1$, rather than adiabatic. For the current discussion, this distinction is not significant.

[14] The dielectric constant of air at atmospheric pressure is sufficiently close to that of a vacuum that the difference can usually be ignored and the free-space value of the permittivity can be used to calculate the capacitance. If the space between the membrane and the backplate is filled with an insulating material, then the capacitance of Eq. (6.65) must be increased by the relative dielectric constant of the material, such as the Teflon used for electret microphones in Sect. 6.3.3.

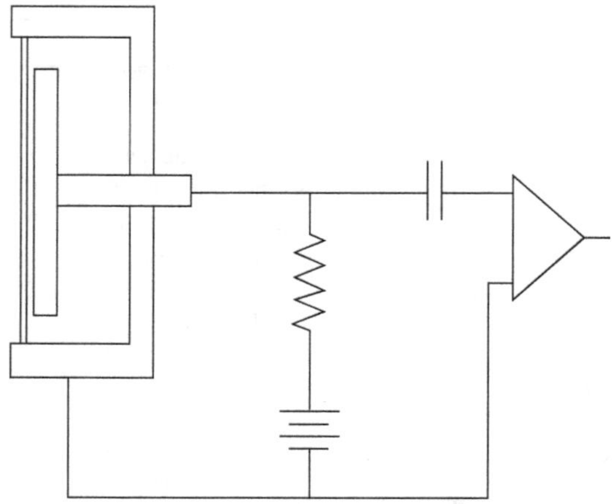

Fig. 6.15 Schematic diagram of an electrical circuit that provides a polarizing voltage, V_{bias} (*battery symbol*), to the condenser microphone's backplate through an electrical resistor with very large resistance. That DC voltage, V_{bias}, is blocked by a large-value capacitor acting as a high-pass filter to allow only the time-varying voltage, $V(t)$, produced by the changing capacitance of the condenser microphone, to reach the pre-amplifier, while blocking V_{bias} from the input to the first-stage pre-amplifier that is usually located in close proximity to the condenser microphone, typically in the microphone's handle. (Figure courtesy of J. D. Maynard)

averaged variation in the gap, $\langle h(t)\rangle = \Re e\left[\widehat{\mathbf{h}}e^{j\omega t}\right]$, produced by the forcing pressure differential, $\delta p(t) = \Re e\left[\widehat{\mathbf{p}}e^{j\omega t}\right]$, if the static pressure of the air trapped back chamber that remains close to p_m.[15]

Logarithmic differentiation (see Sect. 1.1.3) of Eq. (6.64) relates the area-averaged variation in the gap to the variation in the capacitance of the membrane and backplate.

$$\frac{\delta C(t)}{C_o} = -\frac{\delta\langle g(t)\rangle}{h_o} = -\frac{\Re e\left[\widehat{\mathbf{h}}e^{j\omega t}\right]}{h_o} \tag{6.65}$$

In ordinary operation, an electronic circuit has to be added to convert the changes in capacitance to changes in voltage (or current) that can be amplified, detected, displayed, and recorded by one or more electronic instrument (e.g., oscilloscope, dynamic signal analyzer, digital multimeter, sound level meter, etc.).

This production of a time-dependent voltage,[16] $V(t) = \Re e\left[\widehat{\mathbf{V}}e^{j\omega t}\right]$, which can be sensed electronically, is typically accomplished by providing the condenser microphone with a *polarizing voltage*, V_{bias}, through an electrical resistor, R_{el}, having a high electrical resistance. A circuit for providing the DC-bias voltage is shown schematically in Fig. 6.15. The exponential time constant, $\tau = R_{el}\, C_o$, for charge to move on or off the microphone would be about one-half second if $R_{el} = 10\ \text{G}\Omega$. For frequencies at the lower end of the audio spectrum, defined by the traditional limits of human hearing

[15] The capillary vent shown in Fig. 6.14 is designed so that it will let the mean pressure equilibrate over longer times while blocking oscillatory flow of air at the lowest operational design frequency of a particular microphone.

[16] Note that $\widehat{\mathbf{V}}$ should be considered to be a complex number (phasor) since there can be a phase difference between $\delta p(t)$ and $V(t)$.

(20 Hz < *f* < 20 kHz), the charge, Q_o, stored in the capacitor formed by the membrane and backplate is effectively constant.[17] If $V_{bias} = 200\ V_{dc}$, then $Q_o = C_0 V_{bias} \cong 10$ nanocoulombs (10 nC).[18]

6.3.1 Optimal Backplate Radius

With a constant charge, Q_o, on the condenser microphone, we can again invoke logarithmic differentiation to relate changes in capacitance, δC, to changes in the voltage, $V(t)$, that would appear across the terminals of the pre-amplifier stage in Fig. 6.15, which we assume has an infinite input electrical impedance.

$$Q_o = CV \quad \Rightarrow \quad \frac{\delta V}{V_{bias}} = \frac{V(t)}{V_{bias}} = -\frac{\delta C}{C_o} = \frac{\langle h(t) \rangle}{h_o} \tag{6.66}$$

The final term on the right-hand side of Eq. (6.66) incorporates the result of Eq. (6.65).

Before continuing to calculate an open-circuit sensitivity, $\mathbf{M_{oc}} = \widehat{\mathbf{V}}/\widehat{\mathbf{p}}$, for the microphone capsule based on Eq. (6.66), it will be instructive to think about our choice of a backplate area, πb^2, for a given diaphragm area, πa^2. Since the motion of the membrane is greatest at its center and the sensitivity is proportional to $\delta C/C_o$, making b as small as possible would provide the largest value of $\delta C/C_o$ by making $\langle h(t) \rangle / h_o$ have its largest value, thus making $V(t)$ as large as possible. It would also make the electrical output impedance, $|\mathbf{Z_{el}}| = (\omega\ C_o)^{-1}$ very high, suggesting that the voltage, $V(t)$, that is produced will supply a very tiny electrical current.

For optimum performance and minimum electronic noise [9], the electrical power output should be maximized, not just the voltage. The electrical power is proportional to the product of current and voltage. The time-averaged electrical power, $\langle \Pi_{el} \rangle_t$, delivered to a load by a capacitor is related to the electrostatic energy, $(PE)_{el}$, stored in a capacitor that is charging and discharging at a frequency, ω, with a voltage, $V(t) = \Re e\left[\widehat{\mathbf{V}} e^{j\omega t}\right]$, across its terminals.

$$\langle \Pi_{el} \rangle_t = \left| \frac{d(PE)_{el}}{dt} \right| = \omega\, C_o \frac{\left|\widehat{\mathbf{V}}\right|^2}{2} \tag{6.67}$$

To calculate $\widehat{\mathbf{V}}$ from Eq. (6.66), the area-averaged, pressure-driven displacement, $\langle h(t) \rangle$, has to be determined by integration of the membrane's displacement over the backplate area, πb^2. Since we seek a result for the low-frequency sensitivity, the membrane's displacement as a function of radius, $z(r)$, can be represented by Eq. (6.58).

---

[17] To use a condenser microphone to detect infrasonic pressures, a high-frequency "carrier" is used rather than a DC-bias voltage because the output impedance of a capacitance goes to infinity as the frequency approaches zero. Another way to detect the diaphragm's displacement down to DC is to resonate the microphone's capacitance, C, with an inductance, L, and use the changes in resonance frequency, $\omega_o = (LC)^{-1/2}$, of the electrical tank circuit, to determine the diaphragm displacement. Another way to extend the response down to DC is to make the microphone be one arm of a capacitive Wheatstone bridge circuit that uses a high-frequency AC-bias voltage [20]. Of course, careful attention must also be paid to the pressure equilibration through the capillary vent to ensure that frequency response of the microphone itself will not be reduced by air leakage through the capillary at very low frequencies.

[18] The SI abbreviation for coulomb is [C]. For temperature in degrees Celsius, it is [°C] to distinguish it from electrical charge.

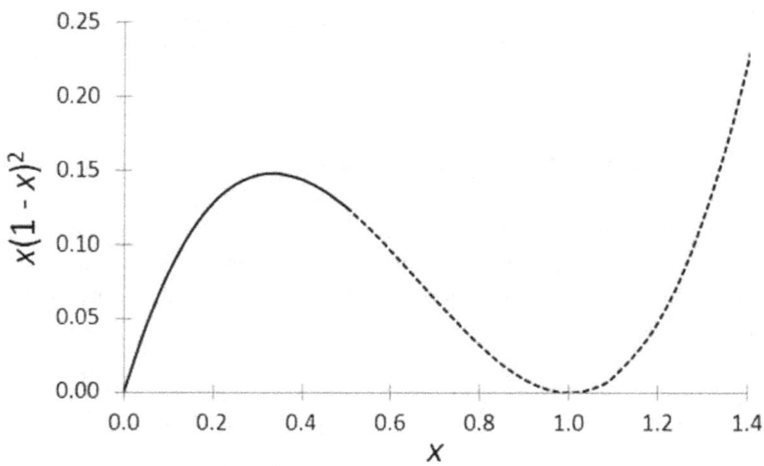

Fig. 6.16 Graph of Eq. (6.80) showing the local maximum at $x = 1/3$ with the maximum value at that point equal to $4/27$. The values of $x \geq 0.5$ that are not physically realizable are plotted with the dashed line

$$\langle h(t)\rangle = \Re e\left[\frac{\widehat{\mathbf{p}}e^{j\omega t}}{4\pi\Im}\right]\frac{a^2}{b^2}\int_0^b\left(1-\frac{r^2}{a^2}\right)2\pi r\,dr = \frac{a^2}{4\Im}\left(1-\frac{b^2}{2a^2}\right)\Re e\left[\widehat{\mathbf{p}}e^{j\omega t}\right] \tag{6.68}$$

This result is reasonable. For a given pressure differential across the membrane, a larger membrane will have a larger average deflection, and a lower tension per unit length will also lead to larger average deflection. Both increases in sensitivity (with increasing area or decreasing tension) come at the cost of a lower fundamental resonance frequency, hence, a more limited useful frequency bandwidth. Substituting this result into Eq. (6.66) produces the time-dependent variation in output voltage caused by the acoustic pressure.

$$V(t) = V_{bias}\frac{\langle h(t)\rangle}{h_o} \cong V_{bias}\frac{a^2}{4h_o\Im}\left(1-\frac{b^2}{2a^2}\right)\Re e\left[\widehat{\mathbf{p}}e^{j\omega t}\right] \tag{6.69}$$

The signal voltage, $V(t)$, increases as b^2 decreases, as postulated earlier, but the capacitance, C_o, increases as b^2 increases.

$$C_o = \varepsilon_o\frac{\pi b^2}{h_o} = \varepsilon_o\frac{\pi a^2}{h_o}\left(\frac{b^2}{a^2}\right) \tag{6.70}$$

Letting $x^2 = b^2/a^2$, Eqs. (6.69) and (6.70) can be substituted into Eq. (6.67) to calculate the time-averaged electrical power, $\langle\Pi_{el}\rangle_t$, produced by the pressure-driven microphone in terms of x.

$$\langle\Pi_{el}\rangle_t = \Lambda x^2\left(1-\frac{x^2}{2}\right)^2; \quad \Lambda = \frac{\varepsilon_o\pi\omega a^6}{2h_o^3}\left(\frac{V_{bias}|\widehat{\mathbf{p}}|}{4\Im}\right)^2 \tag{6.71}$$

The optimum ratio for $b/a = x_{opt}$ can be found by differentiation of Eq. (6.71) with respect to x.

$$\frac{d\langle\Pi_{el}(x)\rangle_t}{dx} = \Lambda\left(\frac{3x^5}{2} - 4x^3 + 2x\right) = 0 \quad\Rightarrow\quad x_{opt}^2 = \frac{4\pm 2}{3} \tag{6.72}$$

Since $b/a \leq 1$, the optimum backplate-to-membrane radius ratio is $(b/a)_{opt} = \sqrt{2/3} = 0.8165$. The radius ratio for the 1-inch microphone in Fig. 6.13 is slightly smaller: $b/a = 0.744$.

To determine the theoretically maximum sensitivity of a condenser microphone, the pre-amplifier in Fig. 6.16 will be assumed to have infinite electrical input impedance, so that the full pressure-driven voltage produced by the microphone will appear across the pre-amplifier's input terminals. To remind ourselves of this approximation, the microphone's sensitivity will be designated the *open-circuit microphone sensitivity*, $\mathbf{M_{oc}}$.

$$|\mathbf{M_{oc}}| \equiv \left|\frac{\widehat{\mathbf{V}}}{\widehat{\mathbf{p}}}\right| = V_{bias}\frac{\langle h(t)\rangle}{h_o} = V_{bias}\frac{a^2\left(1-\frac{b}{2a^2}\right)}{4h_o\Im} \tag{5.73}$$

$$(|\mathbf{M_{oc}}|)_{max} = \frac{a^2}{6h_o\Im}V_{bias}$$

At sufficiently low frequencies, the voltage and pressure will be in-phase, so M_{oc} is usually represented by a real number. In reality, $\mathbf{M_{oc}}$ must be a complex number to incorporate phase differences between pressure and output voltage. Using the parameters in the table on the right in Fig. 6.14 and applying the typical bias voltage for condenser microphones, $V_{bias} = 200\ V_{DC}$, the optimal open-circuit sensitivity of that microphone would be $(M_{oc})_{max} = 46.3$ mV/Pa or $-$ 26.7 dB *re:* 1.0 V/Pa.

6.3.2 Limits on Polarizing Voltages and Electrostatic Forces

As shown in Eq. (6.73), the open-circuit sensitivity of a condenser microphone is directly proportional to the polarization voltage, V_{bias}. In principle, it appears that the larger the value of V_{bias}, the better. This section will address two limitations on the possible value for V_{bias}. The first is "arcing," which occurs if the voltage is large enough to cause a spark to jump from the diaphragm to the backplate. The second is electrostatic collapse. Since the electrostatic force between the diaphragm and the backplate increases with decreasing separation, the application of the bias voltage causes the electrostatic force to bring the diaphragm closer to the backplate. As the separation decreases, the electrostatic attractive force increases. At some point, the gap becomes small enough that the elastic restoring force due to the tension in the diaphragm is insufficient to counteract the electrostatic force that is increasing quadratically with decreasing gap.

It is worthwhile to reflect momentarily on the fact that a 200 V_{DC} voltage difference across a 20-micron gap corresponds to an electric field strength of 10×10^6 V/m = 10 MV/m. The accepted value for the electric field strength necessary to cause *dielectric breakdown* (i.e., sparking) in air, at atmospheric pressure, is about 3 MV/m = 3 V/μm [10].

The reason that a condenser microphone can exceed the breakdown threshold by a factor of three is that sparking relies on an avalanche process where one electron is accelerated to an energy that is sufficient to ionize another atom prior to its final collision (in this case with the microphone's backplate or diaphragm). For nitrogen, the ionization energy is about 15 eV, and the distance between collisions is the order of one mean free path.[19] For gaps on the order of 10 microns and above, the gap dependence of the breakdown voltage, V_{BD}, is given by an empirical relation known as Paschen's law [11].

$$V_{BD} = \frac{a p_m g}{\ln(p_m g) + b} \tag{6.74}$$

In Eq. (6.74), p_m is in atmospheres (1 atm ≡ 101,325 Pa), and the gap, g, is in meters, with $a = 4.36 \times 10^7$ V/(atm-m) and $b = 12.8$ [12]. At atmospheric pressure, Eq. (6.74) yields V_{BD}/

[19] The mean free path of an atom in air at atmospheric pressure is about 100 nanometers (see Sect. 9.5.1). The mean free path of electrons in air is about 500 nanometers.

m = 3.4 MV/m over 1 m, and at 20 microns, $V_{bias} = 440\ V_{DC}$. The suppression of breakdown in very small devices has generated renewed contemporary interest due to the development of microelectromechanical systems (MEMS) [13].

The voltage difference between the membrane and the backplate will produce an electrostatic attraction. Since the gap between the backplate and membrane is so small, it is prudent to calculate the static displacement of the membrane due to this electrostatic attraction. The *electrostatic potential energy*, $(PE)_{el}$, stored in a parallel-plate capacitor, depends upon the product of the square of the voltage difference and the capacitance, C_o.

$$(PE)_{el} = \frac{C_o V_{bias}^2}{2} = \frac{\varepsilon_o \pi b^2}{2h_o} V_{bias}^2 \tag{6.75}$$

Since the polarization voltage, V_{bias}, is held constant, the electrostatic force, F_{el}, is just the negative of the derivative of the potential energy (see Sect. 1.2.1) with respect to the gap height, h_o.

$$F_{el} = -\left(\frac{\partial (PE)_{el}}{\partial h_o}\right)_{V_{bias}} = \frac{\varepsilon_o \pi b^2}{2h_o^2} V_{bias}^2 = \frac{C_o V_{bias}^2}{2h_o} \tag{6.76}$$

If we ignore the fact that the electrostatic force acts over the area of the backplate and not the entire membrane, thus assuming that $b \cong a$, then the electrostatic pressure, P_{el}, can be approximated well enough to determine the displacement of the membrane's center due to the polarization voltage.

$$P_{el}|_{b=a} \cong \frac{\varepsilon_o}{2h_o^2} V_{bias}^2 \tag{6.77}$$

The deflection of the membrane's center, $z\,(0)$, using the microphone parameters in Fig. 6.14 and $V_{bias} = 200\ V_{DC}$, is determined using Eq. (6.58).

$$z(0)|_{b=a} = \frac{P_{el} a^2}{4\mathfrak{T}} = \frac{\varepsilon_o}{8} \frac{a^2}{h_o^2} \frac{V_{bias}^2}{\mathfrak{T}} = 2.8 x 10^{-6} \text{m} = 2.8 \text{ microns} \tag{6.78}$$

The electrostatic force reduces the minimum gap by about 11%. The error in the electrostatic deflection introduced by assuming that $a = b$ is negligible for our purposes [14].

$$\frac{z(0)|_{b<a}}{z(0)|_{b=a}} = \frac{b^2}{a^2}\left[1 + 2\ln\left(\frac{a}{b}\right)\right] \tag{6.79}$$

For the microphone parameters in Fig. 6.4, this corresponds to a reduction in the displacement below that calculated in Eq. (6.78) resulting in an electrostatic deflection of 2.46 microns.

There is another intrinsic limit on the polarization voltage that is a result of this electrostatic deflection of the membrane. As V_{bias} increases, Eq. (6.77) shows that the electrostatic pressure increases.[20] The restoring force provided by the diaphragm increases linearly with reduction in gap, h, but the electrostatic attraction increases inversely as the gap squared, h^{-2}. There will be some critical

[20] The electrostatic pressure is used as a means to make absolute calibrations of condenser microphones using a grid placed above the diaphragm called an "electrostatic actuator." Because the gap between such a grid and the microphone's diaphragm must be small to produce significant sinusoidal electrostatic pressures, uncertainty in that gap is the most significant source of uncertainty in such calibrations. For further discussion, see H. Miura, "Laboratory Calibration Methods at Electrotechnical Laboratory, Japan," in *AIP Handbook of Condenser Microphones: Theory, Calibration and Measurements*, G. S. K. Wong and T. F. W. Embleton, Editors (Am. Inst. Phys., 1995); ISBN 1-56396-284-5. See §8.7, "Electrostatic Actuator Measurements."

value of the polarization voltage, $(V_{bias})_{Crit}$ that, when exceeded, will cause a catastrophic collapse of the membrane.

The critical value of the polarization voltage can be calculated by substitution of P_{el}, from Eq. (6.77), into the expression for the membrane's average (effective) static ($\omega = 0 \Rightarrow ka = 0$) deflection, $\langle z \rangle_{eff}$, given in Eq. (6.63), that will provide the reduction in the gap due to the polarization voltage. When we allow the gap, h, to be reduced by that average deflection, so the polarization-dependent gap, $h(V_{bias}) = h_o - \langle z \rangle_{eff}$, then we obtain an expression that includes the effective deflection on both sides of the equation.

$$\frac{\langle z \rangle_{eff}}{h_o} \cong \frac{\varepsilon_o}{16} \frac{V_{bias}^2}{\Im} \frac{b^2}{h_o^3} \frac{1}{\left(1 - \frac{\langle z \rangle_{eff}}{h_o}\right)^2} \quad \Rightarrow \quad x(1-x)^2 = \frac{\varepsilon_o}{16} \frac{V_{bias}^2}{\Im} \frac{b^2}{h_o^3} \tag{6.80}$$

In the right-hand version of Eq. (6.80), we have let $x = \langle z \rangle_{eff}/h_o < ½$. That limit on x is based on the fact that the average deflection is half the maximum deflection of the center.

Figure 6.16 is a plot of $x(1-x)^2$ vs. x. Differentiation of that function shows that there is a local maximum at $x = ⅓$. At that point, the function has a value of $^4/_{27} \cong 0.148$. If the dimensionless quantity on the right-hand side of Eq. (6.80) exceeds $^4/_{27}$, then no solution is possible. This result provides the maximum magnitude of the polarization voltage $(V_{bias})_{Crit}$.

$$(V_{bias})_{Crit} = |V_{bias}|_{max} = \sqrt{\frac{64}{27} \frac{\Im h_o^3}{\varepsilon_o b^2}} \tag{6.81}$$

For the 1-inch microphone parameters in Fig. 6.13, $(V_{bias})_{Crit} \cong 450$ V_{DC}. Coincidentally, the polarization voltage limits imposed by both dielectric breakdown of the air and electrostatic collapse are nearly the same.

6.3.3 Electret Condenser Microphone

The large polarizing voltage required for operation of the condenser microphone, described in the previous section, is readily available in a laboratory or recording studio that has dedicated microphone power supplies capable of providing stable polarizing voltages in the range of 48–200 V_{DC}. For portable battery-operated electronics, like telephones or small tape/digital recorders, the required high voltages are not easy to produce. Fortunately, in addition to the nonstick properties of Teflon® that make it a popular coating for cooking utensils, Teflon also has the ability to trap electrical charges [15].[21]

Charge can be deposited in Teflon by exposure of an aluminized Teflon film to break down electrical fields [16], by ion implantation using electron beams [17] or by rubbing an alcohol-soaked cotton swab over the Teflon if there is a voltage difference as small as 300 V_{DC} between the moist swab and the conducting surface of the Teflon film [18, 19].

[21] This property is not unique to Teflon. Other polymers, like Mylar®, also exhibit the ability to trap electrical charges. Teflon is the most common electret material for microphones because it has the highest volume electrical resistivity, which is on the order of 10^{18} Ω cm.

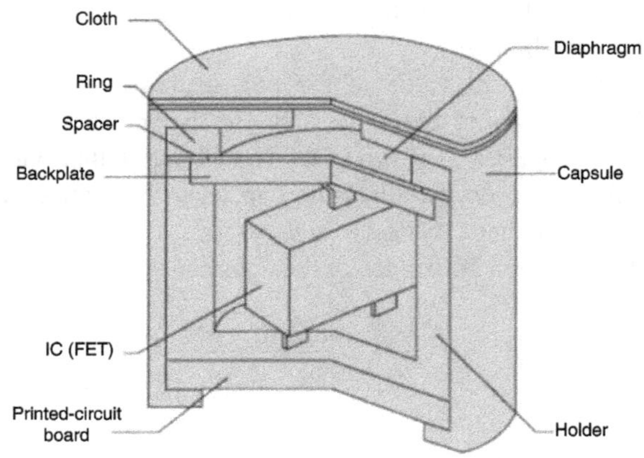

Fig. 6.17 An electret condenser microphone can be designed so that a field-effect transistor (FET) can be included in a package that is less than a centimeter in diameter and a few millimeters in height. (*Left*) The microphone backplate can be attached directly to the gate electrode of the FET, and the source and drain terminals of the FET can be exposed. (*Right*) The aluminized Teflon membrane (diaphragm), with the aluminized surface facing away from the backplate, is typically mounted to a thin spacer (washer), and the package is usually covered by a thin sheet of porous fabric that keeps dust and other particulates off of the membrane. The membrane will be attracted to the backplate by the electrostatic forces of Eq. (6.76). [Drawing courtesy of Hosiden Corp]

There are two strategies for using charge trapped in Teflon to provide the polarizing voltage for a condenser microphone. For laboratory-quality microphones, the Teflon is bonded to the backplate, a configuration known as a "back electret." The other option is to use a thin Teflon film (usually between 6 and 12 microns thick) as both the membrane and the electret.

The cost of an *electret microphone* can be made very low (on the order of pennies each), and the packaging can include a field-effect transistor (FET) [20]. The FET will convert the output electrical impedance $(j\omega C_o)^{-1}$ of the microphone's capacitance of only a few picofarads to about 1 kΩ. Such a package can be just as small as a TO-92 transistor case. As shown in Fig. 6.17, the gate electrode of the FET can be attached directly to the microphone backplate, and the source and drain electrodes are available for transmission of the microphone signal through modest lengths of cable.

This approach has led to the sale of such microphones for use in telephones, tape recorders, hearing aids, etc., at around two billion units per year [21]. By 2010, the annual sales of electret condenser microphones had decreased, while the sale of microelectromechanical systems (MEMS) has increased to 1.1 billion units in 2013 (see Problem 15). With improvements in the MEMS devices, that trend will most likely continue.

Although the trapped charge will escape as the temperature increases, Teflon electret microphones have been used down to temperatures approaching absolute zero [22].[22] Electret transducers are very easy to assemble in a laboratory because the charged membrane will be held in place by the electrostatic attraction, given in Eq. (6.76), between the membrane and the conductive backplate. The sensitivity of such an electret microphone depends primarily upon the ambient air pressure and the equivalent polarization voltage, $V_{bias} = Q_o/C_o$, produced by the trapped charge, Q_o.

The microphone can be modeled as two capacitors that are placed in series electrically. One capacitor is formed by the conductive (usually aluminized) coating and the Teflon film. If the thickness

[22] The thermal activation of the trapped charge is used to stabilize the electret microphone's sensitivity. Typically, the temperature of the assembled unit will be raised to about 60 °C so that the charge, and hence the sensitivity, will be stable for use at temperatures less than 60 °C.

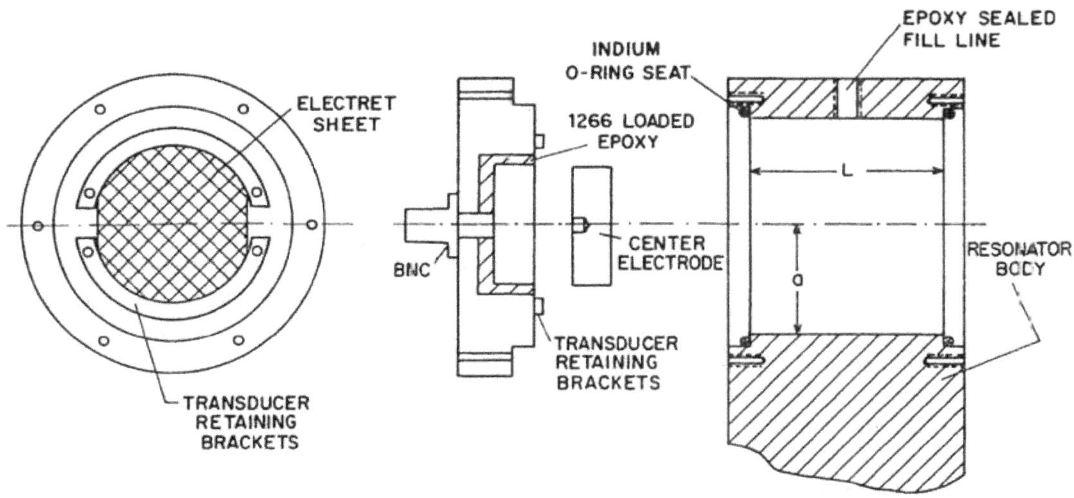

Fig. 6.18 Cylindrical resonator of length, L, and diameter, $2a$, with rigid endcaps formed by electret transducers that consist of an insulated "center electrode" covered by an electrically charged, thin, aluminized Teflon® film that can either excite or detect acoustic resonances within the resonator [25].

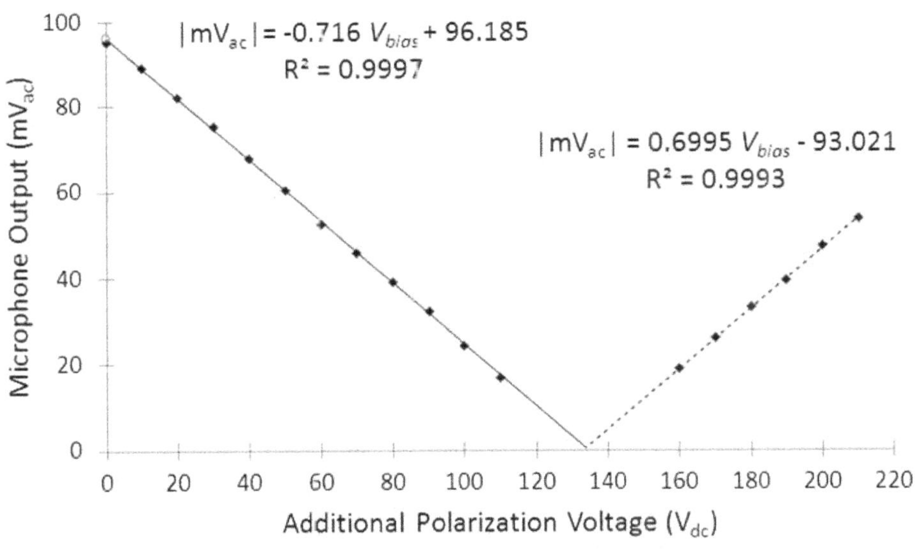

Fig. 6.19 Change in an electret microphone output voltage vs. additional applied polarization voltage, V_{bias}. The applied polarization is opposing the polarization provided by the trapped charge. Straight-line fits to the descending (*solid*) and ascending (*dashed*) data extrapolate to $V_{bias} = 133.7 \pm 0.7\ V_{DC}$. The data near zero microphone output have lower signal-to-noise ratios and are not plotted

of the Teflon is t, then the Teflon's capacitance per unit area, C_{Teflon}/A, is a constant, since the thickness is not a function of the acoustic pressure on the conducting surface of the film: $C_{Teflon}/A = \varepsilon_{Teflon}\varepsilon_o/t$. The relative dielectric constant is $\varepsilon_{Teflon} = 2.1$. For ½-mil thick ($t = 12.7$ micron) Teflon film, $C_{Teflon}/A \cong 150\ \text{pF/cm}^2$. The other series capacitor is formed by the bottom surface of the Teflon film and the backplate that are separated by a thin film of air. That average air gap, h_{Air}, will be very thin due to the electrostatic attraction making, $h_{Air} \ll t$.[23] For that reason, the net capacitance per unit area is just that of the Teflon film. Since the Teflon® film is far less compressible than the air trapped between the Teflon and the backplate, the sound impinging on the microphone will vary the average thickness of the air gap and will produce a voltage, $V(t) = \Re e\left[\widehat{\mathbf{V}}e^{j\omega t}\right]$, that is proportional to the equivalent polarization voltage, $V_{bias} = Q_o/C_o$, produced by the trapped charge, Q_o, and the time-varying thickness of the air gap, expressed in Eq. (6.66).

It is actually quite simple to accurately measure the equivalent polarization voltage, V_{bias}, using a *vibrating reed electrometer* technique [23, 24]. The data provided in Fig. 6.19 was obtained by driving a cylindrical resonator with two electret transducers on either end, similar to that shown in Fig. 6.18 [25], at its fundamental resonance and connecting the receiving electret to a circuit like that in Fig. 6.15, which has a variable polarization supply voltage. Keeping the amplitude of the acoustical resonance constant, the amplitude of the received signal can be plotted against the additional externally applied polarization voltage. Data for one such electret microphone is shown in Fig. 6.19.

The reverse bias provided by the additional polarization voltage decreased the microphone's output until the reverse bias voltage equaled the effective electret polarization voltage. When the reverse bias exceeded the effective electret polarization voltage, the microphone signal returned. The stored charge can be estimated by extrapolation of the two straight lines in Fig. 6.19 to zero microphone output. Based on that extrapolation, $V_{bias} = 133.7 \pm 0.7\ V_{DC}$, so $Q_o = C_o V_{bias} = 2.7 \times 10^{-7}$ coulombs. The capacitance of that electret microphone was about 2000 pF, corresponding to a backplate diameter of 4.25 cm.

Calculation of the electret microphone's sensitivity depends upon a relationship between the air gap thickness, $h_{Air}(t) = h_o + \Re e\left[\widehat{\mathbf{h}}e^{j\omega t}\right]$, and the excess pressure, $\delta p(t)$, on its surface. We can use the Adiabatic Gas Law in Eq. (6.45), but since the equilibrium thickness of that air layer, h_o, is so small, the gas compresses isothermally rather than adiabatically, as it did in our tympani example in Sect. 6.2.6. This isothermal behavior can be represented in Eq. (6.45) by setting $\gamma = 1$.[24] It is also worthwhile to remember that the stiffness of the gas trapped between the film and the backplate provides the restoring force, since there is no tension applied to the membrane. That means that the membrane will move uniformly, like a piston, and no effective area need be calculated.

Following Eq. (6.66), the electret microphone's open-circuit sensitivity, M_{oc}, can be expressed in terms of the mean gas pressure, p_m.

$$\frac{V(t)}{V_{bias}} = \frac{\delta h_{Air}(t)}{h_o} \Rightarrow |\mathbf{M_{oc}}| = \frac{V(t)}{\delta p(t)} = \frac{\left|\widehat{\mathbf{V}}\right|}{|\widehat{\mathbf{p}}|} = \frac{Q_o}{p_m C_o} = \frac{V_{bias}}{p_m} \tag{6.82}$$

The effective polarization voltage for a Teflon electret mic is typically about 100 V_{DC}, and the $p_m \cong 100$ kPa, so for the electret microphone, $|\mathbf{M_{oc}}| \cong 1.0$ mV/Pa.

[23] When I construct an electret transducer, I usually sandblast the backplate's surface to create microscopic bumps that ensure a compressible trapped air cushion. At the microscopic level, this produces a "mountain range" with the Teflon film resting on the peaks.

[24] As will be demonstrated in Sect. 9.3.2, the temperature of the air trapped between the Teflon film and the microphone's backplate will be held constant by the larger heat capacity of the film and backplate since the thickness of the gas layer is much smaller than the thermal penetration depth, δ_κ (see Sect. 9.3.1). In air at 10 kHz, $\delta_\kappa \cong 27$ microns and grows larger at as the frequency decreases. Since $2\delta_\kappa \gg h_o$, the air cannot change its temperature adiabatically as calculated in Eq. (1.19).

6.4 Vibrations of Thin Plates

The vibration of thin plates has the same relationship to the vibration of membranes as the flexural vibration of bars has to the transverse vibrations of a limp string. However, the plate vibration is complicated by Poisson's ratio, ν, in a way that was not a problem with bars because when a bar is bent, the bar's cross-section could "bulge." For a plate, the bulge is constrained by the adjacent plate material. As with the bar, the bending of a plate compresses material above the neutral plane and stretches the material below. The sideways spreading caused when $\nu \neq 0$ will create a tendency for the plate to curl downward in a direction perpendicular to the upward bend, and vice versa. The derivation of Eq. (6.83) is treated in books devoted to the theory of elasticity [26].

$$\nabla^4 z + \frac{12(1-\nu^2)}{Et^2}\rho\frac{\partial^2 z}{\partial t^2} = 0 \tag{6.83}$$

In this equation of motion, t in the Et^2 product in the denominator of Eq. (6.83) is the thickness of the plate and E is the Young's modulus of the plate's material. Of course, the "t" in the vertical acceleration, $(\partial^2 z/\partial t^2)$, represents time, not thickness. As before, ν is Poisson's ratio. The similarity of Eq. (6.83) to that for the flexural vibrations of a bar in Eq. (5.31) suggests that waves in plates will also be dispersive (see Sect. 5.3.1). The differential bi-harmonic operator, ∇^4, is difficult to separate in most coordinate systems, but in polar coordinates, it is easy enough to justify the analysis of the vibration of thin circular and annular disks.

As before, we assume time-harmonic vibration at frequency, ω, so $z(r, \theta, t) = \Re e[z(r, \theta)e^{j\omega t}]$. We can also factor the resulting operator, as we did for the wave equation in Eq. (3.6), and reintroduce the square of longitudinal wave speed in bars, $c_B{}^2 = E/\rho$.

$$\left(\nabla^2 - \gamma^2\right)\left(\nabla^2 + \gamma^2\right)z(r,\theta) = 0 \quad \text{where} \quad \gamma^4 = \frac{12\omega^2(1-\nu^2)}{t^2 c_B^2} \tag{6.84}$$

This equation is satisfied when either $(\nabla^2 z + \gamma^2 z) = 0$ or $(\nabla^2 z - \gamma^2 z) = 0$.

Since we are expressing both ∇^2 and z in polar coordinates, we can separate $z(r, \theta) = R(r)\Theta(\theta)$, as we did for the circular membrane in Eq. (6.32), to provide a solution for $(\nabla^2 z - \gamma^2 z) = 0$.

$$z(r,\theta) = C_{m,n} J_m(\gamma r)\begin{Bmatrix} \cos(m\theta) \\ \sin(m\theta) \end{Bmatrix} \tag{6.85}$$

Our disk has no hole in the center, so we can reject the Neumann functions and again impose periodic boundary conditions to force m to assume only integer values, including zero.

To solve $(\nabla^2 z + \gamma^2 z) = 0$ and thus obtain the other two solutions to our fourth-order differential equation, we can simply use Eq. (6.84) but substitute $\gamma = j\gamma$ to impose the requisite sign change. This introduces the *modified Bessel functions*, $I_m(r)$ and $K_m(r)$, that are the equivalent of the hyperbolic trigonometric functions, sinh (x) and cosh (x), that provided the third and fourth solutions required to solve the fourth-order differential equation that described the flexural modes of bars.[25] Those hyperbolic solutions were generated when Eq. (5.36) required both real and imaginary values for the wavenumbers.

[25] We can choose to consider the modified Bessel functions to be the hyperbolic Bessel functions in the same way that sinh (x) and cosh (x) are the hyperbolic trigonometric functions.

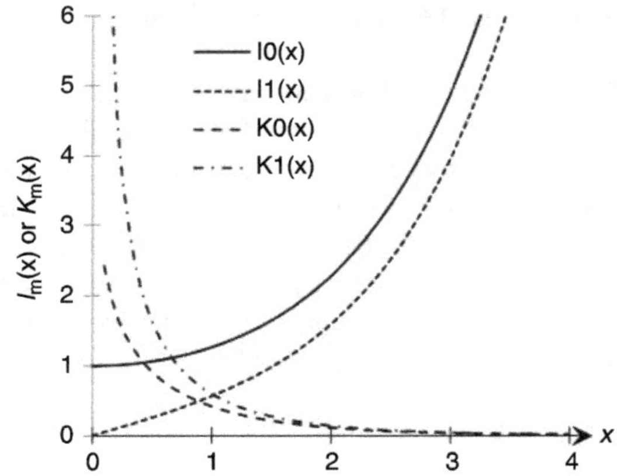

Fig. 6.20 Modified Bessel functions of the first kind, $I_m(x)$, are the radial solutions to $(\nabla^2 - \gamma^2) I_m(x) = 0$ and the second kind, $K_m(x)$, that are solutions to $(\nabla^2 + \gamma^2) K_m(x) = 0$. Unlike the Bessel functions in Fig. 6.8 or the Neumann functions in Fig. 6.9, the values of the modified Bessel functions are not oscillatory functions of their arguments

The modified Bessel functions are related to the Bessel functions with purely imaginary arguments: $I_m(x) = j^{-m} J_m(jx)$. Bessel functions with imaginary arguments are the solution to a Bessel equation like Eq. (6.28), but with different signs [2].

$$\frac{d^2R}{dr^2} + \frac{1}{r}\frac{dR}{dr} - \left(k^2 + \frac{m^2}{r^2}\right)R = 0 \tag{6.86}$$

Unlike ordinary Bessel functions, the modified Bessel functions are not oscillatory but are monotonically increasing or decreasing functions of their arguments. This is also true of the hyperbolic trigonometric functions. Graphs of $I_0(x)$, $I_1(x)$, $K_0(x)$, and $K_1(x)$ are provided in Fig. 6.20.

The second set of solutions to Eq. (6.86) is represented as $K_m(x)$, and their values, like those of the Neumann functions, are infinite at the origin. For a solid disk, the $K_m(x)$ solutions will be rejected, but they will be required for calculation of the displacement and normal mode frequencies of annular disks [27]. The properties of the $I_m(x)$ are similar to those of $J_m(x)$ in Appendix C. Some of those most useful properties are reproduced below:

$$\begin{aligned} I_{m-1}(x) - I_{m+1}(x) &= \frac{2m}{x} I_m(x) \\ \frac{d}{dx} I_m(x) &= \frac{1}{2}[I_{m-1}(x) + I_{m+1}(x)] \quad \text{and} \quad \frac{d}{dx} I_0(x) = I_1(x) \\ \int x I_0(x)\, dx &= x I_1(x) \quad \text{and} \quad \int I_1(x)\, dx = I_0(x) \end{aligned} \tag{6.87}$$

It is worthwhile to notice that although these are similar to the Bessel function relations, they are not identical. The sign of the terms in the recurrence relation and most of the derivative and integral relations are reversed when applied to the modified Bessel functions.

6.4.1 Normal Modes of a Clamped Circular Plate

The general solution to Eq. (6.84) for a thin disk will be the superposition of the Bessel and the modified Bessel solutions.

Table 6.6 Quantized values of $\gamma_{m,n}\, a$ that determine the normal mode frequencies of a thin disk, $t \ll a$, that is clamped at its outer radius, $r = a$. For larger values of n, $\gamma_{m,n}\, a \cong \pi[(m/2) + n]$

m	$n = 1$	$n = 2$	$n = 3$	$n = 4$
0	3.19622	6.30644	9.43949	12.57713
1	4.61090	7.79927	10.95806	14.10862
2	5.90568	9.19688	12.41188	15.57949
3	7.14353	10.53667	13.66009	17.00529
4	8.34662	11.83672	15.14987	18.39595

$$\mathbf{z}(r,\theta,t) = [AJ_m(\gamma r) + BI_m(\gamma r)]\begin{Bmatrix} \cos(m\theta) \\ \sin(m\theta) \end{Bmatrix} e^{j\omega t} \tag{6.88}$$

For a disk that is clamped at $r = a$, $z(a) = 0$ and $[dz/dr]_{r=a} = 0$. Imposition of the first boundary condition sets the ratio of contributions of the $J_m(\gamma r)$ and $I_m(\gamma r)$ solutions.

$$\frac{B}{A} = -\frac{J_m(\gamma a)}{I_m(\gamma a)} \tag{6.89}$$

This ratio remains finite because $I_m(x) > 0$ for all $x > 0$.

The restriction on the slope at $r = a$ will quantize the values of γ.

$$I_m(\gamma_{m,n}a)\left[\frac{d}{dr}J_m(\gamma_{m,n}r)\right]_a - J_m(\gamma_{m,n}a)\left[\frac{d}{dr}I_m(\gamma_{m,n}r)\right]_a = 0 \tag{6.90}$$

That equation can be transformed by use of the relationships in Appendix C for the Bessel functions and for the modified Bessel functions in Eq. (6.87). The result for the radially symmetric, $m = 0$, modes is particularly simple.

$$\begin{aligned} \frac{J_0(\gamma_{0,n}a)}{J_1(\gamma_{0,n}a)} &= -\frac{I_0(\gamma_{0,n}a)}{I_1(\gamma_{0,n}a)} \quad \text{for} \quad m = 0 \\ \frac{J_m(\gamma_{m,n}a)}{I_m(\gamma_{m,n}a)} &= \frac{J_{m-1}(\gamma_{m,n}a) - J_{m+1}(\gamma_{m,n}a)}{I_{m-1}(\gamma_{m,n}a) + I_{m+1}(\gamma_{m,n}a)} \quad \text{for} \quad m \geq 1 \end{aligned} \tag{6.91}$$

Solutions to Eq. (6.91) for small values of m and n are provided in Table 6.6 to five-digit accuracy, following the recommendation of Gabrielson [27].

$$f_{m,n} = \frac{(\gamma_{m,n}a)^2}{2\pi a^2\sqrt{12(1-\nu^2)}} c_B t \tag{6.92}$$

The normal mode frequencies for other values of m and n and for plates of other shapes (e.g., rectangular plates) are available in the excellent compendium by Leissa [28].

The complete solution for the transverse displacements of a thin disk vibrating in one of its normal modes can be written based on Eq. (6.88) using Eq. (6.89).

$$\mathbf{z_{m,n}}(r,\theta,t) = C_{m,n}\left[J_m(\gamma_{m,n}r) - \frac{J_m(\gamma_{m,n}a)}{I_m(\gamma_{m,n}a)} I_m(\gamma_{m,n}r)\right]\begin{Bmatrix} \cos(m\theta) \\ \sin(m\theta) \end{Bmatrix} e^{j\omega_{m,n}t} \tag{6.93}$$

Following our previous focus on the fundamental (0, 1) normal mode, due to its dominance for coupling to a surrounding fluid medium, the effective piston area, A_{eff}, can be calculated by integration of Eq. (6.93) over the disk's surface.

$$A_{eff} = \int_0^a \left[J_0(\gamma_{0,1}r) - \frac{J_0(\gamma_{0,1}a)}{I_0(\gamma_{0,1}a)} I_0(\gamma_{0,1}r) \right] 2\pi r\, dr \tag{6.94}$$

Again, letting $x = \gamma_{0,n}\, r$, $r = x/\gamma_{0,n}$, and $dr = dx/\gamma_{0,n}$, the integration is simplified by use of Eqs. (6.87) and (6.42).

$$\frac{A_{eff}}{\pi a^2} = \frac{2}{(\gamma_{0,1}a)} \left[J_1(\gamma_{0,1}a) - \frac{J_0(\gamma_{0,1}a)}{I_0(\gamma_{0,1}a)} I_1(\gamma_{0,1}a) \right] = 0.3289 \tag{6.95}$$

For the (0, 1) mode, $\gamma_{0,1}a = 3.19622$, $J_1\,(\gamma_{0,1}a) = 0.262861$, $J_0\,(\gamma_{0,1}a)/I_0\,(\gamma_{0,1}a) = -0.05571$, and $I_m\,(\gamma_{0,1}a) = 4.71815$, making $A_{eff}/\pi a^2 = 0.3289$. This is less than the result for a circular membrane because the disk must approach the boundary at $r = a$, with zero slope as well as zero displacement. The equivalent displacement that would produce the same volume velocity for a rigid piston of area, πa^2, can be expressed in terms of the maximum displacement at the disk's center, z_{max}. The displacement of the disk's center, using Eq. (6.93), for the (0, 1) mode is $z(0) = 1.05571 C_{0,1}$.

$$\frac{\langle z \rangle_{eff}}{z_{max}} = \frac{1}{1.0557} \frac{A_{eff}}{\pi a^2} = 0.312 \tag{6.96}$$

6.5 Flatland

This chapter has provided a glimpse into the similarities and differences between one-dimensional and two-dimensional vibrating systems. A two-dimensional membrane or plate requires two integer indices to specify each normal mode of vibration. The number of such modes within a fixed frequency interval increases as the center frequency of that interval increases, even though the width of the interval remains constant. As with a string, a membrane with negligible flexural rigidity that is placed under tension was described by a second-order differential equation, but a rigid bar or a plate, with displacements that were restored by stiffness, obeyed a fourth-order differential equation.

The string and membrane were also driven in different ways. The string was driven by the application of a force or a displacement at a point. The membrane was driven by the application of a uniform pressure difference across its surface. Solutions for the pressure-driven membrane facilitated the analysis of the condenser microphone. The effect of gas stiffness in providing an additional restoring force to that of the membrane's tension was incorporated by assuming that the gas trapped behind the membrane was compressing and expanding adiabatically or isothermally, depending upon the proximity of solid surfaces. A much more systematic analysis of the ideal gas laws will be presented in the next chapter.

Talk Like an Acoustician

Laplacian operator	Gas stiffness
Helmholtz equation	Polytropic coefficient
Separation condition	Inhomogeneous differential equation
Degenerate modes	Particular solution
Mode splitting	Condenser microphone
Modal density	Permittivity of free space
Periodic boundary condition	Polarizing voltage
Bessel functions	Electrostatic potential energy
Neumann functions	Paschen's law
k-space	Open-circuit microphone sensitivity

Fig. 6.21 Nobel Prize-winning physicist Richard P. Feynman, playing his favorite alto Solomongo [31]

Reciprocal space	Electret microphone
Compactness criterion	Vibrating reed electrometer
Volume velocity	Modified Bessel functions

Exercises

1. **Mass-loaded square membrane.** A square membrane with $L_x = L_y = 30.0$ cm has a mass of 10.0 gm and its fundamental normal mode frequency, $f_{1,1} = 100.0$ Hz.
   (a) *Tension per unit length.* What is the tension per unit length to which the membrane was stretched?
   (b) *Split degeneracy.* If a mass, $M = 1.0$ gm, is placed as shown in Fig. 6.3, what are the frequencies of the $f_{1,2}$ and the $f_{2,1}$ modes.
2. **Modes of the baritone Solomongo.** Biblical legend has it that King Solomon, while walking through his kingdom, came across two young boys who were arguing over a drum that both wanted to play and neither wanted to share. Buoyed by his recent success in the resolution of a dispute between two women claiming to be the mother of one baby, the wise king told the boys that he would split the drum and each could keep half. The boys responded: "לכי תזדייני."[26] They then laughed uncontrollably. Dismayed by the boys' response, the wise king instructed one of his guards to cut the boys' drum in half. After a respectable interval, the boys attached planks to the half-drum bodies to close them and carefully smoothed the junction between the half-drum and the sealing plank to remove any rough edges that could rip the membrane. They then stretched a goat skin membrane over the half-drums.
   Such percussion instruments are now known as Solomongos (rhymes with bongos) and have become popular worldwide because they allow the hands to be positioned in the same way as one

[26] Loosely translated, the boys said, "Your royal highness, please leave us in peace and as you depart perform a miracle that would otherwise be anatomically impossible."

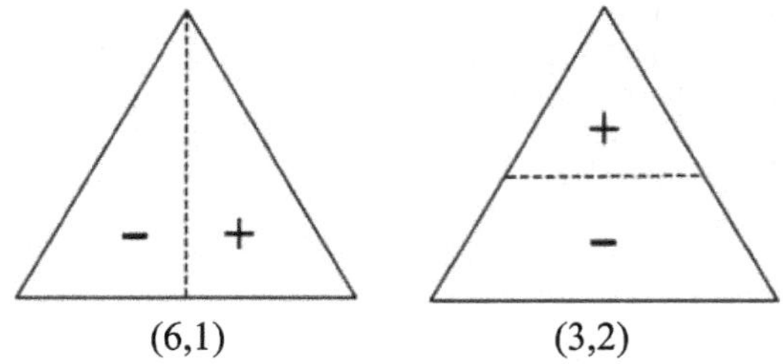

Fig. 6.22 Two sample mode shapes for the equilateral triangular membrane

would play a keyboard instrument as shown in Fig. 6.21. The baritone, with membrane diameter, $D = 60$ cm, is particularly popular, since it can radiate intricate rhythmic patterns generated by tapping of finger tips but when struck with the palm of the hand produces heart-throbbing bass that is radiated from the half-drum's opening at the bottom.

(a) *Fundamental mode frequency.* If the thickness of the goat skin is 1.2 mm and the average mass density of the skin is 1200 kg/m^3, what is the frequency of the fundamental mode if the total tension applied by the tuning straps is 500 N?

(b) *Modal frequencies.* What are the mode numbers and frequencies of the next five lowest-frequency modes using mode numbering of a full circular membrane?

(c) *Effective piston area.* Express the equivalent piston area, $\langle A \rangle$, for a Solomongo vibrating in its fundamental mode in terms of the maximum transverse displacement of the membrane, z_{max}. At what radial distance from the center of the Solomongo's straight edge does the maximum displacement occur?

3. **Mode number approximation**. Determine the number of modes with normalized frequencies, $f_{m,n}/f_{1,1} \leq 5$, predicted by Eq. (6.15) for the rectangular membrane with $L_x = 3L_y/2$. Compare your result with the exact result in Table 6.1, $N\ (f_{5,5}) = 34$, by calculating the percent difference between two determinations.
4. **Density of modes**. A rigidly clamped rectangular membrane is 10.0 cm wide and 14.1 cm long, with a transverse wave speed of 50 cm/s.
   (a) How many modes could be excited by a band-limited noise source which has a frequency range of $70 < f < 80$ Hz?
   (b) How many modes could be excited by a band-limited noise source which has a frequency range of $145 < f < 155$ Hz?
5. **Relative radii of nodal circles**. Calculate the relative radii, r/a, for the nodal circles for the six modes which have nodal circles in Fig. 6.10. The modes in the middle column will have one nodal circle in addition to the node that exists at $r/a = 1$. The modes in the right-hand column will have two nodal circles.
6. **Triangular membrane mode shapes**. Using diagrams like those in Fig. 6.22, sketch the approximate mode shapes for the ten lowest-frequency modes given in Table 6.4 for the membrane that is an equilateral triangle.
7. **Number of modes for a triangular membrane**. Equation (6.15) approximated the number of modes below a given frequency $N_{rect}\ (f_{m,n})$ based on the area and half the perimeter of the rectangular membrane. Using the area and the half-perimeter of a circular membrane, that expression was modified to estimate the number of modes of a circular membrane, $N_{circ}\ (f_{m,n})$, in Eq. (6.34). Both expressions were successful in predicting the number of modes to within a few percent. Modify Eq. (6.15) to accommodate the area and half-perimeter of the equilateral triangle, and compare the predictions of this approximation to the ten lowest-frequency normal modes for the equilateral triangle provided in Table 6.4.

8. **Kinetic energy of vibration**. Show that the maximum kinetic energy of a circular membrane of radius, a, oscillating in its fundamental (0, 1) mode at frequency, $f_{0,1}$, is correctly expressed in terms of the maximum displacement of its center, z_{max} and ρ_S, in Eq. (6.97).

$$(KE)_{\max} = 0.135\ \pi a^2 \rho_S \left(2\pi f_{0,1} z_{\max}\right)^2 \tag{6.97}$$

9. **Modes of a latex diaphragm**. A rigidly clamped 20.0-cm-diameter latex sheet is 0.010″ = 254 micron thick. When vibrating in its fundamental mode, $f_{0,1} = 50.0$ Hz.
   (a) *Mass*. If the density of latex is $\rho = 960$ kg/m^3, what is the mass of the membrane?
   (b) *Tension*. What is the membrane's tension per unit length?
   (c) *Unperturbed modal frequencies*. What are the frequencies and mode numbers of the next five normal modes?
   (d) *Small mass perturbation*. If a small piece of putty with mass, $M = 0.20$ gm, is stuck to the center of the membrane, by how much is each of the frequencies of the fundamental and the next five modes changed? Report your results for parts (c) and (d) in a table with the mode numbers, unperturbed frequencies, and perturbed frequencies.
   (e) *Large lumped-mass loading*. If the putty of part (d) above is removed and two magnetic disks, each with a mass of 10.0 gm and a diameter of 4.0 cm, are attached to the top and bottom of the membrane, what will be the frequency of the lowest-frequency normal mode?
10. **Pressure-driven rectangular membrane**. Calculate the area-averaged displacement of a rectangular membrane driven by a time-harmonic excess pressure acoustic, $\delta p(t) = \Re e[\widehat{\mathbf{p}} e^{j\omega t}]$, in a fluid that is applied to one surface of the membrane, assuming that $c_{fluid} \gg c = (\Im/\rho_S)^{½}$ and $\omega \ll 2\pi f_{1,1}$.
11. **Kettledrum**. The circular membrane of a tympani has a diameter of 50.0 cm and a total mass of 31.4 gm and is stretched to a tension per unit length of $\Im = 10{,}000$ N/m.
    (a) *Transverse wave speed*. What is the speed of transverse waves on the membrane, ignoring the gaseous restoring force?
    (b) *Fundamental frequency*. Determine the fundamental frequency, $f_{0,1}$, for the membrane if the volume of the kettle were infinite (i.e., calculate the frequency assuming there is no gas stiffness).
    (c) *Kettle volume*. If the kettle volume, V_o, raises the frequency by a factor of 1.27, what is the volume of the kettle? How does this volume compare to the volume of a hemisphere with a diameter that is the same as the membrane?
12. **Condenser microphone lumped-element model**. Using the low-frequency parabolic approximation for the deflection of the microphone's membrane, subject to a uniform pressure difference, given in Eq. (6.58), the transverse deflection can be expressed in terms of the maximum deflection, $z_{\max} = z(0) = a^2(\delta p)/4\Im$.

$$z(r) = z(0)\left(1 - \frac{r^2}{a^2}\right) \quad \text{for} \quad 0 \le r \le a \tag{6.98}$$

    (a) *Effective area*. Calculate the effective piston area, A_{eff}, of the membrane, and express your result as a fraction of the membrane's area, πa^2.
    (b) *Effective mass*. Determine the effective moving mass, m_{eff}, of the membrane by calculating the membrane's kinetic energy. Compare the effective mass to the membrane's physical mass.
    (c) *Effective stiffness*. The potential energy, PE, stored in the microphone's membrane can be calculated by integrating the work done by the force produced by the excess pressure times

the membrane's displacement over the membrane's area, $dW = (\delta p\ 2\pi r\ dr)\ dz$. Show that $K_{eff} = 2\pi\ \mathfrak{I}$ by setting $PE = ½[K_{eff}\, z(0)^2]$.

(d) *Fundamental mode frequency*. Using the effective mass and the effective stiffness, calculate the lumped-element approximation to the frequency of the lowest-frequency normal mode, $\omega_{0,1} = \sqrt{K_{eff}/m_{eff}}$, and then compare your result to the exact result using the Bessel function result of Eq. (6.33).

(e) *Damping*. Determine the mechanical resistance, R_m, necessary to make the membrane's quality factor $Q = 2$ using the lumped-element model.

(f) *Resonance response*. What is the response relative to the low-frequency value for the microphone at $f_{1,0}$ with $Q = 2$?

(g) *Bandwidth*. Determine the frequency below $f_{1,0}$ at which the microphone's response exceeds the low-frequency value by 1.0 dB. Report your answer in terms of $f_{1,0}$.

(h) *High-frequency extended response*. Determine the frequency relative to $f_{1,0}$ at which the microphone's response is −1.0 dB relative to the low-frequency value above $f_{1,0}$. Report your answer in terms of $f_{1,0}$.

13. **Condenser microphone with a stretched stainless-steel diaphragm**. The 0.2% yield strength of stainless-steel alloy 316 at room temperatures is 290 MPa, and its mass density is $\rho = 7990\ kg/m^3$. A condenser microphone diaphragm made from that alloy has a diameter of 3.0 cm and is 0.001″ (25 microns) thick.

(a) *Maximum tension/length*. Calculate the tension per unit length that would stretch the stainless to its 0.2% yield limit.

(b) *Transverse wave speed*. Calculate the transverse wave speed of the diaphragm when stretched to its 0.2% limit.

(c) *Fundamental frequency*. Calculate the fundamental normal mode frequency $f_{0,1}$ for that diaphragm.

(d) *Open-circuit sensitivity*. If a backplate of optimum radius is placed 25 microns from the diaphragm and is polarized by $V_{bias} = 200\ V_{dc}$, what will be the low-frequency (i.e., $ka < 1$) sensitivity of this microphone? Report your result in both V/Pa and in dB *re:* 1 V/Pa.

14. **Altec-Lansing Model 21C microphone**. This condenser microphone used a ½″ diameter, 50 μm thick glass plate (fused silica, SiO_2) as the diaphragm, instead of a stretched membrane. The glass plate has a sputtered gold coating to make it electrically conductive. If the Young's modulus of glass is 73.1 GPa and its shear modulus is 31.2 GPa, what is the lowest resonance frequency of the plate if it is clamped at its boundary?

15. **MEMS microphone**. Knowles Acoustics marketed the first commercial microelectromechanical systems (MEMS) silicon diaphragm miniature surface-mount SiSonic® microphone, shown schematically in Fig. 6.23, in 2003 [29].

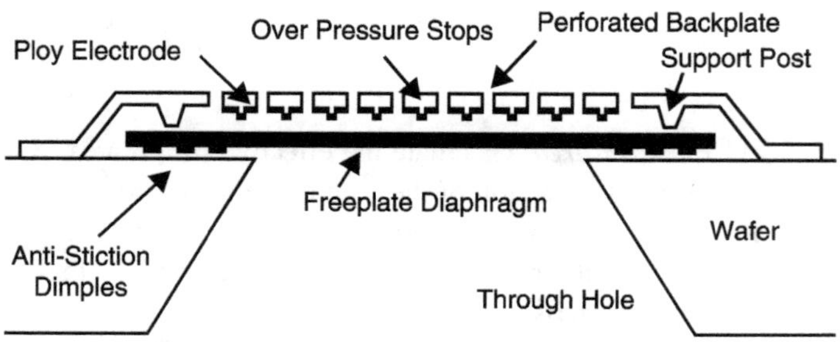

Fig. 6.23 Schematic cross-sectional diagram of a MEMS microphone [30]

The diaphragm is $t = 1.0$ μm thick, has an effective diameter of 500 μm, and is made of polycrystalline silicon (polysilicon) with $E = 169$ GPa and $\nu = 0.22$ [30].

(a) *Fundamental mode frequency*. Calculate the frequency, $f_{0,1}$, of the lowest purely radial normal mode, assuming that the boundary is clamped.

(b) *Diaphragm thickness*. Based on the density of polysilicon, $\rho_{Si} = 2.33$ gm/cm^3, and the atomic mass of silicon, $M_{Si} = 28.08$ a.m.u., express t in terms of the number of atoms.

References

1. E. Jahnke, F. Emde, *Tables of Functions with Formulae and Curves*, 4th edn. (Dover, New York, 1945)
2. M. Abramowitz, I.A. Stegun, *Handbook of Mathematical Functions with Formulas, Graphs and Mathematical Tables*, Nat'l. Bur. Standards – Applied Mathematics Series #55 (U.S. Department of Commerce, National Bureau of Standards, Washington, DC, 1964)
3. M. Greenspan, Simple derivation of the Boltzmann-Ehrenfest adiabatic principle. J. Acoust. Soc. Am. **27**(1), 34–35 (1955)
4. E.C. Wente, A condenser transmitter as a uniformly sensitive instrument for the absolute measurement of sound intensity. Phys. Rev. **10**, 39–63 (1917)
5. H.L.F. Helmholtz, *On the Sensation of Tone as a Physiological Basis for the Theory of Music* (Longmans & Co, London, 1885). See Ch. III – Analysis of musical tones by sympathetic resonance. Reprinted Dover, 1954; ISBN 0-486-60753-4
6. J. Tyndall, *Sound*, 7th edn. (Longmans, Green, and Co, London, 1898), pp. 236–282, see Lecture VI – Singing Flames
7. J. Zuckerwar, Theoretical response of condenser microphones. J. Acoust. Soc. Am. **64**(5), 1278–1285 (1978)
8. J. Zuckerwar, Principles of operation of condenser microphones, in *AIP Handbook of Condenser Microphones*, ed. by G. S. K. Wong, T. F. W. Embleton, (American Institute of Physics, New York, 1995). See Chapter 3: ISBN 1-56396-284-5
9. H.W. Baerwald, The absolute noise level of microphones. J. Acoust. Soc. Am. **12**(1), 131–139 (1940)
10. J.D. Combine, *Gaseous Conductors: Theory and Engineering Application* (McGraw-Hill, New York, 1941), reprinted (Dover, 1958)
11. F. Paschen, Ueber die zum Funkenübergang in Luft, Wasserstoff und Kohlensäure bei verschiedenen Drucken erforderliche Potentialdifferenz (On the potential difference required for spark initiation in air, hydrogen, and carbon dioxide at different pressures). Ann. Phys. **273**(5), 69–75 (1889)
12. T.W. Dakin, D. Berg, Theory of gas breakdown, in *Progress in Dielectrics*, ed. by J. B. Berks, J. Hart, vol. 4, (Wiley, New York, 1962), pp. 152–198
13. R.S. Dhariwal, J.-M. Torres, M.P.Y. Desmulliez, Electric field breakdown at micrometre separations in air and nitrogen at atmospheric pressure. IEE Proc. Sci. Meas. Technol. **147**(5), 261–265 (2000)
14. M.S. Hawley, F.F. Romanow, J.E. Warren, The Western electric 640AA capacitance microphone: Its history and theory of operation, in *AIP Handbook of Condenser Microphones*, ed. by G. S. K. Wong, T. F. W. Embleton, (American Institute of Physics, New York, 1995). See Chapter 2; ISBN 1-56396-284-5
15. G.M. Sessler, J.E. West, D.A. Berkley, G. Morgenstern, Determination of spatial distribution of charges in thin dielectrics. Phys. Rev. Lett. **38**(7), 368–371 (1977)
16. G.M. Sessler, J.E. West, Production of high quasi permanent charge densities on polymer foils by application of breakdown fields. J. Appl. Phys. **43**(2), 922–926 (1972)
17. G.M. Sessler, J.E. West, Electret transducers: A review. J. Acoust. Soc. Am. **53**(6), 1589–1600 (1973)
18. P.W. Chudleigh, Mechanism of charge transfer to a polymer surface by conducting liquid contact. J. Appl. Phys. **47** (10), 4475–4483 (1976)
19. G.M. Sessler, Physical principles of electrets, in *Electrets*, ed. by G. M. Sessler, (Springer, New York, 1987).; See §2.2.4 (Charging with liquid contact). ISBN 978-3-540-17335-9
20. P. Horowitz, W. Hill, *The Art of Electronics*, 2nd edn. (Cambridge University Press, New York, 1989). See Chapter 3
21. G.W. Elko, K.P. Harney, A history of consumer microphones: The electret condenser microphone meets the micro-electro-mechanical systems. Acoust. Today **5**(2), 4–13 (2009)
22. S.L. Garrett, S. Adams, S. Putterman, I. Rudnick, Resonant nonlinear mode conversion in He II. Phys. Rev. Lett. **41** (6), 413–416 (1978)
23. H. Palevsky, R.K. Swank, R. Grenchik, Design of dynamic condenser electrometers. Rev. Sci. Instrum. **18**(5), 298–314 (1947)

24. G. Rietveld, H.E. van den Brom, Vibration reed electrometer for accurate measurement of electrical currents below 10 pA. IEEE Trans. Instrum. Meas. **56**(2), 559–563 (2007)
25. E. Polturak, S.L. Garrett, S.G. Lipson, Precision acoustic gas analyzer for binary gas mixtures. Rev. Sci. Instrum **57** (11), 2837–2841 (1986)
26. L.D. Landau, E.M. Lifshitz, *Elasticity* (Pergamon, 1970). See §12; ISBN 0 08 006465 5
27. T.B. Gabrielson, Frequency constants for transverse vibration of annular disks. J. Acoust. Soc. Am. **105**(6), 3311–3317 (1999)
28. W. Leissa, *Vibration of Plates*, NASA Report SP-160 (U.S. Government Printing Office, Washington, DC, 1969). reprinted by the Acoustical Society of America, 1993; ISBN 1-56396-294-2
29. P. Loeppert and S. B. Lee SiSonic™ – The first commercialized MEMS microphone. Solid-State Sensors, Actuators, and Microsystems Workshop, Hilton Head, June 4–8, 2006
30. W.N. Sharpe Jr., B. Yuan, R. Vaidyanathan, R.L. Edwards, Measurement of Young's modulus, Poisson's ratio, and tensile strength of polysilicon, in *Proceedings of the 10th International Workshop on Microelectromechanical Systems*, (Nagoya, Japan, 1997), pp. 424–429
31. R.P. Feynman, R.B. Leighton, M. Sands, Introduction, in *The Feynman Lectures on Physics*, vol. I, (Addison-Wesley, Reading, 1963)

Permissions

All chapters in this book were first published by Springer by Steven L. Garrett; hereby published with permission under the Creative Commons Attribution License or equivalent. Every chapter published in this book has been scrutinized by our experts. Their significance has been extensively debated. The topics covered herein carry significant information for a comprehensive understanding. They may even be implemented as practical applications or may be referred to as a beginning point for further studies.

We would like to thank the editorial team for lending their expertise to make the book truly unique. They have played a crucial role in the development of this book. Without their invaluable contributions this book wouldn't have been possible. They have made vital efforts to compile up to date information on the varied aspects of this subject to make this book a valuable addition to the collection of many professionals and students.

This book was conceptualized with the vision of imparting up-to-date and integrated information in this field. To ensure the same, a matchless editorial board was set up. Every individual on the board went through rigorous rounds of assessment to prove their worth. After which they invested a large part of their time researching and compiling the most relevant data for our readers.

The editorial board has been involved in producing this book since its inception. They have spent rigorous hours researching and exploring the diverse topics which have resulted in the successful publishing of this book. They have passed on their knowledge of decades through this book. To expedite this challenging task, the publisher supported the team at every step. A small team of assistant editors was also appointed to further simplify the editing procedure and attain best results for the readers.

Apart from the editorial board, the designing team has also invested a significant amount of their time in understanding the subject and creating the most relevant covers. They scrutinized every image to scout for the most suitable representation of the subject and create an appropriate cover for the book.

The publishing team has been an ardent support to the editorial, designing and production team. Their endless efforts to recruit the best for this project, has resulted in the accomplishment of this book. They are a veteran in the field of academics and their pool of knowledge is as vast as their experience in printing. Their expertise and guidance has proved useful at every step. Their uncompromising quality standards have made this book an exceptional effort. Their encouragement from time to time has been an inspiration for everyone.

The publisher and the editorial board hope that this book will prove to be a valuable piece of knowledge for students, practitioners and scholars across the globe.

Index

www.ingramcontent.com/pod-product-compliance
Lightning Source LLC
LaVergne TN
LVHW062221200926
848484LV00006B/1447